AF323439

Analytic Elements in p-Adic Analysis

Alain Escassut

Université Blaise Pascal, France

Analytic Elements in p-adic Analysis

World Scientific

Singapore • New Jersey • London • Hong Kong

Published by

World Scientific Publishing Co. Pte. Ltd.

P O Box 128, Farrer Road, Singapore 9128

USA office: Suite 1B, 1060 Main Street, River Edge, NJ 07661

UK office: 57 Shelton Street, Covent Garden, London WC2H 9HE

British Library Cataloguing-in-Publication Data
A catalogue record for this book is available from the British Library.

ANALYTIC ELEMENTS IN p-ADIC ANALYSIS

Copyright © 1995 by World Scientific Publishing Co. Pte. Ltd.

ISBN 981-02-2234-3

This book is printed on acid-free paper.

Printed in Singapore by Uto-Print

INTRODUCTION

This book is aimed at all graduate students, Mathematicians and Physicists interested in p-adic analysis. The theory of p-adic analytic functions in domains other than simple disks is not very well known yet, although such kind of functions happens to intervene in questions linked to p-adic functional analysis, number theory, and others. Actually, as we will see, this is first a generalization and a deepening of the properties of rational functions in ultrametric algebraically closed complete fields, as the theory of holomorphic functions may also be considered like a generalization of the properties of rational functions in $\mathbb{C}$. No knowledge in p-adic analysis is required from the reader: all the basic results on ultrametric fields are given and shown when they appear to be necessary. The reader is only supposed to have the classical background of a master in pure mathematics, in algebra and topology, in particular when using methods involving filters.

In p-adic analysis, it hasn't been easy to construct a theory of holomorphic functions provided with strong properties, like those in complex analysis. Indeed, given a power series, its radius of convergence does exist and is computed with the same formula as in Archimedean analysis. But the difference comes next, when one wants to extend the domain of definition by means of another disk, the series is defined inside: the new disk, actually, is the same as the previous, because given two disks whose intersection is not empty, one is included in the other! This immediate remark was frequently described by Marc Krasner who found this situation very funny! Thus, except in disks, it first appeared very hard to define a relevant kind of holomorphic functions, in non Archimedean fields.

However, by Runge's Theorem the holomorphic functions in an open connected subset D of $\mathbb{C}$ also are limits of rational functions with no poles in D, with respect to the topology of uniform convergence on compact subsets of D. This is why Marc Krasner had the idea of considering the set $R(D)$ of rational functions with no poles in a given set D of an algebraically closed complete ultrametric field K, and then defined analytic elements in D as the uniform limits of rational functions with no poles in D. This definition, given by Krasner during the fifties, first concerned "quasi-connected" sets D, and was designed to construct a theory of holomorphic functions by connecting quasi-connected sets, and analytic elements defined on these sets. In fact, it soon appeared that analytic elements were provided with very interesting properties mainly based on the existence of multiplicative semi-norms whose definition is very clear for a rational function and for a power series, or a Laurent series in its set of convergence. These properties were often stated in terms of valuation function, because it is piecewise linear. All these properties of the valuation functions, first extended to the quasi-connected sets by Krasner, actually still hold in a much more general class of sets: the infraconnected sets. Besides, this class is the biggest such that no proper subset of a set D in this class admits a characteristic function among the analytic elements. Later, Bernard Guennebaud systematically studied the multiplicative semi-norms on a K-normed algebra and in particular considered those on the rational functions: this was specialized by G. Garandel in terms of circular filters. This concept of circular filter is essential in this book.

A famous result on the analytic elements is the so-called Mittag-Leffler theorem, proven by Krasner, showing a space of analytic elements to be a direct topological sum of elementary subspaces. This theorem actually is much stronger than the original Mittag-Leffler theorem for holomorphic functions in $\mathbb{C}$ and should certainly be called "Krasner's Theorem". Here again, the class of the infraconnected sets is the largest in which the Mittag-Leffler theorem holds.

Many fine properties on the behaviour of analytic elements soon appeared to be linked to the following question: can an analytic element vanish along some filter on D less thin than a sequence a_n such that $|a_n - a_{n+1}|$ is a strictly monotonous sequence of limit different from 0. The answer is "yes" whenever the filter satisfies a certain (very involved) metric property based on the holes of D. Such filters were named T-filters. In particular, when there is no T-filter on D, provided $\overline{D}$ is open and bounded, every analytic element is the product of a polynomial (whose zeros lie in D) by an invertible analytic element, and then the set $H(D)$ of analytic elements is a principal ideal ring, while any element $f \in H(D)$ satisfying $f(x) = 0$ in a disk included in D is identically zero on all D (the sets satisfying this property for every $f \in H(D)$ are called "analytic sets", introduced by Elhanan Motzkin and Philippe Robba). But as soon as a T-filter exists on D, very pathological analytic elements appear, and make the algebra very complicated. Thus, it was possible to characterize the K-algebras that are noetherian, integrally closed, those with divisors of zero, the maximal ideals of infinite codimension and the analytic sets. We could also give an answer to the elementary question whether "$f'(x) = 0$ for all x in an open infraconnected set D", implies "$f = cst$". If $\overline{D}$ is open, the answer is "yes". This way, infraconnectedness is characteristic of this implication in the class of the open sets such that $\overline{D}$ is open. However, examples show that the answer is "no" for certain infraconnected open sets such that $\overline{D}$ is not open.

Many proofs of the results on T-filters necessarily involve very hard calculations, and understanding them is not easy. However, it is useful to remember that most of the time they are suggested by the behaviour of the valuation function of a rational function.

Analytic elements obviously had applications in general analytic functions theory, like Abdelbaki Boutabaa's work, and have been frequently used in the theory of differential equations mainly initiated by Bernard Dwork, Philippe Robba, Gilles Christol. For example, certain polynomials whose coefficients are rational functions happen to admit a factorization whose factors belong to $H(D)[Y]$, but not to $R(D)[Y]$. Here we only give quite elementary results about the equation $y' = fy$, on an infraconnected set D, with $y,\ f \in H(D)$. Yet, we get surprising results on the dimension of the space of solutions when the field is $\mathbb{C}_p$.

Analytic elements are also useful in spectral theory (we haven't found space to include it in this book), and they let us obtain a solution (in a common work with Yvette Amice) to the enigma of the p-adic Fourier transform asked by Bernard de Mathan (while another solution was found by de Mathan and Fresnel). Interpolation of analytic functions in the disk $|x| < 1$ is a natural problem. With Jesus Araujo we give this problem a solution by using infinite van der Monde matrices previously introduced by Marie-Claude Sarmant. Meanwhile, this provides us with examples to see how complicated the composition of linear mappings is in spaces of sequences.

The meromorphic products introduced by Marie-Claude Sarmant at first looked like very special analytic elements, whose behaviour was obviously linked to problems on T-filters. In fact, they increasingly showed very astonishing and deep links with the general theory of the analytic elements: any Laurent series converging in a set $|x| \geq r$ has a continuation to a meromorphic product $\prod_{n=0}^{\infty} \dfrac{x - a_n}{x - b_n}$, with b_n a sequence in the disk $|x| < r$, and a_n a sequence such that $\lim\limits_{n \to \infty} (a_n - b_n) = 0$. Meanwhile, Motzkin discovered a factorization into singular factors, for invertible analytic elements. Actually, this factorization is directly connected with the Mittag-Leffler series, while Marie-Claude Sarmant showed that given $\epsilon > 0$, each singular factor f^T is equal to a meromorphic product satisfying $\max\limits_{n \in \mathbb{N}} |a_n - b_n| \leq \|f^T - 1\| + \epsilon$. Finally, such factorizations into meromorphic products make the group of the invertible analytic elements look like a Meccano whose elementary pieces are homographic functions. (This has also been useful in differential equations).

This surprising property (all based on the T-filters) allowed us to show very particular equivalent sufficient conditions for an analytic element to be injective, involving Mittag-Leffler series, Motzkin factorization, and meromorphic products: there are reasons to think that these conditions are also necessary provided the set D satisfies minimal conditions. If this conjecture was confirmed, the injective analytic elements would be "very close" (topologically speaking) to the homographic functions. This recalls an even stronger result obtained by Yvette Perrin in hypervalued fields. It is interesting to note that these sufficient conditions of injectivity are rather basic and do not involve T-filters, although one doesn't see how to obtain them without intervention of meromorphic products, whose properties come from T-filters.

I would like to thank very sincerely all those who have helped me write this book, in particular Jesus Araujo who read many chapters, and gave me many pieces of advice, Bertin Diarra, who gave me a very nice proof of the existence of a spherically complete extension, and clear generalizations in characteristic different from zero, Labib Haddad, whose ideas are each time original, like the finite increasing properties, and Marie-Claude Sarmant whose meromorphic products have deeply renewed the theory.

CONTENTS

Analytic Elements in p-Adic Analysis

1. ABSOLUTE VALUES AND NORMS

Let E be a field provided with an absolute value $|\,.\,|$. It is convenient and useful to define the valuation v associated to the absolute value $|\,.\,|$. Let $\omega \in\,]1, +\infty[$ and let log be the real logarithm function of base ω. We put $v(x) = -\log|x|$, and v is named *the valuation associated to the absolute value* $|\,.\,|$.

Lemmas 1.1 and 1.2 are classical, and proven in the same way no matter what the absolute value of E.

Lemma 1.1: *Let E be a field provided with two absolute values whose associated valuations are v and w, respectively. They are equivalent if and only if there exists $r > 0$ such that $w(x) = rv(x)$ whenever $x \in E$.*

Proof: If such a r exists, the two absolute values are seen to be equivalent. Reciprocally, we assume them to be equivalent, and take $a \in E$ such that $v(a) \geq 0$. It is seen that $w(a) \geq 0$. Besides, for all $x \in E$, and for all $m, n \in \mathbb{N}$, we have $v(\frac{x^m}{a^n}) > 0$ if and only if $w(\frac{x^m}{a^n}) > 0$. Therefore, we see that $\dfrac{v(x)}{v(a)} > \dfrac{n}{m}$ is equivalent to $\dfrac{w(x)}{w(a)} > \dfrac{n}{m}$. Then, since $\mathbb{Q}$ is dense in $\mathbb{R}$, we have $\dfrac{v(x)}{v(a)} = \dfrac{w(x)}{w(a)}$ whenever $x \in E$, and therefore $\dfrac{w(x)}{v(x)} = \dfrac{w(a)}{v(a)}$.

Lemma 1.2: *Let E be a field provided with an absolute value and let V be an E-vector space of finite dimension provided with two norms. These two norms are equivalent in each one of these two cases:*
   The dimension of V is one;
   E is complete.

Proof: Let $\|\,.\,\|$ and $\|\,.\,\|'$ be the two norms on V. First we suppose that $V = E$. So, we have $\|ab\| = |a|\|b\| = |b|\|a\|$, hence $\dfrac{\|x\|}{\|a\|}$ is constant in E, and therefore $\|\,.\,\|$ is obviously equivalent to $|\,.\,|$. This is immediately generalized to the case when V has dimension 1 because then, by isomorphism, the two norms $\|\,.\,\|$ and $\|\,.\,\|'$ on V define on E two norms that are equivalent to the absolute value.

Now, we suppose that E is complete. We will proceed by induction on the dimension of V and we assume the equivalence true for subspaces of dimension $n < q$. Let V have dimension q. Let $e_1, ...e_q$ be a base of V. Let us suppose that the two norms are not equivalent on V. Then there exists a sequence $(u_n)_{n\in\mathbb{N}}$

of the form $u_n = \sum\limits_{j=1}^{q} a_{j,n} e_j$, with $\|u_n\| \geq 1$, such that $\lim\limits_{n\to\infty} \|u_n\|' = 0$. Let

S be the subspace of V generated by $\{e_1, ..., e_{q-1}\}$. For every $n \in \mathbb{N}$, we put

$$v_n = \sum_{j=1}^{q-1} a_{j,n} e_j.$$

First, we suppose that

$$(1) \quad \lim_{n\to\infty} |a_{q,n}| = 0.$$

Since $\lim\limits_{n\to\infty} \|u_n\|' = 0$, we have $\lim\limits_{n\to\infty} \|v_n\|' = 0$. By hypothesis, the restrictions

of the two norms to S are equivalent, hence we have $\lim\limits_{n\to\infty} \|v_n\| = 0$. But since

$\|u_n\| \geq 1$ for all $n \in \mathbb{N}$, this contradicts (1).

Now, since (1) is not true, there exists a subsequence of the sequence $(|a_{q,n}|)_{n\in\mathbb{N}}$ that admits a strictly positive lower bound, and therefore, without loss of generality, we can clearly assume that there exists $r > 0$ such that $|a_{q,n}| \geq r$ for all $n \in \mathbb{N}$. Let $(x_n)_{n\in\mathbb{N}}$ be the sequence defined as $x_n = \dfrac{u_n}{a_{q,n}}$. It

is seen that

$$(2) \quad \lim_{n\to\infty} \|x_n\|' = 0.$$

The two norms $\| \, . \, \|$ and $\| \, . \, \|'$ are equivalent on S, and they both are equivalent

to the product norm $\| \, . \, \|''$ defined as $\|\sum\limits_{j=1}^{q-1} b_j e_j\|'' = \max\limits_{1\leq j\leq q-1} |b_j|$. Since E is

complete, S is complete with respect to $\| \, . \, \|''$, hence S is closed in V with respect to the two norms $\| \, . \, \|$ and $\| \, . \, \|'$. Hence by (2) e_q belongs to S, which is absurd and finishes the proof.

Theorem 1.3: *Let E be a field complete for a non trivial absolute value and let F be an algebraic extension of E, provided with two absolute values extending the one E. These absolute values are equal.*

Proof: Let v, w be the valuations associated to these absolute values. Let $a \in F$. By Lemma 1.5, the two absolute values are equivalent on $E[a]$. Hence by lemma 1.1, there exists $r > 0$ such that $w(x) = rv(x)$ whenever $x \in E[a]$. But since $v(x) = w(x)$ whenever $x \in E$, and since there exists $u \in E$ such that $v(u) \neq 0$, we have $r = 1$.

Notations: Henceforth L will denote a field provided with a non trivial ultrametric absolute value $| \, . \, |$. The set of the $x \in L$ such that $|x| \leq 1$ will be denoted by U_L, and the set of the $x \in L$ such that $|x| < 1$ will be denoted by M_L.

Then Lemma 1.4 is immediate:

Lemma 1.4: U_L *is a local subring of* L *whose maximal ideal is* M_L.

Definitions and notations: Henceforth U_L is called *the valuation ring of* L. The maximal ideal M_L of U_L is called the *valuation ideal* and the field $\mathcal{L} = \dfrac{U_L}{M_L}$ is called *the residue class field of* L. For any $a \in L$, the residue class of a will be denoted by $\bar{a}$. If D is a set in L we put $|D| = \{|x| \ \mid \ x \in D\}$, and
$v(D) = \{v(x) \ \mid \ x \in D\}$
The characteristic of $\mathcal{L}$ is named *the residue characteristic of* L and will be denoted by p.
We put $L^* = L \setminus \{0\}$. Then $|L^*|$ is a subgroup of the multiplicative group $(\mathbb{R}_+^*, .)$. The image of L^* by the valuation v associated to L is then a subgroup of the additive group $(\mathbb{R}, +)$ called *the valuation group of* L. The valuation of L is said to be *discrete* if its valuation group is a discrete subgroup of $\mathbb{R}$. Else, the valuation group is dense in $\mathbb{R}$, and then the valuation is said to be *dense*.

Lemma 1.5: *If* L *is algebraically closed, its valuation group is dense in* $\mathbb{R}$.
Proof : Given $\alpha \in L$ such that $0 < |\alpha| < 1$ and $\beta \in L$ such that $\beta^q = \alpha^s$ we have $v(\beta) = \dfrac{q}{s} v(\alpha)$ whenever $s \in \mathbb{N}^*$ and $q \in \mathbb{Z}$.

When $p \neq 0$ it is useful to take $\omega = p$. The most classical example of an ultrametric complete algebraically closed field is the field $\mathbb{C}_p$ and it will be described in Chapter 8.

Definitions and notations: In this chapter A *denotes a commutative* L-*algebra with a unity denoted by* 1. Let us recall that *a semi-norm* (resp. *a norm*) *of* L- *algebra is a semi-norm* (resp. *a norm) of* K-linear space φ that satisfies $\varphi(x.y) \leq \varphi(x)\,\varphi(y)$. Besides φ is said to be *semi-multiplicative* if $\varphi(x^n) = \varphi(x)^n$ whenever $x \in A$ and φ is said to be *multiplicative* if $\varphi(x.y) = \varphi(x)\,\varphi(y)$ whenever $x, y \in A$. In particular, a multiplicative norm is just an absolute value extending that of L (when identifying the unity with 1 in L).

Given a semi-norm of L-algebra φ we will denote by $Ker(\varphi)$ the set of the $x \in A$ such that $\varphi(x) = 0$.

Lemma 1.6: *Let* φ *be a semi-norm of* L- *algebra of* A. *Then* $Ker(\varphi)$ *is an ideal of* A. *If* φ *is multiplicative,* $Ker(\varphi)$ *is a prime ideal of* A.

Notations : $Max(A)$ will denote the set of the maximal ideals of A, and $Max_1(A)$ will denote the subset of the maximal ideals of codimension 1. We will denote by $SM(A)$ the set of the semi-multiplicative semi-norms of A, by $Mult(A)$ the set of the ultrametric multiplicative semi-norms of A, by $Mult_m(A)$ the set of the $\varphi \in Mult(A)$ such that $Ker(\varphi) \in Max(A)$ and by $Mult^*(A)$ the set of the multiplicative norms of A [41], [42].

Lemma 1.7:   *Every element of $SM(A)$ is ultrametric.*

Proof: Let $\varphi \in SM(A)$ and let $a, b \in A$ satisfy $\varphi(a) \geq \varphi(b)$. We just have to show that $\varphi(a + b) \leq \varphi(a)$. Obviously we have $\varphi((a + b)^n) = \varphi\left(\sum_{k=0}^{n} C_n^k a^k b^{n-k}\right)$.

For each $k = 0, ...n$ we have $\varphi(C_n^k a^k b^{n-k}) = |C_n^k|\varphi(a^k b^{n-k}) \leq \varphi(a)^k \varphi(b)^{n-k} \leq \varphi(a)^n$ hence $\varphi((a + b)^n) \leq (n + 1)\varphi(a)^n$ and therefore $\varphi(a + b) \leq \sqrt[n]{n + 1} \, \varphi(a)$ for all $n \in \mathbb{N}^*$. Finally we obtain $\varphi(a + b) \leq \varphi(a)$.

Notations :   Let A be a L-algebra provided with a topology T of L-vector space. We will denote by $SM(A, T)$ (resp. $Mult(A, T)$, resp. $Mult^*(A, T)$, resp. $Mult_m(A, T)$) the set of the $\varphi \in SM(A)$ (resp. $\varphi \in Mult(A)$, resp. $\varphi \in Mult^*A$, resp. $\varphi \in Mult_m A$) that are continuous with respect to the topology T.

In particular when A is a normed L-vector space the norm of which is denoted by $\| \, . \, \|$ we will denote by $SM(A, \| \, . \, \|)$ (resp. $Mult(A, \| \, . \, \|)$, resp. $Mult^*(A, \| \, . \, \|)$, resp. $Mult_m(A, \| \, . \, \|)$) the set $SM(A, T)$ (resp. $Mult(A, T)$, resp. $Mult^*(A, T)$, resp. $Mult_m(A, T)$).

The idea of studying systematically the multiplicative semi-norms is due to Bernard Guennebaud and was much helpful in several domains [41], [42]. The following Lemmas 1.8 and 1.9 are obvious.

Lemma 1.8:   *Let S be a subset of $Mult A$ (resp. of $Mult(A, T)$). Then the mapping ϕ defined on A by $\phi(x) = \sup\limits_{\varphi \in S} \varphi(x)$ belongs to $SM(A)$ (resp. $SM(A, T)$).*

Lemma 1.9:   *Both $Mult(A)$ and $SM(A)$ are closed in $(\mathbb{R}_+)^A$ with respect to the topology of simple convergence.*

Lemma 1.10:   *Let A be provided with a norm $\| \, . \, \|$ of L-algebra and let $\varphi \in SM(A)$. Then $\varphi \in SM(A, \| \, . \, \|)$ if and only if $\varphi(x) \leq \| \, x \, \|$ whenever $x \in A$. Besides, if A has an identity u, and if φ is not identically 0, then $\varphi(\lambda u) = |\lambda|$ whenever $\lambda \in L$.*

Proof: Suppose that for some $x \in A$ we have $\varphi(x) > \| \, x \, \|$. Since the valuation group of L is not trivial, it contains a subgroup of the form $a\mathbb{Z}$, with $a > 0$. Let $q \in \mathbb{N}$ be such that $q(\log(\varphi(x)) - \log(\|x\|)) > a$. Then there clearly exists $\lambda \in L$ satisfying $\| \, x \, \|^q < |\lambda| < \varphi(x)^q$. With greater reason we have $\| \, x^q \, \| < |\lambda| < \varphi(x^q)$. Let $t = x^q$. Then we have

$\lim_{n \to \infty} \left(\dfrac{t}{\lambda}\right)^n = 0$ but $\lim_{n \to \infty} \varphi\left(\left(\dfrac{t}{\lambda}\right)^n\right) = +\infty$, and then φ is not continuous. Now let u be a unity in A. Either $\varphi(u) = 0$ and then $\varphi(x) = 0$ whenever $x \in A$, or $\varphi(u) = 1$ and then we have $\varphi(\lambda u) = |\lambda|\varphi(u) = |\lambda|$ whenever $\lambda \in L$. This ends the proof of Lemma 1.10.

Theorem 1.11: *Let A be provided with a norm of L-algebra $\| \cdot \|$. Then $Mult(A, \| \cdot \|)$ is compact with respect to the topology of simple convergence.*

Proof: Let U be the unit ball of A. By Lemma 1.10, each $\varphi \in Mult(A, \| \cdot \|)$ has a restriction $\widehat{\varphi}$ to U which satisfies $\widehat{\varphi}(U) \subset [0, 1]$. Hence $Mult(U, \| \cdot \|)$ is a closed subset of $[0, 1]^U$ provided with the topology of simple convergence on U. But by Tykhonov's Theorem, $[0,1]$ is compact for this topology and then so is $Mult(U, \| \cdot \|)$. Besides the mapping $\varphi \to \widehat{\varphi}$ from $Mult(A, \| \cdot \|)$ into $Mult(U, \| \cdot \|)$ is a bijection. Indeed it is clearly injective and it is surjective because given $\psi \in Mult(U, \| \cdot \|)$, we may extend ψ to A by putting $\overline{\psi}(x) = |\lambda|\psi(\dfrac{x}{\lambda})$ with $\lambda \in L$, $|\lambda| \geq \|x\|$. Finally this bijection is bicontinuous with respect to the simple convergence on both $Mult(A, \| \cdot \|)$ and $Mult(U, \| \cdot \|)$ and this ends the proof of Lemma 1.11.

In order to introduce the spectral semi-norm $\| \cdot \|_{si}$ in Theorem 1.11 we need a very elementary (but not trivial) little lemma in classical analysis.

Lemma 1.12 : *Let $(x_n)_{n \in \mathbb{N}}$ be a sequence in $\mathbb{R}$ such that $x_{p+q} \leq x_p + x_q$ whenever $p, q \in \mathbb{N}$. Then the sequence $\left(\dfrac{x_n}{n}\right)_{n \in \mathbb{N}}$ admits a finite limit.*

Proof : Let $u_n = \dfrac{x_n}{n}$. The sequence u_n is obviously upper bounded by u_1 because $x_n \leq nx_1$. Besides, we notice that $u_{tq} \leq u_t$ whenever $t, q \in \mathbb{N}^*$. Let $B \in \mathbb{R}$ be such that $\liminf_{n \to \infty}(u_n) \leq B$ and let ϵ be > 0. We will show that there exists $n \in \mathbb{N}$ such that $u_n \leq B + 2\varepsilon$ whenever $n \geq N$ so that $\liminf_{n \to \infty}(u_n)$ will be proven to be equal to $\limsup_{n \to \infty}(u_n)$. Indeed there exists $t \in \mathbb{N}$ such that $u_t < B + \varepsilon$.

Now let $s \in \mathbb{N}$ be such that $\dfrac{1}{s} \max_{0 \leq i < t} u_i < \varepsilon$ for $n \geq st$. Let us divide n in the form $n = qt + r$ with $q \in \mathbb{N}, q \geq s, r \in \mathbb{N}, 0 \leq r < t$. Then we have

$$u_n = \frac{x_{qt+r}}{qt + r} \leq \frac{x_{qt} + x_r}{qt} \leq u_{qt} = u_{qt} + \frac{x_r}{qt} \leq$$

$$\leq u_{qt} + \frac{u_r}{q} \leq (B + \varepsilon) + \frac{1}{s}u_r \leq B + 2\varepsilon.$$

Thus we have proven that $u_n \leq B + 2\epsilon$ as soon as $n \geq st$, and that ends the proof of Lemma 1.12.

Theorem 1.13 :   *Let $\| . \|$ be a norm of L-algebra on A. Then for every $x \in A$ the sequence $\left(\| x^n \|^{\frac{1}{n}} \right)_{n \in \mathbb{N}}$ has a limit denoted by $\| x \|_{si}$ and the mapping f defined in A as $f(x) = \| x \|_{si}$ belongs to $SM(A, \| . \|)$. Besides $\| x \|_{si} < 1$ if and only if $\lim_{n \to \infty} x^n = 0$.*

Proof :   The existence of $\lim_{n \to \infty} \| x^n \|^{\frac{1}{n}}$ is an obvious consequence of Lemma 1.12 by putting $x_n = \log \| x^n \|$. It is then easily seen that $\| \lambda x \|_{si} = |\lambda| \| x \|_{si}, \| xy \|_{si} \leq \| x \|_{si} \| y \|_{si}$ and $\| x^q \|_{si} = \| x \|_{si}^q$ whenever $q \in \mathbb{N}$. Thus, to show that $\| . \|_{si}$ belongs to $SM(A, \| . \|)$ we just have to show that $\| x + y \|_{si} \leq \max \left(\| x \|_{si}, \| y \|_{si} \right)$. Obviously we have

$$(1) \quad \| (x + y)^n \| \leq (n + 1) \max_{0 \leq i \leq n} \left(\| x^i \| \; \| y^{n-i} \| \right).$$

Let $M = \max \left(\| x \|, \| y \| \right)$. We may obviously assume $\| x \|_{si} \geq \| y \|_{si}$, let $\varepsilon > 0$ and let $N \in \mathbb{N}$ be such that $\| x^n \|^{\frac{1}{n}} \leq \| x \|_{si} + \varepsilon$ and $\| y^n \|^{\frac{1}{n}} \leq \| y \|_{si} + \varepsilon$ for $n \geq N$. Henceforth we assume (2) $n \geq N^2$. Let $q_n \in \mathbb{N}$, satisfy $\sqrt{n} \leq q_n < \sqrt{n} + 1$ and let $s_n = n - q_n$. By (2) both q_n, s_n are greater than N. We first suppose $i \leq q_n$. We have $\| y^{n-i} \| \leq \left(\| y \|_{si} + \varepsilon \right)^{n-i}$ hence,

$\| x^i \| \; \| y^{n-i} \| \leq \| x \|^i \left(\| y \|_{si} + \varepsilon \right)^{n-i}$. Since $\| y \|_{si} \leq \| x \|_{si}$ we obtain

$\| x^i \| . \| y^{n-i} \| \leq \left(\| x \| + \varepsilon \right)^i \left(\| x \|_{si} + \varepsilon \right)^{n-i}$.

But since $\| x \|_{si} \leq \| x \|$ we have

$$\left(\| x \| + \varepsilon \right)^i \left(\| x \|_{si} + \varepsilon \right)^{n-i} \leq \left(\| x \| + \varepsilon \right)^{q_n} \left(\| x \|_{si} + \varepsilon \right)^{s_n} \leq$$

$(M + \varepsilon)^{q_n} \left(\| x \|_{si} + \varepsilon \right)^{s_n}$ hence finally

$$(3) \quad \| x^i \| \; \| y^{n-i} \| \leq \left(\frac{M + \varepsilon}{\| x \|_{si}} \right)^{q_n} \left(\| x \|_{si} + \varepsilon \right)^n.$$

In the same way when $i \geq s_n$ we obtain

$$\| x^i \| \; \| y^{n-i} \| \leq \left(\| x \|_{si} + \varepsilon \right)^i \left(\| y \| + \varepsilon \right)^{n-i} \leq \left(\| x \|_{si} + \varepsilon \right)^{s_n} (M + \varepsilon)^{n-s_n}$$

hence again (3) $\| x^i \| \; \| y^{n-i} \| \leq \left(\| x \|_{si} + \varepsilon \right)^n \left(\frac{M + \varepsilon}{\| x \|_{si}} \right)^{q_n}$.

Now when $q_n < i < s_n$ we have at the same time $\| x^i \| \leq \left(\| x \|_{si} + \varepsilon \right)^i$ and $\| y^{n-i} \| \leq \left(\| y \|_{si} + \epsilon \right)^{n-i}$. Finally, since $\| y \|_{si} \leq \| x \|_{si}$, we see that

$$\| \, x^i \, \| \, \| \, y^{n-i} \, \| \; \leq \; \Big(\| \, x \, \|_{si} + \varepsilon \Big)^n \; \text{for } q_n < i < s_n.$$

Thus Relation (3) holds for every $i = 0, ..., n$, hence

$$\|(x+y)^n \, \|^{\frac{1}{n}} \; \leq \; \sqrt[n]{n+1} \Big(\frac{M+\varepsilon}{\| \, x \, \|_{si}} + \varepsilon \Big)^{\frac{q_n}{n}} \, (\, \| \, x \, \|_{si} + \varepsilon).$$

But since $\lim\limits_{n \to \infty} \dfrac{q_n}{n} = 0$ we see that $\| \, (x+y) \, \|_{si} \leq (\| \, x \, \|_{si} + \varepsilon)$. But ε was taken arbitrary small, hence finally we have $\| \, x+y \, \|_{si} \leq \max(\| \, x \, \|_{si} \, , \; \| \, y \, \|_{si})$.

If $\lim\limits_{n \to +\infty} x^n = 0$ we have $\|x^n\| < 1$ for n big enough, hence $\|x^n\|_{si} < 1$, and therefore $\|x\|_{si} < 1$. Now, reciprocally, if $\| \, x \, \|_{si} < 1$, there exist $\rho < 1, N \in \mathbb{N}$ such that $\| \, x^n \, \|^{\frac{1}{n}} < \rho$ for $n \geq N$, hence $\| \, x^n \, \| \leq \rho^n$ and then $\lim\limits_{n \to \infty} x^n = 0$. This finishes proving Theorem 1.13.

Theorem 1.14: *Let A be a L-Banach Algebra. For every $x \in A$ satisfying $\|x\|_{si} < 1$, $1 - x$ is invertible in A. For every algebra homomorphism ψ from A onto L, the mapping $|\psi|$ from A into $\mathbb{R}_+$ defined as $|\psi|(x) = |\psi(x)|$ belongs to $Mult(A, \| \, . \, \|_{si})$.*

Proof : Let $x \in A$ satisfy $\|x\|_{si} < 1$. By Theorem 1.13 we have $\lim\limits_{n \to \infty} x^n = 0$ and then the series $\sum\limits_{n=0}^{\infty} x^n$ converges in A to a limit S that obviously satisfies $S(1 - x) = 1$. Hence for every $\lambda \in K$ satisfying $\|x\|_{si} < |\lambda|, 1 - \dfrac{x}{\lambda}$ is invertible in A. Now $|\psi|$ obviously belongs to $Mult(A)$. Assume that $|\psi(x)| > \| \, x \, \|_{si}$ for some $x \in A$. Let $\lambda = \psi(x) \in L$, and let $y = 1 - \dfrac{x}{\lambda}$. Then $\psi(y) = 0$, hence $1 - \dfrac{x}{\lambda} \in Ker(\psi)$. But $|\lambda| > \| \, x \, \|_{si}$ hence $\|\dfrac{x}{\lambda}\|_{si} < 1$, so $1 - \dfrac{x}{\lambda}$ is invertible. This just contradicts $1 - \dfrac{x}{\lambda} \in Ker(\psi)$. Finally $|\psi(x)| \leq \| \, x \, \|_{si}$ for every $x \in A$, hence $|\psi| \in Mult(A, \| \, . \, \|_{si})$.

Theorem 1.15 was given in several works [38], [49] . This proof mainly is given in [49] .

Theorem 1.15: *Let F be a field extension of L provided with a non zero semi-norm of L-algebra $\| \, . \, \|$. Then $\| \, . \, \|$ is a norm of L-algebra, and there exists an absolute value φ on F extending that of L, such that $\varphi(x) \leq \|x\|$ whenever $x \in A$.*

Proof: By Lemma 1.6, it is seen that $\| \, . \, \|$ is a norm because $Ker\| \, . \, \| = \{0\}$. In the same way, so is the spectral semi-norm $\| \, . \, \|_{si}$ associated to $\| \, . \, \|$. Now $SM(A, \| \, . \, \|_{si})$ is easily checked to be inductive with respect to the order $\geq$, i.e. given a totally ordered subset W of $SM(A, \| \, . \, \|_{si})$, the mapping ψ defined in A by $\psi(x) = \inf\{\theta(x)|\theta \in W\}$ belongs to $SM(A, \| \, . \, \|_{si})$. Then by Zorn's Lemma, $SM(A, \| \, . \, \|_{si})$ admits a minimal element φ. As we just saw, φ is a norm of L-algebra, and by Lemma 1.10 we have $\varphi(x) \leq \|x\|_{si}$ whenever $\;x \in A$. We will prove that $\varphi(ab) = \varphi(a)\varphi(b)$ whenever $\;a, \; b \in A$. Let $a \in A \setminus \{0\}$. For every $x \in A$, we put $u_n(x) = \dfrac{\varphi(a^n x)}{\varphi(a)^n}$. The sequence $(u_n(x))_{n \in \mathbb{N}}$ is seen to be decreasing. We put $\sigma(x) = \lim\limits_{n \to +\infty} \dfrac{\varphi(a^n x)}{\varphi(a)^n}$ whenever $\;x \in A$.

First we will check that σ is a norm of L-algebra. Obviously, it is seen that for every $n \in \mathbb{N}, \;u_n$ is a norm of L-vector space, hence so is σ. Next, we have

$$u_n(x)u_n(y) = \frac{\phi(a^n x)\phi(a^n y),}{\phi(a^n)\phi(a^n)} = \frac{\phi(a^n x)\phi(a^n y)}{\phi(a^{2n})} \geq \frac{\phi(a^{2n} xy)}{\phi(a^{2n})} \geq \sigma(xy), \text{ whenever}$$

$x, \; y \in A$, hence $\sigma(x)\sigma(y) \geq \sigma(xy)$. So, σ is a norm of L-algebra. Now, we check that σ is semi-multiplicative, because:

$$\lim_{n \to +\infty} \left(\frac{\varphi(a^n x^q)}{\varphi(a^n)} \right) = \lim_{n \to +\infty} \left(\frac{\varphi(a^{qn} x^q)}{\varphi(a^{qn})} \right) = \lim_{n \to +\infty} \left(\frac{\varphi(a^n x)}{\varphi(a^n)} \right)^q = \sigma(x)^q$$

Then, since σ satisfies (1) $\sigma(x) \leq \varphi(x) \leq \|x\|_{si}$ whenever $\;x \in A$, it clearly belongs to $SM(A, \| \, . \, \|_{si})$. But since φ is minimal in $SM(A, \| \, . \, \|_{si})$, actually φ is equal to σ. Now, as the sequence $(u_n)_{n \in \mathbb{N}}$ is decreasing, we have $\sigma(x) \leq \dfrac{\varphi(ax)}{\varphi(a)} \leq \varphi(x)$, and finally by (1), $\varphi(ax) = \varphi(a)\varphi(x)$ whenever $a, x \in A$. This ends the proof of Theorem 1.15.

Theorem 1.16 is an obvious consequence.

Theorem 1.16:   *Let A be a L-Banach algebra . For every maximal ideal $\mathcal{M}$ of A, there exists $\varphi \in Mult(A, \| \, . \, \|)$ such that $Ker(\varphi) = \mathcal{M}$.*

Proof: Let $\mathcal{M}$ be a maximal ideal of A, and let F be the field $\dfrac{A}{\mathcal{M}}$. Since A is complete, $\mathcal{M}$ is closed, and therefore F is provided with the quotient norm. By Theorem 1.15, F admits an absolute value $| \, . \, |$ which extends that of L. Let ψ be the canonical surjection from A to F. On A we put $\varphi(x) = |\psi(x)|$. Then φ is an element of $Mult(A, \| \, . \, \|)$ such that $Ker(\varphi) = \mathcal{M}$.

Notations :   Given $x \in \mathbb{R}$ we will denote by $Int(x)$ the integral part of x (i.e., the unique $n \in \mathbb{N}$ such that $n \leq x < n+1$).

Proposition 1.17:   *Let A be a L-Banach algebra whose norm $\|\cdot\|$ is ultra-metric , let $r \in]0, +\infty[$ and let B be the set of the series $\sum_{n=0}^{\infty} a_n x^n$ such that*

$$\lim_{n\to\infty} \|a_n\| r^n = 0.$$

Then B, provided with the multiplication of series , is a commutative algebra which contains A and admit 1 for unity. Let $\|\cdot\|_r$ be defined on B by $\|\sum a_n x^n\|_r = \sup_{n\in\mathbb{N}} \|a_n\| r^n$. Then $\|\cdot\|_r$ is a norm of L-algebra on B and B is complete for this norm.

Proof: Since B obviously is a L-vector space we first check that it is ring. Let

$$f = \sum_{n=0}^{\infty} a_n x^n, \quad g = \sum_{n=0}^{\infty} b_n x^n$$

belong to B. For every $m \in \mathbb{N}$ let $s_m = \sup_{n\geq m} \|a_n\| r^n$, $t_m = \sup_{n\geq m} \|b_n\| r^n$. Hence by hypothesis we have $\lim_{m\to\infty} s_m = \lim_{m\to\infty} t_m = 0$.

For each $q \in \mathbb{N}$ we put $\ell(q) = Int(\frac{q}{2})$. Then it is easily seen that we have

$$(1) \quad \left\| \sum_{j=0}^{q} a_j b_{q-j} \right\| r^q \leq \max\left(\|f\|_r\, t_{\ell(q)}, \|g\|_r\, s_{\ell(q)} \right).$$

Now , putting $c_n = \sum_{j=0}^{n} a_j b_{n-j}$ $(n \in \mathbb{N})$ we see that $\lim_{n\to\infty} \|c_n\| r^n = 0$. As a consequence the series $\sum_{n=0}^{\infty} c_n x^n$ does belong to B, and therefore, B is a commutative L-algebra admitting A as a subalgebra and 1 as a unity.

Then $\|\cdot\|_r$ obviously is a norm of L-vector space. Actually by (1) each coefficient c_n satisfies $\|C_n\| r^n \leq \max\left(\|f\|_r, \|g\|_r \right)$ and therefore $\|\cdot\|$ is a norm of L-algebra on B.

Finally we check that B is complete for this norm. Indeed let $(f_m)_{m\in\mathbb{N}}$ be a Cauchy sequence in B and for each $m \in \mathbb{N}$ let $f_m = \sum_{n=0}^{\infty} a_{m,n} x^n$. For each fixed $n \in \mathbb{N}$, it is seen that the sequence $(a_{m,n})_{m\in\mathbb{N}}$ is a Cauchy sequence. Let $a_n = \lim_{n\to\infty} a_{m,n}$. Now, putting $f = \sum_{n=0}^{\infty} a_n x^n$, we will check that $f \in B$ and that $\lim_{m\to\infty} \|f_m - f\|_r = 0$.

Indeed , let $\epsilon \in]0, +\infty]$, and let $N \in \mathbb{N}$ be such that $\|f_m - f_q\|_r \leq \epsilon$ whenever $m, q \geq N$. Fixing $m = N$, we see that for every $n \in \mathbb{N}$ we have

(2) $\|a_{N,n} - a_n\| r^n \leq \epsilon$. Hence , as soon as $\|a_{N,n}\| r^n \leq \epsilon$, we have $\|a_n\| r^n \leq \epsilon$,

and this shows f belongs to B. Besides by (2) it is seen that $\|f_N - f\|_r \leq \epsilon$. Thus B is complete and this ends the proof.

Theorem 1.18 is given in [52]. We give the same kind of proof.

Theorem 1.18: *Let A be a L-Banach algebra whose norm $\| \cdot \|$ belongs to $SM(A)$. For every $t \in A$, there exists $\varphi \in Mult\left(A, \| \cdot \|\right)$ such that $Ker\varphi \in Max(A)$, satisfying $\varphi(t) = \|t\|$.*

Proof: Let $t \in A$ be $\neq 0$, and let $r = \dfrac{1}{\|t\|}$. Let B be the Banach algebra of the series $\displaystyle\sum_{n=0}^{\infty} a_n x^n$ such that $\displaystyle\lim_{n\to\infty} \|a_n\| r^n = 0$, with respect to the norm $\| \cdot \|_r$ defined in Proposition 1.17. We will show that $1 - tx$ is not invertible in B. Indeed, we suppose it has an inverse $\displaystyle\sum_{n=0}^{\infty} a_n x^n$. Hence we have $\displaystyle\sum_{n=0}^{\infty} a_n x^n = 1 + \sum_{n=0}^{\infty} t a_n x^{n+1}$ and therefore $a_{n+1} = t a_n$ for every $n \geq 0$. By induction we have $a_n = t^n$ for all $n \geq 1$, and therefore $\|a_n\| r^n = 1$ for all $n \geq 1$. This obviously contradicts the hypothesis $\displaystyle\sum_{n=0}^{\infty} a_n x^n \in B$, and finally shows $1 - tx$ is not invertible. Let $\mathcal{M}$ be a maximal ideal of B which contains $1 - tx$ and let ψ be the canonical surjection from B onto the field $E = \dfrac{B}{\mathcal{M}}$. Then E is a L-Banach algebra with respect to the quotient norm $\| \cdot \|_q$ of $\dfrac{B}{\mathcal{M}}$. By Theorem 1.15, E admits an absolute value ϕ extending that of L, such that $\phi(x) \leq \|x\|_q$ whenever $x \in E$. In particular we have

(1) $\phi(\psi(t)) \leq \|\psi(t)\|_q \leq \|t\|$, and

(2) $\phi(\psi(x)) \leq \|x\|_q \leq \|x\| = r = \dfrac{1}{\|t\|}$.

But on the other hand $\psi(tx) = \psi(1) = 1$, hence $\phi(\psi(tx)) = 1$ and therefore $\phi(\psi(x)) = \dfrac{1}{\varphi(\psi(t))}$. Hence by (2) we obtain $\phi(\psi(t)) \geq \|t\|$, and by (1) we have $\phi(\psi(t)) = \|t\|$.

Now, putting $\varphi = \phi \circ \psi$, we see that φ belongs to $Mult_m(A, \| \cdot \|)$ and satisfies $\varphi(t) = \|t\|$. This ends the proof of Theorem 1.18.

2. INFRACONNECTED SETS

Infraconnected sets [11] were introduced to provide a class of sets as wide as possible, aimed to play the same role as the connected sets do in complex analysis towards holomorphic functions. We will see that for many things, this class of sets is quite satisfying.

Notations and definitions: In all this chapter L is an ultrametric field whose valuation is dense. Let $a \in L$ and let $r \in \mathbb{R}_+$. We will denote by $d(a,r)$ the disk $\{x \in L| \ |x - a| \ \leq r\}$, by $d(a,r^-)$ the disk $\{x \in L| \ |x - a| \ < r\}$ and we will call *circle of center a, of radius r* the set $C(a,r) = d(a,r) \setminus d(a,r^-)$.

Given r_1 and r_2 such that $0 < r_1 < r_2$ we will denote by $\Gamma(a,r_1,r_2)$ the annulus $\{x \in L| \ r_1 < |x-a| \ < r_2\}$ and by $\Delta(a,r_1,r_2)$ the annulus $\{x \in K| \ r_1 \leq |x - a| \ \leq r_2\}$.

We know that if $b \in d(a,r)$ then $d(b,r) = d(a,r)$. In the same way if $b \in d(a,r^-)$ then $d(b,r^-) = d(a,r^-)$. Moreover given two disks T and T' such that $T \cap T' \neq \emptyset$ then either $T \subset T'$ or $T' \subset T$.

Of course the three following statements are seen to be equivalent :

i) $d(a,r) = d(a,r^-)$

ii) $C(a,r) = \emptyset$

iii) $r \notin |\, L\,|$

Besides the disks $d(b,r^-)$ included in $C(a,r)$ (resp. in $d(a,r)$) are the disks $d(b,r^-)$ such that $b \in C(a,r)$ (resp. in $d(a,r)$). They are called *the classes* of $C(a,r)$ (resp. of $d(a,r)$).

Remark : In particular, all these notations will apply to a complete algebraically closed field.

Notations: *Henceforth D will denote a set in the field L .*

The closure of D is denoted by $\overline{D}$ and the interior of D is denoted by $\overset{\circ}{D}$.

We put $diam(D) = \sup\{|x - y| \ |x \in D, y \in D\}$ and $diam(D)$ is named the *diameter* of D.

If D is bounded of diameter R we denote by $\widetilde{D}$ the disk $d(a,R)$ for any $a \in D$.

If D is not bounded we put $\widetilde{D} = L$.

Given a point $a \in L$ we put $\delta(a,D) = \inf\{|x - a| \ |x \in D\}$. Then $\delta(a,D)$ is named *the distance of a to D.*

Lemma 2.1: $\widetilde{D} \setminus \overline{D}$ *admits a unique partition of the form* $(T_i)_{i \in I}$, *whereas each* T_i *is a disk of the form* $d(a_i, r_i^-)$ *with* $r_i = \delta(a_i, D)$.

Proof : For every $a \in \widetilde{D} \setminus \overline{D}$ let $r(a) = \delta(a, D)$. Let α and β be two points in $\widetilde{D} \setminus \overline{D}$ such that $|\beta - \alpha| < r(\alpha)$. It is easily seen that for every $x \in D$, we have $|x - \beta| = |x - \alpha|$, and then the family of the disks $T(\alpha) = d(\alpha, r(\alpha)^-)\,(\alpha \in \widetilde{D} \setminus \overline{D})$ makes a partition of $\widetilde{D} \setminus D$ because given α and $\beta \in \widetilde{D} \setminus \overline{D}$, either $|\alpha - \beta| < r(\alpha)$ and then $T(\alpha) = T(\beta)$, or $|\alpha - \beta| \geq r(\alpha)$ and then $|\alpha - \beta| \geq r(\beta)$ hence $T(\alpha) \cap T(\beta) = \emptyset$.

Definition: Such disks $d(a_i, r_i^-)$ are called *the holes* of D.

Example 1 : The holes of a disk $d(a, r^-)$ with $r \in |\,L\,|$ are the classes of $C(a, r)$.
Example 2 : The only one hole of $L \setminus d(0, 1^-)$ is $d(0, 1^-)$.
Example 3 : The holes of $L \setminus d(0, 1)$ are the disks $d(a, 1^-)$ with $a \in d(0, 1)$.

Definitions: A set D is said to be *infraconnected* if for every $a \in D$, the mapping I_a from D to $\mathbb{R}_+$ defined by $I_a(x) = |x - a|$ has an image whose closure in $\mathbb{R}_+$ is an interval. In other words, a set D is not infraconnected if and only if there exist a and $b \in D$ and an annulus $\Gamma(a, r_1, r_2)$ with $0 < r_1 < r_2 < |a - b|$ such that $\Gamma(a, r_1, r_2) \cap D = \emptyset$.
D is said to be *strongly infraconnected* if for every hole $T = d(a, r^-)$ with $r \in |\,L\,|$, there exists a sequence $(x_n)_{n \in \mathbb{N}}$ in D such that $|x_n - a| = |x_n - x_m| = r$ whenever $n \neq m$.

Lemma 2.2 is obvious:
Lemma 2.2: *If* D *is infraconnected of diameter* $R \in \mathbb{R}$ *(resp.* $+\infty$*) then* $\overline{I_a(D)} = [0, R]$ *(resp.* $\overline{I_a(D)} = [0, +\infty[$ *).*

Lemma 2.3: *Let* D *be strongly infraconnected. Let* $a \in D$ *and let* $r \in |\,L\,|$ *be such that* $r < diam(D)$. *There exists a sequence* $(x_n)_{n \in \mathbb{N}}$ *in* D *such that* $|x_n - a| = |x_n - x_m| = r$ *whenever* $n \neq m$.

Proof: Without loss of generality we may clearly assume D to be closed. Suppose Lemma 2.3 is false. Let $b \in C(a, r)$. If $b \notin D$ then b belongs to a hole $T = d(b, \rho^-)$. If $\rho = r$ then by definition of the strongly infraconnected sets, there exists a sequence $(a_n)_{n \in \mathbb{N}}$ in $C(b, r) \cap D$ such that $|a_n - a_m| = |a_n - b| = r$ whenever $n \neq m$, and therefore $|a_n - a| = r$ for every $n \in \mathbb{N}$ except maybe for one index, which is false by hypothesis. Hence we have $\rho < r$, and then there exists $\beta \in d(b, r^-) \cap D$. Now let $(b_n)_{n \in \mathbb{N}}$ be a sequence in $C(a, r)$ such that $|b_n - b_m| = r$ whenever $n \neq m$. The last reasoning applied to each b_n shows the existence of a point $\beta_n \in D$ such that $|\beta_n - b_n| < r$ and then we have $|\beta_n - \beta_m| = |\beta_n - a| = r$ whenever $n \neq m$.

Corollary 2.4:   *A strongly infraconnected set is infraconnected.*

Now we will study thoroughly the infraconnected sets. The following Lemma 2.5 gives a point of view from a hole of D.

Lemma 2.5:   *Let D be an infraconnected set and let α belong to a hole T of diameter ρ. The closure of the set $\{|x - \alpha| \mid x \in D\}$ is an interval whose lower bound is ρ.*

Proof: We just have to show that for every r and r' such that $\rho < r < r' < diam(D)$, there exists $\beta \in D$ such that $r < |\beta - \alpha| < r'$. By definition of the holes there exists $b \in D$ such that $|\alpha - b| < r$ and then, since D is infraconnected, there exists $\beta \in D$ such that $r < |b - \beta| < r'$. But it is seen that $|\beta - \alpha| = |b - \beta|$.

Given two infraconnected sets A and B we may prove $A \cup B$ to be infraconnected in the two following hypothesis (Th. 2.6 and 2.8).

Theorem 2.6:   *Let A and B be two infraconnected sets such that $A \cap B \neq \emptyset$. Then $A \cup B$ is infraconnected.*

Proof: If A and B are not bounded, the statement is obvious because for every $a \in A$, $\overline{I_a(A)} = \mathbb{R}_+$ and for every $a \in B$ we have $\overline{I_a(B)} = \mathbb{R}_+$. Now we may assume A to be bounded, of diameter R, while B has diameter $R' \geq R$ (resp. is not bounded). Then $A \cup B$ has diameter R' (resp. is not bounded). Let $c \in A \cap B$, let $a \in A \cup B$, and let us show that $\overline{I_a(A \cup B)} = [0, R']$ (resp. $[0, +\infty[$).

For convenience we first assume B to be bounded. Since $c \in A \cap B$ we see that $|x - a| \leq \max(|x - c|, |c - a|) \leq R'$ whenever $x \in A \cup B$ hence $\overline{I_a(A \cup B)} \subset [0, R']$. Hence we just have to show that $\overline{I_a(A \cup B)} \supset [0, R']$. Obviously $\overline{I_a(A \cup B)} = \overline{I_a(A)} \cup \overline{I_a(B)} = [0, R] \cup \overline{I_a(B)}$. Hence we just have to show that $\overline{I_a(B)} \supset [R, R']$. But when $x \in B$ with $|x - a| > R$, we see that $|x - a| = |x - c|$ (because $|c - a| \leq R$) hence $I_a(B) \cap]R, R'] = I_c(B) \cap]R, R']$ hence finally $\overline{I_a(B)} \supset [R, R']$ because $\overline{I_c(B)} \supset [R, R']$.

When B is not bounded, in the same way it is seen that $\overline{I_a(A \cup B)} = [0, +\infty[$. This finishes showing that $A \cup B$ is infraconnected.

Corollary 2.7:   *Let D be a set in K. The relation $\mathcal{R}$ defined by $x\mathcal{R}y$ if there exists an infraconnected subset of D that contains x and y, is an equivalence relation.*

Proof: $\mathcal{R}$ is obviously reflexive and symmetric. It is transitive by Theorem 2.6.

Definition: The equivalence classes with respect to this relation are called *the infraconnected componants.*

Examples :   1) $d(0,1^-) \cup d(1,1^-)$ is infraconnected. Its holes are the disks $d(\alpha,1^-)$ with $|\alpha| = |\alpha - 1| = 1$.
2)   Let $r \in]0,1[$ and let $D = d(0,1^-) \cup d(1,r)$. Then D is not infraconnected, its infraconnected components are $d(0,1^-)$ and $d(1,r)$. The holes of D are the disks $d(\alpha,1^-)$ with $|\alpha| = |\alpha - 1| = 1$ and the disks $d(\alpha,|\alpha - 1|^-)$ with $r < |\alpha - 1| < 1$.

Theorem 2.8:   *Let A and B be infraconnected sets such that $\widetilde{A} = \widetilde{B}$. Then $A \cup B$ is infraconnected.*

Proof: Obviously $\widetilde{A \cup B} = \widetilde{A}$. If A is bounded let $\widetilde{A} = d(\alpha, R)$ and otherwise let $\widetilde{A} = L$. First let us assume A to be bounded. For $a \in A$, the set $\{|x-a| \mid x \in A\}$ is dense in $[0,R]$ hence with greater reason so is the set $\{|x-a| \mid x \in A \cup B\}$. In the same way B plays the same role hence this still holds for $a \in B$. Finally if A is not bounded we just replace $[0,R]$ by $[0,+\infty[$. That finishes proving Theorem 2.8.

Definition: We will call *an empty annulus of D* an annulus $\Gamma(a,r_1,r_2)$ such that

$$
\begin{aligned}
\text{i)} \qquad r_1 &= \sup\{|x-a| \quad |x \in D, |x-a| \le r_2\} \\
\text{ii)} \qquad r_2 &= \inf\{\,|x-a| \quad |x \in D, |x-a| \ge r_1\}
\end{aligned}
$$

The set $d(a,r_1) \cap D$ will be denoted by $\mathcal{I}_D(\Gamma(a,r_1,r_2))$ while the set $(L \backslash d(a,r_2^-)) \cap D$ will be denoted by $\mathcal{E}_D(\Gamma(a,r_1,r_2))$. When there is no risk of confusion about the set D we will just write $\mathcal{I}(\Gamma(a,r_1,r_2))$, (resp. $\mathcal{E}(\Gamma(a,r_1,r_2))$) , instead of $\mathcal{I}_D(\Gamma(a,r_1,r_2))$, (resp. $\mathcal{E}_D(\Gamma(a,r_1,r_2))$) .

Remark 1:   By definition, D is not infraconnected if and only if it admits an empty annulus.

Remark 2:   By definition $\{\mathcal{I}(\Gamma(a,r_1,r_2)), \mathcal{E}(\Gamma(a,r_1,r_2))\}$ is a partition of D.

Examples: Let $r \in]0,1[$, let $D = d(0,r) \cup d(1,1^-)$ and let $D' = d(0,r^-) \cup d(1,r)$. Then $\Gamma(0,r,1)$ is an empty annulus of D and also of D'. In the same way $\Gamma(1,r,1)$ is also an empty annulus of D'.

Notation: Let $\mathcal{X}(D)$ be the set of the empty annuli of D. Given Λ_1 and $\Lambda_2 \in \mathcal{X}(D)$, it is easily seen that $\mathcal{I}(\Lambda_1) \subset \mathcal{I}(\Lambda_2)$ is equivalent to $\mathcal{E}(\Lambda_1) \supset \mathcal{E}(\Lambda_2)$.

We will denote by $\leq$ the relation defined on $\mathcal{X}(D)$ by $\Lambda_1 \leq \Lambda_2$ if $\mathcal{I}(\Lambda_1) \subset \mathcal{I}(\Lambda_2)$. It is easily seen that $\leq$ is a relation of order on $\mathcal{X}(D)$.

We will denote by $<$ the relation defined by $\Lambda_1 < \Lambda_2$ if $\Lambda_1 \leq \Lambda_2$ and $\Lambda_1 \neq \Lambda_2$.

The following Lemmas 2.9 and 2.10 are easily seen.

Lemma 2.9: *Let Λ_1 and Λ_2 be two empty annuli of D. The following assertions are equivalent :*

  i) *Λ_1 and Λ_2 are not comparable with respect to the order $\leq$*

  ii) *$\mathcal{I}(\Lambda_1) \subset \mathcal{E}(\Lambda_2)$*

  iii) *$\mathcal{I}(\Lambda_2) \subset \mathcal{E}(\Lambda_1)$*

  iv) *$\mathcal{I}(\Lambda_1) \cap \mathcal{I}(\Lambda_2) = \emptyset$.*

Lemma 2.10: *Let $\Lambda \in \mathcal{X}(D)$ and let $x \in \mathcal{I}(\Lambda)$ (resp. $x \in \mathcal{E}(\Lambda)$). The infraconnected component of x is included in $\mathcal{I}(\Lambda)$ (resp. in $\mathcal{E}(\Lambda)$). If $\Lambda' \in \mathcal{X}(D)$ is such that $\Lambda < \Lambda'$ then $\mathcal{I}(\Lambda') \cap \mathcal{E}(\Lambda) \neq \emptyset$.*

The following Lemma 2.11 is a direct consequence of Lemmas 2.9 and 2.10 .

Lemma 2.11: *Let Σ be an empty annulus of D. The family of the empty annuli $\Lambda \geq \Sigma$ is totally ordered.*

Proof: Let Λ_1 and $\Lambda_2 \in \mathcal{X}(D)$ satisfy $\Lambda_1 \geq \Sigma, \Lambda_2 \geq \Sigma$. Then $\mathcal{I}(\Lambda_1) \cap \mathcal{I}(\Lambda_2) \supset \mathcal{I}(\Sigma) \neq \emptyset$ hence $\mathcal{I}(\Lambda_1)$ is not included in $\mathcal{E}(\Lambda_2)$ hence Λ_1 and Λ_2 are comparable.

Lemma 2.12: *Let Σ be a minimal element of $\mathcal{X}(D)$ for the order $\leq$. Then $\mathcal{I}(\Sigma)$ is an infraconnected component of D.*

Proof: Suppose that $\mathcal{I}(\Sigma)$ is not infraconnected. By definition $\mathcal{I}(\Sigma)$ is of the form $d(a, R) \cap D$ hence there exists an empty annulus $\Lambda = \Gamma(\alpha, r_1, r_2)$ of $\mathcal{I}(\Sigma)$ with $\alpha \in d(a, R), r_1 < r_2 \leq R$ and some $\beta \in \mathcal{I}(\Sigma)$ such that $r_2 \leq |\alpha - \beta| \leq R$. Since $\Lambda \subset d(a, R)$ we see that $\Lambda \cap D = \emptyset$ hence Λ is an empty annulus of D and therefore $\Lambda < \Sigma$. This ends the proof of Lemma 2.12.

Theorem 2.13: *D has finitely many infraconnected components if and only if it has finitely many empty annuli. Moreover if so does D then one of the infraconnected components is $A_0 = \displaystyle\bigcap_{\Sigma \in \mathcal{X}(D)} \mathcal{E}(\Sigma)$ while the others are of the form*

$$A_i = \mathcal{I}(\Lambda_i) \cap \left(\bigcap_{\Sigma < \Lambda_i} \mathcal{E}(\Sigma) \right), \quad \text{with } \Lambda_i \in \mathcal{X}(D).$$

Proof: We will first assume $\mathcal{X}(D)$ to be finite and we will prove that the infraconnected components are in the form A_i, above, so that there will be finitely many ones.

Let $\Lambda_1, ..., \Lambda_n$ be these empty annuli of D and for every $i = 0, ..., n$, let A_i be the subsets of D defined from $\Lambda_1, ..., \Lambda_n$ as above. For every $x \in D$, for every $i = 1, ..., n$, either $x \in \mathcal{I}(\Lambda_i)$ or $x \in \mathcal{E}(\Lambda_i)$ hence it is easily seen that x belongs to one of the A_i, hence $D = \bigcup_{i=0}^{n} A_i$. We check that $A_i \cap A_j = \emptyset$ whenever $i \neq j$. First we assume $i = 0, j > 0$. Hence $A_0 \subset \mathcal{E}(\Lambda_j)$ while $A_j \subset \mathcal{I}(\Lambda_j)$ hence $A_0 \cap A_j = \emptyset$. Now we suppose $i > 0, j > 0$. If $\Lambda_i < \Lambda_j$ then $a_j \subset \mathcal{E}(\Lambda_i)$ while $A_i \subset \mathcal{I}(\Lambda_i)$ and then $A_i \cap A_j = \emptyset$. Hence we may assume that Λ_1 and Λ_2 are not comparable and then by Lemma 2.9 we have $\mathcal{I}(\Lambda_i) \cap \mathcal{I}(\Lambda_j) = \emptyset$ hence $A_i \cap A_j = \emptyset$. Then the family $(A_i)_{0 \leq i \leq n}$ makes a partition of D.

Now we will show that each A_i is infraconnected. Suppose that a certain A_h is not infraconnected for some $h > 0$ (resp. $h = 0$). Then it admits an empty annulus $\Lambda = \Gamma(a, r_1, r_2)$. First we notice that if $h = 0$ then $\lambda_h > \lambda$ because both a, b are centers of Λ_h. Now, if $h = 0$ (resp. $h > 0$), let $\Sigma \in \mathcal{X}(D)$ (resp. let $\Sigma \in \mathcal{X}(D)$ be such that $\Sigma < \Lambda_h$). Since both a, b belong to $\mathcal{E}(\Sigma)$, it is seen that all Λ is included in $\mathcal{E}(\Sigma)$, and therefore is included in A_h. This contradicts the hypothesis and finishes proving that A_h is infraconnected.

Next we check that each A_j is maximal in the set of the infraconnected subsets of D. Indeed let B be a subset of D that strictly contains a certain A_h and let $a \in B \setminus A_h$. If $h = 0$, there exists $\Sigma \in \mathcal{X}(D)$ such that $a \in \mathcal{I}(\Sigma)$, but $A_h \subset \mathcal{E}(\Sigma)$ and therefore Σ is included in an empty annulus of B. If $h > 0$, either a belongs to $\mathcal{E}(\Lambda_0)$ whereas $A_h \subset \mathcal{I}(\Lambda_0)$, and then Λ_0 is included in an empty annulus of B, or there exists $\Sigma \in \mathcal{X}(D)$ satisfying $\Sigma < \Lambda_0$ and $a \in \mathcal{I}(\Sigma)$, but then $A_h \subset \mathcal{I}(\Sigma)$ and therefore Σ is included in an empty annulus of B. Thus in each case B is not infraconnected, and this finishes showing that each A_i is maximal in the set of the infraconnected subsets of D. As a consequence, the infraconnected components of D are the A_i.

Now reciprocally we assume D to have infinitely many empty annuli. First let us suppose that D has a sequence of empty annuli $(\Lambda_n)_{n \in \mathbb{N}}$ such that $\Lambda_n < \Lambda_{n+1}$ (resp. $\Lambda_n > \Lambda_{n+1}$) for all $n \in \mathbb{N}$. By Lemma 2.8 , for every $n \in \mathbb{N}$ there exists $x_n \in \mathcal{E}(\Lambda_n) \cap \mathcal{I}(\Lambda_{n+1})$ (resp. $x_n \in \mathcal{I}(\Lambda_n) \cap \mathcal{E}(\Lambda_{n+1})$) and then the infraconnected component X_n of x_n satisfies $X_n \subset \mathcal{E}(\Lambda_n) \cap \mathcal{I}(\Lambda_{n+1})$ (resp. $X_n \subset \mathcal{I}(\Lambda_n) \cap \mathcal{E}(\Lambda_{n+1})$) hence $X_n \cap X_m = \emptyset$ for all $n \neq m$, hence D has infinitely many infraconnected components.

Finally we see that we may assume that every totally ordered set of empty annuli is finite. Hence there exists a sequence of empty annuli Λ_n that are minimal elements for the order $\leq$ on $\mathcal{X}(D)$, and then $\mathcal{I}(\Lambda_n) \cap \mathcal{I}(\Lambda_m) = \emptyset$ whenever $n \neq m$. By Lemma 2.12 , $\mathcal{I}(\Lambda_n)$ is an infraconnected component D_n of D such that $D_n \cap D_m = \emptyset$ whenever $n \neq m$. This finishes proving that D has infinitely many infraconnected components and this ends the proof of Theorem 2.13 .

Notations: $\widehat{L}$ will denote an extension of L provided with an absolute value that extends the one of L.

Given $a \in \widehat{L}$, $r > 0$, $\widehat{d}(a,r)$ (resp. $\widehat{d}(a,r^{-})$) will denote the disk $\{x \in \widehat{L} \mid |x-a| \leq r\}$ (resp. $\{x \in \widehat{L} \mid |x-a| < r\}$).

Let D be a set in L, of diameter $R \in \mathbb{R}$ (resp. $+\infty$), whose holes form a family $\left(d(a_i, r_i^{-})\right)_{i \in \mathbb{N}}$. Let $a \in D$. We will denote by $\widehat{D}$ the set $\widehat{d}(a,R) \setminus$

$$\left(\left(\bigcup_{i \in I} \widehat{d}(a_i, r_i^{-})\right) \bigcup (\overline{D} \setminus D)\right) \quad (\text{resp. } \widehat{L} \setminus \left(\bigcup_{i \in I} \widehat{d}(a_i, r_i^{-})\right).$$

Proposition 2.14 is easy:

Proposition 2.14 : *If D is closed (resp. infraconnected) in L, so is $\widehat{D}$ in $\widehat{L}$. Besides if L is complete and if D is open, so is $\widehat{D}$.*

Proof: If D is infraconnceted, it is obvious that so is $\widehat{D}$. Now assume that D is closed, and let a be a point of $\overline{\widehat{D}} \setminus \widehat{D}$. Since a does not belong to a hole of $\widehat{D}$, it belongs to $\overline{D} \setminus D$, but D is closed, and therefore a belongs to $\widehat{D}$.

Now we suppose L complete and D open. Let $a \in \widehat{D} \setminus \overset{\circ}{\widehat{D}}$. As L is complete, by definition of $\widehat{D}$, it is seen that a lies in $\overline{D}$. However, since D is open, a does not belong to D, because then it would belong to a disk $\widehat{d}(a,r)$ included in $\widehat{D}$. This contradicts the hypothesis, and finishes the proof.

3. MONOTONOUS AND CIRCULAR FILTERS

L denotes a complete ultrametric field whose valuation is dense. Monotonous and circular filters play an important role in ultrametric functional analysis, mainly because any rational function, in absolute value, admits a limit along any one of these filters [11], [14], [16], [17], [40].

Definitions: A filter $\mathcal{F}$ on L will be said to be *thinner* than a filter $\mathcal{G}$ if every element of $\mathcal{G}$ belongs to $\mathcal{F}$. In such a case, $\mathcal{G}$ will also said to be *less thin than* $\mathcal{F}$.

A sequence $(u_n)_{n\in\mathbb{N}}$ in L will be said to be *thinner* than a filter $\mathcal{G}$ if so is the filter defined by the sets $A_q = \{u_n | n \geq q\}$ $(q \in \mathbb{N})$. In such a case, $\mathcal{G}$ will also said to be *less thin than* the sequence $(u_n)_{n\in\mathbb{N}}$.

A sequence $(u_n)_{n\in\mathbb{N}}$ in L will be said to be *an increasing distances sequence* (resp. *a decreasing distances sequence*) if the sequence $|u_{n+1} - u_n|$ is strictly increasing (resp. decreasing) and has a limit $\ell \in \mathbb{R}_+^*$.

The sequence $(u_n)_{n\in\mathbb{N}}$ will be said to be *a monotonous distances sequence* if it is either an increasing distances sequence or a decreasing distances sequence.

A sequence $(u_n)_{n\in\mathbb{N}}$ in L will be said to be *an equal distances sequence* if $|u_n - u_m| = |u_m - u_q|$ whenever $n, m, q \in \mathbb{N}$ such that $n \neq m \neq q$.

Theorem 3.1:   *Let $(u_n)_{n\in\mathbb{N}}$ be a bounded sequence in L. Either we may extract a convergent subsequence or we may extract a monotonous distances subsequence or we may extract an equal distances subsequence from the sequence $(u_n)_{n\in\mathbb{N}}$.*

Proof: We suppose Theorem 3.1 to be false. For every $q \in \mathbb{N}$ the set of the circles $C(u_q, r)$ that contain some u_n is then finite.

Suppose that we have already defined integers n_q for $q \leq t$ satisfying

(1) $|u_{n_q} - u_{n_{q-1}}| < |u_{n_{q-1}} - u_{n_{q-2}}|$ for $2 \leq q \leq t$

and such that $d(u_{n_q}, |u_{n_q} - u_{n_{q-1}}|^-)$ contains infinitely many terms of the sequence (u_n). For every $q = 2, ..., t$, let $r_q = |u_{n_q} - n_{n_{q-1}}|$. Obviously, at least one of the circles $C(u_{n_t}, r)$, with $r < r_t$ contains infinitely many terms of the sequence $(u_n)_{n\in\mathbb{N}}$. Let $C(u_{n_t}, r_{t+1})$ be such a circle. It is seen that at least one class Λ of this circle contains infinitely many terms of the sequence because otherwise we would have a sequence of classes (Λ_j) each one containing at least one term $u_{\tau(j)}$ and then they should satisfy $|u_{\tau(j)} - u_{\tau(i)}| = r_{t+1}$ whenever $i \neq j$. Hence the sequence $(u_n)_{n\in\mathbb{N}}$ should admit an equal distances subsequence. Then we may pick up one term $u_{n_{t+1}}$ in Λ and we have constructed the finite subsequence up to the rank $t + 1$, satisfying the properties mentionned above. In the same way we may initiate the induction by defining n_2 from arbitrary n_0, n_1. The sequence $(u_{n_t})_{t \in \mathbb{N}}$ is then defined for every $t \in \mathbb{N}$ and satisfies (1) for $t > 1$. Let $\ell = \lim_{t \to \infty} |u_{n_t} - u_{n_{t+1}}|$. If $\ell = 0$ the subsequence $(u_{n_t})_{t \in \mathbb{N}}$ is convergent. If $\ell > 0$ this is a decreasing distances subsequence. Finally the hypothesis "Theorem 3.1 is false" is proven to be false and therefore Theorem 3.1 is proven.

Definitions and notations :   Let $a \in \tilde{D}$ and $R \in \mathbb{R}_+^*$ be such that $\Gamma(a, r, R) \cap D \neq \emptyset$ whenever $r \in]0, R[$ (resp. $\Gamma(a, R, r) \cap D \neq \emptyset$ whenever $r > R$). We call an *increasing* (resp. *a decreasing*) *filter of center a and diameter R, on D* the filter $\mathcal{F}$ on D that admits for base the family of sets $\Gamma(a, r, R) \cap D$ (resp. $\Gamma(a, R, r) \cap D$) . For every sequence $(r_n)_{n \in \mathbb{N}}$ such that $r_n < r_{n+1}$ (resp. $r_n > r_{n+1}$) and $\lim_{n \to \infty} r_n = R$, it is seen that the sequence $\Gamma(a, r_n, R) \cap D$ (resp. $\Gamma(a, R, r_n) \cap D$) is a base of $\mathcal{F}$ and such a base will be called *a canonical base* .

We call *a decreasing filter with no center of canonical base* $(D_n)_{n \in \mathbb{N}}$ *and diameter* $R > 0$, *on D* a filter $\mathcal{F}$ on D that admits for base a sequence $(D_n)_n \in \mathbb{N}$ in the form $D_n = d(a_n, r_n) \cap D$ with $D_{n+1} \subset D_n$, $r_{n+1} < r_n,\ \lim_{n \to \infty} r_n = R$, and

$$\bigcap_{n \in \mathbb{N}} d(a_n, r_n) = \emptyset.$$

Given an increasing (resp. a decreasing) filter $\mathcal{F}$ on D of center a and diameter r we will denote by $\mathcal{P}_D(\mathcal{F})$ the set $\{x \in D|\ |x - a| \geq r\}$ (resp. the set $\{x \in D|\ |x - a| \leq r\}$ and by $\mathcal{C}_D(\mathcal{F})$ the set $\{x \in D|\ |x - a| < r\}$ (resp. the set $\{x \in D|\ |x - a| > r\}$. When there is no risk of confusion we will only write $\mathcal{P}(\mathcal{F})$ instead of $\mathcal{P}_D(\mathcal{F})$, and $\mathcal{C}(\mathcal{F})$ instead of $\mathcal{C}_D(\mathcal{F})$. Besides $\mathcal{C}_D(\mathcal{F})$ will be named *the body of* $\mathcal{F}$ and $\mathcal{P}_D(\mathcal{F})$ will be named *the beach of* $\mathcal{F}$.

We call *a monotonous filter on D* a filter which is either an increasing filter or a decreasing filter (with or without a center).

Given a monotonous filter $\mathcal{F}$ we will denote by $diam(\mathcal{F})$ its diameter.

The field L is said to be *spherically complete* if every decreasing filter on L has a center in L. The field $\mathbb{C}_p$ for example is not spherically complete. However, every algebraically closed complete ultrametric field admits a spherically complete algebraically closed extension and this will be recalled in chapter 7.

Lemma 3.2:   *Let* $(a_n)_n \in \mathbb{N}$ *be an increasing distances (resp. a decreasing distances) sequence in D. There exists a unique increasing (resp. decreasing) filter $\mathcal{F}$ on D such that the sequence* $(a_n)_n \in \mathbb{N}$ *is thinner than $\mathcal{F}$.*

Proof: Let $r_n = |a_{n+1} - a_n|$ and let $R = \lim_{n \to \infty} r_n.$

We first suppose $(a_n)_{n\in\mathbb{N}}$ to be an increasing distances sequence. The increasing filter $\mathcal{F}$ of center a_0 , of diameter R is obviously less thin than the sequence $(a_n)_{n\in\mathbb{N}}$. We will show that $\mathcal{F}$ is unique. Let $\mathcal{G}$ be an increasing filter of center a, of diameter R', less thin than the sequence $(a_n)_{n\in\mathbb{N}}$. For every $r < R'$, there exists $q \in \mathbb{N}$ such that $a_n \in \Gamma(a, r, R')$ whenever $n \geq q$. If $a \in d(a_0, R^-)$ this clearly requires that $R = R'$ and then $\mathcal{G} = \mathcal{F}$. Let us suppose that $a \notin d(a_0, R^-)$. Then we have $|a - a_n| = |a_n - a_m| = C$ whenever $n \neq m$ so $R' > R$ and then $\Gamma(a, r, R')$ does not contain the a_n whenever $r > R$. Finally $\mathcal{G} = \mathcal{F}$.

We now suppose the sequence $(a_n)_{n\in\mathbb{N}}$ to be a decreasing distances sequence with a point a such that $|a - a_n| = |a_{n+1} - a_n|$ whenever $n \in \mathbb{N}$. Then the decreasing filter of center a, of diameter R is a decreasing filter less thin than the sequence $(a_n)_{n\in\mathbb{N}}$. We will show it to be the only decreasing filter less thin than the sequence $(a_n)_{n\in\mathbb{N}}$. Indeed given a decreasing filter $\mathcal{G}$ less thin than the sequence $(a_n)_{n\in\mathbb{N}}$, it must have a center because if it had no center, the sequence $d(a_{n+1}, |a_{n+1} - a_n|)$ would be a canonical base of it, but by definition it has an intersection that contains a. Then, symmetrical to the case when $\mathcal{F}$ is increasing, it is easily seen that $\mathcal{F}$ is unique.

Now we suppose that the sequence $(a_n)_{n\in\mathbb{N}}$ is a decreasing distances sequence and that there does not exist $a \in L$ such that $|a - a_n| = |a_{n+1} - a_n|$ whenever $n \in \mathbb{N}$. We put $|a_{n+1} - a_n| = r_n$. Hence the sequence of disks $d(a_{n+1}, r_n)$ has empty intersection and then the filter $\mathcal{F}$, a base of which is the sequence $(D_n)_{n\in\mathbb{N}}$ with $D_n = d(a_{n+1}, r_n) \cap D$, is a decreasing filter with no center, of diameter R. There is no decreasing filter with center $a \in L$, less thin than the sequence (a_n) because we should have $|a - a_n| = r_n$ whenever $n \in \mathbb{N}$. Hence it just remains to show that $\mathcal{F}$ is the only decreasing filter with no center less thin than the sequence (a_n). Let us suppose that there exists another decreasing filter $\mathcal{G}$ of diameter R' with no center, of canonical base $(D'_m)_{m\in\mathbb{N}}$ less thin than the sequence (a_n). If $R' > R$, since every D'_m contains points a_n, it is seen that all the a_n lie in $D \cap d(a_0, R) \subset D'_m$ whenever $m \in \mathbb{N}$, and this contradicts that $\mathcal{G}$ has no center. Hence we have $R' \leq R$. But symmetrically we have $R \leq R'$. Hence $R = R'$. We will show that $\mathcal{G} = \mathcal{F}$. For every $m \in \mathbb{N}$, let ρ_m be the diameter of D'_m and let $a_q \in D'_m$ be such that $r_q \leq \rho_m$. Clearly $a_n \in D'_m$ whenever $n \geq q$ hence $D_n \subset D'_m$ whenever $n > q$. In the same way, let $n \in \mathbb{N}$ and $t \in \mathbb{N}$ be such that $\rho_m < r_n$ whenever $m \geq t$. Then D'_m contains some a_s which belongs to $d(a_{n+1}, r_n) \cap D = D_n$, hence $D'_m \subset D_n$ whenever $m \geq t$. That finishes showing that $\mathcal{G} = \mathcal{F}$, and that ends the proof of Lemma 3.2 .

Lemma 3.3:   *Let D be infraconnected. Let $\mathcal{F}$ be an increasing filter (resp. a decreasing filter) on L, of center $a \in \tilde{D}$ and diameter $R \leq diam(D)$ (resp.*

$R < diam(D)$) *such that a does not belong to a hole of diameter $\rho \geq R$ (resp. $\rho > R$). Then $\mathcal{F}$ is secant with D and induces on D an increasing filter (resp. a decreasing filter) of center a and diameter R, on D.*

Proof: We just have to check that $\Gamma(a, r, R) \cap D \neq \emptyset$ whenever $r \in]0, R[$ (resp. $\Gamma(a, R, r) \cap D \neq \emptyset$ whenever $r > R$), and this is obvious when $a \in \overline{D}$ because D is infraconnected, and $R \leq diam(D)$ (resp. $R < diam(D)$). Now let us assume a to belong to a hole T of diameter $\rho < R$ (resp. $\rho \leq R$). By Lemma 2.3 for every $r < R$ (resp. $r > R$), D has points α such that $r < |a - \alpha| < R$ (resp. $R < |a - \alpha| < r$) and this ends the proof.

Definition: Let $\mathcal{F}$ be an increasing (resp. a decreasing) filter of center a and diameter R on D . $\mathcal{F}$ is said *to be pierced* if for every $r \in]0, R[$, (resp. $r < R$) , $\Gamma(a, r, R)$ (resp. $\Gamma(a, R, r)$) contains some hole T_m of D.
A decreasing filter with no center $\mathcal{F}$ on D is said *to be pierced* if for every $m \in \mathbb{N}$, $\widetilde{D}_m \setminus \widetilde{D}_{m+1}$ contains some hole T_m of D.

Remarks: The definition of a pierced filter with no center also applies to a deacreasing filter with a center and then is equivalent to the one given just above for such a filter .
If $\mathcal{F}$ is an increasing (resp. a decreasing) filter of center a, of diameter R , $\mathcal{F}$ is pierced if and only if there exists a sequence of holes $(T_n)_{n \in \mathbb{N}}$ of D such that $\delta(a, T_n) < \delta(a, T_{t+1})$, (resp. $\delta(a, T_n) > \delta(a, T_{n+1})$), $\lim_{n \to \infty} \delta(a, T_n) = R$.

Definitions: Given a Cauchy filter $\mathcal{F}$ on D , of limit a in L, we will call *a canonical base of $\mathcal{F}$* a sequence D_m in the form $d(a, r_m) \cap D$ with $0 < r_m < r_{m+1}$ and $\lim_{m \to \infty} r_m = 0$. The filter $\mathcal{F}$ is said *to be pierced* if for every $m \in \mathbb{N}, \widetilde{D}_m$ contains some hole of D.

Let $a \in \widetilde{D}$. Let $(T_{m,i})_{\substack{1 \leq i \leq s(m) \\ m \in \mathbb{N}}}$ be a sequence of holes of D which satisfies
$\delta(a, T_{m,i}) = d_m$ $(1 \leq i \leq h_m)$, $d_m < d_{m+1}$ (resp. $d_m > d_{m+1}$), $\lim_{m \to \infty} d_m = S > 0$.
The sequence $(T_{m,i})_{\substack{1 \leq i \leq s(m) \\ m \in \mathbb{N}}}$ is called *an increasing (resp. a decreasing) distances holes sequence that runs the increasing (resp. decreasing) filter of center a, of diameter R.*
Now let $(T_{m,i})_{\substack{1 \leq i \leq s(m) \\ m \in \mathbb{N}}}$ be a sequence of holes of D that satisfies
$\delta(a_m, T_{m,i}) = d_m$ $(1 \leq i \leq s(m))$, $d_m > d_{m+1}$, $\lim_{m \to \infty} d_m = R > 0$, where the filter $\mathcal{F}$ of base $D_m = d(a_m, d_m) \cap D$ is a decreasing filter with no center. The

sequence $(T_{m,i})_{\substack{1 \leq i \leq s(m) \\ m \in \mathbb{N}}}$ is called *a decreasing distances holes sequence that runs*
$\mathcal{F}$.

Summarizing these definitions, an increasing (resp. decreasing) distances holes sequence that runs an increasing (resp. decreasing) filter $\mathcal{F}$ will be just named *an increasing (resp. decreasing) distances holes sequence* and the filter $\mathcal{F}$ will be named *the increasing (resp. decreasing) filter associated to the sequence* $(T_{m,i})_{\substack{1 \leq i \leq s(m) \\ m \in \mathbb{N}}}$. The diamater of $\mathcal{F}$ will be called *the diameter of the sequence* $(T_{m,i})_{\substack{1 \leq i \leq s(m) \\ m \in \mathbb{N}}}$. If $\mathcal{F}$ has a center a , a will be named *the center of the sequence* $(T_{m,i})_{\substack{1 \leq i \leq s(m) \\ m \in \mathbb{N}}}$. If $\mathcal{F}$ has no center, the sequence $(T_{m,i})$ will be called *a decreasing distances holes sequence with no center.*

Finally, an increasing (resp. decreasing) distances holes sequence will be called *a monotonous distances holes sequence* and the sequence $(d_m)_{m \in \mathbb{N}}$ is called *the monotony* of the monotonous distances holes sequence.

Let $(T_{m,i})_{\substack{1 \leq i \leq s(m) \\ m \in \mathbb{N}}}$ be a monotonous distances holes sequences and for every $(m,i)_{\substack{1 \leq i \leq s(m) \\ m \in \mathbb{N}}}$ let $\rho_{m,i} = diam(T_{m,i})$. The number $\inf_{\substack{1 \leq i \leq s(m) \\ m \in \mathbb{N}}} \rho_{m,i}$ will be called

piercing of the sequence $(T_{m,i})_{\substack{1 \leq i \leq s(m) \\ m \in \mathbb{N}}}$.

If a monotonous holes sequence has a piercing $\rho > 0$, it will be said to be *well pierced*. If a monotonous filter $\mathcal{F}$ is run by a well pierced monotonous holes sequence, $\mathcal{F}$ will be said to be well pierced.

In each case the sequence of circles $C(a, d_m)$ when $\mathcal{F}$ has center a (resp. $C(a_{m+1}, d_m)$ when $\mathcal{F}$ has no center) will be said *to run the filter $\mathcal{F}$*, and *to carry* the monotonous distances holes sequence $(T_{m,i})_{\substack{1 \leq i \leq s(m) \\ m \in \mathbb{N}}}$.

A monotonous distances holes sequences $(T_{m,i})_{\substack{1 \leq i \leq s(m) \\ m \in \mathbb{N}}}$ will be said to be *simple* if $s(m) = 1$ for all $m \in \mathbb{N}$

Besides, a sequence of holes $(T_m)_{m \in \mathbb{N}}$ of D will be called *a Cauchy sequence of holes of limit* $a \in L$ if $\lim_{m \to \infty} \delta(a, T_m) = 0$. Such a sequence will be said *to run* the Cauchy filter of base $\{d(a, r) \cap D | r > 0\}$.

Notations: In all the propositions, Theorem, Corollaries below, γ is the homographic function $b + \dfrac{1}{x - a}$ with $a, b \in L$.

Proposition 3.4: *Let $\alpha \in D, r > 0$ be such that $|a - \alpha| < t$. Then $\gamma(C(\alpha, r)) = C(b, \frac{1}{r})$.*

Proof: We may assume $b = 0$ and then the proof is immediate.

Corollary 3.5:   Let $\alpha \in L, r_1, r_2 \in]0, +\infty[$ with $|a - \alpha| < r_1 < r_2$. Then
$$\gamma(\Gamma(\alpha, r_1, r_2)) = \Gamma(b, \frac{1}{r_2}, \frac{1}{r_1}).$$

Corollary 3.6:   Let $\mathcal{F}$ be the increasing (resp. decreasing) filter of center α and diameter $R > |a - \alpha|$, on $L \setminus \{a\}$. Then $\gamma(\mathcal{F})$ is the decreasing (resp. increasing) filter of center b and diameter $\dfrac{1}{R}$.

Lemma 3.7:   Let $\alpha \in L$ be such that $|\alpha - a| \neq r$. Then
$$\gamma(C(\alpha, r)) = C\left(\gamma(\alpha), \frac{r}{|a - \alpha|^2}\right).$$

Proof: When x belongs to $C(\alpha, r)$ we have
$$\gamma(x) - \gamma(\alpha) = \left|\frac{\alpha - x}{(x - a)(a - \alpha)}\right| = \frac{r}{|a - \alpha|^2}$$

hence $\gamma(C(\alpha, r)) \subset C\left(\gamma(\alpha), \frac{r}{|a - \alpha|^2}\right)$. Now let $\xi(u) = \gamma^{-1}(u) = a + \dfrac{1}{u - b}$. We see that $C\left((\gamma(\alpha), \frac{r}{|a - \alpha|^2}\right) \subset C(\alpha, r)$. Since γ and ξ are injective we see that γ must be a surjection onto $C\left(\gamma(\alpha), \frac{r}{|a - \alpha|^2}\right)$.

Corollary 3.8:   Let $\alpha \in L$ and $r, r' \in]0, +\infty[$ be such that $0 < r < r' < |a - \alpha|$. Then we have $\gamma\left(\Gamma(\alpha, r, r')\right) = \Gamma\left(\gamma(\alpha), \frac{r}{|a - \alpha|^2}, \frac{r'}{|a - \alpha|^2}\right)$,
$$\gamma(d(\alpha, r)) = d\left(\gamma(\alpha), \frac{r}{|a - \alpha|^2}\right), \quad \gamma(d(\alpha, r^-)) = d\left(\gamma(\alpha), \left(\frac{r}{|a - \alpha|^2}\right)^-\right).$$

Corollary 3.9:   Let $\mathcal{F}$ be the increasing (resp. decreasing) filter of center α and diameter R on $L \setminus \{a\}$ with $|a - \alpha| > R$. Then $\gamma(\mathcal{F})$ is an increasing (resp. a decreasing) filter of center $\gamma(\alpha)$, of diameter $\dfrac{R}{|a - \alpha|^2}$ on $L \setminus \{b\}$.

Corollary 3.10:   Let $\mathcal{F}$ be a decreasing filter with no center , of center base $(D_n)_{n \in \mathbb{N}}$ on $L \setminus \{a\}$ such that $a \notin D_0$. Then $\gamma(\mathcal{F})$ is a decreasing filter with no center, of canonical base $\left(\gamma(D_n)\right)_{n \in \mathbb{N}}$ on $L \setminus \{b\}$.

Theorem 3.11: *We suppose $a \in D$. Let $D' = \gamma(D)$. Let $\mathcal{F}$ be a filter on D which is either a monotonous filter or a Cauchy filter. Then $\mathcal{F}$ is pierced if and only if $\gamma(\mathcal{F})$ is a pierced filter on D'.*

Proof: For example we suppose first $\mathcal{F}$ to be a monotonous filter. By definition, $\mathcal{F}$ is the intersection with D of a monotonous filter $\mathcal{G}$ of $L \setminus \{a\}$. Hence $\mathcal{F}$ is pierced if and only if $\mathcal{G}$ is secant with $(L \setminus \{a\}) \setminus \overline{D}$. Since γ is bicontinuous in $L \setminus \{a\}$ we see that $\gamma(\mathcal{G})$ is secant with $(L \setminus \{b\}) \setminus \overline{D'}$ if and only if $\mathcal{G}$ is secant with $(L \setminus \{a\}) \setminus \overline{D}$ because $\gamma(\overline{D}) \setminus \{a\} = \overline{D'} \setminus \{b\}$. Hence the conclusion is clear. In the same way if $\mathcal{F}$ is a Cauchy filter of limit $\alpha \in L$, we consider the filter $\mathcal{G}'$ of the neighbourhoods of $\gamma(\alpha)$ in $L \setminus \{b\}$, and we see that $\mathcal{G}$ is secant with $(L \setminus \{a\}) \setminus \overline{D}$ if and only if $\mathcal{G}'$ is secant with $(L \setminus \{b\}) \setminus \overline{D'}$.

We are now going to define the circular filters, which roughly characterize the absolute values on $L(x)$.

Lemma 3.12: *Let $a \in \tilde{D}$, let ρ be the distance from a to D and let R be such that $\rho \leq R \leq diam(D)$. For $j = 1, ..., q$ let $\alpha_j \in d(a, R)$ and let $r'_j, r''_j \in \mathbb{R}_+$ be such that $r'_j < R < r''_j$. Then $\bigcap\limits_{j=0}^{q} (\Gamma(\alpha_j, r'_j, r''_j) \cap D) \neq \emptyset$.*

Proof: If $\rho < R$ we put $r' = \max\limits_{1 \leq j \leq q} r'_j$ and we see that $\Gamma(a, r', R) \cap D$ is not empty (because D is infraconnected) and is included in every set $\Gamma(\alpha_j, r'_j, r''_j) \cap D$. If $R < diam(D)$ we put $r'' = \min\limits_{1 \leq j \leq q} r''_j$ and in the same way, $\Gamma(a, R, r'') \cap D$ is not empty and is included in every set $\Gamma(\alpha_j, r'_j, r''_j) \cap D$.

Now if $\rho = R = diam(D)$, let $b \in D$, and let $r' = \max\limits_{1 \leq j \leq q} r'_j$. Then $\Gamma(b, r', R) \cap D$ is not empty and is included in every set $\Gamma(\alpha_j, r'_j, r''_j) \cap D$.

Definition: Let $a \in \tilde{D}$, let $\rho = \delta(a, D)$ and let $R \in]0, +\infty[$ be such that $\rho \leq R \leq diam(D)$. We call *circular filter of center a and diameter R on D* the filter $\mathcal{F}$ which admits as a generating system the family of sets $\Gamma(\alpha, r', r'') \cap D$ with $\alpha \in d(a, R), r' < R < r''$, i.e. $\mathcal{F}$ is the filter which admits for base the family of sets of the form $D \cap (\bigcap\limits_{i=1}^{q} \Gamma(\alpha_i, r'_i, r''_i))$ with $\alpha_i \in d(a, R), r'_i < R < r''_i$ $(1 \leq i \leq q , q \in \mathbb{N})$.

For reasons that will appear when characterizing the absolute values of $L(x)$, a decreasing filter with no center, of canonical base $(D_n)_{n \in \mathbb{N}}$ will also be called *circular filter on D with no center, of canonical base $(D_n)_{n \in \mathbb{N}}$.*

Finally the filter of the neighbourhoods of a point $a \in D$ will be called *circular filter of the neighbourhoods of a on D*. It will also be named *circular filter of center a and diameter* 0.

A circular filter on D will be said to be *large* if it has diameter different from 0. Given a circular filter $\mathcal{F}$, its diameter will be denoted by $diam(\mathcal{F})$.

The following Proposition 3.13 is just a translation of the definitions.

Proposition 3.13:   *Let $\mathcal{F}$ be an increasing filter (resp. a decreasing filter) of center a and diameter R, on D . Then the circular filter of center a and diameter R on L is secant with D, and is the only circular filter on D less thin than $\mathcal{F}$.*

Reciprocally let $\mathcal{F}$ be a circular filter of center a and diameter R, on D secant with $d(a, R^-)$ (resp. $L \setminus d(a, R)$). Then the increasing filter (resp. decreasing filter) of center a and diameter R on L is secant with D and thinner than $\mathcal{F}$.

Proposition 3.14:   *Let D be infraconnected, let $a \in \tilde{D}$, let Θ be the closure of $\{|x - a| \,\, |x \in D\}$ in $\mathbb{R}$. For every $r \in \Theta$ the circular filter $\mathcal{F}$ of center a and diameter r is secant with D.*

Proof: Let $a \in \tilde{D}$. We first suppose that $C(a,r) \cap D = \emptyset$. Then either there exists a sequence $(x_n)_{n \in \mathbb{N}}$ in D such that $r < |x_{n+1} - a| < |x_n - a|$, $\lim_{n \to \infty} |x_n - a| = r$, or there exists a sequence $(x_n)_{n \in \mathbb{N}}$ in D such that $|x_n - a| < |x_{n+1} - a| < r$, $\lim_{n \to \infty} |x_n - a| = r$. In both cases the circular filter $\mathcal{F}$ of center a and diameter R is clearly secant with D.

Now we may suppose that $C(a,r) \cap D \neq \emptyset$. Let $b \in C(a,r) \cap D$. We see that b is also a center of $\mathcal{F}$. Since D is infraconnected and since $|a - b| = r \leq diam(D)$ there does exist a sequence $(x_n)_{n \in \mathbb{N}}$ in D such that $\lim_{n \to \infty} |x_n - b| = r$. Hence $\mathcal{F}$ (that has center b and diameter r), is secant with D.

Proposition 3.15:   *Let $(a_n)_{n \in \mathbb{N}}$ be a sequence in D that is either a monotonous distances sequence or a constant distances sequence. Then there exists a unique circular filter on D less thin than the sequence (a_n).*

Proof: First we suppose that the sequence $(a_n)_{n \in \mathbb{N}}$ is an increasing (resp. a decreasing) distances sequence. By Lemma 3.2 there exists a unique increasing (resp. decreasing) filter $\mathcal{F}$ on D less thin than the sequence $(a_n)_{n \in \mathbb{N}}$. If $\mathcal{F}$ has center a and diameter R, by Proposition 3.13, $\mathcal{F}$ is less thin than the circular filter of center a , of diameter R on D. If $\mathcal{F}$ is decreasing with no center, $\mathcal{F}$ is a circular filter.

Now we suppose that $(a_n)_{n \in \mathbb{N}}$ is a constant distances sequence. We put $R = |a_n - a_m|$ for $n \neq m$ and $a = a_0$. The circular filter of center a of diameter R on

L is clearly secant with D because each set $\Lambda_n = \Gamma(a_n, r', r'')$ with $r' < R < r''$ belongs to its canonical generating system and contains a_m for every $m > n$ hence its intersection with D is a circular filter $\mathcal{C}$ on D less thin than the sequence $(a_n)_{n \in \mathbb{N}}$. That ends the proof.

Notations: Let L' be an extension of L provided with an absolute value that extends the one of L. Let D be a set in L, let $\mathcal{F}$ be a monotonous filter on D, and let D' be a set in L' that contains D. Let $(a_n)_{n \in \mathbb{N}}$ be a monotonous distances sequence that runs $\mathcal{F}$. In $\widehat{D}$, there is a unique monotonous filter less thin than the sequence $(a_n)_{n \in \mathbb{N}}$. This filter will be denoted by $\widehat{\mathcal{F}}$.

In the same way, let $\mathcal{G}$ be a circular filter of center a and diameter r on D. We will denote by $\widehat{\mathcal{G}}$ the filter of center a and diameter r on D'. Finally let $\mathcal{G}$ be a circular filter with no center. Then it is a decreasing filter, hence we have already previously defined $\widehat{\mathcal{G}}$.

4. ULTRAMETRIC ABSOLUTE VALUES

AND VALUATION FUNCTIONS $v(h, \mu)$ ON $K(x)$

Notations: As in Chapter 1, and in the following ones, F denotes an algebraically closed field provided with an ultrametric absolute value, and L is a subfield of F. Besides $\mathcal{L}$ is the residue class field of L and for every $x \in L$ such that $|x| \leq 1$, $\overline{x}$ is the residue class of x in $\mathcal{L}$.

We will show the absolute values on $F(x)$ to be characterized by the circular filters on F. Actually all the properties of the Analytic Elements are depending on this. At the same time we will define the valuation functions $v(h, \mu)$ that are convenient mainly because they are piecewise linear. They just translate the properties of the abolute values on $F(x)$ into terms of piecewise linear functions.

Let $L[x_1, ..., x_q]$ be an algebra of polynomials in q indeterminates, with coefficients in L. For each $P(x_1, ..., x_q) = \displaystyle\sum_{j_1 + ... + j_q \leq t} a_{i_1, ..., i_q} x_1^{j_1} ..... x_q^{j_q}$ we put

$$\overline{P(x_1, ..., x_q)} := \sum_{j_1 + ... + j_q \leq t} \overline{a_{i_1, ..., i_q}} x_1^{j_1} ..... x_q^{j_q}.$$

On $L[x_1, ..., x_q]$, we put $\|P\| := \displaystyle\sup_{j_1 + ... + j_q \leq t} |a_{i_1, ..., i_q}|$.

Lemma 4.1: $\| \, . \, \|$ *is a multiplicative norm of L-algebra.*

Proof :   Let $B = L[x_1, ..., x_q]$. $\| \, . \, \|$ is clearly seen to be an ultrametric norm of L-vector space on B. We will check that $\|PQ\| = \|P\|\|Q\|$ whenever $P, Q \in B$. Both $\|P\|$, $\|Q\|$ belong to $|L|$. Hence, without loss of generality, we may clearly assume $\|P\| = \|Q\| = 1$. Thus we have $\overline{P} = \overline{Q} = \overline{1}$. But since $\mathcal{L}[x_1, ..., x_q]$ is a ring without divisors of zeros, we have $\overline{PQ} \neq \overline{0}$ and therefore $\|PQ\| = 1$. This ends the proof.

Definition: The norm $\| \, . \, \|$ on $L[x_1, ..., x_q]$ is named *the Gauss norm*.

Lemma 4.2:   *Let $P(x) = \displaystyle\sum_{j=0}^{n} a_j x^j \in F[x] \backslash \{0\}$, and let $r \in \mathbb{R}_+$. Then $|P(x)|$*

admits a limit $|P|(r)$ equal to $\displaystyle\max_{0 \leq j \leq n} |a_j| r^j$ when $|x|$ approaches r but remains *different from r. Let $x \in d(0,r)$. Then $|P(x)| \leq |P|(r)$. If P has no zero in the class of x in $d(0,r)$ then $|P(x)| = |P|(r)$. If P has at least one zero in this class, then $|P(x)| < |P|(r)$.*

Proof: Let $r' < r$ and $r'' > r$ be such that P has no zero in $\Gamma(0,r',r) \cup \Gamma(0,r,r'')$. We may obviously assume P to be monic. Let $P(x) = \displaystyle\prod_{i=1}^{n}(x - \alpha_i)$ with

$$|\alpha_i| < r' \text{ for } i \leq h$$
$$|\alpha_i| > r'' \text{ for } i \geq \ell$$
$$|\alpha_i| = r \text{ for } i = h, ..., \ell.$$

Now let $x \in \Gamma(0,r',r)$. Clearly $|x - \alpha_i| = |x|$ whenever $i \leq h$ while $|x - \alpha_i| = |\alpha_i|$ whenever $i > h$ hence $|P(x)| = |x|^h \displaystyle\prod_{i=h+1}^{n} |\alpha_i|$ hence $\displaystyle\lim_{|x| \to r^-} |P(x)| = r^h \displaystyle\prod_{i=h+1}^{n} |\alpha_i|$.

Symmetrically we show that $\displaystyle\lim_{|x| \to r^+} |P(x)| = r^{\ell-1}\displaystyle\prod_{i=\ell}^{n}|\alpha_i| = r^h \displaystyle\prod_{i=h+1}^{n} |\alpha_i|$. But the terms $|a_j x^j|$ are all different for every $|x|$ except for finitely many values so that there exist $\rho' \in [r',r[$ and $\rho'' \in]r,r'']$ such that the $|a_j x^j|$ are all different when $x \in \Gamma(0,\rho',r) \bigcup \Gamma(0,r,\rho'')$. Then we have $|P(x)| = \displaystyle\max_{0 \leq j \leq n} |a_j||x|^j$ hence $\displaystyle\lim_{\substack{|x| \to r \\ |x| \neq r}} |P(x)| = \displaystyle\max_{0 \leq j \leq n} |a_j| r^j$.

Now let $x \in d(0,r)$ and let us assume P to have no zero in the class Λ of x in $d(0,r)$. This means that $|x - \alpha_i| = r$ whenever $i = 0, ..., \ell - 1$ and $|x - \alpha_i| = |\alpha_i|$ whenever $i \geq \ell$. Thus $|P(x)| = r^{\ell-1}\displaystyle\prod_{i=\ell}^{n}|\alpha_i|$. If P has at least one zero α_{h+1} in the class of x we see that $|x - \alpha_{h+1}| < r$ while $|x - \alpha_i| \leq r$ for every $i = h+2, ..., \ell-1$, hence $|P(x)| < r^{\ell-1}\displaystyle\prod_{i=\ell}^{n}|\alpha_i|$ and finally $|P(x)| < |P|(r)$.

Lemma 4.3: *For every $r > 0$ the mapping from $F[x]$ into $\mathbb{R}_+$ defined by $P \to |P|(r)$ is an absolute value on $F[x]$.*

Proof: Since $|P|(r) = \displaystyle\lim_{\substack{|x| \to r \\ |x| \neq r}} |P(x)|$ we obviously have

$$|P + Q|(r) \leq \max(|P|(r), |Q|(r))$$

and $|PQ|(r) = |P|(r)\,|Q|(r)$. Let $P(x) = \sum_{j=0}^{n} a_j x^j$. Since $|P|(r) = \max_{0 \le j \le n} |a_j| r^j$ we see that $|P|(r) = 0$ if and only if $P = 0$.

Lemma 4.4: *Let $r \in \mathbb{R}_+$ and let $a \in F$ be such that $|a| \le r$. Then $|P(x)|$ has a limit when $|x - a|$ approaches r but remains different from r.*

Proof: We set $x = a + u$ and $P_a(u) = P(a + u)$. For every $P \in F[x]$ we have $\lim\limits_{\substack{|u| \to r \\ |u| \ne r}} |P_a(u)| = |P_a|(r)$. In particular $|P_a|(r) = \lim\limits_{\substack{|u| \to r+ \\ |u| < r}} |P_a(u)|$. But for every $\rho > r$, $C(0, \rho) = C(a, \rho)$, hence $\lim\limits_{|x| \to r+} |P(x)| = \lim\limits_{|u| \to r+} |P(a + u)| = \lim\limits_{|u| \to r+} |P_a(u)|$. That ends the proof of Lemma 4.4.

Theorem 4.5: *Let $P(x) = \sum_{j=0}^{q} a_j x^j \in F[x]$ be a monic polynomial such that $a_j \in d(0, 1)$ whenever $j = 0, ..., q$. Then the q zeros of P belong to $d(0, 1)$.*

Proof: Let ψ be the absolute value defined on $F[x]$ by $\psi(P) = \lim\limits_{|u| \to 1, |u| \ne 1} |P(u)|$ and let $P(x) = \prod_{j=1}^{q} (x - c_j)$. By Lemma 4.2, for each $j = 1, ..., q$, it is seen that $\psi(x - c_j) \ge 1$ while $\psi(P) = 1$. Hence $\psi(x - c_j) = 1$ for every $j = 1, ..., q$ and therefore by Lemma 4.2 again, we have $c_j \le 1$ whenever $j = 1, ..., q$.

These ultrametric absolute values defined on $F[x]$ are immediately extended to ultrametric absolute values defined on $F(x)$ by $\left|\dfrac{P}{Q}\right|(r) := \dfrac{|P|(r)}{|Q|(r)}$. Their properties are summarized in Lemma 4.6 that is an immediate application of Lemmas 4.2 , 4.3 , 4.5.

Lemma 4.6: *Let $h \in F(x)$ and let $r \in \mathbb{R}_+$. For every $a \in d(0, r)$ we have $\lim\limits_{\substack{|x-a| \to r \\ |x-a| \ne 0}} |h(x)| = |h|(r)$. Let $x \in C(0, r)$. If h has no zero and no pole in the class of x in $C(0, r)$ then $|h(x)| = |h|(r)$.*

Notations: For $\mu \in \mathbb{R}$ we put $v(h, \mu) = -\log(|h|(\omega^{-\mu}))$. The translation of Lemmas 4.1, 4.2, 4.3, 4.4, 4.5, 4.6 into terms of valuation allows us to obtain the following Lemmas 4.7 and 4.8.

Lemma 4.7:   Let $P(x) = \sum_{j=0}^{n} a_j x^j \in L[x] \setminus \{0\}$. For every $\mu \in \mathbb{R}$, we have $v(P,\mu) = \inf_{0 \leq j \leq n} v(a_j) + j\mu$. For all $x \in L$, we have $v(P(x)) \geq v(P, v(x))$ and if $L = F$, the equality $v(P(x)) = v(P, v(x))$ holds if and only if P has no zero α such that $v(x) < v(x - \alpha)$.

Lemma 4.8:   Let $h \in F(x) \setminus \{0\}$. We have $v(h(x)) = v(h, v(x))$ for every $x \in F$ such that h has no zero α satisfying $v(x - \alpha) > v(x)$ and no pole β satisfying $v(x - \beta) > v(x)$.

Lemma 4.9:   Let $h_1, h_2 \in L(x) \setminus \{0\}$. Then we have $v(h_1 + h_2, \mu) \geq \min\big(v(h_1, \mu), v(h_2, \mu)\big)$. When $v(h_1, \mu) < v(h_2, \mu)$, then we have $v(h_1 + h_2, \mu) = v(h_1, \mu)$. Besides $v(h_1.h_2, \mu) = v(h_1, \mu) + v(h_2, \mu)$.

Notations :   In order to perform easily any change of origin, for every $a \in F$ and $h \in F(x) \setminus \{0\}$ we put $v_a(h, \mu) = v(h_a, \mu)$ with $h_a(u) = h(a + u)$. Thus if $\mathcal{F}$ denotes the circular filter of center a, and diameter $\omega^{-\mu}$ then $v_a(h, \mu) = -\log(\varphi_{\mathcal{F}}(h))$.

We now consider again a polynomial $P(x) = \sum_{j=0}^{n} a_j x^j \neq 0$. We denote by $N^+(P, \mu)$ (resp. $N^-(P, \mu)$) the biggest (resp. the smallest) index j such that $v(a_j) + j\mu = v(P, \mu)$.

Let f be a function defined in an interval I, provided with a right side (resp. left side) derivative at a point $a \in I$. We will denote by $f'^r(a)$ (resp. $f'^l(a)$) its right side (resp. left side) derivative at a.

For convenience, on $L[x]$ we denote by $v(\, . \,)$ the valuation associated to the norm $\| \, . \, \|$ i.e. $v(P) = v(P, 0)$ whenever $P \in L[x]$.

Lemma 4.10 is a consequence of Lemma 4.6:

Lemma 4.10:   Let $h \in F(x) \setminus \{0\}$ and let $a, b \in F$. For every $\mu \leq v(a - b)$, we have $v_a(h, \mu) = v_b(h, \mu)$.

Theorem 4.11:   Let $P(x) = \sum_{j=0}^{n} a_j x^j \in F[x]$. For every $\mu \in \mathbb{R}$, $N^+(P, \mu) - N^-(P, \mu)$ is equal to the number of zeros admitted by P in the circle $C(0, \omega^{-\mu})$ in F. The function $N^+(P, .)$ (resp. $N^-(P, .)$) is decreasing and continuous on the left (resp. on the right).

The function $v(P,.)$ is continuous, piecewise linear, increasing, concave, and has a left side derivative (resp. a right side derivative) equal to $N^+(P,\mu)$ (resp. $N^-(P,\mu)$).

Proof: It is easily seen that the equality (1) $N^+(P,\mu) = N^-(P,\mu)$ holds for every μ but finitely many values, at most n. It is also clear that the functions $N^+(P,.)$ and $N^-(P,.)$ are decreasing. By continuity we see that the function $N^+(P,\mu)$ is continuous on the left at each point while $N^-(P,\mu)$ is continuous on the right at each point. Finally if (1) holds in an interval $]\mu',\mu''[$, then the functions $N^+(P,.)$ and $N^-(P,.)$ are constant and equal. Consider an interval $]\mu',\mu''[$ such that $N^+(P,\mu) = N^-(P,\mu)$ for all $\mu \in]\mu',\mu''[$, and let $i = N^+(P,\mu)$ whenever $\mu \in]\mu',\mu''[$. Then $v(P,\mu) = v(a_i) + i\mu$ so that the function $v(P,.)$ is in the form $A + i\mu$ in this interval.

Now let ν be such that $N^+(P,\nu) > N^-(P,\nu)$. We see that $v(P,.)$ is still continuous at ν and has a right side derivative equal to $N^-(P,\nu)$ and a left side derivative equal to $N^+(P,\nu)$. Finally the function $v(P,.)$ is continuous, piecewise linear, concave, and largely increasing.

If P and $Q \in F(x) \setminus \{0\}$ then $N^+(PQ,\mu)$ is the left side derivative of the function $v(PQ,.)$. But $v(PQ,.) = v(P,.) + v(Q,.)$, hence its left side derivative at μ is just $N^+(P,\mu) + N^+(Q,\mu)$. In the same way we have $N^-(PQ,\mu) = N^-(P,\mu) + N^-(Q,\mu)$ by considering right side derivatives.

Then, to prove that $N^+(P,\mu) - N^-(P,\mu)$ is the number of zeros of P in $C(0,\omega^{-\mu})$, it is sufficient to show it when P is a binomial $x - a$. But then it is obvious because $N^+(P,\mu) = N^-(P,\mu) = 0$ whenever $\mu > v(\alpha)$, $N^+(P,\mu) = N^-(P,\mu) = 1$ whenever $\mu < v(\alpha)$, while $N^+(P,v(\alpha)) = 1$, $N^-(P,v(\alpha)) = 0$. So all the statements of Theorem 4.11 have been proven.

Applying Lemma 4.7 and Theorem 4.11 to the numerator and the denominator of a rational function, we obtain Corollary 4.12.

Corollary 4.12: *Let $h \in F(x) \setminus \{0\}$. The function in μ $v(h,.)$ is continuous and piecewise linear.*

If μ is such that $d(0,\omega^{-\mu})$ contains s zeros and t poles of h, (taking multiplicities into account), but neither any zero nor any pole in $C(0,\omega^{-\mu})$ then $v(h,.)$ has a derivative equal to $s - t$ at μ .

If μ is such that $C(0,\omega^{-\mu})$ contains s zeros and t poles of h, (taking multiplicities into account), then we have $v'^l(h,\mu) - v'^r(h,\mu) = s - t$. Besides, if the function $v(f,\mu)$ is not derivable at μ, then μ lies in $v(K)$.

Corollary 4.13: *Let $h \in F(x) \setminus \{0\}$ have s zeros and t poles in $d(0,r)$ and have neither any zero nor any pole in $\Gamma(0,r,r')$. Then in $\Gamma(0,r,r')$, $v(h(x))$ is* *of the form $A + (s - t)v(x)$.*

Now we will characterize all the absolute values on $F(x)$ [40], [42].

Theorem 4.14: (B.Guennebaud)   *For every large circular filter $\mathcal{F}$ on F, for every rational function $h(x) \in F(x), |h(x)|$ has a limit $\varphi_{\mathcal{F}}(h)$ along the filter $\mathcal{F}$. The mapping $\mathcal{F} \to \varphi_{\mathcal{F}}$ from the set of the large circular filters of F into the set of the multiplicative norms of $F(x)$ is a bijection. If $\mathcal{F}$ has center 0 and diameter r, then $\varphi_{\mathcal{F}}(h) = |h|(r)$.*

Proof: We first suppose that $\mathcal{F}$ has center $a \in F$ and diameter $r > 0$. With no loss of generality we may obviously assume $a = 0$ by means of the change of variable $x = a + u$. Then by Lemma 4.2, $|h(x)| = |h|(r)$ holds in every class of $C(0,r)$ but finitely many ones $\Lambda_1, ..., \Lambda_q$. For every $j = 1, ..., q$ let $\alpha_j \in \Lambda_j$ and let $\alpha_0 = 0$. Let ϵ be > 0. By Lemma 4.2 there exist $\rho' \in]0, r[$ and $\rho'' > r$ such that $|\, |h(x)| - |h|(r)\, | \le \epsilon$ for every $x \in \bigcap_{j=0}^{q} \Gamma(\alpha_j, \rho', \rho'')$, so $\lim_{\mathcal{F}} |h(x)| = |h|(r)$.

Now we suppose that $\mathcal{F}$ has no center in F. It admits a canonical base $(D_n)_{n \in \mathbb{N}}$ and then given $h \in R(D)$, there exists $q \in \mathbb{N}$ such that h has neither any zero nor any pole inside D_q. Hence $|h(x)|$ is equal to a constant l in D_q, and therefore we have $\lim_{\mathcal{F}} |h(x)| = l$. By the same kind of reasonning as in Lemma 4.3 it is easily seen that $\varphi_{\mathcal{F}}$ is an absolute value on $F(x)$. We will check that the mapping $\mathcal{F} \to \varphi_{\mathcal{F}}$ is injective. Indeed let $\mathcal{F}_1, \mathcal{F}_2$ be two different circular filters and let r_1 (resp. $r_2 \ge r_1$) be the diameter of $\mathcal{F}_1$ (resp. $\mathcal{F}_2$). We first suppose that we may find disks $\Lambda_1 = d(a_1, \rho_1)$ and $\Lambda_2 = d(a_2, \rho_2)$ such that $\Lambda_1 \cap \Lambda_2 = \emptyset$, and such that $\mathcal{F}_1$, (resp.$\mathcal{F}_2$) is secant with Λ_1, (resp.Λ_2) . Then it is seen that $|a_1 - a_2| > \rho_1 \ge r_1$ hence $\varphi_{\mathcal{F}_1}(x - a_1) \le r_1$ while $\varphi_{\mathcal{F}_2}(x - a_1) = |a_1 - a_2| > r_1$ and therefore $\varphi_{\mathcal{F}_1} \ne \varphi_{\mathcal{F}_2}$.

We now suppose that we cannot find disks Λ_1, Λ_2 defined as above. Since $r_1 \le r_2$, any disk Λ which belongs to $\mathcal{F}_1$ is included in any disk that belongs to $\mathcal{F}_2$ and therefore any point of Λ_1 is a center of $\mathcal{F}_2$. Thus $\mathcal{F}_2$ admits a center $a \in \Lambda$ and then, $\mathcal{F}_1$ is secant with $d(a, r_2)$. Hence we have $r_1 < r_2$ because otherwise $\mathcal{F}_1$ would be equal to $\mathcal{F}_2$. In particular $\mathcal{F}_2$ is secant with one class $d(\alpha, r_2^-)$ of $d(a, r_2)$. Then we have $\varphi_{\mathcal{F}_1}(x - \alpha) \le r_1$ while $\varphi_{\mathcal{F}_2}(x - \alpha) = r_2$. This finishes showing that the mapping $\mathcal{F} \to \varphi_{\mathcal{F}}$ is injective.

Now we will show that this mapping is also surjective. Indeed let ψ be an absolute value on $F(x)$, and let $r = \inf_{\lambda \in F} \psi(x - \lambda)$.

We first suppose that there exists $a \in F$ such that $\psi(x - a) = r$. Since ψ is an absolute value, we check that $r > 0$ because if $r = 0$, we have $\psi(h) = h(a)$ for every $h \in F(x)$ and then ψ is not an absolute value. Hence we can assume

$r > 0$. Let $\mathcal{F}$ be the circular filter of center a, of diameter r. By lemma 1.5, we know that ψ is ultrametric and then for every $b \in F$, we have $\psi(x - b) \le \max$ $(\psi(x-a), |a-b|) = \max(|a-b|, r)$. But by definition we have $\psi(x-b) \ge r$, hence (1) $r \le \psi(x - b) \le \max(r, |a - b|)$.

If $|a - b| > r$, then it is seen that both $\psi(x - b), \varphi_{\mathcal{F}}(x - b)$ are equal to $|a - b|$. If $|a - b| \le r$, then b is another center of $\mathcal{F}$ and then we have $\varphi_{\mathcal{F}}(x - a) = \varphi_{\mathcal{F}}(x - b) = r$. But by (1) we see that $\psi(x - b) = r$. So we have shown that $\varphi_{\mathcal{F}}(x - b) = \psi(x - b)$ for all $b \in F$ and since F is algebraically closed, this finishes proving that $\psi = \varphi_{\mathcal{F}}$.

We now suppose that there does not exist $a \in F$ such that $r = \psi(x - a)$. There exists $\alpha_n \in F$ such that $r < \psi(x - \alpha_n) < r + \dfrac{1}{n}$. Let $\rho_n = \psi(x - \alpha_n)$. For $b \in K \setminus d(\alpha_n, \rho_n)$ clearly we have (2) $\psi(x - b) = |b - \alpha_n| > \rho_n$ because $\psi(x - \alpha_n) < |\alpha_n - b|$. Further, if $\mathcal{F}_n$ is the circular filter of center α_n and diameter ρ_n we have (3) $\psi(x - b) = |b - \alpha_n| = \varphi_{\mathcal{F}_n}(x - b)$. However there exists $\alpha_{n+1} \in F$ such that $r < \psi(x - \alpha_{n+1}) < \min(\rho_n, r + \dfrac{1}{n + 1})$. Hence by (2) we see that $\alpha_{n+1} \in d(\alpha_n, \rho_n)$. That way, we may define a decreasing sequence of disks $D_n = d(\alpha_n, \rho_n)$ such that $r < \rho_n < r + \dfrac{1}{n}$ and $\psi(x - \alpha_n) = \rho_n$. Let $D'_n = D_n \cap D$. Then the decreasing filter $\mathcal{F}$ of base $(D'_n)_{n \in \mathbb{N}}$ satisfies $\lim_{\mathcal{F}} \psi(x - \alpha_n) = r$. It is easily seen that $\mathcal{F}$ has no center because if α is a center of $\mathcal{F}$ then $\psi(x - \alpha) \le \max (|\psi - \alpha_n|, |\alpha_n - \alpha|)$ hence $\psi(x - \alpha) = r$. We will show that $\psi = \varphi_{\mathcal{F}}$. Let $b \in F$ and let $q \in \mathbb{N}$ be such that $b \notin D_q$. Then by (3) for $n \ge q$ we have $\psi(x - b) = \varphi_{\mathcal{F}_n}(x - b)$. On the other hand it is easily seen that $\varphi_{\mathcal{F}_n}(x - b) = \varphi_{\mathcal{F}}(x - b)$. Finally $\psi(x - b) = \varphi_{\mathcal{F}}(x - b)$ whenever $b \in F$ and then $\psi = \varphi_{\mathcal{F}}$. This ends the proof of Theorem 4.14.

Notation :   When $\mathcal{F}$ is the circular of center a, of diameter r, we will also denote by $\varphi_{a,r}$ the absolute value $\varphi_{\mathcal{F}}$. Hence by definition we have $\varphi_{a,r}(h) = \lim\limits_{\substack{|x-a| \to r \\ |x-a| \ne a}} |h(x)|$. In particular we notice that $\varphi_{0,r}(h) = |h|(r)$.

5.   HENSEL LEMMA

Notations :   Given a field E and $g, h \in E[x]$, (g, h) will denote the monic greatest common divisor of g and h.

Given $Q(x) = \sum_{j=0}^{n} b_j x^j \in U_L[x]$, we will denote by $\overline{Q}$ the polynomial

$\sum_{j=0}^{q} \overline{a_j}\, x^j \in \mathcal{L}[x]$. In this chapter $P(x) = \sum_{j=0}^{q} a_j x^j \in L[x]$ will denote a polynomial of degree q.

The "Hensel Lemma" is a strong result that roughly says : "if L is complete and if $\overline{P}$ splits in the form $\gamma\eta$ with $(\gamma, \eta) = \overline{1}$, then P also splits in $L[x]$ in the form gh with $\overline{g} = \gamma$, $\overline{h} = h$, $deg(\overline{g}) = deg(\gamma)$". The proof is not very easy and requires a serious preparation. Here we will roughly follow the same process as in [1], but Lemmas will follow the logic order.

Lemma 5.1. is immediate :

Lemma 5.1:   *The quotient ring* $\dfrac{U_L[x]}{M_L[x]}$ *is isomorphic to* $\mathcal{L}[x]$.

Lemma 5.2:   *For all $\alpha \in L$ we have $v(P(\alpha)) \geq v(P, 0) + \min(0, qv(\alpha))$.*

Proof :   By Lemma 4.7 we have $v(P(\alpha)) \geq \min_{0 \leq j \leq q} (v(a_j) + j\, v(\alpha)) \geq$ $\min_{0 \leq j \leq q} v(a_j) + \min_{0 \leq j \leq q} jv(\alpha)$. But $\min_{0 \leq j \leq q} v(a_j) = v(P)$ and $\min_{0 \leq j \leq q} jv(\alpha) = \min(0, qv(\alpha))$.

Definition :   A polynomial $\sum_{j=0}^{q} a_j x^j \in L[x]$ will be said to be *quasi-monic* if $|a_q| = 1$.

Lemma 5.3:   *Let $P, E \in U_L[x]$ with E quasi-monic. Let $Q, R \in U_L[x]$ satisfy $P = EQ + R$ and $deg(R) < deg(E)$. Then we have $v(Q) \geq v(P)$ and $v(R) \geq v(P)$.*

Proof: We can clearly assume $P \neq 0$. Then, by multiplying P by a suitable constant λ, we can also assume $v(P) = 0$. Since E is quasi-monic, the Euclidean division of P by E is clearly possible in $U_L[x]$, and therefore Q is the quotient,

R is the rest of this division, due to the fact that $deg(R) < deg(E)$. So we have $v(Q) \geq 0$, $v(R) \geq 0$ because both Q, R belong to $U_L[x]$.

Lemma 5.4:   Let $g, h \in U_L[x]$ be quasi-monic, such that $(\overline{g}, \overline{h}) = \overline{1}$ and $deg(P) < deg(g) + deg(h)$. There exist $V, W \in L[x]$ satisfying $v(Vg + Wh - P) > v(P)$, $v(V) \geq v(P)$, $v(W) \geq v(P)$, $deg(V) < deg(h)$, $deg(W) < deg(g)$.

Proof: Since $(\overline{g}, \overline{h}) = \overline{1}$, by Bezout's Theorem there exists φ and $\psi \in \mathcal{L}[x]$ such that $\varphi \overline{g} + \psi \overline{h} = \overline{1}$, $deg(\varphi) < deg(\overline{h})$, $deg(\psi) < deg(\overline{g})$. Let $s, t \in U_L[x]$ satisfy $\overline{S} = \varphi$, $\overline{T} = \psi$, $deg(S) = deg(\varphi)$, $deg(T) = deg(\psi)$. Thus we have $\overline{Vg + Wh - 1} = \overline{0}$ i.e. (1) $v(Sg + Tg - 1) > 0$.

We now consider the Euclidean division of SP by h and TP by G, respectively. We obtain $SP = S_0 h + V$ and $TP = T_0 g + W$. By Lemma 6.2, it is seen that $\min(v(V), v(S_0)) \geq v(SP) \geq v(P)$. Besides, by hypothesis we have

(1) $deg(V) < deg(h)$ and

(2) $deg(W) < deg(g)$

Let $B = Sg + Th - 1$. Then we have $BP = (S_0 + T_0)gh + Vg + Wh - P$. Since $deg(P) < deg(g) + deg(h)$, by (1) and (2) we see that $deg(Vg + Wh - P) < deg(g) + deg(h)$ and therefore $Vg + Wh - P$ is just the remainder of the Euclidean division of BP by gh. But then, by Lemma 6.3, we have $v(Vg + Wh - P) \geq v(BP) = v(B) + v(P)$, and therefore by (1) and by definition of B it is seen that $v(B) > 0$. This finishes proving that $v(Vg + Wh - P) > v(P)$, and this ends the proof of Lemma 5.4.

Lemma 5.5:   Let $g, h \in U_L[x]$, be monic and satisfy $(\overline{g}, \overline{h}) = \overline{1}$. There exists a constant $\lambda(g, h) \in \mathbb{R}$ such that, for every polynomial $Q \in L[x]$ satisfying $deg(Q) < deg(g) + deg(h)$, there exist $V, W \in L[x]$ satisfying $v(Vg + Wh - Q) \geq v(Q) + \lambda(g, h)$, $v(V) \geq v(Q)$, $v(W) \geq v(Q)$, $deg(V) < deg(h)$, $deg(W) < deg(g)$.

Proof: We can apply Lemma 6.4 to each polynomial $Q_n = x^n$ for every $n = 0, \ldots, d - 1$. Thus, we have polynomials V_n, W_n satisfying $v(V_n g + W_n h - x - n) > 0$, $v(V_n) \geq 0$, $v(W_n) \geq 0$, $deg(V_n) < deg(h)$, $deg(W_n) < deg(g)$. We put

$$\lambda_n = v(V_n g + W_n h - x^n), \quad (0 \leq n \leq d - 1). \text{ Now let } Q = \sum_{n=0}^{d-1} a_n x^n, \text{ let } V =$$

$$\sum_{n=0}^{d-1} a_n V_n, \quad W_n = \sum_{n=0}^{d-1} a_n W_n \text{ and let } \lambda(g, h) = \min_{0 \leq n \leq d-1} \lambda_n. \text{ Clearly we have}$$

$$v(Vg + Wh - Q) \geq \min_{0 \leq n \leq d-1} (v(a_n) + \lambda_n) \geq \min_{0 \leq n \leq d-1} v(a_n) + \min_{0 \leq n \leq d-1} \lambda_n = v(Q) + \lambda(g, h).$$

But trivially $v(V) \geq \min\limits_{0 \leq n \leq d-1} v(a_n)$, $v(W) \geq v(Q)$, $deg(V) \leq \max\limits_{0 \leq n \leq d-1} (deg(V_n)) <$
$deg(h)$, $deg(W) \leq \max\limits_{0 \leq n \leq d-1} (deg(W_n)) < deg(h)$. This ends the proof of Lemma 5.5.

Lemma 5.6: $\;$ Let $Q \in L[x]$ and let $g, h \in U_L[x]$ be quasi-monic and satisfy $(\overline{g}, \overline{h}) = \overline{1}$. There exist monic polynomials V, $W \in L[x]$ satisfying
$v(Vg + Wh - Q) \;\geq\; \lambda(g,h) + v(Q)$,
$deg(W) \;<\; deg(g)$, $deg(V) \;\leq \max\;(deg(h), deg(Q) - deg(g))$,
$v(V) \geq v(Q), v(W) \geq v(Q)$.

Proof : $\;$ We consider the Euclidean division of Q by gh : $Q = \ell gh + Q_1$. Hence $deg(Q_1) \;<\; deg(g) + deg(h)$. By Lemma 5.3 we have

(1) $\;v(Q_1) \;\geq\; v(Q)$,

(2) $\;v(\ell) \;\geq\; v(Q)$.

By lemma 5.5. there exist $V_1, W_1 \in L[x]$ satisfying

(3) $\;v(V_1 g + W_1\, h - Q_1) \;\geq v(Q_1) + \lambda(g,h)$,

(4) $\;v(V_1) \;\geq\; v(Q_1)$,

(5) $\;v(W_1) \;\geq\; v(Q_1)$,

(6) $\;deg(V_1) \;<\; deg(h)$,

(7) $\;deg(W_1) \;<\; deg(g)$,

$\qquad$ Now we put $V = V_1 + \ell h$, $W = W_1$. So we have $Vg + Wh - Q = V_1 + \ell gh + W_1 h - \ell gh - Q_1$ and therefore by (3) we obtain $v(Vg + Wh - Q) \;\geq\; v(Q_1) + \lambda(g,h)$. Hence by (1) we obtain (8) $\;\;v(Vg + Wh - Q) \;\geq\; v(Q) + \lambda(g,h)$.

By (1) , (5) it is seen that

(9) $\;v(W) \;\geq\; v(Q)$. We will check

(10) $\;v(V) \;\geq v(Q)$. Indeed we have $v(h) = 0$, hence by (2) we see

(11) $\;v(\ell h) \;\geq\; v(Q)$.

But by (4) we have $v(V_1) \;\geq\; v(Q)$ and therefore by (11) we obtain (10). Finally, by definition we have $deg(\ell) = deg(Q) - deg(gh)$ and therefore

(12) $\;deg(V) \;\leq \max\;(deg(V_1),\; deg(\ell h)) \;\leq\; \max\;(deg(h),\; deg(Q) - deg(g))$.

Thanks to (7), (8) , (9) , (10), (12), Lemma 5.6 is now proven.

Theorem 5.7: $\;$ **(Hensel Lemma)** $\;$ L is supppposed to be complete. Let $P \in U_L[x]$ be such that $\overline{P}$ splits in $\mathcal{L}[x]$ in the form $\gamma\eta$ with $(\gamma, \eta) = \overline{1}$. There exists $g, h \in U_L[x]$ such that $P = gh, \overline{g} = \gamma, \overline{h} = \eta$, $deg\,(g) = deg(\gamma)$.

Proof : $\;$ We can obviously take quasi-monic polynomials $g_0, h_0 \in U_L[x]$ such that $\overline{g_0} = \gamma$, $\overline{h_0} = \eta$. We put $\nu = v(P - g_0 h_0)$ and $\tau = \min(\nu, \lambda(g_0, h_0))$. We will construct sequences $(g_n)_{n \in \mathbb{N}}, (h_n)_{n \in \mathbb{N}}$ in $L[x]$ satisfying for all $n \geq 0$:

$\quad$ i) $\;v(P - g_n h_n) \;\geq\; (n+1)\tau$

ii) $v(g_{n+1} - g_n) \geq (n+1)\tau$, $v(h_{n+1} - h_n) \geq (n+1)\tau$

iii) $deg(h_n) \leq deg(P) - deg(g_0)$, $deg(g_n) = deg(g_0)$

iv) $\overline{g_n} = \gamma$, $\overline{h_n} = \eta$.

First we put $P_1 = P - g_0 h_0$. By Lemma 5.6, there exist $V_1, W_1 \in L[x]$ satisfying

(1) $deg(W_1) < deg(g_0)$,

(2) $deg(V_1) < deg(P) - deg(g_0)$,

(3) $v(V_1) \geq \tau$,

(4) $v(W_1) \geq \tau$,

(5) $v(V_1 g_0 + W_1 h_0 - P) \geq \lambda(g_0, h_0) + v(Q) \geq 2\tau$,

Next we put $g_1 = g_0 + W_1$, $h_1 = h_0 + V_1$. We will check that i), ii) , iii), iv) are satisfied for $n = 0$. Indeed i) , iii), iv) are satisfied for $n = 0$ by definition of g_0, h_0, τ. Besides by (3) and (4), ii) is clearly satisfied for $n = 0$.

Now we suppose we have already constructed the couples (g_j, h_j) satisfying i), ii), iii), iv) for every $j = 0, \ldots, n$. Then we put $P_{n+1} = P - g_n h_n$. By i), true at the order n, we have (6) $v(P_{n+1}) \geq (n+1)\tau$. We can apply Lemma 5.6 to the case when (Q, g, h) is equal to (P_{n+1}, g_n, h_n). So we obtain $V_{n+1}, W_{n+1} \in L[x]$ satisfying

(7) $v(W_{n+1} g_n + V_{n+1} h_n - P_{n+1}) \geq \lambda(g_n, h_n) + v(P_{n+1})$

(8) $deg(W_{n+1}) < deg(g_n)$,

(9) $deg(V_{n+1}) \leq \max(deg(h_n), deg(P_{n+1}) - deg(g_n))$

(10) $v(V_{n+1}) \geq v(P_{n+1})$, $v(W_{n+1}) \geq v(P_{n+1})$.

By (6) and (7) we obtain

(11) $v(W_{n+1} g_n + V_{n+1} h_n - P_{n+1}) \geq (n+2)\tau$.

Now we put $g_{n+1} = g_n + W_{n+1}$, $h_{n+1} = h_n + W_{n+1}$. We check that $P - g_{n+1} h_{n+1} = (P_{n+1} - h_n W_{n+1} - g_n V_{n+1}) - V_{n+1} W_{n+1}$. By (10) and (6) it is seen that $v(V_{n+1} W_{n+1}) \geq (n+2)\tau$ and therefore by (11) we obtain i) at the order $n+1$. By (10) and (6), we notice that ii) is clear at the order $n+1$. Besides, iv) at the order $n+1$ clearly comes from iv) at the order n together with (6) and (10). By (8) it is seen that $deg(g_{n+1}) = deg(g_n) = deg(\gamma)$. Finally by (9) we have $deg(h_{n+1}) \leq \max(deg(h_n), deg(P_{n+1}) - deg(g_n))$ and therefore by iii) at the order n we have $deg(h_{n+1}) \geq \max(deg(P) - deg(g_0), \; deg(P_{n+1}) - deg(g_n))$. But $deg(P_{n+1}) \leq \max(deg(P), deg(g_n h_n))$ and then by iii) at the order n, we see that $deg(g_n h_n) \leq deg(P)$. Hence, as $deg(g_n) = deg(g_0)$, we have $deg(P_n g_n) \leq deg(P) - deg(g_0)$ and finally we have proven that $deg(P_{n+1}) \leq deg(P) - deg(g_0)$. This finishes showing iii) at the order $n + 1$, and therefore the sequences $(g_n)_{n\in\mathbb{N}}$, $(h_n)_{n\in\mathbb{N}}$ satisfying i), ii), iii), iv) are now constructed. Since L is complete, the vector space $L_q[x]$ of the polynomial of degree $m \leq q$ is obviously complete with respect to the Gauss norm $\|\cdot\|$ which is characterized by $-\log\|Q\| = v(Q)$.

Then by ii) it is seen that the sequences $(g_n)_{n\in\mathbb{N}}$, $(h_n)_{n\in\mathbb{N}}$ converge in $L_q[x]$. We put $g = \lim\limits_{n\to\infty} g_n$, $h = \lim\limits_{n\to\infty} h_n$. Clearly by iii) we have $deg(g) = deg(g_0) = deg(\gamma)$. By iv) we have $\overline{g} = \gamma$, $\overline{h} = \eta$, and finally by i) we have $v(P - gh) = +\infty$ hence $P = gh$. This ends the proof of Theorem 5.7.

6. ULTRAMETRIC FIELD EXTENSIONS

Several problems on Analytic Elements require one to consider an ultrametric algebraically closed extension of K which is spherically complete, in order to give every circular filter a center. Others require one to have a complete algebraically closed extension which admits a non countable residue class field. Although these very basic results are well known in the mathematical world, they cannot be easily found in books of p-adic analysis. So, we are going to recall why K does admit a spherically complete algebraically closed extension whose residue class field is not countable.

Notations: As in the previous chapters, L will denote an ultrametric field whose absolute value is not trivial. F will denote an algebraically closed ultrametric field whose absolute value is not trivial. If a belongs to an algebraic extension of L, $irr(a, L)$ will denote the minimal polynomial of a over L.

Lemma 6.1 is classical in algebra:

Lemma 6.1: *Let E be a field, let A be a finite algebraic extension of E, let $q = [A : E]$, and let N be the norm of A over E. Let $a \in A$, let $P_a = irr(a, E)$, and let $d = deg(P_a)$. Then N satisfies $N(a) = \left((-1)^d P_a(0)\right)^{\frac{q}{d}}$, and $N(ab) = N(a)N(b)$, whenever $b \in A$.*

Lemma 6.2: *Let L' be an algebraic extension of L, let $a \in L'$ and let $P = irr(a, L)$. Then a is integral over U_L if and only if $|P(0)| \leq 1$.*

Proof: First we assume a integral over U_L. Then there exists a monic polynomial $Q \in U_L[x]$ such that $Q(a) = 0$. Therefore, P divides Q in $L[x]$. Let $Q(x) = P(x)T(x)$. Since both P, Q are monic, so is T, hence we have $v(P) \leq 0, v(T) \leq 0$. But since $v(Q) = 0$, P, T must satisfy $v(P) = v(T) = 0$, and thefore $|P(0)| \leq 1$.

Now we assume $|P(0)| \leq 1$. Suppose $v(P) < 0$. There exists $b \in M_L$ such that $v(bP) = 0$, and then we have $|bP(0)| < 1$, hence $\overline{0}$ is a zero of $\overline{bP}$. Besides we notice (1) $deg(\overline{bP}) < deg(P)$. Let $\overline{bP} = x^d \phi$ with $\phi(\overline{0}) \neq \overline{0}$. then we have $(x^d, \phi) = \overline{1}$, and therefore, by Theorem 5.7 there exist g, $h \in U_L[x]$ such that $\overline{g} = x^d$, $\overline{h} = \phi$, $deg(g) = d$, and $P = gh$. But since P is irreducible in $L[x]$, and since $d > 0$, h must be a constant and therefore $deg(P) = d$. But this contradicts (1) and finishes the proof.

Theorem 6.3:   *Let L be complete and let Ω be an algebraic closure of L. There exists a unique absolute value φ on Ω that extends the one of L. Further, this absolute value is ultrametric and defined as follows:*

given $a \in \Omega$, $Q = irr(a, L)$ and $t = deg(P)$, then $\varphi(a) = \sqrt[t]{|Q(0)|}$.

Proof: We first notice that $\varphi(a) = |a|$ whenever $a \in L$. We will show that φ is an ultrametric absolute value on Ω. Clearly we have $\varphi(a) \neq 0$ whenever $a \in \Omega$. By Lemma 6.1 it is easily seen that we have $\varphi(ab) = \varphi(a)\varphi(b)$ whenever $a, b \in \Omega$, and therefore $\varphi(a)^{-1} = \varphi(a^{-1})$. So it remains to show the ultrametric inequality. For this, we will show $\varphi(1 + z) \leq 1$ for every $z \in U_\Omega$. For convenience, we put again $P_z = irr(z, E)$ whenever $z \in \Omega$. Let $z \in U_\Omega$. So we have $|P_z(0)| \leq 1$ and then by lemma 6.2, z is integral over U_L, hence so is $1 + z$. Hence by Lemma 6.2 we have $|P_{1+z}(0)| \leq 1$, and therefore $\varphi(1 + z) \leq 1$. Now, the ultrametric inequality will be easily deduced. Let $a, b \in \Omega$ satisfy $0 < |a| \leq |b|$. We have $\varphi(a + b) = \varphi\big(b(1 + \frac{a}{b})\big) = \varphi(b)\varphi(1 + \frac{a}{b})$. But $\varphi(1 + \frac{a}{b}) \leq 1$, hence finally $\varphi(a + b) \leq \varphi(b)$. Thus we have now proven φ to be an ultrametric absolute value that extends the one of L. Then by Theorem 1.3, this absolute value on Ω is unique. This ends the proof.

Corollary 6.4:   *Let L be complete and let Ω be an algebraic closure of L provided with the unique absolute value $| \, . \, |$ that extends the one of L. Then U_Ω is equal to the integral closure of U_L. Besides $|\Omega| = \{ \sqrt[n]{r} \, | \, r \in |L| \}$.*

     Lemma 6.5 is immediate:

Lemma 6.5:   *Let B be a subfield of L. The residue class field of B is a subfield of $\mathcal{L}$.*

Lemma 6.6:   *Let L be complete, and let Ω be an algebraic closure of L provided with the unique absolute value extending the one of L. Then the residue class field of Ω is an algebraic closure of $\mathcal{L}$.*

Proof: Let $\mathcal{T}$ be the residue class field of Ω. Let $u \in U_\Omega$ and let $P = irr(u, L)$. By Corollary 6.4, P belongs to $U_L[x]$, and obviously satisfies $\overline{P}(\overline{u}) = \overline{0}$, hence $\overline{u}$ is algebraic over $\mathcal{L}$. So $\mathcal{T}$ is an algebraic extension of $\mathcal{L}$. Now let $q \in \mathcal{T}[x]$, and let $Q \in U_L[x]$ be a polynomial such that $\overline{Q} = q$. Then Q has at least one zero u in Ω. By Theorem 4.5 u belongs to U_Ω, and therefore $\overline{u}$ is a zero of q. But by definition $\overline{u}$ belongs to $\mathcal{T}$. Finally we have shown that $\mathcal{T}$ is an algebraic extension of $\mathcal{L}$ such that every polynomial $q \in \mathcal{L}[x]$ has at least one zero: hence $\mathcal{T}$ is equal to the algebraic closure of $\mathcal{L}$.

Corollary 6.7:   *$\mathcal{K}$ is algebraically closed .*

Corollary 6.8: *Let L be complete and let M be an algebraic extension of L provided with the unique absolute value that extends the one of L. Then the residue class field of M is an algebraic extension of the residue class field of L.*

Lemma 6.9: *Let $P(x) = \sum_{j=0}^{t} a_j x^j$, $Q(x) = \sum_{j=0}^{t} b_j x^j$ be monic and belong to $F[x]$. For each zero α of P, Q admits at least one zero β such that $|\alpha - \beta|^t \leq \max_{0 \leq j \leq t} |a_j - b_j|$.*

Proof: Let $s = \max_{0 \leq j \leq t} |a_j - b_j|$, and let α be a zero of P. By Lemma 4.2 we have $|(P - Q)(x)| \leq s$ whenever $x \in d(0,1)$, hence in particular $|Q(\alpha)| \leq s$. Let $\beta_1, ... \beta_t$ be the zeros of Q (taking multiplicities into account). So we have $\prod_{j=1}^{t} |\beta_j - \alpha| \leq s$ and then it is seen that at least one of the β_j satisfies $|\beta_j - \alpha| \leq s$.

Theorem 6.10: *The completion of F is algebraically closed.*

Proof: Let E be the completion of F, and let $P(x) = \sum_{j=0}^{t} a_j x^j \in E[x]$ be monic. Let Ω be an algebraic closure of E, provided with the unique absolute value extending the one of E, and let $\alpha_1, ..., \alpha_t$ be the zeros of P in Ω. Let $\varepsilon \in]0, 1[$ and let $Q(x) = \sum_{j=0}^{t} b_j x^j \in K[x]$ be such that $\max_{0 \leq j \leq t} |a_j - b_j| \leq \varepsilon^t$. By Lemma 6.9 Q admits a zero β such that $|\alpha - \beta|^t \leq \varepsilon$. Since $Q \in F[x]$, obviously β belongs to F and therefore we see that α belongs to E.

Theorem 6.11: (M. Krasner) *Let L be complete and have characteristic zero. Let Ω be an algebraic closure of L provided with the unique absolute value extending the one L. Let $a \in \Omega$, let $a_2, ... a_n$ be the conjugates of a in Ω, and let $b \in \Omega$ satisfying $|b - a| < |b - a_j|$ for every $j = 2, ... n$, and $[L[b] : L] \leq n$. Then we have $L[a] = L[b]$.*

Proof: Let $a_1 = a$, and let $P(x) = irr(a, L)$. In $\Omega[x]$ P splits in the form $\prod_{j=1}^{n} (x - a_j)$. Let $Q(x) = irr(a, L[b])$. Then Q divides P. Let $t = deg(Q)$ and suppose the a_j ranged in such a way that $Q(x) = \prod_{j=1}^{t} (x - a_j)$. Let $R(y) = Q(b+y)$. Then $R(y)$ is seen to be irreducible in $L[b][y]$ like $Q(x)$ in $L[b][x]$. Besides the zeros of R are just the $a_j - b$, with $1 \leq j \leq t$. Thus we have $R = irr(a-b, L[b])$. But since L is complete, by Theorem 6.3 we have $|a_j - b| = \sqrt[q]{R(0)}$ for every $j = 1, ..., t$. In particular, for $j = 2, ...t$, we have $|a_j - b| = |a_b|$, and this contradicts the hypothesis. Finally, we have $t = 1$, and therefore a lies in $L[b]$. But since $[L[b] : L] \leq [L[a] : L]$, we see that $L[a] = L[b]$.

7. ULTRAPRODUCTS AND

SPHERICALLY COMPLETE EXTENSIONS

Several problems on Analytic Elements require one to consider an ultrametric algebraically closed extension of K which is spherically complete, in order to give every circular filter a center. Others require one to have a complete algebraically closed extension which admits a non countable residue class field. Proving the existence of a spherically complete algebraically closed extension of the ground field K isn't easy, most of the ways involving basic considerations in logic. Here we will follow the very original way proposed by Bertin Diarra in [9], that is only based on the notion of ultraproducts.

Notations and definitions :   In this chapter $(E_i)_{i \in I}$ denotes an infinite family of field extensions of L, provided each with an utrametric absolute value $| \, . \, |_i$ extending the one of L. Besides $\mathcal{U}$ will denote an ultrafilter on I. We remember that $\mathcal{U}$ is said to be *principal* if there exists $\alpha \in I$ such that $\mathcal{U}$ is the set of the subsets of I that contain α. $\mathcal{U}$ is said to be *incomplete* if there exists a decreasing sequence $(X_n)_{n \in \mathbb{N}}$ of elements of $\mathcal{U}$, such that $\bigcap_{n \in \mathbb{N}} X_n = \emptyset$. Since I is infinite, there obviously exist incomplete ultrafilters on I. In particular, any incomplete ultrafilter is non principal.

$\mathcal{R}$ will denote the subring of $\prod_{i \in I} E_i$ that consists of the set $(a_i)_{i \in I} \in \prod_{i \in I} E_i$ such that $\sup_{i \in I} |a_i|_i < +\infty$. Of course $\mathcal{R}$ is a L-algebra.

We will denote by φ the mapping from $\mathcal{R}$ into $\mathbb{R}_+$ defined as $\varphi((a_i)_{i \in I}) = \lim_{\mathcal{U}} |a_i|_i$. Then φ is seen to be a multiplicative semi-norm of the L-algebra $\mathcal{R}$.

We put $\mathcal{J} = Ker(\varphi)$, and $\mathcal{S} = \dfrac{\mathcal{R}}{\mathcal{J}}$, and we denote by ψ the canonical surjection from $\mathcal{R}$ onto $\mathcal{S}$. Then $\mathcal{S}$ is obviously provided with an absolute value $| \, . \, |$ defined as
$|\psi(a)| = \varphi(a), \ (a \in \mathcal{R})$.

On the other hand $\mathcal{R}$ is seen to be provided with a norm of L-algebra $\| \, . \, \|$ defined as $\|(a_i)_{i \in I}\| = \sup_{i \in I} |a_i|$. Next we denote by $\|| \, . \, \||$ the semi-norm quotient of the norm of L-algebra by the ideal $\mathcal{J}$, defined on $\mathcal{R}$ as $\||a\|| = \inf_{t \in \mathcal{J}} \|a - t\|$.

We remember that given a field E, $\alpha \in E$ and $r > 0$, $d_E(\alpha, r)$ denotes the disk $\{x \in E | \quad |x - \alpha| \leq r\}$.

Theorem 7.1: (B.Diarra) $\mathcal{S}$ *is a field extension of L, and its absolute value $|\,.\,|$ extends the one of L. Moreover, if $\mathcal{U}$ is non principal and if each E_i has a dense valuation group, then $\mathcal{S}$ has a valuation group equal to $\mathbb{R}$. Further, if each E_i is algebraically closed, then so is $\mathcal{S}$.*

Proof: Let $\alpha \in \mathcal{S} \setminus \{0\}$, and let $a = (a_i)_{i \in I} \in \mathcal{R}$ be such that $\psi(a) = \alpha$. By definition we have $\lim_{\mathcal{U}} |a_i|_i \neq 0$. Hence there exists $J \in \mathcal{U}$ such that for every $i \in J$ we have $\dfrac{\varphi(a)}{2} < |a_i|_i < \dfrac{3\varphi(a)}{2}$, hence $a_i \neq 0$, and therefore

$$(1) \qquad \frac{2}{3\varphi(a)} < |a_i^{-1}|_i < \frac{2}{\varphi(a)} \quad \text{whenever} \quad i \in J.$$

Now let $b = (b_i)_{i \in I} \in \displaystyle\prod_{i \in I} E_i$ be defined as $b_i = a_i^{-1}$ whenever $i \in J$, and $b_i = 1$ whenever $i \in I \setminus J$. By (1) it is seen that b does belong to $\mathcal{R}$. But now, $ab - 1$ is an element $(c_i)_{i \in I}$ of $\mathcal{R}$ that satisfies $c_i = 0$ whenever $i \in J$, hence $\lim_{\mathcal{U}} |c_i| = 0$. Therefore $ab - 1$ belongs to $\mathcal{J}$, and finally in $\mathcal{S}$ we have $\psi(a)\psi(b) = 1$. This shows $\mathcal{S}$ to be a field.

Next we suppose that for each $i \in I$ the valuation group of E_i is dense. Let $r \in]0, +\infty[$. We can obviously find a family $(\varepsilon_i)_{i \in I}$ in $]0, +\infty[$ such that $\lim_{\mathcal{U}} \varepsilon_i = 0$. For every $i \in I$, let $a_i \in E_i$ satisfy $r - \varepsilon_i < |a_i| < r$ and let $a = (a_i)_{i \in I}$. Of course, a belongs to $\mathcal{R}$, and satisfies $\lim_{\mathcal{U}} |a_i|_i = r$, hence r belongs to $|\mathcal{S}|$. This shows that the valuation group of $\mathcal{S}$ is equal to $\mathbb{R}$.

Finally we suppose that each field E_i is algebraically closed. Let $P(x) = \displaystyle\sum_{n=0}^{q} \lambda_n x^n \in \mathcal{S}[x]$, with $\lambda_q = 1$, and $q > 0$. We will show that P admits at least one zero in $\mathcal{S}$. For every $n = 0, 1, ..., q-1$, let $(a_{i,n})_{i \in I} \in \mathcal{R}$ satisfy $\psi((a_{i,n})_{i \in I}) = \lambda_n$, and let $a_{i,q} = 1$ whenever $i \in I$. So, we have $\psi((a_{i,n})_{i \in I}) = \lambda_n$ whenever $n = 0, ..., q$. For every $i \in I$, we put $T_i(x) = \displaystyle\sum_{n=0}^{q} a_{i,n} x^n$. Since E_i is algebraically closed, and since $a_{i,q} = 1$ for every $i \in I$, at least one of the zeros α_i of T_i in E_i satisfies $|\alpha_i|^q \leq |a_{i,0}|$. But by hypothesis $(a_{i,0})_{i \in I}$ belongs to $\mathcal{R}$, hence so does $(\alpha_i)_{i \in I}$. Hence we can put $\alpha = \psi((\alpha_i)_{i \in I})$, and then we have $P(\alpha) = 0$. This finishes showing that $\mathcal{S}$ is algebraically closed and this ends the proof of Theorem 7.1.

Lemma 7.2: (B.Diarra) *Let $a = (a_i)_{i \in I} \in \mathcal{R}$. Then we have $|||a||| = \varphi(a)$.*

Proof: Let $J \in \mathcal{U}$ and let $e = (e_i)_{i \in I} \in \mathcal{R}$ be defined as $e_i = 0$ whenever $i \in J$, and $e_i = 1$ whenever $i \notin J$. For convenience we put $b = ae$. Clearly b belongs to $\mathcal{J}$, hence, we have $|||a||| = \inf_{t \in \mathcal{J}} \|a - t\| \leq \|a - b\|$. But now, we check that $\|a - b\| = \sup_{i \in J} |a_i|_i$. Besides, this is true for every $J \in \mathcal{U}$. Hence we obtain $|||a||| \leq \inf_{J \in \mathcal{U}} \left(\sup_{i \in J} |a_i|_i \right) = \lim_{\mathcal{U}} |a_i|_i = \varphi(a)$. On the other hand, for all $t \in \mathcal{J}$, we have $\varphi(a - t) = \varphi(a) \leq \|a - t\|$, hence $\varphi(a) \leq |||a|||$. This ends the proof of Lemma 7.2.

Theorem 7.3: (B.Diarra) *If $\mathcal{U}$ is incomplete, then $\mathcal{S}$ is spherically complete.*

Proof: Let $(\alpha_n)_{n \in \mathbb{N}}$ be a decreasing distances sequence in $\mathcal{S}$, and for every $n \in \mathbb{N}$, let $a_n \in \mathcal{R}$ be such that $\psi(a_n) = \alpha_n$. By induction we can easily construct another sequence $(b_n)_{n \in \mathbb{N}}$ in $\mathcal{R}$ satisfying

$(\mathcal{V}_n)$ $\quad \psi(b_n) = \alpha_n$ for every $n \in \mathbb{N}$.

$(\mathcal{W}_n)$ $\quad \|b_n - b_{n-1}\| < r_{n-2}$ whenever $n > 1$.

Indeed, let $q \in \mathbb{N}^*$, and suppose that we have defined $b_0, ..., b_q$ satisfying $(\mathcal{V}_n)$ for every $n = 0, ..., q$. Of course we have

$(1) \quad \varphi(b_q - a_{q+1}) = \varphi(a_q - a_{q+1}) = r_q < r_{q-1}$.

By Lemma 7.2 we have $\varphi(b_q - a_{q+1}) = |||b_q - a_{q+1}|||$, hence by (1) there exists $c \in \mathcal{J}$ such that $\|b_q - a_{q+1} - c\| < r_{q-1}$. So, we put $b_{q+1} = a_{q+1} + c$, and then $(\mathcal{V}_{q+1})$, $(\mathcal{W}_{q+1})$ are satisfied. In order to begin the induction, we put $b_0 = a_0$, $\quad b_1 = a_1$ and then, we can define the sequence $(b_n)_{n \in \mathbb{N}}$ for every $n \in \mathbb{N}$, satisfying $(\mathcal{V}_n)$, and $(\mathcal{W}_n)$.

Now for each $n \in \mathbb{N}$ we put $b_n = (b_{i,n})_{i \in I}$. Since $\mathcal{U}$ is incomplete, we can take a decreasing sequence $(X_n)_{n \in \mathbb{N}}$ of elements of $\mathcal{U}$ such that $\bigcap_{n \in \mathbb{N}} X_n = \emptyset$. We put $I_0 = I \setminus X_0$, and for each $n \in \mathbb{N}^*$, $I_{n+1} = X_n \setminus X_{n+1}$. Thus the family $(I_n)_{n \in \mathbb{N}}$ makes a partition of I. Further, for each $q \in \mathbb{N}$, $(I_n)_{n \in \mathbb{N}}$ makes a partition of X_q. Hence we can define a surjective mapping g from I onto $\mathbb{N}$ as $g(i) = n$ whenever $i \in I_n$. Now for every $i \in I$, we put $h_i = b_{i,g(i)}$. By $(\mathcal{W}_n)$, we have $\|b_n\| \leq r_0$ whenever $n \in \mathbb{N}$, hence $|h_i|_i \leq r_0$ for each $i \in I$, and therefore $(h_i)_{i \in I}$ belongs to $\mathcal{R}$. We put $h = (h_i)_{i \in I}$, and $\omega = \psi(h)$. We will show

$(2) \quad |\omega - \alpha_n| \leq r_{n-1}$ whenever $n \in \mathbb{N}^*$.

Let $n \in \mathbb{N}^*$ be fixed. It is seen that for every $m > n$, we have $\|b_n - b_m\| < r_{n-1}$, hence for every $i \in I$ we have $|b_{i,n} - b_{i,m}|_i < r_{n-1}$. Beside, since $(I_m)_{m>n}$ makes a partition of X_{n+1}, for every $i \in X_{n+1}$ there exists $m > n$ such that $i \in I_m$, and then we have $|b_{i,n} - h_i|_i = |b_{i,n} - b_{i,g(i)}|_i = |b_{i,n} - b_{i,m}|_i < r_{n-1}$ whenever $i \in X_{n+1}$. But X_{n+1} belongs to $\mathcal{U}$, and therefore in $\mathcal{S}$ we have $|b_n - h| =$

$\lim\limits_{\mathcal{U}} |b_{i,n} - h_i|_i \leq r_{n-1}$. This is true for every $n \in \mathbb{N}^*$, and finally this shows (2). Hence ω belongs to $\bigcap\limits_{n \in \mathbb{N}} d_{\mathcal{S}}(\alpha_{n+1}, r_n)$ and this finishes proving that $\mathcal{S}$ is spherically complete.

Theorem 7.4:   *K admits a spherically complete ultrametric algebraically closed extension whose residue class field is not countable and whose valuation group is equal to* $\mathbb{R}$.

Proof: First we will construct a complete algebraically closed extension of K whose residue class field is not countable. Let T be a transcendental extension of the form $K((x_j)_{j \in \mathbb{R}})$ provided with the absolute value $|\,.\,|$ defined on $K[(x_j)_{j \in \mathbb{R}}]$ by

$$\left| \sum_{j_1,\ldots,j_q \leq N} a_{j_1,\ldots,j_q} x_{j_1}^{t_1} \ldots x_{j_q}^{t_q} \right| = \max_{j_1,\ldots,j_q \leq N} |a_{j_1,\ldots,j_q}|.$$

It is seen that $|x_j - x_h| = 1$ whenever $j, h \in \mathbb{R}$ such that $j \neq h$ and therefore the residue class field of T is not countable. Let T' be the completion of T, and let E be an algebraic closure of T', provided with the unique absolute value that extends the one of T'. Let E' be the completion of E. By Theorem 6.11, E' is algebraically closed. Obviously, its residue class field contains the one of T and therefore is not countable.

Now, we can construct $\mathcal{S}$ by taking $I = \mathbb{N}$, and $E_i = E'$ for every $i \in \mathbb{N}$. Since E' is algebraically closed, by Theorem 7.1 so is $\mathcal{S}$. Moreover, the valuation group of E' obviously is dense, and therefore, by Theorem 7.1, $\mathcal{S}$ has a valuation group equal to $\mathbb{R}$. Finally, by Theorem 7.3 $\mathcal{S}$ is spherically complete. This ends the proof of Theorem 7.4.

Remark: Actually, one can deduce from "Remarques diverses" in [9] that when the field $\mathcal{S}$ is constructed by taking for each E_i a field with an infinite residue class field, then $\mathcal{S}$ has a non countable residue class field.

Notation: Henceforth, $\widehat{K}$ will denote an algebraically closed spherically complete extension of K whose residue class field is not countable and whose valuation group is $\mathbb{R}$.

For every disk $d(a, r^-)$ (resp. $d(a, r)$) in K, we will denote by $\widehat{d}(a, r^-)$ (resp. $\widehat{d}(a, r)$) the disk of same center and diameter in $\widehat{K}$.

8.   A STUDY IN C_p, THE p^nth ROOTS OF 1

For every $s \in \mathbb{N}^*$, we put $u_s = \dfrac{1}{p^s(p-1)}$, and $r_s = p^{-u_s}$. We will study the p^s-th roots of 1 and will show that they lie in circles of center 1 and radius r_s. First we have to recall the definition of the p-adic fields.

Notation and definition: Let p be a prime number, and let $\omega \in]1, +\infty[$. On $\mathbb{Z}$, the p-adic absolute value is defined as follows: given $n \in \mathbb{Z}^*$, it factorizes in a unique way in the form $p^s q$, with $q \in \mathbb{Z}^*$, prime to p. We put $|n|_p = \omega^{-s}$.

Lemma 8.1 is immediate.

Lemma 8.1: $| \, . \, |_p$ *is an ultrametric absolute value on $\mathbb{Z}$ that has continuation to $\mathbb{Q}$ and defines an ultrametric absolute value on $\mathbb{Q}$, and $\mathbb{N}$ is dense in $\mathbb{Z}$. Then $U_\mathbb{Q} = \mathbb{Z}$, $M_\mathbb{Q} = p\mathbb{Z}$, and the residue characteristic is p. The valuation group of $\mathbb{Q}$ is isomorphic to $\mathbb{Z}$.*

Notation and definition: We denote by $v_p(x)$ the valuation associated to the absolute value $| \, . \, |_p$.

Lemma 8.2:   *Let $| \, . \, |$ be an ultrametric absolute value on $\mathbb{Q}$. If this absolute value is trivial, the residue characteristic is zero. If the absolute value is not trivial, there exists a prime number q such that $| \, . \, |$ is equivalent to $| \, . \, |_q$.*

Proof: If this absolute value is trivial, it is clear that the residue characteristic is zero. So we suppose that the absolute value is not trivial. For every $n \in \mathbb{N}^*$ we have $|n| \leq 1$. Since $| \, . \, |$ is not trivial, there certainly exists $s \in \mathbb{N}^*$ such that $|s| < 1$. Let q be the smallest $s \in \mathbb{N}^*$ such that $|s| < 1$. It is easily checked that q is prime. Since $M_\mathbb{Q}$ is a principal ideal of $\mathbb{Z}$, we have $M_\mathbb{Q} = q\mathbb{Z}$.

Let $\nu = \left| \dfrac{1}{q} \right|$. It is easily checked that given $m \in \mathbb{Z}^*$, of the form $q^s m$, with $n \in \mathbb{Z}^*$, prime to q, we have $n = \nu^{-s}$. Let w be the valuation associated to this absolute value. Then w is clearly proportional to v_q, and by Lemma 1.1 is equivalent to v_q. This ends the proof.

Notation and definition: Now $\mathbb{Q}$ is completed with respect to the p-adic absolute value and its completion is denoted by $\mathbb{Q}_p$. The closure of $\mathbb{Z}$ in $\mathbb{Q}_p$ is denoted by $\mathbb{Z}_p$.

Lemma 8.3 is easily seen.

Lemma 8.3: $\mathbb{N}$ *is dense in* $\mathbb{Z}_p$, *the invertible elements in* $\mathbb{Z}_p$ *are the ones whose absolute value is* 1, $\mathbb{Z}_p$ *is compact, equal to* $U_{\mathbb{Q}_p}$, $p\mathbb{Z}_p$ *is equal to* $M_{\mathbb{Q}_p}$. *The residue class field of* $\mathbb{Q}_p$ *is equal to the field of* p *elements* $\mathbb{F}_p$.

An algebraic closure Ω_p of $\mathbb{Q}_p$ is provided with the unique extension of the p-adic absolute value defined on $\mathbb{Q}_p$. The valuation group of $\mathbb{Q}_p$ is obviously equal to this of $\mathbb{Q}$. Next, the valuation group of Ω_p is easily seen to be isomorphic to $(\mathbb{Q}, +)$. For convenience, we take $\omega = p$.

By Theorem 6.10 Ω_p has a completion $\mathbb{C}_p$ that is algebraically closed. The valuation group of $\mathbb{C}_p$ is then isomorphic to $(\mathbb{Q}, +)$ like this of Ω_p.

Besides, by Lemma 6.6, the residue class field of Ω_p is an algebraic closure of $\mathbb{F}_p$ and the one of $\mathbb{C}_p$ is seen to be the same.

We will now study the p^s-th roots of 1. We will need certain technical lemmas.

Lemma 8.4: *Let* $s \in \mathbb{N}^*$. *For every* $n \in \mathbb{N}^*$ *such that* $n < p^s$, *we have* $\left|\binom{p^s}{n}\right|_p \le \dfrac{1}{p^s|n|_p} \le \dfrac{1}{p}$. *Further, if* $s = 1$ *then* $\left|\binom{p}{n}\right|_p = \dfrac{1}{p}$ *for every* $n = 0, ...p - 1$.

Proof : We notice that when n is a multiple of p^h for some $h < s$, so is $p^s - n$. Now, let b be the bijection from $\{1, ..., n\}$ onto $\{(p^s - n + 1), .., p^s\}$, defined as $b(j) = p^s - j + 1$. Then when j is divisible by p^h, so is $b(j+1)$, hence $|b(j+1)|_p \le |j|_p$ and therefore we have $|(p^s - 1)(p^s - 2)...(p^s - n + 1)|_p \le |(n-1)!|_p$. As a consequence we see that $\left|\binom{p^s}{n}\right|_p \le \dfrac{|p^s(p^s - 1)(p^s - 2)...(p^s - n + 1)|_p}{|n!|_p} \le \dfrac{p^{-s}}{|n|_p}$.

In particular, since $n < p^s$, we have $\left|\binom{p^s}{n}\right|_p \le \dfrac{1}{p}$. In particular, if $s = 1$, it is seen that $|n!|_p = |(p - n)!|_p = 1$, and that $|p!|_p = \dfrac{1}{p}$.

Notations : Let $s \in \mathbb{N}$. We will denote by A_s the group of the p^s-roots of 1 in $\mathbb{C}_p$ i.e. the $\zeta \in \mathbb{C}_p$ such that $\zeta^{p^s} = 1$ and we will denote by B_s the set $A_s \setminus A_{s-1}$.

F_s will denote the polynomial $\displaystyle\sum_{j=0}^{p-1} x^{jp^{s-1}}$ and we put $G_s(x) = F_s(1 + x)$.

Definition : A monic polynomial $P(x) = \displaystyle\sum_{j=0}^{q} a_j x^j \in L[x]$ will be called *an*

Eisenstein polynomial if it satisfies $a_j \in M_L$ whenever $j = 0, \ldots, q - 1$ and $a_0 \notin M_L^2$.

Theorem 8.5: (Eisenstein) *Let L have a discrete valuation . Let $P(x) = \sum_{j=0}^{q} a_j x^j \in L[x]$ be an Eisenstein polynomial. Then P is irreducible in $L[x]$.*

Proof : We suppose P not irreducible. Then P splits in $L[x]$ in the form $S(x)T(x)$ with $S(x) = \sum_{j=0}^{m} \alpha_j x^j$, $T(x) = \sum_{j=0}^{n} \beta_j x^j$ and $\alpha_m = \beta_n = 1$. Since S, T are monic we have $\|S\| \geq 1$, $\|T\| \geq 1$ and since $\|S\|\|T\|\|ST\| = \|P\| = 1$, we have $\|S\| = \|T\| = 1$. Hence, both S, T belong to $U_L[x]$. First we notice that if α_0 belongs to M_L then β_0 does not, because $a_0 \notin (M_L)^2$. Hence we may assume $\alpha_0 \in M_L$ and $\beta_0 \notin M_L$. Besides we have $\alpha_j \in M_L$ for every $j = 0, \ldots, m - 1$. Indeed let ℓ be the smallest of the integers h such that $|\alpha_h| = 1$. Then we have

$$|\alpha_h| = |\beta_0 \alpha_\ell + \sum_{j=1}^{\ell} \beta_j \alpha_{\ell-j}| = 1 \text{ because } |\beta_0 \alpha_\ell| = 1 \text{ and } \sum_{j=1}^{\ell} \beta_j \alpha_{\ell-j} \in M_L. \text{ Thus}$$

we have $\ell = m$. But then $\overline{S}$ is equal to x^m and therefore $\overline{P} = x^m \overline{T} = \sum_{j=0}^{n} \overline{\beta_j} x^{j+m}$.

But by hypothesis we assume $\overline{P} = x^q$. Hence, $x^q = \sum_{j=0}^{n} \overline{\beta_j} x^{j+m}$ and therefore $|\beta_j| < 1$ whenever $j < n$. This contradicts the relation $|\beta_0| = 1$ and finishes the proof.

Lemmas below will be useful in the sequel.

Lemma 8.6: *Let G be a subgroup of the multiplicative group $(L^*, \cdot)$ included in $C(0,1)$ and let $u \in G$. The bijection ψ from G onto G defined as $\psi(x) = ux$ is isometric.*

Lemma 8.7: *Let $(j,n) \in \mathbb{N} \times \mathbb{N}^*$ be such that $j < n$. Then we have*

$$\binom{n}{j+1} = \sum_{h=j}^{n-1} \binom{h}{j}.$$

Lemma 8.8: *For every $s \in \mathbb{N}$, G_s is an Eisenstein polynomial.*

Proof : First we suppose $s = 1$. We have

$$G_1(x) = \sum_{h=0}^{p-1} \left(\sum_{j=0}^{h} \binom{h}{j} x^j \right) = \sum_{j=0}^{p-1} \left(\sum_{h=j}^{p-1} \binom{h}{j} \right) x^j.$$

Hence by Lemma 8.7, we have $G_1(x) = \sum_{j=0}^{p-1} \binom{p}{j+1} x^j$. Besides by Lemma 8.4, we have $\left| \binom{p}{j+1} \right|_p = \dfrac{1}{p}$ for every $j = 0, \dots, p-2$ and therefore G_1 is an Eisenstein polynomial.

Now we consider the general case $s \in \mathbb{N}^*$. First we put $T_s(x) = (1+x)^{p^s}$. By Lemma 8.4, it is seen that $T_1(x)$ is of the form $1 + x^p + \tau_1(x)$ with $\tau_1(x) \in xp\mathbb{Z}_p[x]$ and $deg(\tau_1) = p-1$. Then by an immediate induction, we see that $T_s(x)$ is of the form $1 + x^{p^s} + \tau_s(x)$ with $\tau_s(x) \in xp\mathbb{Z}_p[x]$ and $deg(\tau_s) = p^s - 1$.

As a consequence it is easily seen that G_s is an Eisenstein polynomial if and only if so is the polynomial $g_s(x) = \sum_{j=0}^{p-1} (1 + x^{p^s})^j$. But we have $g_s(x) = G_1(x^{p^s})$. Since G_1 is an Eisenstein polynomial so is g_s. This ends the proof.

Theorem 8.9:   *For each $s \in \mathbb{N}^*$, B_s consists of $p^s - p^{s-1}$ roots of 1 of order p^s that lie in $C(1, r_s)$. For every $t \in B_s$, $irr(t, \mathbb{Q}_p)$ is equal to F_s and $B_s \cap d(t, r_s^-)$ is equal to tA_{s-1}.*

Proof :   Let $t \in B_s$, and let $F(x) = irr(t, \mathbb{Q}_p)$. Then F divides $x^{p^s} - 1$ and has a degree $t > p^{s-1}$. Since $x^{p^s} - 1 = (x^{p^{s-1}} - 1)F_s$, F divides F_s. But by Lemma 8.8, and Theorem 8.5, G_s is irreducible over $\mathbb{Q}_p$, hence so is F_s and therefore $F = F_s$. Then by Theorem 6.3 we have $v_p(t-1) = \dfrac{v_p(G_s(0))}{p^s - p^{s-1}} = u_s$. Hence B_s is included in $C(1, r_s)$, and obviously consists of $p^s - p^{s-1}$ different points in this circle. The equality $B_s \cap d(t, r_s^-) = tA_{s-1}$ comes from Lemma 8.6.

Corollary 8.10:   *For every ξ, $\zeta \in A_1$ of $\xi \neq \zeta$, we have $v_p(\xi - \zeta) = \dfrac{1}{p-1}$.*

Theorem 8.11:   *Let $n \in \mathbb{N}^*$ and let ζ be a n-th root of 1 of order n. Then ζ belongs to $d(1, 1^-)$ if and only if n is of the form p^s $(s \in \mathbb{N})$.*

Proof :   By Theorem 8.9 we know that if n is of the form p^s then ζ belongs to $d(1, 1^-)$. Now we suppose $\zeta \in d(1, 1^-)$ and put $n = q\, p^s$ with q prime to p. Let $\xi = \zeta^{p^s}$. It is seen that ξ also belongs to $d(1, 1^-)$, because $|\xi - 1|_p = |\zeta - 1|_p \left| \sum_{j=0}^{p^s-1} \zeta^j \right|_p$. Let $P(x) = x^q - 1$. If $\xi \neq 1$, then $\overline{P}(x)$ admits 1 as a zero of order $t \geq 2$. But $\overline{P}'(\overline{1}) = \overline{q} \neq \overline{0}$, hence $\xi = 1$. Therefore we have $q = 1$ and this ends the proofs.

9. ALGEBRAS $R(D)$

Notations: From now D denotes an infinite set in K, and K^D is provided with the topology $\mathcal{U}_D$ of uniform convergence on D.

$R(D)$ will denote the K-algebra of the rational functions $h(x) \in K(x)$ with no pole in D. Since D is infinite, $R(D)$ is clearly a K-subalgebra of K^D, and then is provided with the topology induced by $\mathcal{U}_D$, that makes it a topological subgroup of K^D. Algebraically, $R(D)$ is a K-subalgebra of $K(x)$ and more precisely, is of the form $S(D)^{-1}K[x]$ with $S(D)$ the multiplicative set of the polynomials whose zeros do not belong to D.

$R_b(D)$ will denote the K-subalgebra of $R(D)$ that consists of the $f \in R(D)$ which are bounded in D. Finally, if D is not bounded, $R_0(D)$ will denote the K-subagebra of $R(D)$ that consists of the $f \in R(D)$ such that $\lim\limits_{|x|\to+\infty,\ x\in D} f(x) = 0$.

For every $f \in K^D$ we put $\|f\|_D = \sup\limits_{x\in D}|f(x)| \in [0, +\infty]$.

Let E be an algebraically closed extension of K and let A be a set in E. We will denote by $R_E(A)$ the E-algebra of the rational functions $h \in E(x)$ with no pole in A.

The following Lemma 9.1 is then immediate.

Lemma 9.1: *$R(D)$ is a principal ideal ring. Every ideal is of the form $P(x)R(D)$ with P a polynomial whose zeros belong to D.*

Lemma 9.2 is an immediate application of the general properties of the supremum, once the set $[0, +\infty]$ is provided with the classical extensions of the addition and the multiplication :

$$a + (+\infty) \quad = \quad +\infty \qquad \text{for every } a \in [0, +\infty]$$
$$a.(+\infty) \quad = \quad +\infty \qquad \text{for every } a \in]0, +\infty]$$

Lemma 9.2 *For every $g, h \in R(D)$ we have*

i) $\|h\|_D = 0$ *if and only if* $h = 0$

ii) $\|\lambda h\|_D = |\lambda|\,\|h\|_D$ *for every* $\lambda \in K^*$

iii) $\|h + g\|_D \leq \max(\|h\|_D, \|g\|_D)$

iv) *If* $(\|f\|_D, \|g\|_D)$ *is different from* $(0, +\infty)$ *and from* $(+\infty, 0)$, *then* $\|hg\|_D \leq \|h\|_D.\|g\|_D$.

v) $\|h^n\|_D = (\|h\|_D)^n$ *whenever* $n \in \mathbb{N}$.

Theorem 9.3:　　$R_b(D) = R(D)$ *if and only if D is closed and bounded. Besides, if D is closed and bounded, then $\| \cdot \|_D$ is a semi-multiplicative ultrametric norm of K-algebra.*

Proof:　　We first suppose D to be bounded. By Lemma 9.2 we just have to show that $\|h\|_D < +\infty$ for every $h \in R(D)$ in order to show that $\| \cdot \|_D$ is a norm of K-algebra such that $\|h^n\|_D = \|h\|_D^n$. Since D is bounded, obviously every polynomial P satisfies $\|P\|_D < +\infty$, hence by Lemma 9.2 iv) we just have to check that $\|\frac{1}{Q}\|_D < +\infty$. To show this, it is sufficient to prove that

$\|\frac{1}{x-a}\|_D < +\infty$ for every $a \in K \setminus D$. Since D is closed the distance r from a to D is not zero, hence $\|\frac{1}{x-a}\|_D \le \frac{1}{r}$.

Now if D is not bounded, obviously $\|x\|_D = +\infty$. If D is not closed there exists at least one point $a \in K \setminus D$ with a sequence $(a_n)_{n \in \mathbb{N}}$ in D which converges to a, hence $\|\frac{1}{x-a}\|_D = +\infty$. That ends the proof of Theorem 9.3 .

Theorem 9.4: (G.Garandel)　　*Let $\mathcal{F}$ be a large circular filter on K of diameter $s > 0$. The three following assertions are equivalent*
i) $\varphi_{\mathcal{F}}(h) \le \|h\|_D$ *whenever $h \in R(D)$*
ii) $\varphi_{\mathcal{F}}$ *is an ultrametric continuous multiplicative norm on $R(D)$ with respect to the topology of uniform convergence.*
iii) $\mathcal{F}$ *is secant with D.*

Proof:　　First i) and ii) are obviously equivalent. Next, iii) clearly implies i) because if $\mathcal{F}$ is secant with D then $\lim_{\mathcal{F}}|h(x)| = \lim_{\mathcal{F} \cap D}|h(x)| \le \|h\|_D$.

Hence we just have to show that i) implies iii). For this, we assume iii) to be false and will prove that i) is false. We first assume $\mathcal{F}$ to have center a. There exist annuli $\Gamma(a_i, r_i', r_i'')$ $(1 \le i \le q)$ with $|a_i - a_j| = s$ whenever $i \ne j$ and $r_i' < s < r_i''$, such that the set $B = \bigcap_{i=1}^{q} \Gamma(a_i, r_i', r_i'')$ belongs to $\mathcal{F}$ and satisfies $B \cap D = \emptyset$. We put $r' = \max_{1 \le i \le q} r_i'$ and $r'' = \min_{1 \le i \le q} r_i''$. Let $\rho' \in]r', s[$, let $\rho'' \in]s, r''[$, and for every $i = 1, ..., q$ let $b_i \in \Gamma(a_i, \rho', s)$ and let $b \in \Gamma(a, s, \rho'')$.

We put $h(x) = \left(\prod_{i=1}^{q} \left(\frac{x - a_i}{x - b_i}\right)\right)\left(\frac{\lambda}{x - b}\right)$ with $|\lambda| = b$. We first notice that

$\varphi_{\mathcal{F}}(h) = 1$ because $\varphi_{\mathcal{F}}\left(\frac{x - a_i}{x - b_i}\right) = 1$ whenever $i = 1, ..., q$ and $\varphi_{\mathcal{F}}\left(\frac{\lambda}{x - b}\right) = 1$.

Besides, it is easily seen that $\|h\|_D \leq \max\left(\dfrac{r'}{\rho'}, \dfrac{\rho''}{r''}\right)$. Indeed $|h(x)| = |\dfrac{x}{b_i}| \leq \dfrac{r'}{\rho'}$ when $|x - a_i| \leq r'$ and $|h(x)| = |\dfrac{\lambda}{x - b}| \leq \dfrac{\rho''}{r''}$ when $|x - a| \geq r''$. Hence we have $\|h\|_D < 1$ while $\varphi_{\mathcal{F}}(h) = 1$ and that contradicts the assertion i).

We now suppose that $\mathcal{F}$ has no center. Let $d(a, r)$ belong to $\mathcal{F}$ such that $d(a, r) \cap D = \emptyset$ and let $\rho \in]s, r[$. There still exists a disk $d(\alpha, r) \in \mathcal{F}$ such that $d(\alpha, \rho) \subset d(a, r)$. Let us take $b \in \Gamma(\alpha, \rho, r)$ and $\lambda \in K$ such that $|\lambda| = |b|$. We just put $h(x) = \dfrac{\lambda}{x - b}$ and we have $\varphi_{\mathcal{F}}(h) = 1$ because $|h(x)| = 1$ whenever $x \in d(\alpha, \rho)$, while $|h(x)| \leq \dfrac{|b - \alpha|}{r} < 1$ whenever $|x - \alpha| \geq r$, hence finally $\|h\|_D < 1$. That ends the proof of Theorem 9.4.

The characterization of the continuous multiplicative norms of $(R(D), \| \cdot \|_D)$ by means of the large circular filters secant with D suggests us extending this characterization to the multiplicative semi-norms of $R(D)$.

Notation. $\Phi(D)$ will denote the set of the circular filters on D and $\Phi^*(D)$ will denote the set of the large circular filters on D .

Remark:. Let $\mathcal{F} \in \Phi^*(D)$. Then by definition $\mathcal{F}$ has diameter different from 0.

Theorem 9.5: *The mapping Ψ from $\Phi(D)$ into $Mult(R(D))$ defined as $\Psi(\mathcal{F}) = \varphi_{\mathcal{F}}$ is a bijection from $\Phi(D)$ onto $Mult(R(D), \mathcal{U}_D)$. Besides $\varphi_{\mathcal{F}}$ is an absolute value if and only if $\mathcal{F}$ does not converge in D.*

Proof: On the first hand, by Theorem 4.7 and Theorem 9.4, it is clearly seen that the mapping defined in $\Phi^*(D)$ by $\mathcal{F} \longrightarrow \varphi_{\mathcal{F}}$ is a bijection from this set onto the set of continuous multiplicative norms on $R(D)$.

On the other hand, it is seen that every $a \in D$ defines a multiplicative semi-norm ψ by $\psi(h) = |h(a)|$, the kernel of which is the maximal ideal $(x - a)R(D)$. Thus we have a mapping from the set of the convergent circular filters on D into the set of the multiplicative semi-norms which are not norms: this mapping is obviously injective.

Finally let ψ be a multiplicative semi-norm whose kernel is not zero. Then $Ker(\psi)$ is a prime ideal hence a maximal ideal of $R(D)$, and therefore it is of the form $(x - a)R(D)$ with $a \in D$. Then $\psi(x - a) = 0$, hence $\psi(x - b) = |a - b|$ whenever $b \in K$ and therefore ψ is of the form $\varphi_{\mathcal{F}_a}$ with $\mathcal{F}_a$ the filter of the neighbourhoods of a in D. That ends the proof of Theorem 9.5 .

10. THE ANALYTIC ELEMENTS

Due to the fact that any disk $d(a,r)$ is exactly the same as $d(b,r)$ for every $b \in d(a,r)$, it is easily seen that a power series $\sum_{n=0}^{\infty} a_n(x-a)^n$ which admits the disk $d(a,r)$ for disk of convergence, may not be extended outside its convergence disk as it is done in complex analysis, by means of a change of origin.

However by Runge's Theorem we remember that a holomorphic function in a compact subset D of $\mathbb{C}$ is equal to the limit of a sequence of rational functions with respect to the uniform convergence on D. This is why Marc Krasner introduced the analytic elements on a subset D of K directly by considering the limits of the sequences of rational functions with respect to uniform convergence on D [44], [48] .

Actually Marc Krasner constructed a theory of analytic functions f defined on a quasi-connected set D equal to the union of a chained family of quasi-connected sets $(D_i)_{i \in I}$ such that the restriction of f to each D_i is an analytic element on D_i. (This construction was widen to the analytic infraconnected sets by Philippe Robba [51]). But the analytic elements themselves have appeared to be so rich and useful that we have focused on them. Sometimes analytic functions are considered mainly in a disk $d(a, r^-)$, or when deriving an analytic element.

Notations and definitions: We will denote by $H(D)$ the completion of $R(D)$ for the topology $\mathcal{U}_D$ of uniform convergence on D. The elements of $H(D)$ are called the *analytic elements on D* [44], [48].

The set $H(D)$ is then provided with the topology of uniform convergence on D for which it is complete, and every $f \in H(D)$ defines a function on D which is the uniform limit (on D) of a sequence $(h_n)_{n \in \mathbb{N}}$ in $R(D)$. Thus, given two infinite sets D, D' such that $D \subset D'$, the restriction to D of elements of $H(D')$ enables us to consider that $H(D')$ is included in $H(D)$.

Besides $H(D)$ is a K-vectorial space and a complete topological group with respect to the topology $\mathcal{U}_D$. The question whether the product of two analytic elements on D is an analytic element on D will be studied later. However it is easily seen that given $f \in H(D)$, the function f^n also belongs to $H(D)$. Lemma 9.2 has an obvious extension to the space $H(D)$ in the following way.

We will sometimes need to consider analytic elements defined in an extension of K. Let E be a complete ultrametric algebraically closed extension of K and let A be a set in E. We will denote by $H_E(A)$ the completion of $R_E(A)$ with respect to the topology of the uniform convergence on A.

Lemma 10.1:   *For every $f, g \in H(D)$ we have*

   i) $\|f\|_D = 0$ *if and only if* $f = 0$

   ii) $\|\lambda f\|_D = |\lambda|\, \|f\|_D$ *whenever* $\lambda \in K^*$

   iii) $\|f + g\|_D \leq \max(\|f\|_D, \|g\|_D)$

   iv) *If* $(\|f\|_D, \|g\|_D)$ *is different from* $(0, +\infty)$ *and from* $(+\infty, 0)$ *then the function fg satisfies* $\|fg\|_D \leq \|f\|_D\, \|g\|_D$.

   v) $\|f^n\|_D = \|f\|_D^n$ *whenever* $n \in \mathbb{N}^*$.

Notations: We will denote by $H_b(D)$ the set of the elements $f \in H(D)$ bounded on D. Then $H_b(D)$ is clearly a K-vectorial subspace of $H(D)$ and is closed in $H(D)$. Besides $\| \, . \, \|_D$ is a norm on $H_b(D)$ that makes it a Banach algebra. If D is unbounded, we will denote by $H_0(D)$ the set of the $f \in H(D)$ such that $\displaystyle \lim_{\substack{|x| \to +\infty \\ x \in D}} f(x) = 0$.

Theorem 10.2 is an immediate consequence of Theorem 9.3.

Theorem 10.2: $H_b(D)$ *is a Banach K-subalgebra of K^D. The three following conditions are equivalent*

   i) $H_b(D) = H(D)$

   ii) $H(D)$ *is topological K-vector space,*

   iii) $(H(D), \| \, . \, \|_D)$ *is a K-Banach algebra*

   iv) D *is closed and bounded.*

If these conditions are satisfied, then $\| \, . \, \|_D$ is a semi-multiplicative norm.

Definition: Let $f \in H(D)$ have no zero in D. f will be said to be *invertible in* $H(D)$ if the function $\dfrac{1}{f}$ (also denoted by f^{-1}) belongs to $H(D)$. This definition holds even if $H(D)$ is not a ring.

The following Lemma is classical and easy.

Lemma 10.3:   *Let $f \in H(D)$ be such that $\displaystyle \inf_{x \in D} |f(x)| > 0$. Then f^{-1} belongs to $H(D)$. Besides if D is closed and bounded, f^{-1} belongs to $H(D)$ if and only if $\displaystyle \inf_{x \in D} |f(x)| > 0$.*

Let $g \in H(D)$ satisfy $|g(x)| = 1$ for all $x \in D$ and $\|f - 1\|_D > \|g - 1\|_D$. Then we have $\|fg - 1\|_D = \|f - 1\|_D$

Proof: We suppose $\inf_{x \in D} |f(x)| = \lambda > 0$. Let $(h_n)_{n \in \mathbb{N}}$ be a sequence in $R(D)$ such that $\lim_{n \to \infty} \|h_n - f\|_D = 0$. For n big enough we have $|h_n(x)| \geq \lambda$ whenever $x \in D$ hence $\left| \dfrac{1}{h_n(x)} - \dfrac{1}{f(x)} \right| = \left| \dfrac{f(x) - h_n(x)}{h_n(x)f(x)} \right| \leq \dfrac{\|f - h_n\|_D}{\lambda^2}$, hence the sequence $\dfrac{1}{h_n}$ converges to $\dfrac{1}{f}$.

Reciprocally if D is closed and bounded and if $\dfrac{1}{f} \in H(D)$, $\dfrac{1}{f}$ has to be bounded by some $M \in \mathbb{R}_+$ hence $|f(x)| \geq \dfrac{1}{M}$ whenever $x \in D$.

Now let $g \in H(D)$ satisfy $|g(x)| = 1$ for all $x \in D$ and $\|f - 1\|_D > \|g - 1\|_D$. For every $x \in D$, we have $|f(x) - 1||g(x)| > |g(x) - 1|$, and therefore $|f(x)g(x) - 1| = |f(x) - 1||g(x)| = |f(x) - 1|$. This finishes proving Lemma 10.3.

Theorem 10.4: *Let $f \in H(D)$ and let $h \in R_b(D)$. Then fh belongs to $H(D)$.*

Proof: Let ε be > 0 and let $g \in R(D)$ satisfy $\|f - g\|_D < \varepsilon$. Then we have $\|hf - hg\|_D < \varepsilon \|h\|_D$ and this clearly shows that $fh \in H(D)$.

When D is not closed or is not bounded, we will show how to split an element $f \in H(D)$.

Theorem 10.5: *The vector space $H(D)$ is equal to the direct sum $R_0(K \setminus (\overline{D} \setminus D)) \oplus H(\overline{D})$. Moreover if D is not bounded, then $H(\overline{D})$ is equal to the direct sum $K[x] \oplus H_0(\overline{D})$.*

Proof: Let $(f_n)_{n \in \mathbb{N}}$ be a sequence in $R(D)$ such that $\lim_{n \to \infty} \|f_n - f\|_D = 0$. In particular there exists $N \in \mathbb{N}$ such that $f_n - f_N$ is bounded when $n \geq N$. We put $g_n := f_n - f_N$. The sequence g_n converges in $H(D)$ to $f - f_N$. Let $g := f - f_N$. Besides each $f_n - f_N$ belongs to $R(D)$ and is bounded in D, hence $f_n - f_N$ belongs to $R(\overline{D})$. Obviously $\|f_n - f_N\|_D = \|f_n - f_N\|_{\overline{D}}$, hence finally g belongs to $H_b(\overline{D})$. Now we may obviously split f_N in the form $E(x) + h_1(x) + h_2(x)$ with $E(x) \in K[x], h_1 \in R_0(\overline{D}), h_2 \in R_0(K \setminus (\overline{D} \setminus D))$. We put $f^* = h_2$ and $\overline{f} = E + h_1 + g$. We have clearly split f in the form $f^* + \overline{f}$ with $f^* \in R_0(K \setminus (\overline{D} \setminus D))$, $\overline{f} \in H(\overline{D})$. Hence we have proven that $H(D) = R_0(K \setminus (\overline{D} \setminus D)) + H(\overline{D})$.

This sum is easily seen to be direct. Indeed, suppose that we have

$h \in R_0(K \setminus (\overline{D} \setminus D))$ and $g \in H(\overline{D})$ such that $h + g = 0$, with $h \neq 0$. Then h has a pole $\alpha \in \overline{D} \setminus D$ and it may be written in the form $\sum_{i=1}^{q} \dfrac{\lambda_i}{(x - \alpha)^i} + h_\alpha$ with $\lambda_i \neq 0 \ (1 \leq i \leq q)$ and $h_\alpha \in R_0(K \setminus (\overline{D} \setminus (D \cup \{\alpha\})))$. But g is obviously bounded around α, hence h has to be bounded when x approaches α, hence finally α does not exist. This shows that the sum is direct.

Now, we suppose D is unbounded. First we will prove that every element $f \in H_b(D)$ admits a limit when $|x|$ tends to $+\infty$. Let $\epsilon \in \mathbb{R}_+^*$, and let $h \in R(D)$ satisfy $\|f - h\|_D < \epsilon$. Since f is bounded in D, so is h. But then h is of the form $\dfrac{P}{Q}$, with $deg(P) = deg(Q)$, and therefore h has a limit λ when $|x|$ tends to $+\infty$. Let $\epsilon > 0$ be such that $|h(x) - \lambda| < \epsilon$ whenever $x \in D \setminus d(0, r)$. Clearly we have $|f(x) - \lambda| < \epsilon$ whenever $x \in D \setminus d(0, r)$. This proves that f does converge along the filter $\mathcal{F}$ which admits as a base the family of sets $D \setminus d(0, r) \ \ (r > 0)$.

Let $f \in H(\overline{D})$ be unbounded. Let $(x_n)_{n \in \mathbb{N}}$ be a sequence such that $\lim_{n \to \infty} |f(x_n)| = +\infty$. Suppose the sequence $(x_n)_{n \in \mathbb{N}}$ does not tend to $+\infty$. Then there exists a bounded subsequence $(x_{n_q})_{q \in \mathbb{N}}$ such that $\lim_{q \to \infty} |f(x_{n_q})| = +\infty$, but this is impossible due to Theorem 10.2, because such a sequence lies in a closed bounded set D' included in D. Now there exists $h \in R(\overline{D})$ such that $f - h$ is bounded, and therefore, we have $\lim_{n \to \infty} |h(x_n)| = +\infty$. Let $h(x) = P(x) + \ell(x)$ with $P \in K[x]$, and $\ell \in R_0(\overline{D})$. Since $f - h$ is bounded, clearly $f - P$ belongs to $H_b(\overline{D})$, hence we have proven that $H(\overline{D}) = K[x] + H_b(\overline{D})$. Moreover, since ℓ has a limit when $|x|$ tends to $+\infty$ in D, we have $H_b(\overline{D}) = H_0(\overline{D}) + K$, and therefore $H(\overline{D}) = K[x] + H_0(\overline{D})$. Finally, by considering elements when $|x|$ tends to $+\infty$, this sum is easily seen to be direct, and this ends the proof.

Notations and definitions: Let $\alpha \in \overline{D} \setminus D$ and $f \in H(D)$, and let $f = f^* + \overline{f}$, with $f^* \in R_0(K \setminus (\overline{D} \setminus D))$, and $\overline{f} \in H(\overline{D})$. Let $a \in \overline{D} \setminus D$ be a pole of f^* and let $f^*(x) = \sum_{j=1}^{q} \dfrac{\lambda_j}{(x - a)^j} + \ell(x)$, with $\ell \in R_0(K \setminus (\overline{D} \setminus (D \cup \{a\})))$. Then λ_1 will be called *the residue of f at a*, and will be denoted by $res(f, a)$.

Besides, each pole α of order q of f^* will be called a *pole of order q of f* . Let $a_i, 1 \leq i \leq n$ be the poles of f and for each i let q_i be the order of a_i. The polynomial $\prod_{i=1}^{n} (x - a_i)^{q_i}$ will be named *polynomial of the poles of f in $\overline{D} \setminus D$*.

Corollary 10.6:   *Let $\alpha \in \overline{D} \setminus D$ and $f \in H(D)$ be such that $|f(x)|$ is bounded* *in $d(\alpha, r) \cap D$ with $r > 0$. Then $f \in H(D \cup \{\alpha\})$.*

Proof: Indeed, as $\overline{f}$ is obviously bounded in $D \cap d(\alpha, r)$, so is f^* and therefore f^* has no pole at α.

Corollary 10.7: $H(D) = H_b(\overline{D}) + R(K \setminus (\overline{D} \setminus D))$, $H_b(D) = H_b(\overline{D}) \subset H(\overline{D})$ and $H_b(D) = H_b(\overline{D})$. *If D is bounded then $H_b(D) = H(\overline{D})$. If D is not bounded, for every unbounded $f \in H(D)$ such that $\lim\limits_{\substack{|x| \to +\infty \\ x \in D}} |f(x)| = +\infty$, there exists a unique $q \in \mathbb{N}^*$ such that $x^{-q} f(x)$ has a non zero limit when x tends to $+\infty$, $x \in D$.*

Corollary 10.8: *If $\overline{D} = K$ then $H(D) = R(D)$.*

Corollary 10.9 comes from the definition of the poles and from Theorem 10.5.

Corollary 10.9: *Let $f \in H(D)$ and let $\alpha \in \overline{D} \setminus D$. Then α is a pole of order $n > 0$ for f if and only if $(x - \alpha)^n f(x)$ has a non zero limit at α. If there exists no $r \in \mathbb{R}_+^*$ such that $|f(x)|$ is bounded in $d(\alpha, r) \setminus \{\alpha\}$ then α is a pole of order $n \geq 1$ for f and $(x - \alpha)^n f(x)$ has a non zero limit at α.*

Theorem 10.10: *Let $f \in H(D)$ and let $\alpha \in \overline{D} \setminus D$. Either f belongs to $H(D \cup \{\alpha\})$ or α is a pole for f.*

Proof : If f does not belong to $H(D \cup \{\alpha\})$, by Corollary 10.6, f is unbounded in any disk $d(\alpha, r)$ whenever $r > 0$. Hence by means of the notations of Theorem 10.5 , α is clearly a pole of f^* and therefore is a pole of f.

Notation: We will denote by $\mathcal{A}$ the family of the sets E such that $H(E)$ is a K-subalgebra of K^E.

Theorem 10.11: *Let $f \in H(D)$. There exists $W \in R_b(D)$, the zeros of which lie in $\overline{D} \setminus D$, and $h \in H(\overline{D})$ such that $f = \dfrac{h}{w}$. Besides, if D is bounded or if $D \in \mathcal{A}$ then there exists $g \in H(\overline{D})$ such that $f = \dfrac{g}{Q}$ with Q the polynomial of the poles of f in $\overline{D} \setminus D$.*

Proof: We may summarize Theorem 10.5 in this way: f is of the form $\widetilde{f}(x) + \widehat{f}(x)$ with $\widetilde{f} \in R(K \setminus (\overline{D} \setminus D))$ and $\widehat{f} \in H_b(\overline{D})$. Indeed if D is bounded we just take $\widehat{f} = \overline{f}$ and if D is not bounded, $\widehat{f}$ is the one defined in Theorem 10.5. Thus $\widetilde{f}(x)$ can be written in the form $\dfrac{P(x)}{Q(x)}$ with $Q(x) = \displaystyle\prod_{i=1}^{n}(x - a_i)^{q_i}$ i.e. the polynomial of the poles of f in $\overline{D} \setminus D$, and $P(x) \in K[x]$. Let $q = \displaystyle\sum_{i=1}^{n} q_i$. Theorem 10.11 is obviously trivial if D has no hole, hence we may assume D to have at least one hole $T = d(a, r^-)$. Let $W(x) = \dfrac{Q(x)}{(x - a)^q}$. We know that $W \in R_b(\overline{D})$ hence $\widehat{f}W \in H_b(\overline{D})$. Besides we see that $W\widetilde{f} \in R(\overline{D})$ hence $fW \in H(\overline{D})$. We just put $h = fW$ and have the factorization $f = \dfrac{h}{W}$.

If D is bounded we see that both Q, h are bounded in D hence $Q\widehat{f}$ belongs to $H_b(\overline{D})$, and then Qf belongs to $H(\overline{D})$. In the same way if $H(D)$ is supposed to be a ring, then $Q\widehat{f}$ belongs to $H(\overline{D})$ and then $Qf = P + Q\widehat{f}$ belongs to $H(\overline{D})$ and this ends the proof.

Corollary 10.12:　　*Let $S(D)$ be the set of polynomials whose zeros belong to $\overline{D} \setminus D$. If $D \in \mathcal{A}$ then $H(D) = S(D)^{-1}H(\overline{D})$.*

Theorem 10.13:　*Let D be closed. Let $g \in H(D)$ and let $P \in K[x]$ be such that Pg belongs to $H(D)$. For every $Q \in K[x]$ such that $\deg(Q) \leq \deg(P)$, Qg also belongs to $H(D)$.*

Proof:　Theorem 10.13 is clearly trivial when D belongs to $\mathcal{A}$. If D has no hole, by Corollary 10.8 $H(D)$ is equal to $R(D)$. Hence without loss of generality, we may assume that D admits holes and that one of the holes is $d(0, r^-)$. Let $q = \deg(P)$. Let $P_q = P$, and let $P_{q-1} = \dfrac{P(x) - P(0)}{x}$. Then P_{q-1} is a polynomial of degree $q - 1$. We see that $(P(x) - P(a))g(x)$ belongs to $H(D)$. But $\left\| \dfrac{1}{x} \right\|_D$ is bounded, and then by Lemma 10.4 we see that $P_{q-1}(x)g(x)$ belongs to $H(D)$. Hence by induction, it is seen that for each $j = 1, ..., q$ there exists a polynomial P_j of degree j such that $P_j g$ belongs to $H(D)$ and this clearly completes the proof.

Now when D is not infraconnected we have to notice an easy result on characteristic functions that shows how rich the algebra $H(D)$ is.

Proposition 10.14:　*Let D have an empty annulus Λ. Let w_1, w_2 be the* *functions defined on D by $w_1(x) = 1, w_2(x) = 0$ if $x \in \mathcal{I}(\Lambda)$ and $w_1(x) = 0, w_2(x) = 1$ if $x \in \mathcal{E}(\Lambda)$. Then w_1 and w_2 belong to $H(D)$.*

Proof: Let $\Lambda = \Gamma(a, r_1, r_2)$, with $a \in D$. With no loss of generality we may obviously assume $a = 0$. Let $\alpha \in \Lambda$ be such that $r_1 < |\alpha| < r_2$ and for each $n \in \mathbb{N}^*$, let $u_n = \dfrac{1}{1 - \left(\dfrac{x}{\alpha}\right)^n}$. Then $\left\|1 - \left(\dfrac{x}{\alpha}\right)^n - 1\right\|_{\mathcal{I}(\Lambda)} \leq \left(\dfrac{r_1}{|\alpha|}\right)^n$ while $\left|1 - \left(\dfrac{x}{\alpha}\right)^n\right| \geq \left(\dfrac{r_2}{|\alpha|}\right)^n$ for every $x \in \mathcal{E}(\Lambda)$ hence finally $\|u_n - w_1\|_D \leq \max\left(\left(\dfrac{r_1}{|\alpha|}\right)^n, \left(\dfrac{|\alpha|}{r_2}\right)^n\right)$. Thus we see that $w_1 = \lim\limits_{n \to \infty} u_n \in H(D)$ and $w_2 = 1 - w_1 \in H(D)$.

Theorem 10.15: *Let E have finitely many infraconnected components $E_1, .., E_q$. For each $i = 1, ..., q$, the characteristic function of E_i belongs to $H(E)$.*

Proof: Let A be one of the infraconnected components of E. By Proposition 2.13, there exist empty annuli $(\Lambda_j)_{0 \leq j \leq n}$ such that A is either of the form

$$\alpha) \quad \mathcal{I}_E(\Lambda_0) \bigcap \left(\bigcap_{j=1}^{n} \mathcal{E}_E(\Lambda_j) \right) \quad \text{or of the form} \quad \beta) \quad \bigcap_{j=1}^{n} \mathcal{E}_E(\Lambda_j).$$

But by Proposition 10.14, the characteristic function u_j of $\mathcal{E}_E(\Lambda_j)$ belongs to $H_b(E)$, $(1 \leq j \leq n)$ and so does the characteristic function u_0 of $\mathcal{I}_E(\Lambda_0)$. Since all the u_j belong to $H_b(E)$, we see that the products $u = \prod\limits_{j=0}^{n} u_j$ and $w = \prod\limits_{j=1}^{n} u_j$ do belong to $H(E)$. Then when A is of the form $\alpha)$ (resp. $\beta)$), its characteristic function is equal to u (resp. w) and therefore belongs to $H(E)$.

11.   COMPOSITION OF ANALYTIC ELEMENTS

Given A and $B \subset K$, $f \in H(A)$ such that $f(A) \subset B$ and $g \in H(B)$ the problem we are looking for is whether $g \circ f \in H(A)$. There is an immediate application to the study of the homomorphisms from an algebra $H(D)$ to another $H(D')$.

We remember that $\mathcal{A}$ denotes the set of the subsets D of K such that $H(D)$ is a K-subalgebra of K^D.

Lemma 11.1 :   *Let A and B be subsets of K and let $f \in H(A)$ be such that $f(A) \subset B$. For every $\lambda \in K \setminus \overline{B}$, $f - \lambda$ is invertible in $H(A)$. Besides, if for every $g \in R(B)$, $g \circ f$ belongs to $H(A)$, then for every $g \in H(B)$, $g \circ f$ belongs to $H(A)$ and for every $\lambda \in \overline{B} \setminus B$, $f - \lambda$ is invertible in $H(A)$.*

Proof :   Let $r = \delta(\lambda, B)$. Let $\lambda \notin \overline{B}$ we have $r > 0$ and then $|f(x) - \lambda| \geq r$ whenever $x \in A$. Hence by Lemma 10. 3, $f - \lambda$ is invertible in $H(A)$.

Now we assume that for every $g \in R(B)$, $g \circ f$ belongs to $H(A)$. Let $h \in H(B)$, let ϵ be > 0, and let $g \in R(B)$ satisfy $\|g - h\|_B \leq \epsilon$. It is seen that $\|g \circ f - h \circ f\|_A \leq \epsilon$. Since $g \circ f$ belongs to $H(A)$, then it is seen that so does $h \circ f$. Now let $\lambda \in \overline{B} \setminus B$ and let $g(u) = \dfrac{1}{u - \lambda}$. Since $g \circ f$ belongs to $H(A)$, $f - \lambda$ is invertible in $H(A)$.

Theorem 11.2 :   *Let A, B be subsets of K and let $f \in H(A)$ satisfy $f(A) \subset B$.*
i) If $f \in R(A)$, then $g \circ f \in H(A)$ whenever $g \in H(B)$.
ii) If $A \in \mathcal{A}$, then $g \circ f \in H(A)$ for all $g \in H(B)$ if and only if $f - \lambda$ is invertible in $H(A)$ for all $\lambda \in \overline{B} \setminus B$.

Proof :   By Lemma 11.1 we just have to show that for every $g \in R(B)$, $g \circ f$ belongs to $H(A)$ in each one of these two hypotheses:

H_1) $f \in R(A)$.

H_2) $f - \lambda$ is invertible in $H(A)$ for all $\lambda \in \overline{B} \setminus B$.

So we take $g(u) = \dfrac{P(u)}{Q(u)} \in R(B)$ and will show that $g \circ f \in H(A)$ in each hypothesis. Let $\lambda_1, ..., \lambda_q$ be the poles of g in $K \setminus B$.

H_1) For every $j = 1, ..., q$, $f - \lambda_j$ is invertible in $R(A)$ because $f - \lambda_j$ has no zero in A. Hence $Q \circ f$ is invertible in $R(A)$ and then $g \circ f$ belongs to $R(A)$.

H_2) For each $j = 1, ..., q$, either λ_j belongs to $\overline{B} \setminus B$, or it belongs to $K \setminus \overline{B}$. In both cases, by Lemma 11.1 each $f - \lambda_j$ is invertible in $H(A)$. Since A belongs to $\mathcal{A}$, $Q \circ f$ is clearly invertible in $H(A)$ and $P \circ f$ belongs to $H(A)$. Hence so does $g \circ f$.

Corollary 11.3 :   *Let $A \in \mathcal{A}$ and let B be a closed subset of K. Let $f \in H(A)$ satisfy $f(A) \subset B$ and let $g \in H(B)$. Then $g \circ f$ belongs to $H(A)$.*

Lemma 11.4 :   *Let $h \in R(D)$ and let $D' = h(D)$. Let $f \in H(D')$. If f is invertible in $H(D')$ then $f \circ h$ is invertible in $H(D)$. If h is a homographic function, f is invertible in $H(D')$ if and only if $f \circ h$ is invertible in $H(D)$.*

Proof :   First we suppose f invertible in $H(D')$. Let $g = \dfrac{1}{f}$. Then by Theorem 11.2 $g \circ h$ belongs to $H(D)$ and is clearly equal to $\dfrac{1}{f \circ h}$. Now we assume that h is a homographic function and we put $\ell = h^{-1}$. If $f \circ h$ is invertible in $H(D)$, $(f \circ h) \circ \ell$ is invertible in $H(D')$ and this ends the proof of Lemma 11.4.

We are now going to study the K-algebra homomorphisms from $H(D)$ into $H(D')$. First we will consider homomorphisms from $R(D)$ into $R(D')$.

Proposition 11.5 :   *Let D, D' be sets in K and let $\gamma \in R(D')$ satisfy $\gamma(D') \subset D$. Let ϕ_γ be the mapping from $R(D)$ into $R(D')$ defined as $\phi_\gamma(f) = f \circ \gamma$ ($f \in R(D)$). Then ϕ_γ is a homomorphism from $R(D)$ into $R(D')$, and this homomorphism is injective if and only if γ is not a constant. Every K-algebra homomorphism is of this form and the mapping $\gamma \to \phi_\gamma$ is a one-to-one mapping from the set of the $\gamma \in R(D')$ such that $\gamma(D') \subset D$ onto the set of the K-algebra homomorphisms from $R(D)$ into $R(D')$.*

Proof: Let $\gamma \in R(D')$ satisfy $\gamma(D') \subset D$. Then it is seen that ϕ_γ takes values in $R(D')$, is a K-algebra homomorphism, and is obviously injective if and only if γ is not a constant.

Reciprocally let ψ be a K-algebra homomorphism from $R(D)$ into $R(D')$ and let $\gamma = \psi(I_{D'})$ with $I_{D'}$ the identical mapping in D'. Then we have $\psi(P) = P \circ \gamma$ for every polynomial P. Besides if $\alpha \notin D$ then $(x - \alpha)$ is invertible in $R(D)$ and $\psi\left(\dfrac{1}{x - \alpha}\right) = (\psi(x - \alpha))^{-1} = (\gamma - \alpha)^{-1}$. Thereby we see that $\psi(h) = h \circ \gamma$ whenever $h \in R(D)$. The mapping $\gamma \to \phi_\gamma$ is obviously one-to-one.

Proposition 11.6 :   *Let D, D' be sets in K and let $\gamma \in H(D')$ satisfy*

$\gamma(D') \subset D$ and $f \circ \gamma \in H(D')$ for all $f \in H(D)$. Let ϕ_γ be the mapping from $H(D)$ into $H(D')$ defined as $\phi_\gamma(f) = f \circ \gamma$. Then ϕ_γ is a linear mapping from $H(D)$ into $H(D')$ continuous with respect to the topology of uniform convergence on D for $H(D)$ and on D' for $H(D')$. Besides, given $f, g \in H(D)$ such that $fg \in H(D)$ we have $\phi_\gamma(fg) = \phi_\gamma(f)\phi_\gamma(g)$. The restriction of ϕ_γ to $H_b(D)$ is a K-Banach algebra homomorphism from $H_b(D)$ into $H_b(D')$.

If γ is a bijection from D' onto D and if $\gamma^{-1} \in H(D)$ then ϕ_γ is a K-vector space isomorphism from $H(D)$ onto $H(D')$ bicontinuous with respect to the topology of uniform convergence on D for $H(D)$ and on $H(D')$ for $H(D')$, satisfying $\left(\phi_\gamma\right)^{-1} = \phi_{\gamma^{-1}}$, and the restriction of ϕ_γ to $H_b(D)$ is a K-Banach algebra isomorphism from $H_b(D)$ onto $H_b(D')$. Further, if $\gamma(D') = D$, then the equality $\|\phi_\gamma(f)\|_{D'} = \|f\|_D$ is true for every $f \in H(D)$, and the restriction of ϕ_γ to $H_b(D)$ is an isometric K-Banach algebra isomorphism from $H_b(D)$ onto $H_b(D')$.

Proof : It is easily seen that ϕ_γ is linear and satisfies $\phi_\gamma(fg) = \phi_\gamma(f)\phi_\gamma(g)$ when $fg \in H(D)$. Besides ϕ_γ is clearly continuous because

$\|\phi_\gamma(f)\|_{D'} = \|f \circ \gamma\|_D = \sup_{x \in D'} |f(\gamma(x))| \leq \sup_{u \in D} |f(u)| = \|f\|_D$. In particular, we notice that if $\gamma(D') = D$, we have $\|\phi_\gamma(f)\|_{D'} = \|f \circ \gamma\|_D = \sup_{x \in D'} |f(\gamma(x))| = \sup_{u \in D} |f(u)| = \|f\|_D$.

If $f \in H_b(D)$ obviously $f \circ \gamma \in H_b(D')$. Now let γ be one-to-one from D' onto D and such that $\gamma^{-1} \in H(D)$. It is seen that $(\phi_{\gamma^{-1}}) \circ \phi_\gamma = I_{H(D)}$ while $\phi_\gamma \circ (\phi_{\gamma^{-1}}) = I_{H(D')}$, hence ϕ_γ is an isomorphism such that $\phi_{\gamma^{-1}} = \left(\phi_\gamma\right)^{-1}$.

We will characterize these subsets. We will first study the K-algebra homomorphisms from $H(D)$ into $H(D')$.

Notations: Given subsets D and D' of K, we will denote by $\Omega(D', D)$ *the set of the $\gamma \in H(D')$ such that $\gamma(D') \subset D$ and such that for every $\lambda \in \overline{D} \setminus D$, $\gamma - \lambda$ is invertible in $H(D')$.*

Given two K-algebras A and B we will denote by $\mathcal{H}om(A, B)$ the set of the K-algebra homomorphisms from A into B.

Remark : In particular $\Omega(D, D')$ contains the set of the $h \in R(D)$ such that $h(D) \subset D'$.

Theorem 11.7 : *Let $D, D' \in A$ and let $\gamma \in \Omega(D', D)$. The mapping ϕ_γ defined in $H(D)$ by $\phi_\gamma(f) = f \circ \gamma$ has values in $H(D')$ and is a K-algebra homomorphism*

from $H(D)$ into $H(D')$. Reciprocally, every K-algebra homomorphism from $H(D)$ into $H(D')$ is continuous and of this form. Further, the mapping $\gamma \to \phi_\gamma$ from $\Omega(D',D)$ onto $\mathcal{H}om(H(D),H(D'))$ is a bijection.

Let $D'' \in \mathcal{A}$ and let $\tau \in \Omega(D'',D')$. Then $\gamma \circ \tau \in \Omega(D'',D)$ and $\phi_{\gamma\circ\tau} = \phi_\gamma \circ \phi_\tau$. Further, a homomorphism ϕ_γ from $H(D)$ into $H(D')$ is an isomorphism if and only if γ is a bijection from D' onto D such that $\gamma^{-1} \in H(D)$ and then, when it is satisfied, we have $\left(\phi_\gamma\right)^{-1} = \phi_{\gamma^{-1}}$.

Proof : By Theorem 11.2 , $f \circ \gamma$ belongs to $H(D')$ whenever $f \in H(D)$ and then by Proposition 11.6 it is seen that ϕ_γ is a K-algebra homomorphism from $H(D)$ into $H(D')$. Let $\psi \in \mathcal{H}om(H(D),H(D'))$ and first let us show that ψ satisfies $\|\psi(f)\|_{D'} \leq \|f\|_D$ whenever $f \in H(D)$. Indeed suppose that for some $f \in H(D)$ we have $\|\psi(f)\|_{D'} > \|f\|_{D'}$. Let $g = \psi(f)$. There exists $\alpha \in D'$ such that

$|g(\alpha)| > \|f\|_{D'}$. Let $\lambda = g(\alpha)$. The series $\dfrac{f}{\lambda}\displaystyle\sum_{n=0}^{\infty}\left(\dfrac{f}{\lambda}\right)^n$ does converge in $H(D)$ to

$(\lambda - f)^{-1}$. Thus $\lambda - f$ is invertible in $H(D)$ and then $\lambda - g = \psi(\lambda - f)$ is invertible in $H(D')$. But by hypothesis α is a zero for $\lambda - g$, hence $\lambda - g$ is not invertible in $H(D')$ and this shows that $\|\psi(f)\|_{D'} \leq \|f\|_D$. Now let $\gamma = \psi(I_D) \in H(D')$ and let us show that $\gamma \in \Omega(D',D)$. Let $\alpha \in K \setminus D$. Since $\psi \in \mathcal{H}om(H(D),H(D')),\psi$ must satisfy $\psi\left(\dfrac{1}{x-\alpha}\right) = \dfrac{1}{\psi(x)-\alpha} = \dfrac{1}{\gamma-\alpha}$ hence $\gamma - \alpha$ has to be invertible in $H(D')$ for every $\alpha \in K \setminus D$ but this just means that $\gamma \in \Omega(D',D)$.

In the same way, we see that for every $h \in R(D)$, we have $\psi(h) = h(\psi(x)) = h \circ \gamma$. Finally since ψ is continuous, the equality $\psi(f) = f \circ \gamma$ holds in all of $H(D)$. Obviously, given $\gamma, \tau \in \Omega(D',D)$, if $\phi_\gamma = \phi_\tau$ then $\phi_\gamma(I_D) = \phi_\tau(I_D)$ hence $\gamma = \tau$. The mapping $\gamma \to \phi_\gamma$ is then a bijection from $\Omega(D',D)$ onto $\mathcal{H}om(H(D),H(D'))$. Now let $D'' \in \mathcal{A}$ and $\tau \in \Omega(D'',D')$. It is seen that $\gamma \circ \tau \in \Omega(D'',D)$ and $\phi_{\gamma\circ\tau}(I_D) = \gamma \circ \tau = \phi_\tau(\gamma) = \phi_\tau(\phi_\gamma(I_D)) = \phi_\tau \circ \phi_\gamma(I_D)$, hence $\phi_{\gamma\circ\tau} = \phi_\tau \circ \phi_\gamma$.

By Proposition 11.6, if γ is a bijection from D' onto D and such that $\gamma^{-1} \in H(D)$ then ϕ_γ is an isomorphism of K-vector space, hence it is an isomorphism of K-algebra, and then by Proposition 11.6 we have $\left(\phi_\gamma\right)^{-1} = \phi_{\gamma^{-1}}$. Reciprocally if ϕ_γ is an isomorphism, then $\left(\phi_\gamma\right)^{-1}$ is in the form ϕ_τ, with $\tau \in \Omega(D,D')$ and $\phi_\tau \circ \phi_\gamma(I_D) = \gamma \circ \tau(I_D) = I_D$ and $\phi_\gamma \circ \phi_\tau(I_{D'}) = \tau \circ \gamma(I_{D'}) = I_{D'}$. Hence γ is a bijection from D' onto D such that $\gamma^{-1} = \tau \in H(D)$. That finishes showing Theorem 11.7.

Around the corner, the following Propositions 11.8 and 11.9 will be often useful to transform unbounded domains into bounded domains.

Theorem 11.8 :   *Let $D \in \mathcal{A}$, and let $h \in R(D)$ be a homographic function. Let $D' = h(D)$. Then D' belongs to $\mathcal{A}$ and $H(D')$ is isomorphic to $H(D)$ with respect to the mapping ψ defined in $H(D')$ as $\psi(f) = f \circ h$.*

Proof :   By Theorem 11.2, for every $f \in H(D')$, $f \circ h$ belongs to $H(D)$ and by Proposition 11.6 this mapping is a K-vector space isomorphism which satisfies $\psi(fg) = \psi(f)\psi(g)$ whenever $f, g \in H(D')$. Hence the space $H(D') = \psi^{-1}(H(D))$ is seen to be a K-algebra isomorphic to $H(D)$. In particular D' belongs to $\mathcal{A}$ and ψ is a K- algebra isomorphism.

Proposition 11.9 :   *Let D be a set with a hole $T = d(a, r^-)$, let $\gamma(x) = \dfrac{1}{x - a}$ and let $D' = \gamma(D)$. Then $\overline{D'} \in \mathcal{A}$ and $H(\overline{D'})$ is isomorphic to $H_b(D)$.*

Proof: Without loss of generality we may clearly assume D to be closed because by Corollary 10.7, $H_b(D)$ is equal to $H_b(\overline{D})$. For every $f \in H(D)$ let $\psi(f) = f \circ \gamma \in H(D')$. Then $\psi(H_b(D))$ is a K-algebra included in $H(D')$. If D is bounded, D' is bounded and closed like D, hence by Proposition 11.8, ψ is an isomorphism from $H(D)$ onto $H(D')$. Now we suppose D unbounded. Then D' is bounded and $\psi(H_b(D))$ is obviously included in $H_b(D')$ which, by Corollary 10.7, is just equal to $H(\overline{D'})$. Besides γ clearly maps $R_b(D)$ onto $R(\overline{D'})$ hence $\psi(H_b(D)) = H(\overline{D'})$.

Theorem 11.10:   *Let $T = d(a, r^-)$ be a hole of D and let $\gamma(x) = \dfrac{1}{x - a}$. Let $D' = \gamma(D)$. The mapping ψ from $H_b(D')$ into $H_b(D)$ defined as $\psi(f) = f \circ \gamma$ is a K-algebra isomorphism.*

Proof : D' is bounded, hence by Corollary 10.7, $H_b(D')$ is equal to the K-Banach algebra $H(\overline{D'})$. Now by Theorem 11.2, we see that $\gamma \in \Omega(D, D')$ and $\gamma^{-1} \in \Omega(D', D)$. Hence ψ is clearly a K-Banach space isomorphism from $H_b(D')$ onto $H_b(D)$. Now , ψ satisfies $\psi(fg) = \psi(f)\,\psi(g)$ whenever $f, g \in H(D)$ such that $fg \in H(D')$. But both $H_b(D)$, $H_b(D')$ are K-Banach algebras, hence ψ is a K-Banach algebra isomorphism.

Theorem 11.11:   *Let a be a point in D which is not isolated. Let $\gamma(x) = \dfrac{1}{x - a}$ and let $D' = \gamma(D \setminus \{a\})$. Then given $f \in H(D')$, $f \circ \gamma$ belongs to $H(D)$ if and only if $f(x)$ has a limit when $|x|$ tends to $+\infty$.*

Proof :   If $f \circ \gamma$ belongs to $H(D)$ then we just have

$$\lim_{|x|\to+\infty} f(x) = \lim_{x\to a} f \circ \gamma(x) = f \circ \gamma(a).$$

Reciprocally, if f has a limit l when $|x|$ tends to $+\infty$, then $f \circ \gamma$ is bounded in certain disks $d(a,r) \setminus \{a\}$. Therefore by Corollary 10.6, $f \circ \gamma$ belongs to $H(D)$.

Corollary 11.12: *Let $D \in \mathcal{A}$, let a be a point in D which is not isolated , such that $(D \setminus \{a\})$ belongs to $\mathcal{A}$. Let $\gamma(x) = \dfrac{1}{x-a}$ and let $D' = \gamma(D \setminus \{a\})$. Then $H(D)$ is isomorphic to the subalgebra of $H(D')$ which consists of the f such that $|f(x)|$ is bounded when $|x|$ approaches $+\infty$.*

<h1 align="center">12.   Mult $(H(D), \mathcal{U}_D)$</h1>

Notations: We will denote by $Mult(H(D), \mathcal{U}_D)$ the set of the continuous semi-norms ψ of the K-vector space $H(D)$ that satisfy $\psi(fg) = \psi(f)\psi(g)$ whenever $f, g \in H(D)$ such that $fg \in H(D)$.

Remark: This notation does not require $H(D)$ to be a K- algebra, though it coincides with the notation already introduced for any topological algebra when $H(D)$ is a normed K-algebra. The multiplicative semi-norms appeared to be the main tool for studying the analytic elements [40].

Theorem 12.1: (G. Garandel)   *For every $\mathcal{F} \in \Phi(D)$, the multiplicative semi-norm $\varphi_\mathcal{F}$ defined on $R(D)$ extends by continuity to $H(D)$ to a continuous semi-norm of K-vector space $_D\varphi_\mathcal{F}$ of $H(D)$ that satisfies $_D\varphi_\mathcal{F}(f.g) = {}_D\varphi_\mathcal{F}(f) \, _D\varphi_\mathcal{F}(g)$ whenever $f, g \in H(D)$ such that $fg \in H(D)$. Besides, the mapping: $\mathcal{F} \to {}_D\varphi_\mathcal{F}$, from $\Phi(D)$ into $Mult(H(D), \mathcal{U}_D)$, is a bijection.*

Proof :   We may obviously extend $\varphi_\mathcal{F}$ by continuity to $_D\varphi_\mathcal{F}$ satisfying $_D\varphi_\mathcal{F}(fg) = {}_D\varphi_\mathcal{F}(f) \, _D\varphi_\mathcal{F}(g)$ whenever $f, g \in H(D)$ such that $fg \in H(D)$. We now check that the mapping $\mathcal{F} \to {}_D\varphi_\mathcal{F}$ from $\Phi(D)$ into $Mult(H(D), \mathcal{U}_D)$ is a bijection. It is obviously injective because if $_D\varphi_{\mathcal{F}_1} = {}_D\varphi_{\mathcal{F}_2}$ then $\varphi_{\mathcal{F}_1} = \varphi_{\mathcal{F}_2}$, hence by Theorem 9.5 $\mathcal{F}_1 = \mathcal{F}_2$. Now let $\psi \in Mult(H(D), \mathcal{U}_D)$. The restriction of ψ to $R(D)$ is an element ψ_0 of $Mult(R(D), \mathcal{U}_D)$, hence by Theorem 9.5, ψ_0 is of the form $\varphi_\mathcal{F}$ and then, by continuity, we have $\psi = {}_D\varphi_\mathcal{F}$.

Remark :   If $\mathcal{F}$ is a large circular filter, we know that $\varphi_\mathcal{F}$ is a norm on $R(D)$. But we don't know if $_D\varphi_\mathcal{F}$ is a norm on $H(D)$. This not trivial question is linked to the problem of T-filters and will be cleared up later.
For the state of convenience we will introduce another way to define multiplicative semi-norms of $H(D)$.

Definition and notations: For convenience, we will generalize the way to define the multiplicative semi-norms. For every $a \in D$, we put $\varphi_a(f) = |f(a)|$ whenever $f \in H(D)$, and so we define the semi-norms $\varphi_a \in Mult(H(D), \mathcal{U}_D)$.
An element $\psi \in Mult(H(D), \mathcal{U}_D)$ will be said to be *punctual* if it is of the form φ_a with $a \in D$.
Let $\mathcal{F}$ be a monotonous filter on D. By Proposition 3.13, on D there exists a unique circular filter $\mathcal{G}$ on D less thin than $\mathcal{F}$. Then we put $_D\varphi_\mathcal{F}(f) = {}_D\varphi_\mathcal{G}(f)$

67

for all $f \in H(D)$.

Now let D be infraconnected. Let $a \in \tilde{D}$ and let r satisfy $\delta(a, D) \leq r \leq diam(D)$. The circular filter $\mathcal{F}$ of center a and diameter r is then secant with D. We put $_D\varphi_{a,r} = {_D\varphi_{\mathcal{F}}}$. Let A be a bounded subset of $\tilde{D}$ and let $\tilde{A} = d(a, r)$. If $\delta(a, D) \leq r \leq diam(D)$ we put $_D\varphi_A = {_D\varphi_{a,r}}$. In particular this notation applies to holes of an infraconnected set D.

Let $\mathcal{F}$ be a circular filter or a monotonous filter on D. We will denote by $\mathcal{I}(\mathcal{F})$ the set of the $f \in H(D)$ such that $\lim_{\mathcal{F}} f(x) = 0$. Hence $\mathcal{I}(\mathcal{F})$ is equal to $ker(_D\varphi_{\mathcal{F}})$, and therefore is a closed prime ideal of $H(D)$.

If $\mathcal{F}$ is a monotonous filter on D, we will denote by $\mathcal{I}_0(\mathcal{F})$ the closed ideal of the $f \in H(D)$ such that $f(x) = 0$ whenever $x \in \mathcal{P}(\mathcal{F})$.

Among many ultrametric properties, we notice the followings.

Lemma 12.2 :   *Let $\mathcal{F}$ be a circular filter or a monotonous filter on D and let $f \in H(D)$. There exists $A \in \mathcal{F}$ such that $|f(x)|$ is bounded in A. Besides, for every sequence $(a_n)_{n \in \mathbb{N}}$ thinner than $\mathcal{F}$, we have $\lim_{n \in \mathbb{N}} \varphi_{a_n} = \varphi_{\mathcal{F}}$.*

Proof: Indeed, there does exist $A \in \mathcal{F}$ such that $|f(x)| \leq {_D\varphi_{\mathcal{F}}}(f) + 1$ for all $x \in A$. The last statement is obvious.

Lemma 12.3:   *Let $f \in H(D)$ be invertible in $H(D)$. Then for every $\psi \in Mult(H(D), \mathcal{U}_D)$ we have $\psi(f) \neq 0$.*

Proof: Indeed we have $\psi(f)\psi(\frac{1}{f}) = 1$.

Lemma 12.4 :   *Let $_D\varphi_{\mathcal{F}} \in Mult(H(D), \mathcal{U}_D)$, let $f \in H(D)$ and $g \in H(D)$ be such that $\|f - g\|_D < {_D\varphi_{\mathcal{F}}}(f)$. Then $_D\varphi_{\mathcal{F}}(f) = {_D\varphi_{\mathcal{F}}}(g)$.*

Proof: Indeed we know that $_D\varphi_{\mathcal{F}}(f - g) \leq \|f - g\|_D$, hence $_D\varphi_{\mathcal{F}}(f - g) < {_D\varphi_{\mathcal{F}}}(f)$ and therefore $_D\varphi_{\mathcal{F}}(g) = {_D\varphi_{\mathcal{F}}}(f)$.

Lemma 12.5:   *Let D be unbounded and let $f \in H_b(D)$. Then $|f(x)|$ has a limit $_D\varphi_D(f)$ when $|x|$ tends to $+\infty$ while x lies in D, and $_D\varphi_D$ belongs to $Mult(H_b(D), \| . \|_D)$.*

Proof: By Theorem 10.5, f admits a limit λ when $|x|$ tends to $+\infty$, $(x \in D)$. Hence $\lim_{|x| \to +\infty, \, x \in D} |f(x)| = |\lambda|$. Thus the mapping φ_∞, defined as $\varphi_\infty(f) = \lim_{|x| \to +\infty} |f(x)|$, belongs to $Mult(R_b(D), \| . \|_D)$, and obviously extends by continuity to an element $_D\varphi_D \in Mult(H_b(D), \| . \|_D)$ which satisfies $_D\varphi_D(f) = \lim_{|x| \to +\infty, \, x \in D} |f(x)|$.

Notations: When there is no risk of confusion about the set D, we will just write $\varphi_{\mathcal{F}}$, (resp. $\varphi_{a,r}$, resp. φ_D, resp. φ_A), instead of $_D\varphi_{\mathcal{F}}$ (resp. $_D\varphi_{a,r}$, resp.

$_D\varphi_D$, resp. $_D\varphi_A$). Besides, when D is unbounded, $_D\varphi_D$ will also be denoted by $_D\varphi_\infty$.

Theorem 12.6: *Let $\psi \in Mult(H_b(D), \| \cdot \|_D) \setminus Mult(H(D),\mathcal{U}_D)$. If D is bounded, ψ is of the form φ_a with $a \in \overline{D} \setminus D$. If D is not bounded, ψ is either of the form φ_a, with $a \in \overline{D} \setminus D$ or of the form $_D\varphi_\infty$.*

Proof: First we suppose D bounded. By Corollary 10.7 we have $H_b(D) = H(\overline{D})$. Hence ψ is equal to some $_{\overline{D}}\varphi_{\mathcal{F}}$, with $\mathcal{F}$ a circular filter on $\overline{D}$. If $\mathcal{F}$ is large, it is a large circular filter secant with D, and then ψ belongs to $Mult(H(D),\mathcal{U}_D)$. Hence $\mathcal{F}$ is the filter of the neighbourhoods of a point $a \in \overline{D}$. But if $a \in D$, obviously φ_a belongs to $Mult(H(D),\mathcal{U}_D)$. Hence $a \in \overline{D} \setminus D$.

Now we suppose D unbounded. If D has no hole we just have $H_b(D) = R_b(D) = K$, hence $Mult(H_b(D), \| \cdot \|_D) = Mult(H(D),\mathcal{U}_D)$. Thus we may assume D to have a hole $T = d(a, r^-)$. Without loss of generality we may assume $a = 0$. Let $\gamma(x) = \dfrac{1}{x}$ and let $D' = \gamma(D)$. Then D' is bounded and, by Proposition 11.9, we know that the algebra $H_b(D)$ is isomorphic to $H(D')$. By Proposition 11.6, the mapping $f \to f \circ \gamma$ defines a K-vector space isomorphism from $H(D)$ onto $H(D')$ and a K-algebra isomorphism from $H_b(D)$ onto $H_b(D') = H(\overline{D'})$. Hence we may define $\psi' \in Mult(H(\overline{D'}))$ by $\psi'(f \circ \gamma) = \psi(f)$ whenever $f \in H_b(D)$. If ψ' belonged to $Mult(H(D'),\mathcal{U}_{D'})$, then we would have $\psi(f) = \psi'(f \circ \gamma)$ whenever $f \in H(D)$ and therefore $\psi \in Mult(H(D),\mathcal{U}_D)$. Hence ψ' does not belong to $Mult(H(D'),\mathcal{U}_{D'})$, and then ψ' is of the form φ_b with $b \in \overline{D'} \setminus D$. If $b \neq 0$ then $\psi = \varphi_{\frac{1}{b}}$. If $b = 0$, then $\psi = {}_D\varphi_D$ and this ends the proof.

Theorem 12.7: *Let $(a_n)_{n \in \mathbb{N}}$ be a bounded sequence in D such that no subsequence converges to any point of $\overline{D} \setminus D$. There exists a subsequence $(a_{n_s})_{s \in \mathbb{N}}$ such that the sequence $(\varphi_{a_{n_s}})_{s \in \mathbb{N}}$ converges in $Mult(H(D),\mathcal{U}_D)$.*

Proof: By Theorem 3.1 we may extract either a convergent subsequence, or a monotonous distances subsequence, or an equal distances subsequence from the sequence $(\alpha_n)_{n \in \mathbb{N}}$. Let $(\alpha_{n_s})_{s \in \mathbb{N}}$ be such a subsequence. If this subsequence converges to a point $\alpha \in K$, then by hypothesis α lies in D, hence $\lim\limits_{s \to +\infty} \varphi_{\alpha_{n_s}} = \varphi_\alpha$. If this subsequence is a monotonous distances subsequence, or an equal distances subsequence, then by Proposition 3.15, on D there exists a large circular filter $\mathcal{F}$ less thin than the sequence (α_{n_s}), and then we see that for every $f \in H(D)$ we have $\lim\limits_{s \to +\infty} |f(\alpha_{n_s})| = {}_D\varphi_{\mathcal{F}}(f)$, hence $\lim\limits_{s \to +\infty} \varphi_{\alpha_{n_s}}(f) = \varphi_{\mathcal{F}}(f)$. Thus, in every case we have proven that the subsequence $\varphi_{\alpha_{n_s}}$ converges in $Mult(H(D),\mathcal{U}_D)$.

13.   POWER SERIES

Let $f = \sum_{n=0}^{\infty} a_n x^n$ be a power series with coefficients in K. As usual, when $\limsup_{n \to \infty} \sqrt[n]{|a_n|} \neq 0$, we call *the radius of convergence of f* the number

$$r = \frac{1}{\limsup_{n \to \infty} \sqrt[n]{|a_n|}} \quad \text{(with } r = 0 \text{ when } \limsup_{n \to \infty} \sqrt[n]{|a_n|} = +\infty\text{)}.$$

When $\limsup_{n \to \infty} \sqrt[n]{|a_n|} = 0$, we say that *$f$ has a radius of convergence equal to* $+\infty$.

The basic properties of the power series are systematically recalled. In a "closed disk" $d(0, r)$, we will see that the analytic elements are exactly the power series converging in this disk. (It is not the same in an "open disk " $d(0, r^-)$.)

Examples: Let $f(x) = \sum_{n=1}^{\infty} n x^n$. The radius of convergence of this series is easily seen to be 1. This function obviously defines the rational function $\dfrac{x}{(1-x)^2}$ in $d(1, 1^-)$.

Notation: For every $a \in K$, $r \in \mathbb{R}_+^*$, we will denote by $A(d(a, r^-))$ the set of the power series in $x - a$ whose radius of convergence is superior or equal to r, and by $A_b(d(a, r^-))$ the set of the bounded power series convergent in $d(a, r^-)$. First we will study $H(d(a, r))$.

Theorem 13.1: *Let $r \in \mathbb{R}_+^*$, let $\mathcal{F}$ be the circular filter of center 0 and diameter r on K, and let $E = d(0, r)$. Then $H(E)$ is the set of the power series*

$$f(x) = \sum_{n=0}^{\infty} a_n x^n \text{ such that } \lim_{n \to \infty} |a_n| r^n = 0 \text{ and we have } \|f\|_E = \max_{n \in \mathbb{N}} |a_n| r^n =$$

$$_E \varphi_{\mathcal{F}}(f) = \|f\|_{C(0,r)}.$$

For every $\alpha \in E$, $H(E)$ is also equal to the set of the series $f(x) = \sum_{n=0}^{\infty} b_n (x - \alpha)^n$

such that $\lim_{n \to \infty} |b_n| r^n = 0.$

Let $B = K \setminus d(0, r^-)$. Then $H(B)$ is the set of the Laurent series $f(x) = \sum\limits_{n=0}^{\infty} \dfrac{a_n}{x^n}$

such that $\lim\limits_{n \to \infty} |a_n| r^{-n} = 0$, and we have $\|f\|_B = \max\limits_{n \in \mathbb{N}} |a_n| r^{-n} = {}_B\varphi_{\mathcal{F}}(f) = \|f\|_{C(0,r)}$.

For every $\alpha \in d(0, r^-)$, $H(B)$ is also equal to the set of the series $f(x) = \sum\limits_{n=0}^{\infty} \dfrac{b_n}{(x - \alpha)^n}$ such that $\lim\limits_{n \to \infty} |b_n| r^{-n} = 0$.

Proof: Let F be the set of the power series $f = \sum\limits_{n=0}^{\infty} a_n x^n$ such that

$\lim\limits_{n \to \infty} |a_n| r^n = 0$. By Proposition 1.15, F is a K-Banach algebra, with respect to the norm $\sup\limits_{n \in \mathbb{N}} |a_n| r^n$.

We will show that $F = H(E)$ and that, for every $f(x) = \sum\limits_{n=0}^{\infty} a_n \, x^n$, we

have $\|f\|_E = \max\limits_{n \in \mathbb{N}} |a_n| r^n = {}_E\varphi_{\mathcal{F}}(f)$. Clearly we have $|f(x)| \leq \max\limits_{n \in \mathbb{N}} |a_n| \, |x^n| \leq \max\limits_{n \in \mathbb{N}} |a_n| \, r^n$, hence $\|f\|_E \leq \max\limits_{n \in \mathbb{N}} |a_n| \, r^n$. At the same time, we see that $F \subset H(E)$

because in $H(E)$, f is the limit of the sequence $P_n = \sum\limits_{i=0}^{n} a_i x^i$. Since $\mathcal{F}$ is also

secant with $C(0, r)$ we have $\|f\|_{C(0,r)} = {}_A\varphi_{\mathcal{F}}(f)$. Now we know that $\varphi_{\mathcal{F}}(P_n) \leq \max\limits_{0 \leq i \leq n} |a_i| r^i$. Since $\varphi_{\mathcal{F}}$ extends continuously to ${}_E\varphi_{\mathcal{F}} \in Mult\big(H(E), \| \cdot \|_E\big)$, we see that for n big enough, we have ${}_E\varphi_{\mathcal{F}}(f) = {}_E \varphi_{\mathcal{F}}(P_n) = |a_j| r^j$ with $j < n$ and $|a_n| r^m < |a_j| r^j$ whenever $m < j$, hence finally ${}_E\varphi_{\mathcal{F}}(f) = |a_j| r^j = \max\limits_{n \in \mathbb{N}} |a_n| r^n$. Hence we have $\|f\|_{C(0,r)} = \|f\|_E \leq |a_j| r^j = \max\limits_{0 \leq n} |a_n| r^n$, and therefore $\|f\|_{C(0,r)} = \|f\|_E = \max_{0 \leq n} |a_n| r^n$. This finishes showing that F is a closed subalgebra of $H(E)$.

Now we will show that $R(E)$ is included in F. For this, we just have to show

that, given any $\alpha \in K \setminus E$, $\left(\dfrac{1}{x - \beta}\right)^q = \left(-\dfrac{1}{\beta} \sum\limits_{n=0}^{\infty} \left(\dfrac{x}{\beta}\right)^n\right)^q$ belongs to F. When

developping $\left(\sum\limits_{n=0}^{\infty} \left(\dfrac{x}{\beta}\right)^n\right)^q$, we see that for every fixed $q \in \mathbb{N}$, the coefficient A_q

of x^q is a sum of terms of the form $\dfrac{\nu}{\beta^q}$, with $\nu \in \mathbb{N}$, hence finally $|A_q| \leq \dfrac{1}{|\beta|^q}$

and therefore $|A_q|r^q \leq \left(\dfrac{r}{|\beta|}\right)^q$. Since $|\beta| > r$, this shows that $\left(\dfrac{1}{x-\beta}\right)^q \in F$. So, we have proven the inclusion $R(E) \subset F \subset H(E)$. Since F is closed, we have $F = H(E)$.

Now let $\alpha \in E$. Since $d(\alpha, r) = d(0, r)$, after the change of variable $x = \alpha + u$, the same reasoning shows that a series $f(x) = \sum_{n=0}^{\infty} a_n\, x^n \in H(E)$ is also of the form $\sum_{n=0}^{\infty} b_n(\alpha)(x-\alpha)^n$ with $\lim_{n\to\infty} |b_n(\alpha)|r^n = 0$. Reciprocally, $H(E)$ is clearly equal to the set of the series $\sum_{n=0}^{\infty} b_n(x+\alpha)^n$ such that $\lim_{n\to\infty} |b_n|r^n = 0$ because any series g of the form $\sum_{n=0}^{\infty} b_n\, u^n$, with $\lim_{n\to\infty} |b_n|r^n = 0$, can be written as $\sum_{n=0}^{\infty} a_n(u+\alpha)^n$.

The statements about $H(B)$ are an obvious consequence of those about $H(E)$ after the change of variable $y = \dfrac{1}{x}$, and more generally, $y = \dfrac{1}{x-\alpha}$.

Corollary 13.2: *Let $\alpha \in D$ and $r \in \mathbb{R}_+^*$ be such that $d(\alpha, r) \subset D$. Let $f \in H(D)$. In $d(\alpha, r)$, $f(x)$ is equal to a power series of the form $\sum_{n=0}^{\infty} a_n(x-\alpha)^n$ such that $\lim_{n\to\infty} |a_n|r^n = 0$. If $f(\alpha) = 0$ and if $f(x)$ is not identically zero in $d(\alpha, r)$, then there exists a unique integer $q \in \mathbb{N}^*$ such that $a_n = 0$ for every $n < q$ and $a_q \neq 0$, and α is an isolated zero of f in $d(\alpha, r)$.*

Proposition 13.3: *Let $r \in \mathbb{R}_+^*$ and let $f(x) = \sum_{n=0}^{\infty} a_n x^n$. The following conditions are equivalent:*

    a) $f \in A(d(0, r^-))$

    b) $f \in \bigcap_{s<r} H(d(0, s))$

    c)   *The series f is convergent in all of $d(0, r^-)$*

Proof: b) and c) are clearly equivalent to the condition $\lim_{n\to\infty} |a_n|s^n = 0$ whenever $s < r$ and, in the same way as in archimedean analysis, it is shortly checked that this is also equivalent to $\limsup\limits_{n\to\infty} \sqrt[n]{|a_n|} \leq \dfrac{1}{r}$.

Remark: If f is convergent for some $\alpha \in C(0,r)$, then $\lim\limits_{n\to\infty} |a_n|r^n = 0$, hence f belongs to $H(d(0,r))$.

Corollary 13.4: *Let $f \in A(d(0,r^-))$ be not identically zero. For every $\alpha \in d(0,r^-)$, $f(x)$ is equal to a power series $\sum\limits_{n=0}^{\infty} b_n(\alpha)(x-\alpha)^n$. If f is not identically zero, and if α is a zero of f in $d(0,r^-)$, α is an isolated zero, and f factorizes in $A(d(0,r^-))$ in the form $(x-\alpha)g(x)$, with $g \in A(d(0,r^-))$.*

Definition: Let $f \in H(D)$, and let $\alpha \in \overset{\circ}{D}$, let $r > 0$ be such that $d(\alpha,r) \subset D$, and let $f(x) = \sum\limits_{n=q}^{\infty} b_n(x-\alpha)^n$ whenever $x \in d(\alpha,r)$, with $b_q(\alpha) \neq 0$, and $q > 0$.
Then α is called *a zero of multiplicity order* q, or more simply, *a zero of order q*. In the same way, q will be named *the multiplicity order of α*.

Remark: In particular, these definitions apply to any element $f \in A(d(a,r^-))$, at any point $\alpha \in d(a,r^-)$.

Notations: We will denote by $A(K \setminus d(0,r))$ the set of the Laurent series $\sum\limits_{n=0}^{+\infty} \dfrac{a_n}{x^n}$ convergent in $K \setminus d(0,r)$. This algebra is obviously isomorphic to $A(d(0,r^-))$. We will denote by $A_b(K \setminus d(0,r))$ the subalgebra of the bounded functions of $A(K \setminus d(0,r))$, which is isomorphic to $A_b(d(0,r^-))$. By Theorem 13.1, we see that $H(K \setminus d(0,r)) \subset A_b(K \setminus d(0,r))$.
For convenience, we will also denote by $A_0(K \setminus d(0,r))$ (resp. $A_{0,b}(K \setminus d(0,r))$) the subalgebra of the elements $f \in A(K \setminus d(0,r))$ (resp. the $f \in A_b(K \setminus d(0,r))$) such that $\lim\limits_{|x|\to+\infty} f(x) = 0$.
Now let $r_1, r_2 \in \mathbb{R}_+^*$ with $r_1 < r_2$. We will denote by $A(\Gamma(0,r_1,r_2))$ the set of the Laurent series $f(x) = \sum\limits_{-\infty}^{+\infty} a_n x^n$ convergent in $\Gamma(0,r_1,r_2)$, and by $A_b(\Gamma(0,r_1,r_2))$ the set of the $f(x) \in A(\Gamma(0,r_1,r_2))$ bounded in $\Gamma(0,r_1,r_2)$.

Theorem 13.5:   *Let $r \in \mathbb{R}_+^*$ and let $f(x) = \displaystyle\sum_{n=0}^{\infty} a_n x^n \in H(d(0,r))$. The series $f'(x) = \displaystyle\sum_{n=1}^{\infty} n a_n x^{n-1}$ also belongs to $H(d(0,r))$ and satisfies $\|f'\|_{d(0,r)} \leq \frac{1}{r}\|f\|_{d(0,r)}$. For every $\alpha \in d(0,r)$, f has a derivative at α, that is equal to $f'(x)$. Besides, if K has characteristic zero, and if α is a zero of multiplicity order q of f, then we have $f^{(j)}(\alpha) = 0$ for every $j < q$, and $f^{(q)}(\alpha) \neq 0$.*

Proof: By hypothesis we have $\displaystyle\lim_{n\to\infty} |a_n| r^n = 0$, hence obviously $\displaystyle\lim_{n\to\infty} |n a_n| r^{n-1} = 0$, and $f'(x) \in H(d(0,r))$. Then $\|f'\|_{d(0,r)} = \displaystyle\sup_{n\in\mathbb{N}}(|n a_n| r^{n-1}) \leq \displaystyle\sup_{n\in\mathbb{N}}(|a_n| r^{n-1}) = \frac{1}{r}\|f\|_{d(0,r)}$. Let $\alpha \in d(0,r)$. By Theorem 13.1, $f(x)$ is of the form $\displaystyle\sum_{n=0}^{\infty} b_n(\alpha)(x-\alpha)^n$, hence f clearly has a derivative at α, equal to $b_1(\alpha)$. Now the change of variable

$$\sum_{n=0}^{\infty} a_n[(x-\alpha)+\alpha]^n = \sum_{n=0}^{\infty} b_n(\alpha)(x-\alpha)^n$$

shows that $b_1(\alpha) = \displaystyle\sum_{n=1}^{\infty} n a_n \alpha^{n-1}$. The last statement about the multiplicity order of the zeros is obvious. So the proof is complete.

Corollary 13.6:   *Let $r \in \mathbb{R}_+^*$ and let $f(x) = \displaystyle\sum_{n=0}^{\infty} \frac{a_n}{x^n} \in H(K \setminus d(0,r^-))$. Then the series $g(x) = -\displaystyle\sum_{n=1}^{\infty} \frac{n a_n}{x^{n+1}}$ belongs to $H(K \setminus d(0,r^-))$ and satisfies $\|f'\|_{K \setminus d(0,r^-)} \leq \frac{1}{r}\|f\|_{K \setminus d(0,r^-)}$. For every $\alpha \in K \setminus d(0,r^-)$, f has a derivative at α equal to $g(\alpha)$.*

Corollary 13.7:   *Let D be open and let $f \in H(D)$. Then f has derivative f' in D .*

Theorem 13.8:   *Let $f(x) = \displaystyle\sum_{n=0}^{\infty} a_n x^n \in A(d(0,r^-))$. The power series*

$\displaystyle\sum_{n=1}^{\infty} n a_n x^{n-1}$ *also belongs to* $A(d(0, r^-))$ *and is equal to the derivative of* f *in* $d(0, r^-)$*, and the radius of convergence of* f' *is superior or equal to the one of* f*. Further, if* K *has characteristic* 0*, the radius of convergence of* f' *is the same as that of* f*.*

Proof: By Corollaries 13.6 and 13.7, the first statement is clear. Now we suppose that K has characteristic zero. If K has residue characteristic zero, then $|n| = 1$ for all $n \in \mathbb{N}^*$, and therefore the last statement is clear. Now, we assume that K has residue characteristic $p \neq 0$. By Lemma 8.2, there exists a constant $b > 0$ such that $\dfrac{b}{n} \leq |n| \leq 1$ for every $n \in \mathbb{N}^*$. As a consequence, we have $\displaystyle\lim_{n \to +\infty} \sqrt[n]{|n|} = 1$. Therefore it is easily seen that $\displaystyle\limsup_{n \to +\infty} \sqrt[n]{|a_n|} = \limsup_{n \to +\infty} \sqrt[n-1]{|n a_n|}$, and finally f' has same radius of convergence as f.

Remark: Unlike in Archimedean analysis, when the characteristic p of K is not zero, there do exist power series f whose derivative have a radius of convergence bigger than that of f. For example, let $f(x) = \displaystyle\sum_{n=0}^{\infty} x^{p^n}$: the radius of convergence of f is 1 while that of f' is $+\infty$.

Theorem 13.9: *Let* $f(x) = \displaystyle\sum_{n=0}^{\infty} a_n x^n \in A(d(0, r^-))$*. Then* f *is bounded in* $d(0, r)$ *if and only if so is the sequence* $(|a_n| r^n)_{n \in \mathbb{N}}$*. Besides if* f *is bounded, then* $\|f\|_{d(0, r^-)} = \sup_{n \in \mathbb{N}} |a_n| r^n$*.*

Proof: For every $s \in]0, r[$, we put $\theta(s) = {}_{d(0,s)}\varphi_{0,s}(f)$. By Theorems 13.1 and 13.2, we have $\theta(s) = \|f\|_{d(0,s)} = \sup_{n \in \mathbb{N}} |a_n| s^n$. As it is an increasing function in s it has a limit in $\overline{\mathbb{R}}$ equal to $\sup_{n \in \mathbb{N}} |a_n| r^n$ and then the conclusion is clear.

Theorem 13.10: $A_b(d(0, r^-))$ *is a* K*- Banach subalgebra of* $A(d(0, r^-))$ *with respect to the norm* $\| \, . \, \|_{d(0, r^-)}$*. Further, this norm is multiplicative.*

Proof: Let $f(x) = \displaystyle\sum_{n=0}^{\infty} a_n x^n \in A_b(d(0, r^-))$. By Theorem 13.9 we have $\|f\|_{d(0,r^-)} = \sup_{n \in \mathbb{N}} |a_n| r^n$. The norm $\| \, . \, \|_{d(0,r^-)}$ is a norm of K-algebra hence $\|f \, g\|_{d(0,r^-)} \leq \|f\|_{d(0,r^-)} \|g\|_{d(0,r^-)}$. On the other hand, by Theorem

13.1, the norm $\| \, . \, \|_{d(0,s)}$ is multiplicative on $H(d(0,s))$ for every $s < r$, hence $\|fg\|_{d(0,r^-)} \geq \|fg\|_{d(0,s)} = \|f\|_{d(0,s)} \|g\|_{d(0,s)}$ whenever $s < r$, and therefore $\| \, . \, \|_{d(0,r^-)}$ is multiplicative on $A_b(d(0,r^-))$. Now let $(f_m)_{m\in\mathbb{N}}$ be a Cauchy sequence in $A_b(d(0,r^-))$. We put $f_m(x) = \sum_{n=0}^{\infty} a_{n,m} x^n$. By hypothesis, for every $\epsilon > 0$ we have an integer $N(\epsilon)$ such that $|a_{n,m} - a_{n,q}| r^n \leq \epsilon$ for every $n \in \mathbb{N}$, whenever $m, q \geq N(\epsilon)$. Thus it is easily seen that each sequence $(a_{n,m})_{m\in\mathbb{N}}$ converges in K to a limit a_n that satisfies $|a_n - a_{n,m}| r^n \leq \epsilon$ whenever $m \geq N(\epsilon)$, and then the series $f(x) = \sum_{n=0}^{\infty} a_n x^n$ satisfies $\|f - f_m\|_{d(0,r^-)} \leq \epsilon$. Obviously f belongs to $H(d(0,s))$ for all $s < r$, and then the sequence (f_m) is proven to converge in $A_b(d(0,r^-))$.

Theorem 13.11:   *For every $r \in \mathbb{R}_+^*$, $H(d(0,r^-))$ is included in $A_b(d(0,r^-))$.*

Proof: Since $A_b(d(0,r^-))$ is complete with respect to the norm $\| \, . \, \|_{d(0,r^-)}$, we just have to show that $R(d(0,r^-)) \subset A_b(d(0,r^-))$, hence finally we just have to show that given $\alpha \in K \setminus d(0,r^-)$, $\dfrac{1}{x - \alpha} \in A_b(d(0,r^-))$. But we have

$$\frac{1}{x - \alpha} = -\frac{1}{\alpha(1 - \frac{x}{\alpha})} = -\frac{1}{\alpha} \sum_{n=0}^{\infty} \left(\frac{x}{\alpha}\right)^n \text{ for all } x \in d(0,r^-) \text{ because } \left|\frac{x}{\alpha}\right| < 1, \text{ hence}$$

$\dfrac{1}{x - \alpha} \in A_b(d(0,r^-))$ and that finishes proving Theorem 13.11 .

Remark: We will see later that $H(d(0,r^-))$ is much smaller than $A_b(d(0,r^-))$. In particular, in Chapter 39, we will see that $\sqrt[q]{1+x}$ belongs to $A_b(d(0,1^-))$, but does not belong to $H(d(0,1^-))$.

Theorem 13.12:   *If K has characteristic 0, then an element $f \in H(d(0,r))$ has a derivative identically equal to 0 if and only if it is equal to a constant.*
If K has a characteristic $p \neq 0$, then an element $f \in H(d(0,r))$ has a derivative identically equal to 0 if and only if there exists $g \in H(d(0,r))$ such that $f(x) = (g(x))^p$.

Proof: By Theorem 13.5, the statement about the case when K has characteristic zero is obvious. Hence we suppose that K has a characteristic $p \neq 0$. If there exists $g \in H(d(0,r))$ such that $f(x) = (g(x))^p$, obviously we have $f'(x)$ identically equal to 0.

Now, we suppose $f'(x)$ identically equal to 0. Hence $f(x)$ is of the form $\sum_{j=0}^{\infty} b_j x^{jp}$, with $\lim_{j \to \infty} |b_j| r^{jp} = 0$. For each $j \in \mathbb{N}$, we can take $c_j \in K$ such that $(c_j)^p = b_j$. Then, it is seen that $\lim_{j \to \infty} |c_j| r^j = 0$. Now, we can put $g(x) = \sum_{n=0}^{\infty} c_n x^n$, and therefore g belongs to $H(d(0,r))$. Since K has characteristeric p, it is seen that $(g(x))^p = f(x)$. This ends the proof.

Corollary 13.13: *Let $r > 0$. If K has characteristic 0, then an element $f \in H(K \setminus d(0,r^-))$ (resp. $f \in H_0(K \setminus d(0,r^-))$) has a derivative identically equal to 0 if and only if it is equal to a constant (resp. to 0).*
If K has a characteristic $p \neq 0$, then an element $f \in H(K \setminus d(0,r^-))$ (resp. $f \in H_0(K \setminus d(0,r^-))$) has a derivative identically equal to 0 if and only if there exists $g \in H(K \setminus d(0,r^-))$ (resp. $g \in H_0(K \setminus d(0,r^-))$) such that $f(x) = (g(x))^p$.

Corollary 13.14: *If K has characteristic 0, then a power series $f(x) \in A(d(0,r^-))$ has a derivative identically equal to 0 if and only if it is equal to a constant.*
If K has a characteristic $p \neq 0$, then a power series $f(x) \in A(d(0,r^-))$ has a derivative identically equal to 0 if and only if there exists $g \in A(d(0,r^-))$ such that $f(x) = (g(x))^p$.

14.   FACTORIZATION OF ANALYTIC ELEMENTS

In $\mathbb{C}$, it is well known that when a (not identically zero) holomorphic function admits a zero at a point α, this zero has a finite order of multiplicity. Actually this is a generalization of a property of the rational functions. In the non Archimedean context, we find again this property among the analytic elements [11], [14], [15].

Theorem 14.1:   *Let α belong to $\overset{\circ}{\overline{D}}$ and let $f \in H(D)$ be such that $f(\alpha) = 0$. Then f has factorization in $H(D)$ in the form $(x - \alpha)g$ with $g \in H(D)$. If there is no neighbourhood V of α such that $f(x) = 0$ whenever $x \in V$, then there exists a unique integer $q \in \mathbb{N}$ and $h \in H(D)$ such that $f(x) = (x - \alpha)^q\, h(x)$ and $h(\alpha) \neq 0$.*

Proof: First we will prove the main factorization in the form $(x - \alpha)g$. We may obviously assume $\alpha = 0$. Next by Theorem 10.5 we see that there exits a disk $d(0, r)$ such that f belongs to $H(D \cup d(0, r))$. So we can assume that 0 is interior to D and that $d(0, r) \subset D$ without loss of generality. Then by Corollary 13.2, f is equal to a power series $\sum\limits_{n=1}^{\infty} \alpha_n x^n$ for all $x \in d(0, r)$.

Now let t_n be a sequence in $R(D)$ such that $\lim\limits_{n \to \infty} t_n = f$. Clearly $|t_n(0)| \leq \|t_n - f\|_D$ because $f(0) = 0$, so we have $\|t_n - t_n(0) - f_n\|_D \leq \|t_n - f\|_D$. We put $h_n = t_n - t_n(0)$ $(n \in \mathbb{N})$. The sequence (h_n) of $R(D)$ approaches f in $H(D)$ and satisfies $h_n(0) = 0$ whenever $n \in \mathbb{N}$, hence h_n has factorization in $R(D)$ in the form $x g_n$. We will show that the sequence $(g_n)_{n \in \mathbb{N}}$ converges in $H(D)$. Indeed, it is seen that when $x \in D \setminus d(0, r)$ we have :

$$(1) \qquad |g_n(x) - g_m(x)| \ \leq \ \frac{1}{r}\, |h_n(x) - h_m(x)| \ \leq \ \frac{1}{r}\, \|h_n - h_m\|_D .$$

On the other hand, in $d(0, r)$ the norm $\| \, . \, \|_{d(0,r)}$ is equal to the multiplicative norm $\varphi_{\mathcal{F}}$ with $\mathcal{F}$ the circular filter of center 0 and diameter r. Therefore $\|h_n - h_m\|_{d(0,r)} \ = \ \|x\|_{d(0,r)}\, \|g_n - g_m\|_{d(0,r)}$ hence

$$(2) \qquad \|g_n - g_m\|_{d(0,r)} \ = \ \frac{1}{r}\, \|h_n - h_m\|_{d(0,r)}.$$

Finally, by (1) and (2) we see that $\|g_n - g_m\|_D \ \leq \ \frac{1}{r}\, \|h_n - h_m\|_D$ and then the sequence is proven to be convergent in $H(D)$. Let g be its limit. We will show that $\lim\limits_{q \to \infty} \|x g_q - x g\|_D = 0$. Let $\varepsilon > 0$ and let $N \in \mathbb{N}$ be such that

$\|h_n - f\|_D \leq \varepsilon$ whenever $n \geq N$. We fix $q \leq N$. We have $\|h_n - h_q\|_D \leq \varepsilon$, hence $|xg_n(x) - xg_q(x)| \leq \varepsilon$ whenever $x \in D$. So, when n tends to $+\infty$, we see that $|xg(x) - xg_q(x)| \leq \varepsilon$ whenever x in D. Thus we have $\|xg - xg_q\|_D \leq \varepsilon$, and therefore $\lim\limits_{q \to \infty} \|xg_q - xg\|_D = 0$. But by hypothesis we have $\lim\limits_{q \to \infty} \|f - xg_q\|_D = 0$ and then we see that $f = xg$.

Now we suppose that f is not identically zero in $d(0,r)$. Then at least one of the coefficients α_n of its power series is not zero. By Corollary 13.4, f admits 0 as a zero of order q, and then q is the smallest integer such that $\alpha_q \neq 0$.

In $d(0,r)$ we have $f(x) = \sum\limits_{n=q}^{\infty} \alpha_n x^n = xg(x)$ hence $g(x) = \sum\limits_{n=q}^{\infty} \alpha_n x^{n-1}$ whenever $x \in d(0,r)$. Suppose that f has been proven to be factorized in the form $x^s g_s$ with $s < q$ and $g_s \in H(D)$. Clearly $g_s(x) = \sum\limits_{n=q}^{\infty} \alpha_n x^{n-s}$ whenever $x \in d(0,r)$ hence $g_s(0) = 0$ and therefore g_s has factorization in the form xg_{s+1} with $g_{s+1} \in H(D)$. Thus by induction we obtain $f = x^q g_q(x)$ with $g_q(x) = \sum\limits_{n=q}^{\infty} \alpha_n x^{n-q}$ and then $g_q(0) = \alpha_q \neq 0$. That finishes proving Theorem 14.1.

Corollary 14.2:   *Let $\overline{D}$ be open , let $f \in H(D)$, let α be a zero of f in D. Either there exists a disk $d(\alpha,r)$ such that $f(x) \neq 0$ whenever $x \in d(\alpha,r) \setminus \{\alpha\}$, or there exists a disk $d(\alpha,r)$ such that $f(x) = 0$ whenever $x \in d(\alpha,r)$.*

Corollary 14.3:   *Let $f \in H(D)$ have a zero of order q at a point $\alpha \in \overset{\circ}{D}$. Then for every $s = 1,...,q$, f factorizes in the form $(x - \alpha)^s g_s$, with $g_s \in H(D)$ having a zero of order $q - s$ at α.*

Corollary 14.4: *Let $\alpha \in \overset{\circ}{D}$ and let $f \in H(D)$. Let $\sum\limits_{n=0}^{\infty} a_n(x - \alpha)^n$ be its power series in a disk $d(\alpha,r) \subset D$. Let $P(x) = \sum\limits_{n=0}^{q} a_n(x - \alpha)^n$ and let $g(x) = f(x) - P(x)$. Then g factorizes in the form $(x - \alpha)^q h(x)$, with $h \in H(D)$.*

Definitions: Let $A \subset D$ be an open subset of K, let $f \in H(D)$ have finitely many zeros $a_1,...,a_n$ in A of multiplicity order of $q_1,...,q_n$ respectively. The

polynomial $\prod_{i=1}^{n}(x - a_i)^{q_i}$ will be named *the polynomial of the zeros of f in A.* We
are now able to give the following Corollary.

Corollary 14.5:   *Let A be a subset of D open in K, let $f \in H(D)$ have finitely
many zeros in A and let P be the polynomial of its zeros in A. Then f has a
factorization in the form $f = Pg$, with $g \in H(D)$ and $g(x) \neq 0$ whenever $x \in A$.*

Definitions: An element $f \in H(D)$ will be said to be *semi-invertible* (resp.
quasi-invertible) if it factorizes in the form $P(x)\, g(x)$, with g invertible in $H(D)$
and with P a polynomial whose zeros belong to D (resp. to $D \cap \overset{\circ}{\overline{D}}$).
An element $f \in H(D)$ will be said to be *quasi-minorated* if for every sequence
$(a_n)_{n \in \mathbb{N}}$ of D such that $\lim_{n \to \infty} f(a_n) = 0$ either we may extract a subsequence that
converges in K or we may extract a subsequence $(a_{n_q})_{q \in \mathbb{N}}$ such that $\lim_{q \to \infty} |a_{n_q}| = +\infty$.

Remarks: 1) If a semi-invertible element of $H(D)$ has no zero in D, it is
invertible in $H(D)$.

2) Let D belong to $\mathcal{A}$. If f_1, f_2 are quasi-invertible (resp. semi-invertible)
elements of $H(D)$, then $f_1 f_2$ is also quasi-invertible, (resp. semi-invertible).
However when D does not belong to $\mathcal{A}$, counter- examples show that the product of two semi-invertible (resp. quasi-invertible) elements is not always semi-
invertible (resp. quasi-invertible). Such counter-examples will be given in a
remark in Chapter 31.
In chapter 22 we will show that a product of quasi-minorated elements is quasi-
minorated.

Lemma 14.6:   *Let $f \in H(D)$ be quasi-minorated and $\lambda \in R(D)$ be a homographic function. Let $D' = \lambda(D)$ and let $g = f \circ \lambda^{-1}$. Then g is also quasi-
minorated .*

Proof: Let $(a_n)_{n \in \mathbb{N}}$ be a sequence in D' such that $\lim_{n \to \infty} g(a_n) = 0$ and let
$b_n = \lambda^{-1}(a_n)$, $(n \in \mathbb{N})$. Then $\lim_{n \to \infty} f(b_n) = 0$. Since f is quasi-minorated, one
can extract a subsequence $(b_{n_q})_{q \in \mathbb{N}}$ that either converges or satisfies $\lim_{q \to \infty} |b_{n_q}| = \infty$. But then it is seen that the sequence (a_{n_q}) either converges or satisfies
$\lim_{q \to \infty} |a_{n_q}| = \infty$. Hence f is quasi-minorated.

Theorem 14.7:   *Let D be open closed bounded and let $f \in H(D)$. If f is* *quasi-minorated then it is quasi-invertible .*

Proof: We suppose f is not quasi-invertible and we will prove that f is not quasi-minorated either. First we suppose f to have finitely many zeros. Since D is open, by Theorem 14.1, f has factorization in the form $P(x)\, g(x)$, with P a polynomial whose zeros are interior to D and g an element of $H(D)$ which has no zero in D, but is not invertible in $H(D)$. Hence there exists a sequence $(\alpha_n)_{n\in\mathbb{N}}$ in D such that $\lim_{n\to\infty} g(\alpha_n) = 0$. If f was quasi-minorated we could extract a convergent subsequence from the sequence $(\alpha_n)_{n\in\mathbb{N}}$ whose limit would belong to D and would be a zero of g. Hence f is not quasi-minorated when it has finitely many zeros in D.

Now we suppose that f has a sequence of (different) zeros (α_n) in D and that f is quasi-minorated. Hence we may extract a convergent subsequence of limit α. Obviously α is another zero of f, hence by Corollary 14.2 $f(x)$ is equal to zero inside a disk $d(\alpha, r)$ and then f is not quasi-minorated. This ends the proof of Theorem 14.7 .

Theorem 14.8:   *Let D be closed and bounded. Let $f \in H(D)$ be quasi-minorated and have no zero in D. Then f is invertible in $H(D)$.*

Proof: Assume that $\inf_{x\in D} |f(x)| = 0$ and let $(a_n)_{n\in\mathbb{N}}$ be a sequence in D such that $\lim_{n\to\infty} f(a_n) = 0$. Since D is bounded we can extract a subsequence (a_n) which converges in K to a point $a \in \overline{D}$. Since D is closed, a belongs to D and satisfies $f(a) = 0$, which contradicts the hypothesis. Thus there exists $\lambda > 0$ such that $|f(x)| \geq \lambda$ whenever $x \in D$, and then by Theorem 10.3, f is invertible in $H(D)$.

15. THE MITTAG-LEFFLER THEOREM

The wonderful Mittag-Leffler Theorem for analytic elements is due to Marc Krasner who showed it on quasi-connected sets [47], [48]. The same proof holds on infraconnected sets as it was shown by Philippe Robba [52].

In this chapter, D is supposed to be infraconnected.

We remember that if D is unbounded, $H_0(D)$ denotes the set of the $f \in H(D)$ such that $\lim\limits_{\substack{|x|\to+\infty \\ x\in D}} f(x) = 0$.

Theorem 15.1: (M.Krasner)   *Let D be closed and bounded (resp. unbounded) and let $f \in H_b(D)$. There exists a unique sequence of holes $(T_n)_{n\in\mathbb{N}^*}$ of D and a unique sequence $(f_n)_{n\in\mathbb{N}}$ in $H(D)$ such that $f_0 \in H(\tilde{D})$ (resp. $f_0 \in K$), $f_n \in H_0(K \setminus T_n)$ $(n > 0)$, $\lim\limits_{n\to\infty} f_n = 0$ and*

$$(1) \quad f = \sum_{n=0}^{\infty} f_n \text{ and } \|f\|_D = \sup_{n\in\mathbb{N}} \|f_n\|_D .$$

Moreover for every hole $T_n = d(a_n, r_n^-)$, we have

$$(2) \quad \|f_n\|_D = \|f_n\|_{K\setminus T_n} = {}_D\varphi_{a_n,r_n}(f_n) \leq {}_D\varphi_{a_n,r_n}(f) \leq \|f\|_D .$$

If D is bounded and if $\tilde{D} = d(a,r)$ we have

$$(3) \quad \|f_0\|_D = \|f_0\|_{\tilde{D}} = {}_D\varphi_{a,r}(f_0) \leq {}_D\varphi_{a,r}(f) \leq \|f\|_D .$$

If D is not bounded then $|f_0| = \lim\limits_{\substack{|x|\to\infty \\ x\in D}} |f(x)| \leq \|f\|_D .$

Let $D' = \tilde{D} \setminus \left(\bigcup\limits_{n=1}^{\infty} T_n \right)$. Then f belongs to $H(D')$ (resp. $H_b(D')$) and its decomposition in $H(D')$ is given again by (1) and then f satisfies $\|f\|_{D'} = \|f\|_D .$

Proof: Since $f \in H_b(D)$, by Corollary 10.7 we know that $f \in H(\overline{D})$. Hence without loss of generality we may assume that D is closed. Obviously we may also assume $0 \in \tilde{D}$. First we suppose $f \in R(D)$. Then f has decomposition in the form $E(x) + \sum\limits_{j=1}^{t} \dfrac{\lambda_j}{(x - \alpha_j)^{q_j}}$ with $E(x) \in K[x]$, and $\alpha_j \in K \setminus D$. Now for each j, either α_j belongs to a hole T or α_j belongs to $K \setminus \tilde{D}$. Let $T_1, \ldots T_s$ be the holes that contain some α_j. Then $\displaystyle\sum_{j=1}^{t} \frac{\lambda_j}{(x-\alpha_j)^{q_j}}$ is of the form $\displaystyle\sum_{n=1}^{s} f_n + h_0$ with

$f_i \in H_0(K \setminus T_i)$ and $h_0 \in H_b(\widetilde{D})$. Finally we put $f_0 = E(x) + h_0$ and we have

the announced decomposition: $f = \displaystyle\sum_{n=0}^{s} f_i$ with $f_i \in H_0(K \setminus T_i)$ and $f_0 \in H_b(\widetilde{D})$.

In the case when D is unbounded, f_0 is just a constant.

For each $i = 1, ..., s$, $f - f_i$ clearly belongs to $H_b(D \cup T_i)$ and obviously f

belongs to $H_b\Big(\widetilde{D} \setminus \Big(\displaystyle\bigcup_{i=1}^{s} T_i\Big)\Big)$.

First we will show that for any $n \in \mathbb{N}^*$, we have $\|f_n\|_D = \|f_n\|_{K \setminus T_n}$. Let $\mathcal{F}_n$ be
the circular filter on K of center α_n, and diameter r_n. By Theorem 13.1. we
have

$\qquad$ (4) $\quad \|f_n\|_{K \setminus T_n} = \lim_{\mathcal{F}_n \cap (K \setminus T_n)} |f_n(x)|.$

But by Lemma 3.14, $\mathcal{F}_n$, is secant with D hence

$\qquad$ (5) $\quad \lim_{\mathcal{F}_n \cap (K \setminus T_n)} |f_n(x)| = \lim_{\mathcal{F}_n \cap D} |f_n(x)|$ and obviously

$\qquad$ (6) $\quad \lim_{\mathcal{F}_n \cap D} |f_n(x)| \leq \|f_n\|_D \leq \|f_n\|_{K \setminus T_n}$

Finally by (4), (5), (6) we obtain

$\qquad$ (7) $\quad \|f_n\|_{K \setminus T_n} = \|f_n\|_D = {}_D\varphi_{\mathcal{F}_n}(f_n)$

In the same way, when D is bounded we consider the circular filter $\mathcal{F}_0$ of center
0 of diameter r, in order to prove that

$\qquad$ (8) $\quad \|f_0\|_D = \|f_0\|_{\widetilde{D}} = {}_D\varphi_{\mathcal{F}_0}(f).$

Now let us show that $\|f\|_D \geq \|f_n\|_D$ for any $n \in \mathbb{N}^*$. Since $f \in R(D)$, there
exists an annulus $\Gamma(a_n, r'_n, r_n)$ such that f has neither any zero nor any pole
inside $\Gamma(a_n, r'_n, r_n)$. We put $I =] - \log(r_n), -\log(r'_n)[$. By hypothesis f_n has
no pole in $K \setminus d(0, r'_n)$. Hence, as $\lim_{|x| \to \infty} f_n(x) = 0$, by Corollary 4.12. we see

that $v'_{a_n}(f_n, \mu) < 0$ whenever $\mu \in I$. Let $g_n = f - f_n \in R(D \cup T_n)$. Since g_n

has no pole inside T_n, by Corollary 4.12 we see that $v'^r_{a_n}(g_n, \mu) \geq 0$ whenever

$\mu > -\log(r_n)$.

Therefore the equation $v_{a_n}(f_n, \mu) = v_{a_n}(g_n, \mu)$ has at most one solution in
I, and then $v_{a_n}(f, \mu)$ is equal to $\max(v_{a_n}(f_n, \mu), v_{a_n}(g_n, \mu))$ whenever $\mu \in I$,
hence $v_{a_n}(f, \mu) \leq v_{a_n}(f_n, \mu)$ whenever $\mu \in I$. It follows that the multiplicative
semi-norm $\varphi_{\mathcal{F}_n}$ defined on $R(D)$ satisfies $-\log(\varphi_{\mathcal{F}_n}(f_n)) = v_{a_n}(f_n, -\log(r_n)) \geq$
$v_{a_n}(f, -\log(r_n)) = -\log(\varphi_{\mathcal{F}_n})(f)$ hence

$\qquad$ (9) $\quad \varphi_{\mathcal{F}_n}(f_n) \leq \varphi_{\mathcal{F}_n}(f).$

But $\varphi_{\mathcal{F}_n}(f_n) = \|f_n\|_D$ and $\varphi_{\mathcal{F}_n}(f) \leq \|f\|_D$, hence by (9) we have $\|f\|_D \geq \|f_n\|_D$.

Finally by (7), we see that (2) is clearly proven.

When D is bounded we put $\widetilde{D} = d(0, r)$ and we prove (3) in the same way as above when proving (2) by considering an annulus $\Gamma(0, r, r')$ such that f has neither any zero nor any pole inside $\Gamma(0, r, r')$. Then the element $g_0 = f - f_0$ is of the form $\dfrac{P}{Q}$ with $P, Q \in K[x]$, all the zeros of Q in $\widetilde{D}$ and $deg(P) < deg(Q)$ because $g_0 \in R_0(K \setminus \widetilde{D})$. Hence we have $v'(g_0, \mu) < 0$ while $v'(f_0, \mu) \geq 0$ whenever $\mu \in] - \log r', - \log r \, [$, so we have $_D\varphi_{\mathcal{F}_0}(f_0) \leq \, _D\varphi_{\mathcal{F}_0}(f)$ and then, by (8) we obtain (3).

When D is not bounded the inequality $|f_0| \leq \|f\|_D$ is obvious because $\displaystyle\lim_{\substack{|x| \to \infty \\ x \in D}} f(x) - f_0 = 0$. This finishes proving the Mittag-Leffler Theorem when $f \in R(D)$.

Now let $f \in H(D)$, and let $(h_m)_{m \in \mathbb{N}}$ be a sequence in $R(D)$ that converges to f in $H(D)$. The set of holes of D that contain at least one pole of some h_m is clearly countable. Hence there is a sequence of holes $(T_n)_{n \in \mathbb{N}^*}$ such that, denoting by D' the set $\widetilde{D} \setminus \Big(\displaystyle\bigcup_{n \in \mathbb{N}^*} T_n \Big)$, then h_m belongs to $H(D')$ whenever $m \in \mathbb{N}$. For each $m \in \mathbb{N}$, h_m splits in $H(D')$ in the form $h_m = \displaystyle\sum_{n=0}^{\infty} h_{m,n}$ with $h_{m,0} \in H(\widetilde{D}), h_{m,n} \in H_0(K \setminus T_n)$ and, in particular, for each fixed $n \in \mathbb{N}$, we have $\|h_{m,n} - h_{q,n}\|_D \leq \|h_m - h_q\|_D$. Thus we see that the sequence $(h_{m,n})_{m \in \mathbb{N}}$ converges in $H(K \setminus T_n)$ for $n > 0$ (resp. in $H(\widetilde{D})$ for $n = 0$) to a limit f_n and then we have $f = \displaystyle\sum_{n=0}^{\infty} f_n$ in $H(D')$. Obviously $\|f\|_D = \displaystyle\sup_{n \in \mathbb{N}} \|f_n\|_D$, whereas $\|f_n\|_D = \|f_n\|_{D'}$ whenever $n \in \mathbb{N}$ and so, $\|f\|_D = \|f\|_{D'}$. This ends the proof of Theorem 15.1.

Corollary 15.2:   *Let $(T_i)_{i \in I}$ be the family of holes of D. Let J be a subset of D and let $L = I \setminus J$. Let $E = D \bigcup(\bigcup_{i \in J} T_i)$ and let $F = D \bigcup(\bigcup_{i \in L} T_i)$. Then we have $H(D) = H_0(E) \oplus H(F)$, and for each $g \in H_0(E)$, $h \in H(F)$, we have $\|g + h\|_D = \max(\|g\|_E, \|h\|_F)$.*

The Mittag-Leffler Theorem suggests some new definitions.

Definitions and notations: Let $f \in H_b(D)$. We consider the series $\displaystyle\sum_{n=0}^{\infty} f_n$

obtained in Theorem 15.1, whose sum is equal to f in $H(D)$, with $f_0 \in H(\widetilde{D})$, $f_n \in H(K \setminus T_n) \setminus \{0\}$ and with T_n holes of D. Each T_n will be called a f-hole and f_n will be called the *Mittag-Leffler term of f associated to T_n*, whereas f_0 will be called the *principal term* of f.

For each f-hole T of D, the Mittag-Leffler term of f associated to T will be denoted by $\overline{f}_T$ whereas the principal term of f will be denoted by $\overline{f}_0$. The series $\sum_{n=0}^{\infty} f_n$ will be called *the Mittag-Leffler series of f on the infraconnected set D.*

More generally, let E be an infraconnected set and $f \in H(E)$. According to Theorem 10.5, f is of the form $g + h$ with $g \in R(K \setminus (\overline{E} \setminus E))$, and $h \in H_b(\overline{E})$, and such a decomposition is unique, with respect to an additive constant. For every hole T of $\overline{E}$, we will denote by $\overline{f}_T$ the Mittag-Leffler term of h associated to T, and $\overline{f}_T$ will still be named *the Mittag-Leffler term of f associated to T.*

Corollary 15.3:   *Let $f \in H_b(D)$ let$(T_n)_{n \in \mathbb{N}^*}$ be the sequence of the f-holes and let $f_0 = \overline{f}_0$, and $f_n = \overline{f}_{T_n}$ for every $n \in \mathbb{N}^*$. Let $\widetilde{D} = d(a, s)$,   (resp. $\widetilde{D} = K$). There exists $q \in \mathbb{N}$ such that $\|f\|_D = \|f_q\|_D$. If $q \geq 1$ then $\|f\|_D = {}_D\varphi_{a_q, r_q}(f) = {}_D\varphi_{a_q, r_q}(f_q)$. If $q = 0$ and if D is bounded (resp. is not bounded) then $\|f\|_D = {}_D\varphi_{a,s}(f) = {}_D\varphi_{a,s}(f_0)$ (resp. $\|f\|_D = |f_0|$). Further, given a hole T of D, if f belongs to $H_b(D)$, and if g belongs to $H_0(K \setminus T)$ and satisfies $f - g \in H(D \cup T)$, then $\overline{f}_T$ is equal to g.*

Corollary 15.4:   *Let $f \in H_b(D)$. There exists a large circular filter $\mathcal{F}$ with center $\alpha \in \widetilde{D}$ secant with D such that ${}_D\varphi_{\mathcal{F}}(f) = \|f\|_D$.*

Corollary 15.5:   *Let $f \in H(d(0, 1^-))$ and let $(d(\alpha_m, 1^-))_{m \in \mathbb{N}^*}$ be the family of the f-holes . Then f is of the form*

$$(1) \quad \sum_{n=0}^{\infty} a_{n,0} x^n + \sum_{m,n \in \mathbb{N}^*} \frac{a_{n,m}}{(x - \alpha_m)^n}$$

with $\lim_{n \to \infty} a_n = 0$, $\lim_{n \to \infty} |a_{n,m}| = 0$ whenever $m \in \mathbb{N}^*$ and $\lim_{m \to \infty} \left(\sup_{n \in \mathbb{N}^*} |a_{n,m}| \right) = 0$. Besides f satisfies $\|f\|_{d(0,1^-)} = \sup_{m,n \in \mathbb{N}^*} |a_{n,m}|$. Reciprocally, every function of the form (1), with the α_m satisfying $|\alpha_m| = |\alpha_j - \alpha_m| = 1$ whenever $m \neq j$, belongs to $H(d(0, 1^-))$. The norm $\| \cdot \|_{d(0,1^-)}$ is multiplicative and equal to ${}_{d(0,1^-)}\varphi_{0,1}$.

Theorem 15.6:   *Let $r_1, r_2 \in \mathbb{R}_+$ satisfy $0 < r_1 < r_2$. Then $H(\Delta(0, r_1, r_2))$ is equal to the set of the Laurent series $\displaystyle\sum_{-\infty}^{+\infty} a_n x^n$ with $\displaystyle\lim_{n \to -\infty} |a_n| r_1^n = \lim_{n \to \infty} |a_n| r_2^n = 0$ and we have $\displaystyle\| \sum_{-\infty}^{+\infty} a_n x^n \|_{\Delta(0, r_1, r_2)} = \max\Big(\sup_{n \geq 0} |a_n| r_1^n, \ \sup_{n < 0} |a_n| r_2^n \Big)$.*

Proof: Let $\Lambda = \Delta(0, r_1, r_2)$. We see that Λ has only one hole $T = d(0, r_1^-)$ and the Mittag-Leffler series of f is just $\overline{f}_0 + \overline{f}_1$ with $\overline{f}_0 \in H(d(0, r_2))$ and $\overline{f}_1 \in H_0(K \setminus d(0, r_1^-))$, with $\overline{f}_0 = \displaystyle\sum_{n=0}^{\infty} a_n x^n$, $\overline{f}_1 = \displaystyle\sum_{n < 0} a_n x^n$. Then the conclusions on the norm come from Theorem 15.1.

Theorem 15.7:   *Let $r \in \mathbb{R}_+$. Then $H(C(0, r))$ is equal to the set of the Laurent series $\displaystyle\sum_{-\infty}^{+\infty} a_n x^n$ with $\displaystyle\lim_{|n|_\infty \to \infty} |a_n| r^n = 0$, and we have $\displaystyle\| \sum a_n x^n \|_{C(0, r)} = \sup_{n \in \mathbb{Z}} |a_n| r^n$. Besides the norm $\| \cdot \|_{C(0, r)}$ is multiplicative and equal to $_{C(0, r)}\varphi_{0, r}$.*

Proof: We put $\Lambda = C(0, r)$. We may apply Theorem 15.6 by taking $r_1 = r_2 = r$ and we obtain all the conclusions but the fact that $\| \cdot \|_\Lambda$ is multiplicative. Let us show this. Let $h \in R(\Lambda)$. Hence h is of the form $\dfrac{P}{Q}$, with $P, Q \in K[x]$, and $Q(x)$ has no zero in Λ. Let θ be a class of Λ. By Lemma 4.6 we have $|Q(x)| = \varphi_{0, r}(Q)$ whenever $x \in \theta$ and $|P(x)| \leq \varphi_{0, r}(P)$ whenever $x \in \theta$. Hence we see that $\| \dfrac{P}{Q} \|_\Lambda \leq \varphi_{0, r}\Big(\dfrac{P}{Q} \Big)$ and therefore $\|h\|_\Lambda = \varphi_{0, r}(h)$ whenever $h \in R(\Lambda)$. Hence we have $\|f\|_\Lambda = {}_\Lambda\varphi_{0, r}(f)$ whenever $f \in H(\Lambda)$.

Proposition 15.8:   *Let $r_1, r_2 \in \mathbb{R}_+^*$, with $r_1 < r_2$.*

i) $A(\Gamma(0, r_1, r_2)) = A(d(0, r_2^-)) \oplus A_0(K \setminus d(0, r_1))$ and
$A_b(\Gamma(0, r_1, r_2)) = A_b(d(0, r_2^-)) \oplus A_{0, b}(K \setminus d(0, r_1))$

ii)   *Let $f(x) = \displaystyle\sum_{-\infty}^{+\infty} a_n x^n \in A(\Gamma(0, r_1, r_2))$. Then $f \in A_b(\Gamma(0, r_1, r_2))$ if and only if $\max\Big(\sup_{n \geq 0} |a_n| r_2^n, \sup_{n < 0} |a_n| r_1^n \Big) < +\infty$. Besides, if $f \in A_b(\Gamma(0, r_1, r_2))$ then*

$$\|f\|_{\Gamma(0, r_1, r_2)} = \max\Big(\sup_{n \geq 0} |a_n| r_2^n, \sup_{n < 0} |a_n| r_1^n \Big)$$

iii) $A_b(\Gamma(0, r_1, r_2))$ *is a Banach K-algebra that contains $H(\Gamma(0, r_1, r_2))$.*

Proof: *i)* is obvious. We will show *ii)*. Let $f \in A_b(\Gamma(0, r_1, r_2))$ and let $f = f_1 + f_2$ with $f_2 \in A_b(d(0, r_2^-))$ and $f_1 \in A_{0,b}(K \setminus d(0, r_1))$. We put $\Lambda = \Gamma(0, r_1, r_2)$. It is obviously seen that $\|f\|_\Lambda \leq \max\Big(\|f_1\|_\Lambda, \|f_2\|_\Lambda\Big) \leq \max\Big(\sup_{n \geq 0} |a_n| r_2^n, \sup_{n < 0} |a_n| r_1^n\Big)$. Now for every s_1, s_2 such that $r_1 < s_1 < s_2 < r_2$ we know that f belongs to $H(\Delta(0, s_1, s_2))$ because so do both f_1, $_2$. Then by Theorem 15.1. we have $\|f\|_{\Delta(0, s_1, s_2)} = \max\Big(\|f_2\|_{d(0, s_2)}, \|f_1\|_{K \setminus d(0, s_1^-)}\Big)$. Finally $\|f_2\|_{d(0, s_2)} = \sup_{n \in \mathbb{N}} |a_n| s_2^n$ while $\|f_1\|_{K \setminus d(0, s_1^-)} = \sup_{n < 0} |a_n| s_1^n$. Thus we see that

$$\|f\|_{\Gamma(0, r_1, r_2)} \geq \|f\|_{\Delta(0, s_1, s_2)} = \max\Big(\sup_{n \geq 0} |a_n| s_2^n, \sup_{n < 0} |a_n| s_1^n\Big).$$ This is true for every $s_1, s_2 \in \,]r_1, r_2[$ hence finally $\|f\|_\Lambda = \max\Big(\sup_{n \geq 0} |a_n| r_2^n, \sup_{n < 0} |a_n| r_1^n\Big)$. All the statements in *ii)* are then proven .

We now prove *iii)*. By *ii)* $A_b(\Lambda)$ is just the Banach K-algebra

$$A_b(d(0, r_2^-)) \oplus A_{0,b}(K \setminus d(0, r_1))$$

provided with the norm $\|f_1 + f_2\|_\Lambda = \max\Big(\|f_2\|_{d(0, r_2^-)}, \|f_1\|_{K \setminus d(0, r_1)}\Big)$. We saw that $R(d(0, r_2^-)) \subset A_b(d(0, r_2^-))$ hence $R(d(0, r_2^-)) \subset A_b(\Lambda)$.

In the same way we have $R(K \setminus d(0, r_1)) \subset A_b(K \setminus d(0, r_1))$ and then $A_b(K \setminus d(0, r_1))$ is obviously included in $A_b(\Lambda)$. Since $R(\Lambda) = R(d(0, r_2^-)) + R(K \setminus d(0, r_1))$, $R(\Lambda)$ is included in $A_b(\Lambda)$ which is complete for the norm $\| \, . \, \|_\Lambda$, hence $H(\Lambda) \subset A_b(\Lambda)$. This finishes proving Proposition 15.8 .

Notations: Given a set A in $\widehat{K}$, we will denote by $\widehat{H}(A)$ the set of the analytic elements in A, taking $\widehat{K}$ as ground field.

Now we will apply the Mittag-Leffler Theorem to the analytic extension of analytic elements.

Theorem 15.9: *For all $f \in H(D)$, f has continuation to a unique element $\widehat{f} \in \widehat{H}(\widehat{D})$. Further, if $f \in H_b(D)$ the Mittag-Leffler series of $\widehat{f}$ in $\widehat{D}$ is the same as the one of f in D.*

Proof: By Theorem 10.5 we may easily assume that f belongs to $H_b(\overline{D})$. The Mittag-Leffler series of f on D obviously converges on $\widehat{D}$, to an element of $\widehat{H}(\widehat{D})$. This is unique because so is the Mittag-Leffler series of f .

Theorem 15.10: *Let E be an infraconnected set such that $D \cap E$ is infraconnected and such that every hole of $D \cap E$ is either a hole of D or a hole of E.*

Let $F \in H(D)$, $G \in H(E)$, satisfying $F(x) = G(x)$ whenever $x \in D \cap E$. Then there exists $h \in H(D \cup E)$ such that $h(x) = F(x)$ whenever $x \in D$, $h(x) = G(x)$ whenever $x \in E$, and such that for every h-hole V of $D \cup E$, $\overline{h}_V$ is either of the form $\overline{F}_S$, when V is a F-hole S of D, or of the form $\overline{G}_T$ when V is a G-hole T of E.

Proof: By Theorem 10.5 it is eaily seen that we may assume $F \in H_b(D)$, $G \in H_b(E)$ without loss of generality. Let $A = D \cup E$, $B = D \cap E$. Let h be the restriction of F and G to B. Let $(V_n)_{n \in \mathbb{N}^*}$ be the sequence of h-holes that are holes of D, and let $(W_n)_{n \in \mathbb{N}^*}$ be the sequence of h-holes that are holes of E, but not of D. For each $q \in \mathbb{N}^*$, as $h(x)$ is equal to $F(x)$ in B, $\overline{h}_{W_q}$ is an element of $H_0(D)$ of the form $\sum_{m=1}^{\infty} \overline{F}_{S_m^q}$ with S_m^q F-holes of D included in W_q. We put $f_{q,m} = \overline{F}_{S_m^q}$ for every $(q,m) \in (\mathbb{N}^{*2})$. In the same way, for each $q \in \mathbb{N}^*$, h_{V_q} is an element of $H_0(E)$ of the form $\sum_{m=1}^{\infty} \overline{G}_{T_m^q}$ with T_m^q G-holes of E included in V_q. We put $g_{q,m} = \overline{G}_{T_m^q}$ for every $(q,m) \in (\mathbb{N}^{*2})$. Without loss of generality we may obviously assume $\widetilde{D} \subset \widetilde{E}$, we put $h_0(x) = \overline{G}_0(x)$. We notice that A is clearly included in the set $A' = \widetilde{E} \setminus \left(\left(\bigcup_{(q,m) \in (\mathbb{N}^{*2})} S_m^q \right) \bigcup \left(\bigcup_{(q,m) \in (\mathbb{N}^{*2})} T_m^q \right) \right)$.

Then, it is easily seen that the series $h_0(x) + \sum_{(m,q) \in \mathbb{N}^{*2}} f_{q,m} + \sum_{(m,q) \in \mathbb{N}^{*2}} g_{q,m}$ converges in $H(A')$ because by Corollary 15.2 we have $\|f_{q,m}\|_{A'} = \|\overline{F}_{S_m^q}\|_D$, and $\|g_{q,m}\|_{A'} = \|\overline{G}_{T_m^q}\|_E$, whereas

$\lim_{q+m \to +\infty} \|\overline{F}_{S_m^q}\|_E = \lim_{q+m \to +\infty} \|\overline{G}_{T_m^q}\|_E = 0$. Besides, by construction, $h(x)$ is equal to $F(x)$ and $G(x)$ in B, and is such that for every h-hole V of $D \cup E$, $\overline{h}_V$ is either of the form $\overline{F}_S$, when V is a F-hole S of D, or of the form $\overline{G}_T$ when V is a G-hole T of E. This clearly ends the proof of Theorem 15.10.

16. MAXIMAL IDEALS OF CODIMENSION 1

D denotes a set that belongs to $\mathcal{A}$.

The maximal ideals of codimension 1 of an algebra $H(D)$ are easily characterized by the points of D. This characterization is immediate when D is closed and bounded, but requires some attention in the general case.

Remark: Let A be a commutative ring with unity u, let S a subset of A stable with repect to the multiplication, and let $B = S^{-1}A$. If two ideals J and J' of B satisfy $J \cap A = J' \cap A$, then $J = J'$. As a consequence if $A \cap J$ is maximal in A, then J is maximal in B, because if it were not maximal, then it would be included in a maximal ideal N that should satisfy $A \cap N = A$, hence u would belong to N.

Lemma 16.1: *Let L be field, let A be a commutative L-algebra with unity, and let S a subset of A stable with repect to the multiplication. Let $B = S^{-1}A$ and let M be a maximal ideal of B of codimension 1. Then $A \cap M$ is a maximal ideal of A of codimension 1. Besides M is the only one ideal of B which contains $A \cap M$.*

Proof: Let ϕ be the canonical surjection from B onto $\dfrac{B}{M}$ and let ϕ' be the restriction of ϕ to A. Let u be the unity of A. Of course we have $\phi(u) = 1$, hence $\phi'(\lambda u) = \lambda$ whenever $\lambda \in L$. So $\phi'(A)$ is equal to L and therefore $Ker(\phi')$ is a maximal ideal of A. But $Ker(\phi')$ is equal to $A \cap M$, hence $A \cap M$ has codimension 1. Then by the previous remark, M is the unique ideal J of B such that $J = A \cap M$. Since M is maximal in B, it is the unique ideal of B which contains $A \cap M$. This ends the proof.

Notations: Let $D \in \mathcal{A}$. For every $a \in D$ we will denote by $\mathcal{I}(a)$ the ideal of the $f \in H(D)$ such that $f(a) = 0$.

Theorem 16.2: *The mapping ϕ from D into $Max_1(H(D))$ defined by $\phi(a) = \mathcal{I}(a)$ is a bijection.*

Proof: For each $a \in D$ let ψ_a be the surjective homomorphism of K-algebra from $H(D)$ onto K defined as $\psi_a(f) = f(a)$. Then $\mathcal{I}(a)$ is clearly equal to $Ker(\psi)$, and thereby $\mathcal{I}(a)$ is a maximal ideal of codimension 1. So ϕ, obviously, is an injection from D into $Max_1(H(D))$.

Now let $\mathcal{M}$ be a maximal ideal of codimension 1 of $H(D)$, let ψ be the canonical K-algebra homomorphism from $H(D)$ onto K whose kernel is $\mathcal{M}$ and let $a = \psi(x)$. It is easily seen that $\psi(P(x)) = P(a)$ whenever $P \in K[x]$ and therefore $\psi(h(x)) = h(a)$ whenever $h \in R(D)$.

First we suppose D closed and bounded. Then $H(D)$ is a K-Banach algebra, and then by Theorem 1.14 ψ is continuous. Hence we have $\psi(f) = f(a)$ whenever $f \in H(D)$. Thus Theorem 16.2 is proven when D is closed and bounded.

Now we stop assuming D to be closed and bounded. First we suppose D bounded, but not necessarily closed. Let $S(D)$ be the set of the polynomials whose zeros lie in $\overline{D} \setminus D$. By Corollary 10.12, $H(D)$ is of the form $S(D)^{-1}H(\overline{D})$. Hence $\mathcal{M} \cap H(\overline{D})$ is a maximal ideal of $H(\overline{D})$ and therefore, by the foregoing, there exists $a \in \overline{D}$ such that $\mathcal{M} \cap H(\overline{D})$ is the set of the $f \in H(\overline{D})$ such that $f(a) = 0$. Of course $x - a$ belongs to $\mathcal{M} \cap H(\overline{D})$. Then, if a does not belong to $\overline{D}$, $\dfrac{1}{x - a}$ belongs to $H(D)$, and then 1 belongs to $\mathcal{M}$. So a does not belong to D. Then the ideal $\mathcal{I}(a)$ is a maximal ideal of $H(D)$ and by Lemma 16.1, it is the only one maximal ideal of $H(D)$ that contains $\mathcal{M} \cap H(\overline{D})$, hence $\mathcal{I}(a) = \mathcal{M}$.

Finally we suppose D unbounded. If D has no hole, we have $H(D) = R(D)$, and then by Lemma 9.1 the maximal ideals of $R(D)$ are of the form $(x - a)R(D)$, with $a \in D$. Now we suppose that D has at least one hole $T = d(\alpha, r^-)$. Let $\gamma(x) = \dfrac{1}{(x - \alpha)}$ and let $D' = \gamma(D)$. Then by Theorem 11.5 D' also belongs to $\mathcal{A}$ and there exists an isomorphism ψ from $H(D')$ onto $H(D)$ defined as $\psi(f) = f \circ \gamma$. Since D' is clearly bounded, each maximal ideal of codimension 1 of $H(D')$ is of the form $\{f \in H(D')| \ f(a) = 0\}$ with $a \in D'$. Then it is seen that $\psi^{-1}(\mathcal{M})$ is a maximal ideal of codimension 1 of $H(D')$ of the form $\mathcal{M}' = \{g \in H(D')| \ g(b) = 0\}$. Let $a = \gamma^{-1}(b)$. Then we have $\psi(\mathcal{M}') = \mathcal{I}(a) = \{f \circ \gamma \in H(D)| \ g(b) = 0\} = \{f \in H(D)| \ f(a) = 0\}$. This finishes showing that $\mathcal{M}$ has the form $\mathcal{I}(a)$ and this ends the proof.

Corollary 16.3:   *Every maximal ideal of codimension 1 of $H(D)$ is closed, with respect to $\mathcal{U}_D$.*

17.   DUAL OF A SPACE $H(D)$

Notations: Let E be a K-Banach space. We will denote by $E^\sim$ the K-Banach space of the continuous linear forms of E provided with its usual norm. The dual of a Banach space $H(D)$ was thoroughly studied by Yvette Amice [2].

Theorem 17.1:   **(Y. Amice)** *Let* $r \in \mathbb{R}_+$. *For each* $h(t) = \sum_{n=0}^{\infty} \dfrac{b_n}{t^n} \in$ $A_b(K \setminus d(0,r))$ *there exists a unique* $\phi_h \in H(d(0,r))^\sim$ *satisfying* $\phi_h(x^q) = b_q$, ($q \in$ $\mathbb{N}$). *Besides, on the space* $A_b(K \setminus d(0,r))$ *provided with the norm* $\| \, . \, \|_{K \setminus d(0,r)}$, *the mapping* $h \to \phi_h$ *is an isometric isomorphism from* $A_b(K \setminus d(0,r))$ *onto* $H(d(0,r))^\sim$

Proof: Let $F = K \setminus d(0,r)$. First let $h(t) = \sum_{n=0}^{\infty} \dfrac{b_n}{t^n} \in A_b(F)$ and now let

$$f(x) = \sum_{n=0}^{\infty} a_n x^n \in H(d(0,r)).$$ Since the sequence $\dfrac{|b_n|}{r^n}$ is bounded and

$\lim\limits_{n \to \infty} |a_n| r^n = 0$, it is seen that $\lim\limits_{n \to \infty} a_n b_n = 0$, and then the series $\sum_{n=0}^{\infty} a_n b_n$ is

convergent. Hence we may put $\phi_h(f) = \sum_{n=0}^{\infty} a_n b_n$. Thus, we define a linear form

ϕ_h of $H(d(0,r))$ that is immediately seen to satisfy

$$|\phi_h(f)| \leq \sup_{n \in \mathbb{N}} |a_n b_n| \leq \left(\sup_{n \in \mathbb{N}} |a_n| r^n \right) \left(\sup_{n \in \mathbb{N}} \dfrac{|b_n|}{r^n} \right) = \|f\|_{d(0,r)} \|h\|_F .$$

Therefore, with respect to the norm $\| \, . \, \|$ of $H((d(0,r))^\sim$, we have $\|\phi_h\| \leq$ $\|h\|_F$. Now we check that the equality is satisfied. Indeed let $q \in \mathbb{N}$. We have $\dfrac{|\phi_h(x^q)|}{\|x^q\|_{d(0,r)}} = \dfrac{|\phi_h(x^q)|}{r^q} = \dfrac{|b_q|}{r^q} \leq \sup_{f \neq 0} \dfrac{|\phi_h(f)|}{\|f\|_{d(0,r)}}$ for all $q \geq 0$. Hence we have

$\|h\|_F = \sup_{q \in \mathbb{N}} \dfrac{|b_q|}{r^q} \leq \|\phi_h\|$. So we obtain the announced equality. Thus we have

defined an isometric homomorphism from $A_b(F)$ into $H(d(0,r))^\sim$.

Now we check that this mapping is surjective. Indeed, let $\psi \in H(d(0,r))^\sim$

and for each $n \in \mathbb{N}$, let $b_n = \psi(x^n)$. Obviously we have $\|\psi\| \geq \dfrac{|b_n|}{r^n}$ for every $n \in$

$\mathbb{N}$, Hence the sequence $(|b_n| r^{-n})_{n \in \mathbb{N}}$ is bounded and therefore defines a function $f(t) = \sum_{n=0}^{\infty} \dfrac{b_n}{t^n} \in A_b(F)$. Thus ψ is equal to ϕ_h and therefore the mapping $h \to \phi_h$ is surjective. This ends the proof of Theorem 17.1.

Remark: There obviously exists an isometric homomorphism from $H(d(0,r))$ into $H((d(0,r)))^{\sim}$ that associates to each $f \in H(d(0,r))$ an $\underline{f}$ defined as $\underline{f}(\phi_h) = \phi_h(f)$. The question of whether this homomorphism is surjective depends on the ground field K. If K is spherically complete, this homomorphism is not surjective. If K is not spherically complete, this homomorphism is surjective [52].

Corollary 17.2: *Let $r \in \mathbb{R}_+$. For each $h(t) = \sum_{n=0}^{\infty} b_n t^n \in A_b(d(0,r^-))$, there exists a unique $\phi_h \in H(K \setminus d(0,r^-))^{\sim}$ satisfying $\phi_h(x^{-q}) = b_q$ $(q \in \mathbb{N})$. Besides, the space $A_b(d(0,r^-))$ being provided with the norm $\| \cdot \|_{d(0,r^-)}$, the mapping $h \to \phi_h$ is an isometric isomorphism from $A_b(d(0,r^-))$ onto $H(K \setminus d(0,r^-))^{\sim}$.*

Corollary 17.3: *Let $r \in \mathbb{R}_+$. For each $h(t) = \sum_{n=0}^{\infty} b_n t^n \in A_b(d(0,r^-))$ such that $h(0) = 0$ there exists a unique $\phi_h \in H_0(K \setminus d(0,r^-))^{\sim}$ satisfying $\phi_h(x^{-q}) = b_q$ $(q \in \mathbb{N}^*)$. Besides, this mapping $h \to \phi_h$ from the subspace of the $h \in A_b(d(0,r^-))$ such that $h(0) = 0$ into $H_0(K \setminus d(0,r^-))^{\sim}$, is an isometric isomorphism.*

Now applying Theorem 17.1 to $H(\widetilde{D})$ and Corollary 17.3 to the spaces $H_0(K \setminus T_i)$ for each hole T_i of an infraconnected set D, thanks to Theorem 15.1 we obtain Corollary 17.4.

Corollary 17.4: **(Y. Amice)** *Let D be closed bounded infraconnected and let $(T_i)_{i \in J}$ be the family of the holes of D, and for every $i \in J$, let $a_i \in T_i$. Let $M \in \mathbb{R}_+^*$. Let $h_0 \in A_b(K \setminus \widetilde{D})$ and let $(h_i)_{i \in J}$ be a family such that for each $i \in J$, h_i belongs to $A_b(T_i)$ and satisfies*
> *(1) $h_i(a_i) = 0$ and*
> *(2) $\|h_i\|_{T_i} \leq M$ for all $i \in J$.*
There exists a unique $\psi \in H(D)^{\sim}$ satisfying
> *$\psi(f) = \phi_{h_0}(f)$ for every $f \in H(\widetilde{D})$*
> *$\psi(f) = \phi_{h_i}(f)$ for every $f \in H_0(K \setminus T_i)$, whenever $i \in J$.*

Further, for every element ψ of $H(D)^{\sim}$ there exists a unique $h_0 \in A_b(K \setminus d(0,r))$ and a unique family $(h_i)_{i \in J}$ satisfying (1) and (2) for some $M \in \mathbb{R}_+^*$ such that ψ is defined as above.

Now we will use the continuous linear forms to define the residue of an element on a hole.

Theorem 17.5: *Let $f \in H_b(K \setminus d(a,r^-))$, and for each $\alpha \in d(a,r^-)$, let*

$$f(x) = \sum_{n=0}^{\infty} \frac{b_n(\alpha)}{(x-\alpha)^n}. \quad \text{Then } b_1(\alpha) \text{ does not depend on } \alpha \text{ in } d(a,r^-).$$

Proof: Let $E = K \setminus d(a,r^-))$. We know that $\dfrac{|b_1(\alpha)|}{r} \leq \|f\|_E$, and therefore, fixing α in $d(a,r^-)$, the linear form ψ_α on $H_b(E)$ defined as $\psi_\alpha(f) = b_1(\alpha)$ is obviously continuous. We will show that $\psi_\alpha(f) = \psi_a(f)$. First, for every $q \in \mathbb{N}$, we put $f_q(x) = \dfrac{1}{(x-\alpha)^q}$. We have $f_q(x) = \dfrac{1}{(x-a)^q}\Big(\sum_{j=0}^{\infty}(\dfrac{\alpha-a}{x-a})^j\Big)^q$. Therefore, it is seen that for every $q \in \mathbb{N}$ we have $\psi_a(f_q) = 0$, and that $\psi_a(f_1) = 1$. Hence $\psi_\alpha(f_q) = \psi_a(f_q)$ for every $q \in \mathbb{N}$. This shows that $\psi_\alpha(f) = \psi_a(f)$ for every $f \in H_b(E)$.

Definition and notation: Let $f \in H_b(D)$, let T be a hole of D, and let $a \in T$. Let $\overline{f}_T(x) = \sum_{n=1}^{\infty} \dfrac{b_n(a)}{(x-a)^n}$. By Theorem 17.5, $b_1(a)$ actually does not depend on a in T. We put $res(f,T) := b_1(a)$, and this number $res(f,T)$ will be called *the residue of f on the hole T.*

Remark: Let $f \in H(D)$, let $a \in \overset{\circ}{\overline{D}} \setminus D$, let $r > 0$ be such that f has no pole in $d(a,r) \setminus \{a\}$. Let $D' := D \setminus d(a,r^-)$. Then we have $res(f,a) = res(f,T)$.

By Theorem 15.1, Theorem 17.6 is obvious:

Theorem 17.6: *Let $f \in H(D)$ and let T be a hole of D of diameter r. Then*

$$|res(f,T)| \leq r\|\overline{f}_T\|_{K \setminus T} \leq r\|f\|_D.$$

In Chapter 16 we characterized the maximal ideals of codimension 1 of a K-algebra $H(D)$ with the help of the points of D. Here, we will characterize them among the continuous linear forms.

Theorem 17.7: *Let D be closed bounded infraconnected, let $a \in \widetilde{D}$, let $(T_i)_{i \in J}$ be the family of the holes of D, and for every $i \in J$, let $a_i \in T_i$. Let $M \in \mathbb{R}_+^*$. Let $h_0 \in A_b(K \setminus \widetilde{D})$ and let $(h_i)_{i \in J}$ be a family such that for each $i \in J$, h_i belongs to $A_b(T_i)$ and satisfy Conditions (1) and (2):*

    (1) $h_i(a_i) = 0$ *and*

    (2) $\|h_i\|_{T_i} \leq M$ *for all $i \in J$.*

Let $\psi \in H(D)^\sim$ satisfy

    $\psi(f) = \phi_{h_0}(f)$ *for every $f \in H(\widetilde{D})$*

    $\psi(f) = \phi_{h_i}(f)$ *for every $f \in H_0(K \setminus T_i)$, whenever $i \in J$.*

Then ψ is a homomorphism of K-algebra from $H(D)$ onto K if and only if there exists $\alpha \in D$ such that

$$(3) \quad h_0(t) = \frac{t - a}{t - \alpha},$$

and for every $i \in J$,

$$(4) \quad h_i(t) = \frac{t - a_i}{\alpha - t}.$$

Proof: First we suppose that ψ is a K-algebra homomorphism from $H(D)$ onto K and we put $\psi(x) = a$. As h_0 is of the form $h_0(t) = \displaystyle\sum_{n=0}^{\infty} \frac{b_n}{(t - a)^n}$, here for every $n \in \mathbb{N}$ we have $b_n = \psi((x - a)^n) = (\alpha - a)^n$ and therefore $h_0(t) = \displaystyle\sum_{n=0}^{\infty} \left(\frac{\alpha - a}{t - a}\right)^n = \frac{t - a}{t - \alpha}$. Next, we fix $i \in J$. Then h_i is of the form $h_i(t) = \displaystyle\sum_{n=1}^{\infty} b_{i,n}(t - a_i)^n$. Hence for every $n \in \mathbb{N}^*$, we have $b_{i,n} = \psi\left(\frac{1}{(x - a_i)^n}\right) = \left(\frac{1}{\alpha - a_i}\right)^n$ and therefore $h_i(t) = \displaystyle\sum_{n=1}^{\infty} \left(\frac{t - a_i}{\alpha - a_i}\right)^n = \frac{t - a_i}{\alpha - t}$.

Reciprocally, we suppose (3) and (4) are satisfied. Then it is easily checked that $h_0(t) = \displaystyle\sum_{n=0}^{\infty} \left(\frac{\alpha - a}{t - a}\right)^n$, and therefore for every $n \in \mathbb{N}$, we have $\psi((x - a)^n) = (\alpha - a)^n$. Hence for every $f \in H(\widetilde{D})$, we have $\psi(f) = f(\alpha)$.

In the same way, we check that, fixing $i \in J$, we have $\psi\left(\frac{1}{(x - a_i)^n}\right) = \left(\frac{1}{\alpha - a_i}\right)^n$, hence $\psi(f) = f(\alpha)$ for every $f \in H_0(K \setminus T_i)$. This clearly finishes proving that $\psi(f) = f(\alpha)$ for every $f \in H(D)$.

18. ALGEBRAS $H(D)$

Now we will study what condition a set D has to satisfy for $H(D)$ to be a K-subalgebra of K^D [15].

Proposition 18.1: *Let α belong to $\overset{\circ}{\overline{D}}$ and let $f \in H(D \setminus \{\alpha\})$. For every $q \in \mathbb{N}$, $\dfrac{f}{(x - \alpha)^q}$ belongs to $H(D \setminus \{\alpha\})$.*

Proof: By Theorem 10.5 f is of the form $f^* + \overline{f}$, with $f^* \in R_0(K \setminus (\overline{D} \setminus D))$ and $\overline{f} \in H(\overline{D})$. Since α belong to $\overset{\circ}{\overline{D}}$, there exists a disk $d(\alpha, s)$ included in $\overline{D}$. Besides there exists $r < s$ such that f^* has no pole in $d(\alpha, r) \setminus \{\alpha\}$. Hence by Theorem 10.10 f is of the form $\dfrac{g}{(x - \alpha)^t}$ with $g \in H(D \cup d(\alpha, r))$, and $t \in \mathbb{N}$.

Then $\dfrac{f}{(x - \alpha)^q} = \dfrac{g}{(x - \alpha)^{q+t}}$. Thus without loss of generality we may assume that α belongs to D and that f belongs to $D \cup d(\alpha, r)$.

By Corollary 13.2, in $d(\alpha, r)$, $f(x)$ is equal to a power series $\displaystyle\sum_{n=0}^{\infty} a_n (x - \alpha)^n$.

Let $P(x) = \displaystyle\sum_{n=0}^{q} a_n (x - \alpha)^n$. By Corollary 13.4, $f(x) - P(x)$ factorizes in the form $(x - \alpha)^q h$ with $h \in H(D)$. Hence we see that $\dfrac{f(x)}{(x - \alpha)^q} = \dfrac{P(x)}{(x - \alpha)^q} + h$. Since $\dfrac{P}{(x - \alpha)^q} \in R(D)$ it is clear that $\dfrac{f}{(x - \alpha)^q}$ belongs to $H(D)$.

Corollary 18.2: *Let $f \in H(D)$ and let P be a polynomial whose zeros are interior to D. Then $\dfrac{f}{P}$ belongs to $H(D \setminus \{a_1, ..., a_n\})$.*

Proposition 18.3: *If D is bounded and satisfies $\overline{D} \setminus D \subset \left(\overset{\circ}{\overline{D}}\right)$ then $D \in \mathcal{A}$.*

Proof: Let $f, g \in H(D)$ and let us show that $fg \in H(D)$. By Theorem 10.5, we have $f = f^* + \overline{f}$, $g = g^* + \overline{g}$ with $f^*, g^* \in R_0(K \setminus (\overline{D} \setminus D))$ and $\overline{f}, \overline{g} \in H(\overline{D})$. Since D is bounded, by Corollary 10.6, we have $H(\overline{D}) = H_b(D)$ and then $\overline{f}\overline{g}$

obviously belongs to $H(D)$ while $f^*g^* \in R_0(K \setminus (\overline{D} \setminus D))$. Finally by Corollary 18.2, both $f^*\overline{g}$ and $g^*\overline{f}$ belong to $H(D)$ and therefore so does fg.

Definition: Let $\mathcal{F}$ be a filter in D. An element $f \in H(D)$ will be said to be *vanishing along* $\mathcal{F}$ if $\lim_{\mathcal{F}} f(x) = 0$. Besides f will be said to be *properly vanishing along* $\mathcal{F}$ if $\lim_{\mathcal{F}} f(x) = 0$ and if $\|f\|_A \neq 0$ whenever $A \in \mathcal{F}$.

Lemma 18.4 is a polyvalent result which helps us characterize the sets $D \in \mathcal{A}$ but also find conditions for $H(D)$ not to be a noetherian algebra.

Lemma 18.4:   *Let $\mathcal{F}$ be a pierced filter on D, let $(T_n)_{n\in\mathbb{N}}$ be a sequence of holes of D that runs $\mathcal{F}$ and let $D^* = K \setminus (\bigcup_{n=0}^{\infty} T_n)$. Let $g_1,\ldots,g_q \in H_b(D^*)$ be vanishing along $\mathcal{F}$, with g_1 properly vanishing. For every $x \in D^*$ let $S(x) = \sup_{1\leq i\leq q} |g_i(x)|$ and let F be the ideal generated by $g_1,\ldots,g_q$ in $H_b(D^*)$, and let $\overline{F}$ be its closure in $H_b(D)$.*

There exists a sequence $(z_n)_{n\in\mathbb{N}}$ in D, thinner than $\mathcal{F}$ such that $g_1(z_n) \neq 0$ and an element $T \in \overline{F}$ such that $\lim_{n\to\infty} \dfrac{|T(z_n)|}{S(z_n)} = +\infty$.

Proof: Without loss of generality we may assume $\mathcal{F}$ to be a decreasing filter or a Cauchy filter. Indeed if $\mathcal{F}$ is an increasing filter of center α of diameter R, let $T(b, \rho)$ be a hole of D included in $d(\alpha, R^-)$, let $\gamma(x) = \dfrac{1}{x - b}$, and let $D' = \gamma(D)$. Then by Theorem 3.11 D' admits a decreasing pierced filter $\mathcal{F}'$, image of $\mathcal{F}$, by γ. Besides, D' is clearly bounded. By Proposition 11.9, the mapping ϕ from D onto D' defined by $\phi(f) = f \circ \gamma^{-1}$ is an isomorphism from $H(\overline{D'})$ onto $H_b(D)$. Hence $\mathcal{J}$ is isomorphic to the ideal generated by $\{g_j \circ \gamma^{-1} | 1 \leq j \leq q\}$ in $H(\overline{D'})$. Hence we will assume $\mathcal{F}$ to be a decreasing pierced filter or a Cauchy pierced filter.

Without loss of generality we may clearly assume $D = D^*$. Since the g_j are bounded, we may obviously assume $\|g_j\|_D \leq 1$ whenever $j = 1,\ldots,q$. Let $R = diam(\mathcal{F})$ and let $(x_m)_{m\in\mathbb{N}}$ be a sequence in D thinner than $\mathcal{F}$ such that $g_1(a_m) \neq 0$ whenever $m \in \mathbb{N}$, with $|x_{m+2} - x_{m+1}| < |x_{m+1} - x_m|$. Since $\mathcal{F}$ is pierced , there exists a subsequence $(x_{m_q})_{q\in\mathbb{N}}$ of the sequence (x_m) together with a sequence of holes $(T_q)_{q\in\mathbb{N}}$ of D such that
$$T_q \subset d(x_{m_q+1}, d_{m_q}) \setminus d(x_{m_q+2}, d_{m_q+1}).$$
Hence without loss of generality we may assume that we have a sequence of holes $(T_m)_{m\in\mathbb{N}}$ of D such that $T_m \subset d(x_{m+1}, d_m) \setminus d(x_{m+2}, d_{m+1})$.

We put $D_m = d(x_{m+1}, d_m) \cap D$ and $A_n = D_{2n+1} \setminus D_{2n+3}$. For each n, let $u_n \in A_n$ be such that $|g_1(u_n)| \geq \|g_1\|_{A_n}\left(\dfrac{n}{n+1}\right)$. For each $j = 1, \ldots, t$, let $M_n^j = \|g_j\|_{A_n}$ and let $M_n = \max_{1 \leq j \leq t} M_n^j$. Since $g_1(x_m) \neq 0$ we have $M_n > 0$ whenever $n \in \mathbb{N}$, and since $\|g_j\|_D \leq 1$ for all j, we have $M_n \leq 1$ whenever $n \in \mathbb{N}$.

We will construct a sequence (U_n) in $H_b(D)$ satisfying

(1) $\qquad |U_n(x)| \leq \dfrac{1}{n+1}$ whenever $x \in D \setminus A_n$.

(2) $\qquad \sqrt{M_n}\left(\dfrac{n+1}{n}\right) > \|g_1 U_n\|_{A_n} > \sqrt{M_n}.$

For every $n \in \mathbb{N}$, let $T_n = d(\beta_n, \rho_n^-)$, $u_n = x_{2n+2}$, $a_n = \beta_{n+1}$, $b_n = \beta_{n+2}$, $c_n = \beta_{2n+3}$ and let $\epsilon_n \in d(0, \frac{1}{n})$. Let us fix $n \in \mathbb{N}$. It is seen that $|u_n - a_n| > |u_n - b_n|$, hence there exists $q_n \in \mathbb{N}$ such that

(3) $\qquad |\epsilon_n| \left|\dfrac{u_n - a_n}{u_n - b_n}\right|^{q_n} g(u_n) > \sqrt{M_n}$

and of course there exists q_n' such that

(4) $\qquad \left(\dfrac{d_{2n+1}}{d_{2n+2}}\right)^{q_n} \left(\dfrac{d_{2n+3}}{d_{2n+2}}\right)^{q_n'} < 1$

We put $h_n(x) = \epsilon_n \left(\dfrac{x - a_n}{x - b_n}\right)^{q_n} \left(\dfrac{x - c_n}{x - b_n}\right)^{q_n'}$.

Then by (4) we see that:

when $|x - c_n| > d_{2n+1}$ we have $|h_n(x)| = |\epsilon_n| < \frac{1}{n}$

when $|x - c_n| \leq d_{2n+3}$ we have $|x - a_n| = |a_n - c_n| = d_{2n+1}$ and $|x - b_n| = |b_n - c_n| = d_{2n+2}$ hence $|h_n(x)| \leq |\epsilon_n|\left(\dfrac{d_{2n+1}}{d_{2n+2}}\right)^{q_n} \left(\dfrac{d_{2n+3}}{d_{2n+2}}\right)^{q_n'} < \dfrac{1}{n}.$

But now we notice that x belongs to $D \setminus A_n$ if and only if x satisfies : either $|x - c_n| > d_{2n+1}$ or $|x - c_n| \leq d_{2n+3}$, hence we have proven that $|h_n(x)| < \frac{1}{n}$ whenever $x \in D \setminus A_n$. This shows h_n satisfies (1).

When $x \in A_n$ i.e. when $d_{2n+3} < |x - c_n| < d_{2n+1}$, we see that $\|g_1 h_n\|_{A_n} \geq |g_1(u_n)h_n(u_n)|$ hence by (3) we have $\|g_1 h_n\|_{A_n} \geq \sqrt{M_n}$. Hence there trivially exists $\lambda_n \in d(0, 1)$ such that $\left(\dfrac{n+1}{n}\right)\sqrt{M_n} > |\lambda_n g_1 h_n|_{A_n} > \sqrt{M_n}.$

Now we put $U_n = \lambda_n h_n$, and we see that U_n satisfies (1) and (2). In particular we have $\|g_1 U_n\|_D \leq \max\left(\sqrt{M_n}\left(\dfrac{n+1}{n}\right), \dfrac{\|g_1\|_D}{M}\right)$ hence $\lim_{n \to \infty} \|g_1 U_n\|_D = 0$. Let $T = \displaystyle\sum_{n=0}^{\infty} g_1 U_n$. By definition T belongs to $\overline{F}$ because for every $t \in$

$\mathbb{N}$, $g \sum\limits_{n=0}^{t} U_n$ belongs to F.

By (2) there exists a sequence $(z_n))_{n \in \mathbb{N}}$ in D satisfying $z_n \in A_n$ and

$$(5) \qquad \sqrt{M_n} < |g_1(z_n) U(z_n)| < M_n \left(\frac{n+1}{n} \right)$$

hence we have

$$(6) \qquad |U_n(z_n)| > \frac{\sqrt{M_n}}{|g_1(z_n)|} \geq \frac{1}{\sqrt{M_n}} \text{ because } |g_1(z_n)| \leq M_n.$$

Besides when $j \neq n$, z_n belongs to $D \setminus A_j$ hence by (1) and (6) we have

$$|U_j(z_n)| < \frac{1}{j+1} < \frac{1}{\sqrt{M_n}} < |U_n(z_n)| \text{ whenever } j \neq n.$$ Hence we see that
$|T(z_n)| = |g_1(z_n) U_n(z_n)|$ whenever $n \in \mathbb{N}$. But then, by (5) we see that
$\dfrac{|T(z_n)|}{S(z_n)} = \dfrac{|T(z_n)|}{M_n} > \dfrac{1}{\sqrt{M_n}}$. Thus we have $\lim\limits_{n \to \infty} \dfrac{|T(z_n)|}{S(z_n)} = +\infty$ and this finishes
the proof of Lemma 18.4.

Lemma 18.5:   *Let D have a hole $T = d(a, r^-)$. Let $\gamma(x) = b + \dfrac{\lambda}{x - a}$ with*

$\lambda \in K$, and let $D' = \gamma(D)$. For every $\alpha \in \overline{D}, \alpha$ belongs to $\left(\overset{\circ}{\overline{D}} \right)$ if and only if

$\gamma(\alpha)$ belongs to $\left(\overset{\circ}{\overline{D'}} \right)$. Moreover if D is not bounded then $K \setminus \overline{D}$ is bounded if

and only if b belongs to $\left(\overset{\circ}{\overline{D'}} \right)$.

Proof: γ is obviously a bicontinuous bijection from $K \setminus \{a\}$ onto $K \setminus \{b\}$.

Let $\alpha \in \left(\overset{\circ}{\overline{D}} \right)$. There exists a disk $d(\alpha, r)$ included in $\overline{D}$. Since $a \notin d(\alpha, r)$, γ
is bounded in $d(\alpha, r)$. Since γ is bicontinuous, $\gamma(d(\alpha, r))$ is open in $K \setminus \{a\}$,
hence it is clearly open in K. So $\gamma(\alpha)$ belongs to $\gamma\left(\overset{\circ}{\overline{D}} \right)$. But $\gamma(\overline{D}) \subset \overline{\gamma(D)}$
hence $\gamma(\alpha) \in \left(\overset{\circ}{\overline{D'}} \right)$. Let $\xi = \gamma^{-1}$. Then $\xi(u) = a + \dfrac{\lambda}{u - b}$, and then what is
true for γ is also true for ξ. Hence reciprocally, if $\gamma(\alpha) \in \left(\overset{\circ}{\overline{D'}} \right)$ we see that
$\alpha = \xi(\gamma(\alpha)) \in \overset{\circ}{\overline{D'}}$ because $D = \xi(D')$.

We now suppose D is unbounded. If $K \setminus \overline{D}$ is bounded then $\overline{D}$ contains a set
E of the form $\{|x| \ |x - a| \geq s\}$ with $s > |a - b|$ whose image E' is $d(a, \frac{|\lambda|}{s}) \setminus \{a\}$,
hence $\overline{D'}$ contains $d(a, \frac{|\lambda|}{s})$ and so does $\overset{\circ}{\overline{D'}}$.

Finally we suppose that $K \setminus \overline{D}$ is not bounded. Then $\gamma((K \setminus \{a\}) \setminus \overline{D})$ is an open set Ω in K whose closure contains b. Since $b \in \overline{D'} \setminus D'$, it is easily seen that there is a sequence of holes of $D', (T_n)_{n \in \mathbb{N}}$, that approaches b, (each one is obviously included in Ω) hence $b \notin \left(\overset{\circ}{\overline{D'}} \right)$. This ends the proof of Lemma 18.5.

Corollary 18.6: *Let D have a hole $T = d(a, r^-)$, satisfy $\overline{D} \setminus D \subset \left(\overset{\circ}{\overline{D}} \right)$ and be such that $K \setminus \overline{D}$ is bounded. Let $\gamma = \dfrac{1}{x-a}$ and let $D' = \gamma(D)$. Then $\overline{D'} \setminus D' \subset \left(\overset{\circ}{\overline{D'}} \right)$.*

Lemma 18.7: *The following two conditions are equivalent:*

   A) $\widetilde{D} \setminus \overline{D}$ *is bounded*

   A')   *either D is bounded or $K \setminus \overline{D}$ is bounded.*

Proof: If D is bounded, A) and A') are clearly satisfied hence we have nothing to show. Now we suppose D to be unbounded. Hence $\widetilde{D} = K$ and then $\widetilde{D} \setminus \overline{D} = K \setminus \overline{D}$, so $\widetilde{D} \setminus \overline{D}$ is bounded if and only if $K \setminus \overline{D}$ is bounded. Finally A) and A') are equivalent.

Theorem 18.8: *D belongs to $\mathcal{A}$ if and only if it satisfies the following two conditions.*

   A) $\widetilde{D} \setminus \overline{D}$ *is bounded*

   B) $\overline{D} \setminus D \subset \left(\overset{\circ}{\overline{D}} \right)$.

Proof: By Lemma 18.7, proving Theorem 18.8 consists of showing that D satisfies A') and B) . By Lemma 18.3, if D is bounded and satisfies B) we know that $D \in \mathcal{A}$. Now we suppose $K \setminus \overline{D}$ bounded. We may obviously assume D to have a hole $T = d(a, r^-)$ because if D has no hole then by Corollary 10.8, we have $H(D) = R(D)$. Let $\gamma = \dfrac{1}{x-a}$ and let $D' = \gamma(D)$. The set D' is then bounded and by Corollary 18.6, D' satisfies $\overline{D'} \setminus D' \subset \left(\overset{\circ}{\overline{D'}} \right)$, and hence $D' \in \mathcal{A}$, hence by Proposition 18.3, D belongs to $\mathcal{A}$.

We now suppose that B) is not satisfied and will prove that $D \notin \mathcal{A}$. Indeed let $\alpha \in (\overline{D} \setminus D) \setminus \left(\overset{\circ}{\overline{D}} \right)$. By definitions D has a Cauchy pierced filter $\mathcal{F}$ that converges in K to α. Let $T = d(a, r^-)$ be a hole of D and let $f = \dfrac{x - \alpha}{x - a}$. Then

$f \in R_b(D)$. By Lemma 18.4 there exists $S \in H_b(D)$ such that $S(\alpha) = 0$ together with a sequence $(z_n))_{n \in \mathbb{N}}$ in D such that $\lim\limits_{n \to \infty} z_n = \alpha$, while $\lim\limits_{n \to \infty} \left|\dfrac{S(z_n)}{f(z_n)}\right| = +\infty$.

Let us assume $D \in \mathcal{A}$. Then $\dfrac{S(x)}{f(x)} \in H(D)$ because $\dfrac{x - a}{x - \alpha} \in R(D)$. Since $\left|\dfrac{S(x)}{f(x)}\right|$ is not bounded in any neighbourhood of α, by Theorem 10.10 and Corollary 10.9 there exists an integer $n \geq 1$ such that $(x - \alpha)^n \dfrac{S(x)}{f(x)}$ has a non zero limit ℓ at α. But when $x \in D \cap d(\alpha, |a|)$, we have $\left|(x - \alpha)^n \left(\dfrac{S(x)}{f(x)}\right)\right| = \left|(x - \alpha)^{n-1}(x - a)S(x)\right| = |a| \, |x - \alpha|^{n-1} \, |S(x)|$ hence $\ell = 0$. Thus $D \notin \mathcal{A}$.

Now if D satisfies B) but does not satisfy A) we see that both D and $K \setminus \overline{D}$ are unbounded. Since $K \setminus \overline{D}$ is unbounded, D has a hole $T = d(\alpha, r^-)$ and then by Lemma 18.5, the inversion $\gamma(x) = \dfrac{1}{x - \alpha}$ maps D onto a bounded set D' such that $\alpha \in (\overline{D'} \setminus D') \subset \left(\overset{\circ}{\overline{D'}}\right)$. Hence D' does not belong to $\mathcal{A}$ and then neither does D. This ends the proof of Theorem 18.8 .

Lemma 18.9:   *Let D be open. Then D satisfies Condition B if and only if $\overline{D}$ is open.*

Proof: Since D is open $\overline{D}$ is open if and only if for every $\alpha \in \overline{D} \setminus D$, α is interior to D. This is just equivalent to Condition B).

Notations: Henceforth, Conditions A) and B) will always keep the same meaning as in Theorem 18.8.

Theorem 18.10:   *Let $D \in \mathcal{A}$, and let $a \in D$. If a belongs to $\overset{\circ}{D}$, then $\mathcal{I}(a) = (x - a)H(D)$. Else, $\mathcal{I}(a)$ is not of finite type.*

Proof: First we suppose $a \in \overset{\circ}{D}$. By Theorem 14.1 it is clearly seen that $\mathcal{I}(a) = (x - a)H(D)$. Now let $a \notin \overset{\circ}{D}$. Then the filter of the neighbourhoods of a is a Cauchy pierced filter. We will denote it by $\mathcal{F}$. Suppose that $\mathcal{I}(a)$ is of finite type and let $\{g_1, ..., g_q\}$ be a system of generators. By Theorem 10.11 for each $j = 1, ..., q$, there exists $h_j \in R(D)$, invertible in $R(D)$, such that $h_j g_j$ belongs to $H_b(D)$, and thereby we may clearly assume the g_j to be bounded without loss of generality. Of course, at least one of the g_j is properly vanishing along $\mathcal{F}$, otherwise all the elements of $\mathcal{I}(a)$ would be equal to 0 inside a neighbourhood of a, and then $\mathcal{I}(a)$ couldn't contain $x - a$. For every $x \in D$, we put $S(x) = \max_{1 \le j \le q} |g_j(x)|$. Since $\mathcal{I}(a)$ is closed, by Lemma 18.4, there exists $f \in \mathcal{I}(a)$ together with a sequence $(z_n)_{n \in \mathbb{N}}$ in D, of limit a, such that $S(z_n) \ne 0$ whenever $n \in \mathbb{N}$ and that $\lim_{n \to \infty} \dfrac{|f(z_n)|}{S(z_n)} = +\infty$. This obviously contradicts the fact that f should be of the form $\sum_{j=1}^{q} f_j g_j$ with the f_j in $H(D)$. Thus we have shown that $\mathcal{I}(a)$ is not of finite type, and this ends the proof of Theorem 18.10.

Corollary 18.11: *Let $D \in \mathcal{A}$. If $D \setminus \overset{\circ}{\overline{D}} \ne \emptyset$ then $H(D)$ is not noetherian.*

19.   DERIVATIVE OF ANALYTIC ELEMENTS

The main question we consider here is whether an element f of $H(D)$ has a derivative that belongs to $H(D)$, and when it does, whether its Mittag-Leffler series is obtained by deriving that of f [23].

In all this chapter D is supposed to be open and infraconnected.

Theorem 19.1:   *Let $f \in H_b(D)$, let $\rho = \delta(D,(K \setminus \overline{D}))$. If $\rho > 0$ then f' belongs to $H_b(D)$ and satisfies $\|f'\|_D \leq \dfrac{1}{\rho}\|f\|_D$.*

Proof: Let $(T_n)_{n \in S}$ be the sequence of the f-holes and let $D' = \tilde{D} \setminus \left(\bigcup_{n \in S} T_n \right)$.

By Theorem 15.1 we know that $f \in H_b(D')$ and that (1) $\|f\|_{D'} = \|f\|_D$. By Corollary 13.7, f has a derivative f' in D' and we will first show that the function f' satisfies $\|f'\|_{D'} \leq \dfrac{1}{\rho} \|f\|$. Let $a \in D'$. The disk $d(a,\rho^-)$ is obviously included in D because if a point $b \in d(a,\rho^-)$ belonged $K \setminus D'$, since D is closed there would be a disk $d(b,r^-) \subset K \setminus D'$ with $r < \rho$. Thus, when $x \in d(a,\rho)$, by Corollary 13.2 $f(x)$ is of the form $\sum_{n=0}^{\infty} a_n(x-a)^n$ and by Theorem 13.5 $f'(x)$ is of the

form $\sum_{n=1}^{\infty} na_n(x-a)^{n-1}$ hence $\|f'\|_{d(a,\rho)} = \sup_{n \in \mathbb{N}^*} |na|\rho^{n-1} \leq \left(\dfrac{1}{\rho}\right) \sup_{n \in \mathbb{N}^*} |a_n|\rho^n \leq$

$\dfrac{1}{\rho} \|f\|_{d(a,\rho^-)} \leq \dfrac{1}{\rho} \|f\|_D$ and therefore by (1) we see that $|f'(x)| \leq \dfrac{1}{\rho} \|f\|_D$.

Thus the mapping $\phi : f \to f'$ from $H(D')$ to the algebra $\mathcal{B}(D')$ of the functions bounded in D' is continuous with respect to the norm $\| \cdot \|_{D'}$. Now, obviously we have $\phi(R(D')) \subset R(D')$. Since $H_b(D')$ is closed in $\mathcal{B}(D')$ we see that $\phi(H_b(D')) \subset H_b(D') \subset H_b(D)$ and this finishes the proof of Theorem 19. 1.

Corollary 19.2:   *Let $f \in H(D)$ be such that the set of the diameters of the f-holes has a strictly positive lower bound. Then $f' \in H(D)$.*

Proof: By Theorem 9.5 we know that f is in the form $g + h$ with $h \in R(D)$ and $g \in H_b(\overline{D})$. Obviously h' belongs to $R(D)$ and by Theorem 19.1, we have $g' \in H_b(\overline{D})$.

Corollary 19.3:   *If $\delta(D, K \setminus \overline{D}) > 0$ then for every $f \in H_b(D), f'$ belongs to $H_b(D)$ and $\|f'\|_D \leq \dfrac{\|f\|_D}{\lambda}$.*

Theorem 19.4:   *Let D be closed and open, satisfying Condition A). Let $f \in H_b(D)$, let $(V_n)_{n \in \mathbb{N}^*}$ be the set of the f-holes, and let $\displaystyle\sum_{n=0}^{\infty} f_n$ be its Mittag-Leffler series on D defined as $f_0 = \overline{f}_0$ and for every $n \in \mathbb{N}^*$, $f_n = \overline{f}_{V_n}$. The following three conditions are equivalent*

a) f' belongs to $H(D)$

b) the series $\displaystyle\sum_{n=0}^{\infty} f'_n$ converges in $H(D)$

c) the series $\displaystyle\sum_{n=0}^{\infty} f'_n$ converges to f' in $H_b(D)$.

Proof: We will first prove the equivalence between b) and c). For each $q \in \mathbb{N}$, the sum $\displaystyle\sum_{n=0}^{q} f'_n$ clearly belongs to $H_b(\overline{D})$. Thus we assume this series $\displaystyle\sum_{n=0}^{\infty} f'_n$ converges to an element $h \in H(D)$. Let $\alpha \in D$. There exists a disk $d(\alpha, r) \subset \overline{D}$. We will show that $h(\alpha) = f'(\alpha)$. For every $\psi \in H(\overline{D})$, $\widehat{\psi}$ will denote the restriction of ψ to $d(\alpha, r)$. Since the sequence $\left(\widehat{\displaystyle\sum_{j=0}^{n} f_j} \right)$ converges to $\widehat{f}$, by Theorem 19.1 the sequence of the derivatives $\left(\widehat{\displaystyle\sum_{j=0}^{n} f'_j} \right)$ does converge to $\widehat{f'}$, hence it is clearly seen that $f'(\alpha) = h(\alpha)$, and therefore $f' = h$. We will check that f' is bounded in D. The sequence $\|f'_n\|_D$ has limit 0, hence is obviously bounded and therefore its sum f' is bounded in D. Thus b) and c) are equivalent.

Since c) trivially implies a) we just have to prove that a) implies b). Thus we suppose a) to be true, and will prove b).

First, we suppose D bounded. For each hole T of D that is either a f-hole or a f'-hole, we denote by $\overline{f}_T$ (resp. $\overline{g}_T$) the Mittag-Leffler term of f (resp. f'). Let $\mathcal{G}$ be the set of the holes T such that $(\overline{f}_T)' \neq \overline{g}_T$ and let $\mathcal{F}$ be the set of the f-holes such that $(\overline{f}_T)' = \overline{g}_T$. If we can show that $\mathcal{G} = \emptyset$, then b) is clearly proven.

Hence we suppose $\mathcal{G} \neq \emptyset$. All the $\overline{g}_T$ are equal to zero except maybe a countable family of them. The series $\sum\limits_{T \in \mathcal{F}} \overline{g}_T$ and $\sum\limits_{T \in \mathcal{G}} \overline{g}_T$ obviously converge in $H(D)$, and then we have $f' = \sum\limits_{T \in \mathcal{F}} (\overline{f}_T)' + \sum\limits_{T \in \mathcal{G}} \overline{g}_T$. Since b) implies c), the series $\sum\limits_{T \in \mathcal{F}} (\overline{f}_T)'$ is clearly equal to the derivative of $\sum\limits_{T \in \mathcal{F}} \overline{f}_T$. Let $h = \sum\limits_{T \in \mathcal{G}} \overline{f}_T = f - \sum\limits_{T \in \mathcal{F}} \overline{f}_T$. Then $h' = f' - \sum\limits_{T \in \mathcal{F}} (\overline{f}_T)' = \sum\limits_{T \in \mathcal{G}} \overline{g}_T$. Let $\mathcal{D}$ be the family of the diameters of the holes T that belong to $\mathcal{G}$, and let λ be its lower bound. Suppose $\lambda > 0$. By Theorem 19.1, the series $\sum\limits_{T \in \mathcal{G}} (\overline{f}_T)'$ converges to h', hence $\sum\limits_{T \in \mathcal{G}} (\overline{f}_T)'$ is the Mittag-Leffler series of h' on D, hence $(\overline{f}_T)' = \overline{g}_T$ for all $T \in \mathcal{G}$ and that contradicts the definition of $\mathcal{G}$. Hence $\lambda = 0$.

Now, we will prove that there exists a hole $V = d(a, r^-) \in \mathcal{G}$ with an annulus $\Gamma(a, r, s)$ such that the set $\mathcal{S}$ of the diameters ρ of the f-holes included in $\Gamma(a, r, s)$ has a strictly positive lower bound. Indeed, suppose such a hole V does not exist. Then we can easily construct a sequence of f-holes $(T_n)_{n \in \mathbb{N}^*}$ of the form $T_n = d(a_n, r_n^-)$ with (1) $r_n \leq \dfrac{1}{n}$ and (2) $|a_{n+1} - a_n| \leq \dfrac{2}{n}$. For example, asssume the sequence has just been constructed up to the rank q, satisfying (1) and (2) for $n \leq q$. Since V does not exist, then in $\Gamma(a_q, r_q, \dfrac{2}{q})$ we can find a f-hole $T_{q+1} = d(a_{q+1}, r_{q+1}^-)$ with $r_{q+1} < \dfrac{2}{q+1}$ and then the sequence is clearly constructed by induction by taking first any f-hole $T_1 = d(a_1, r_1^-)$. The sequence $(T_n)_{n \in \mathbb{N}^*}$ clearly converges to a point $w \in \overline{D}$ and that contradicts the hypothesis "D is closed and open". Hence we have now proven the existence of the f-hole V with an annulus $\Gamma(a, r, s)$ and a number $\nu > 0$ such that every f-hole $T \subset \Gamma(a, r, s,)$ satisfies (3) diam $(T) \geq \nu$. Let $\mathcal{L}$ be this family of the f-holes included in $\Gamma(a, r, s)$. Let $l = \sum\limits_{T \in \mathcal{L}} \overline{f}_T$. By Theorem 19.1 the series $\sum\limits_{T \in \mathcal{L}} (\overline{f}_T)'$ converges to l' in $H(D)$. Now let $\psi = h - l - \overline{f}_V$. Clearly ψ belongs to $H(D)$ and no hole T (of D) included in $d(a, s)$ is a ψ-hole. Hence ψ extends to an element of $H(D \cup d(a, s))$. In $d(a, s), \psi(x)$ is equal to a power series $\phi(x) \in H(d(a, s))$, hence $\phi' \in H(d(a, s))$. Thus in $D \cap d(a, s)$, $\psi'(x)$ is equal to the series $\phi'(x)$ and then for every hole T of D included in $d(a, s)$ the Mittag-Leffler term of ψ associated to T (with respect to D) is zero.

On the other hand, we have $\psi' = h' - l' - (\overline{f}_V)' = \sum\limits_{T \in \mathcal{G}} \overline{g}_T - \sum\limits_{T \in \mathcal{L}} (\overline{f}_T)' - (\overline{f}_V)'$

and then the Mittag-Leffler term of ψ' associated to V (with respect to D) is $\overline{g}_V - (\overline{f}_V)' \neq 0$. Hence we have a contradiction with $\psi \in H(d(a, s))$. This finishes proving that b) is true when D is bounded.

Now, we suppose D unbounded. By Condition A), there exists a disk $d(0, S)$ such that all the holes of D are included in this disk. Then for every element $h \in H(D)$, its Mittag-Leffler series in $H(D)$ is the same as in $H(D')$. This is true for both f, f', and therefore b), which is true in $H(D')$, is obviously true in $H(D)$. This ends the proof of Theorem 19.4.

When K has characteristic zero, in most of the cases, we are now able to answer the question "does $f' = 0$ implies $f = ct$". When D is not infraconnected, it admits an empty annulus $\Lambda = \Gamma(a, r', r'')$ and by Proposition 10.14 we know that there exists $w \in H(D)$ such that $w(x) = 1$ whenever $x \in \mathcal{I}(\Lambda)$ while $w(x) = 0$ whenever $x \in \mathcal{E}(\Lambda)$. Thus the condition "$D$ is infraconnected" is certainly necessary to be able to answer "yes" the question above.

The two Theorems that follow show this condition to be sufficient too, provided D satisfies a little extra condition like to be closed or to belong to $\mathcal{A}$.

Theorem 19.5:   *K is supposed to have characteristic zero. Let E be an open set in K such that $\overline{E}$ also is open. Then E is infraconnected if and only if for every $f \in H(E)$ such that $f'(x) = 0$ whenever $x \in E$, we have $f = ct$.*

Proof: If E is not infraconnected it admits at least an empty annulus Λ and then by Proposition 10.14, the characteristic function u of $\mathcal{I}(\Lambda)$ belongs to $H(E)$. Hence there do exist non constant elements $f \in H(E)$ whose derivative is identically 0. Now let E be infraconnected, and let $f \in H(E)$ satisfy $f'(x) = 0$ whenever $x \in E$. We just have to prove that f is a constant.

First we assume that E is bounded. By Theorem 10.5, f is in the form $f^* + \overline{f}$ with $f^* \in R_0(K \setminus (\overline{E} \setminus E))$ and $\overline{f} \in H(\overline{E})$, so $f^{*\prime}(x) + \overline{f}'(x) = 0$ whenever $x \in E$, and therefore $f^{*\prime} = -\overline{f}'$. Hence $\overline{f}' \in R_0(K \setminus (\overline{E} \setminus E)) \cap H(\overline{E})$. Thus f^* is a rational function that has no pole in K and then it is a polynomial. Besides, as an element of $R_0(K \setminus (\overline{E} \setminus E))$ it tends to 0 when $|x|$ goes to $+\infty$, hence $f^* = 0$ and therefore $f^{*\prime}$ is identically 0. Since f^* belongs to $R_0(K \setminus (\overline{E} \setminus E))$ clearly, $f^* = 0$, and therefore f belongs to $H(\overline{E})$.

Let $a \in E$, and let $\displaystyle\sum_{n \in \mathcal{S}} h_n$ be the Mittag-Leffler series of f in $\overline{E}$, with $h_0 = \overline{f}_0$, and for each $n \in \mathcal{S}$, $h_n = \overline{f}_{T_n}$, for any f-hole T_n. Since $\overline{E}$ is open, we can apply Theorem 19.4 to E, and then we have $(h_n)' = 0$ for every $n \in \mathbb{N}$. Since $\overline{E}$ is bounded of diameter r, then by Theorem 13.12 we know that h_0 is a constant. In the same way, by Corollary 13.13, for each $q \in \mathbb{N}$, we know that $h_q = 0$. Hence f is a constant.

Finally let E be unbounded. Then for all $r > 0$ the set $E_r = E \cap d(0,r)$ is such that $\overline{E_r}$ is open hence f is constant in E_r, and therefore in all of E.

Remark: In particular, Theorem 19.5 applies to open closed sets.

Corollary 19.6:   *K is supposed to have characteristic zero. Let E be open and belong to $\mathcal{A}$. Then E is infraconnected if and only if for every $f \in H(E)$ such that $f'(x) = 0$ whenever $x \in E$, f is a constant in E.*

Proof: Indeed, since E belongs to $\mathcal{A}$, it satisfies Condition B) in Theorem 18.8: $\overline{D} \setminus D \subset \left(\overset{\circ}{\overline{D}} \right)$. But then, as it is also open, we check that $\overline{E}$ is open.

We will now study thoroughly the question whether all the analytic elements in a set D have derivative in $H(D)$.

Definition: A set D will be said to be *well pierced* if $\delta(\overline{D}, K \setminus D) > 0$.

Theorem 19.7:   *Let $\overline{D}$ be open . Then D is well pierced if and only if for every $f \in H(D)$, f' also belongs to $H(D)$.*

Proof: If D is well pierced, by Corollary 19.3 we know that for every $f \in H(D)$, f' belongs to $H(D)$. Now let us suppose D has piercing zero and let $T_n = d(\alpha_n, \rho_n^-)$ be a sequence of holes such that $\lim_{n \to \infty} \rho_n = 0$. Let λ_n be a sequence in K such that $\lim_{n \to \infty} \dfrac{|\lambda_n|}{\rho_n} = 0$ while $\lim_{n \to \infty} \dfrac{|\lambda_n|}{\rho_n^2} = +\infty$. It is seen that the series $\displaystyle\sum_{n=1}^{\infty} \dfrac{\lambda_n}{x - \alpha_n}$ converges in $H(D)$ to an element f, while the series

$\displaystyle\sum_{n=1}^{\infty} \dfrac{\lambda_n}{(x - \alpha_n)^2}$ does not. Besides, $\displaystyle\sum_{n=1}^{\infty} \dfrac{\lambda_n}{x - \alpha_n}$ obviously is the Mittag-Leffler series of f. If f' belongs to $H(D)$, by Theorem 19.4 its Mittag-Leffler series must be $\displaystyle\sum_{n=1}^{\infty} \dfrac{\lambda_n}{(x - \alpha_n)^2}$. Since this series does not converge this is just impossible.

Before closing this chapter we will notice the following result that may be sometimes helpful in differential equations.

Theorem 19.8:   *Let $\overline{D}$ be open. We suppose that both f and f' belong to $H(D)$. For every $\epsilon > 0$ there exists $h \in R(D)$ such that $\|f - h\|_D \leq \epsilon$ together with $\|f' - h'\|_D \leq \epsilon$.*

Proof: First we suppose that f belongs to $H_b(\overline{D})$. We have to introduce a few notations. Let $g \in H_b(\widetilde{D})$. If D is bounded, $\widetilde{D}$ is a disk $d(a,r)$ and g is of the form $\sum_{m=0}^{\infty} \lambda_m (x-a)^m$. Then for every $q \in \mathbb{N}$ we put $(g)_q = \sum_{m=0}^{q} \lambda_m (x-a)^m$. If D is unbounded, then g is a constant λ_o and we put $(g)_q = g$ whenever $q \in \mathbb{N}$.

Let $T = d(b, r^-)$ be a hole of D and let $l(x) = \sum_{m=1}^{\infty} \dfrac{\mu_m}{(x-b)^m}$. For each $q \in \mathbb{N}^*$, we put $(l)_q = \sum_{m=1}^{q} \dfrac{\mu_m}{(x-b)^m}$. Now let $\sum_{n=0}^{\infty} f_n$ be the Mittag-Leffler series of f, with $f_0 = \overline{f_0}$, and for each $n \in \mathbb{N}^*$, $f_n = \overline{f_{T_n}}$, for any f-hole T_n. By Theorem 19.4 , the Mittag-Leffler series of f' is $\sum_{n=0}^{\infty} f'_n$ and therefore there exists an integer $N(\epsilon)$ such that

$$(1) \quad \left\| \sum_{n=0}^{N(\epsilon)} f_n - f \right\|_D \leq \epsilon \text{ and } (2) \quad \left\| \sum_{n=0}^{N(\epsilon)} f'_n - f' \right\|_D \leq \epsilon.$$

Obviously we have an integer $Q(\epsilon)$ such that $\|f_n - (f_n)_{Q(\epsilon)}\|_D \leq \epsilon$ whenever $n = 0, \ldots, N(\epsilon)$ and then by (1) and (2) it is easily seen that $\left\| \sum_{n=0}^{N(\epsilon)} (f_n)_{Q(\epsilon)} - f \right\|_D \leq \epsilon$ and $\left\| \sum_{n=0}^{\infty} (f'_n)_{Q(\epsilon)} - f' \right\|_D \leq \epsilon$. By putting $h = \sum_{n=0}^{N(\epsilon)} (f_n)_{Q(\epsilon)}$ we obtain the $h \in R(D)$ we want.

Now we consider the general case. By Corollary 10.7, f is of the form $l + \psi$, with $l \in H_b(\overline{D})$ and $h \in R(D)$. Hence $f' = l' + \psi'$. Since f' belongs to $H(D)$, so does ψ'. Hence by Theorem 19.5, ψ' belongs to $H_b(\overline{D})$. We have just proven that there exists $t \in H_b(\overline{D})$ such that $\|l - t\|_D \leq \epsilon$ and $\|l' - t'\|_D \leq \epsilon$. Hence we just have to consider $h = t + \psi$ and this ends the proof.

In the case when K has a characteristic $p \neq 0$, we have Theorem 19.9.

Theorem 19.9:   *Let K have characteristic $p \neq 0$, let D be closed and let $f \in H_b(D)$. Then $f'(x)$ is identically 0 if and only if there exists $g \in H_b(D)$ such that $f = g^p$.*

Proof: Indeed, if there exists $g \in H_b(D)$ such that $f = g^p$, of course we have $f' = 0$. Now, suppose that $f'(x)$ is identically 0. Let $a \in D$, and let $\displaystyle\sum_{n \in \mathcal{S}} h_n$ be the Mittag-Leffler series of f in D, with $h_0 = \overline{f_0}$, and for each $n \in \mathcal{S}$, $h_n = \overline{f_{T_n}}$, for any f-hole T_n. By Theorem 19.4 we have $(h_n)' = 0$ for every $n \in \mathbb{N}$.

If D is unbounded, h_0 is a constant and then we can find $g_0 \in K$ such that $(g_0)^p = h_0$. If D is bounded of diameter r, then by Theorem 13.12 we can find $g_0 \in H(d(a, r))$ such that $(g_0)^p = h_0$. In the same way, for each $q \in \mathbb{N}$, by Corollary 13.13 we can find $g_q \in H_0(K \setminus T_q)$ such that $(g_q)^p = h_q$ and then, it is seen that $\displaystyle\lim_{n \to \infty} \|g_n\|_D = 0$ because for each $n \in \mathbb{N}^*$, we have $(\|g_n\|_D)^p = \|h_n\|_D$.

So, the series $\left(\displaystyle\sum_{n=0}^{\infty} g_n\right)$ converges in $H_b(D)$ to an element g which clearly satisfies $g^p = f$. This ends the proof of Theorem 19.9.

20. VALUATION FUNCTIONS

FOR ANALYTIC ELEMENTS

In this chapter D is infraconnected.

Notations: For every $a \in \widetilde{D}$, we put $\lambda(a) = -\log \delta(a, D)$ if $\delta(a, D) > 0$ and $\lambda(a) = +\infty$ if $\delta(a, D) = 0$. S denotes the diameter of D, with $S = +\infty$ if D is not bounded.

Let $a \in \widetilde{D}$, and let $\mathcal{F}$ be a circular filter of center a and diameter $r \in [\delta(a, D), S] \cap \mathbb{R}$. By Proposition 3.14, $\mathcal{F}$ is secant with D, and then defines an element $_D\varphi_{\mathcal{F}}$ of $Mult(H(D), \mathcal{U}_D)$.

For every $f \in H(D)$ such that $_D\varphi_{\mathcal{F}}(f) \neq 0$ we put $v_a(f, -\log r) = -\log\left(_D\varphi_{\mathcal{F}}(f)\right)$. Now for an $f \in H(D)$ such that $_D\varphi_{\mathcal{F}}(f) = 0$ we put $v_a(f, -\log r) = +\infty$. When $a = 0$ we just put $v(f, \mu) = v_0(f, \mu)$.

Remark: Let $f \in H(C(0, r))$ for some $r > 0$. By Theorem 15.7 we have
$$v(f, -\log r) = -\log\left(_{C(0,r)}\varphi_{0,r}(f)\right) = -\log \|f\|_{C(0,r)} = \inf_{n \in \mathbb{Z}} v(a_n) - n \log r.$$

Propositions 20.1 is immediate.

Proposition 20.1: *Let $a \in \widetilde{D}$, let $\mu \in [-\log S, \lambda(a)] \cap \mathbb{R}$ and let $f, g \in H(D)$. Then $v_a(f + g, \mu) \geq (v_a(f, \mu), v_a(g, \mu))$ and when $v_a(f, \mu) < v_a(g, \mu)$ then $v_a(f + g, \mu) = v_a(f, \mu)$. Besides $v_a(fg, \mu) = v_a(f, \mu) + v_a(g, \mu)$.*

By Lemma 4.10 and Lemma 12.4 we have Proposition 20.2.

Proposition 20.2: *Let $a \in \widetilde{D}$ and let $[\mu_1, \mu_2] \subset [-\log S, \lambda(a)] \cap \mathbb{R}$. Let $f \in H(D)$ be such that $v_a(f, \mu)$ is bounded in $[\mu_1, \mu_2]$. Then there exists $h \in R(D)$ such that $v_a(h, \mu) = v_a(f, \mu)$ whenever $\mu \in [\mu_1, \mu_2]$. If $v(a - b) \geq \mu$ then $v_a(f, \mu) = v_b(f, \mu)$.*
The function $v_a(f, \mu)$ is continuous and piecewise linear in $[\mu_1, \mu_2]$ and the equality $v_a(f(x)) = v_a(f, v(x - a))$ is satisfied in all of $D \cap \Delta(a, w^{-\mu_2}, w^{-\mu_1})$ except maybe in finitely many classes of finitely many circles $C(a, r)$ with $w^{-\mu_2} \leq r \leq w^{-\mu_1}$.

Let $b \in D$ be such that $|a - b| = r$ and $d(b, r^-) \subset D$, with $w^{-\mu_2} \le r \le w^{-\mu_1}$. Then we have $v(f(x)) \ge v_a(f, -\log r)$ for all $x \in d(b, r^-)$.

Proof: Let $M = \sup\limits_{\mu_1 \le \mu \le \mu_2} v_a(f, \mu)$ and let $h \in R(D)$ satisfy (1) $\|h - f\|_D < w^{-M}$. By Lemma 12.4 we see that (2) $v_a(h, \mu) = v_a(f, \mu)$ whenever $\mu \in [\mu_1, \mu_2]$. Further by Lemma 4.10 the relation $v_a(f, \mu) = v_b(f, \mu)$ when $v(a - b) \ge \mu$ is true for every $f \in R(D)$, hence by (2), is obviously generalized to every $f \in H(D)$.

The function $\mu \to v_a(f, \mu)$ is then continuous and piecewise linear. Now let $x \in D \cap \Delta(a, w^{-\mu_2}, w^{-\mu_1})$ be such that (1) $v(h(x)) = v_a(h, v(x - a))$. Then $v(f(x) - h(x)) \ge -\log \|f - h\|_D > v_a(f, v(x - a)) = v_a(h, v(x - a))$ hence $v(f(x) - h(x)) > v(h(x))$ and therefore $v(f(x)) = v(h(x))$. Hence finally $v(f(x)) = v_a(f, v(x - a))$. By Lemma 4.8 , h satisfies (1) in all of $D \cap \Delta(a, w^{-\mu_2}, w^{-\mu_1})$ except in finitely many classes of finitely many circles $C(a, r)$ with $w^{-\mu_2} \le r \le w^{-\mu_1}$, hence this remains true for f in the same way.

Now, let $h = \dfrac{P}{Q}$ with $P, Q \in K[x]$, and P prime to Q. Since $d(b, r^-)$ is included in D , Q has no zero in $d(b, r^-)$. Hence by Lemma 4.7 in $d(b, r^-)$ we have $v(Q(x)) = v_a(Q, -\log r)$ while P obviously satisfies $v(P(x)) \ge v_a(P, -\log r)$, hence $v(h(x) \ge v_a(h, \log r)$. But by (1) we see that $v(f(x) - h(x)) > v_a(f, -\log r) = v_a(h, -\log r) \le v(h(x))$. Hence finally we have $v(f(x)) \ge v_a(f, -\log r)$ and this ends the proof of Proposition 20.2.

Proposition 20.3: *Let $a \in \widetilde{D}$ and let $f \in H(D)$. The function $v_a(f, .)$ is a continuous function from $[\lambda(a), -\log S] \cap \mathbb{R}$ to $\overline{\mathbb{R}}$. Let $r \in [\delta(a, D), S]$. If $_D\varphi_{a,r}(f) \ne 0$ the equality $|f(x)| = {_D\varphi_{a,r}(f)}$ holds in all the classes of $C(a, r)$ except maybe in finitely many ones.*

Proof: Let $\nu = -\log r \in [\lambda(a), -\log S] \cap \mathbb{R}$. If $v_a(f, \nu) < +\infty$ by Proposition 20.2, $v_a(f, .)$ is continuous at ν and we have $|f(x)| = {_D\varphi_{a,r}(f)}$ in all the classes of $C(a, r)$ except maybe in finitely many, just by taking $\mu_1 = \mu_2 = \nu$. Now suppose $v_a(f, \nu) = +\infty$. By definition of $_D\varphi_{a,r}$, we have $\lim\limits_{|x| \to r, |x - a| \ne r} |f(x)| = 0$ and therefore $\lim\limits_{s \to r, s \ne r} {_D\varphi_{a,s}} = {_D\varphi_{a,r}} = 0$ and finally $\lim\limits_{\mu \to \nu, \mu \ne \nu} v_a(f, \mu) = +\infty$. This ends the proof.

Proposition 20.4: *Let $a \in D$ and let $f \in H(D)$ satisfy $f(a) \ne 0$. There exists $\mu_o \in \mathbb{R}$ such that $v_a(f, \mu) = v(f(a))$ whenever $\mu \ge \mu_o$.*

Proof: Indeed as f is continuous, there exists $r > 0$ such that $|f(x) - f(a)| < |f(a)|$ whenever $x \in d(a, r) \cap D$ hence $|f(x)| = |f(a)|$ whenever $x \in d(a, r) \cap D$ and therefore $v_a(f, \mu) = v(f(a))$ whenever $\mu \ge -\log(r)$.

Lemma 20.5:   *Let $\mu \in \mathbb{R}$ and let $f(x) = \sum\limits_{-\infty}^{+\infty} a_n x^n \in H(C(0, \omega^{-\mu}))$. The set*

of the $j \in \mathbb{N}$ such that $\inf\limits_{n \in \mathbb{Z}} v(a_n) + n\mu = v(a_j) + j\mu$ is finite.

Proof: Indeed by hypothesis we have $\lim\limits_{|n| \to +\infty} v(a_n) + n\mu = +\infty$.

Notation: Let $f(x) = \sum\limits_{-\infty}^{+\infty} a_n x^n \in H(C(0, \omega^{-\mu}))$ $(\mu \in \mathbb{R})$. We will denote
by $N^+(f, \mu)$ (resp. $N^-(f, \mu)$) the biggest (resp. the smallest) integer $j \in \mathbb{Z}$
such that $v(a_j) + j\mu = \inf\limits_{n \in \mathbb{Z}} v(a_n) + n\mu$.

Theorem 20.6:   *Let $r \in \mathbb{R}_+^*$, let $\Lambda = C(0, r)$ and let f and $g \in H(\Lambda)$
satisfy $\|f - g\|_\Lambda < \|f\|_\Lambda$. Then we have $N^+(f, -\log r) = N^+(g, -\log r)$,
$N^-(f, -\log r) = N^-(g, -\log r)$.*

Proof: Let $f(x) = \sum\limits_{-\infty}^{+\infty} a_n x^n$ and let $g(x) = \sum\limits_{-\infty}^{+\infty} b_n x^n$. From the hypothesis
we see that $\|f\|_\Lambda = \|g\|_\Lambda$. By Theorem 15.6 we have (1) $\sup\limits_{n \in \mathbb{Z}} |a_n| r^n = \|f\|_\Lambda =$
$\sup\limits_{n \in \mathbb{Z}} |b_n| r^n$, and $\|f - g\|_\Lambda = \sup\limits_{n \in \mathbb{Z}} |a_n - b_n| r^n$. Let $s = N^-(f, -\log r)$ and let
$t = N^+(f, -\log r)$. We see that $|a_s - b_s| r^s \leq \|f - g\|_\Lambda < \|f\|_\Lambda = |a_s| r^s$ hence
(2) $|b_s| = |a_s|$. In the same way we have (3) $|a_t| = |b_t|$. Now for every $n < s$ and
for every $n > t$ we have $|a_n| r^n < |a_s| r^s = \|f\|_\Lambda$ hence $|b_n| r^n < \|f\|_\Lambda$. Finally
by (1), (2), (3) we see that $N^-(g, -\log r) = s$, $N^+(g, -\log r) = t$.

Proposition 20.7:   *Let $\mu \in \mathbb{R}$ and let $f = \sum\limits_{-\infty}^{+\infty} a_n x^n \in H(C(0, \omega^{-\mu}))$. Then*

*$v(f, \mu)$ is equal to $\inf\limits_{n \in \mathbb{Z}} v(a_n) + n\mu$ and we have $v(f(x)) \geq v(f, \mu)$ for all $x \in$
$C(0, \omega^{-\mu})$. Besides the equality holds in every classes except in finitely many
classes. Further, if $N^+(f, \mu) = N^-(f, \mu)$ then $v(f(x)) = v(f, \mu)$ whenever $x \in$
$C(0, \omega^{-\mu})$.*
*If $h \in H(C(0, \omega^{-\mu}))$ satisfies $v(f - h, \mu) > v(f, \mu)$ then $N^+(f, \mu) = N^+(h, \mu)$
and $N^-(f, \mu) = N^-(h, \mu)$.*

Proof: Let $\Lambda = C(0, \omega^{-\mu})$, let $s = N^-(f, \mu)$ and let $t = N^+(f, \mu)$. By the
Remark above $v(f, \mu)$ is obviously equal to $\inf\limits_{\mu \in \mathbb{Z}} (v(a_n) + n\mu)$. The inequality
$v(f(x)) \geq v(f, \mu)$ is true because $v(f, \mu) = -\log \|f\|_\Lambda \leq v(f(x))$. Finally by

Proposition 20.3 the equality holds in all the classes except in finitely many. If $N^+(f,\mu) = N^-(f,\mu)$ then $v(a_s x^s) = v(a_s) + s\mu < v(a_n x^n)$ whenever $n \neq s$ hence $v(f(x)) = v(f,\mu)$.

Now let $h \in H(\Lambda)$ satisfy $v(f - h, \mu) > v(f,\mu)$ and let $h(x) = \sum_{-\infty}^{+\infty} b_n x^n$. We have $v(a_n - b_n) + n\mu > v(a_s) + s\mu$ whenever $n \in \mathbb{Z}$ hence $v(b_s) = v(a_s)$, $v(b_t) = v(a_t)$, $v(b_n) + n\mu > v(a_s) + s\mu$ whenever $n < s$ and $n > t$, and $v(b_n) + n\mu \geq v(a_s) + s\mu$ whenever $n \in [s,t]$, hence finally $N^+(h,\mu) = N^+(f,\mu)$ and $N^-(h,\mu) = N^-(f,\mu)$.

By applying Proposition 20.2 to a Laurent series on an annulus we have the following Proposition 20.8 that generalizes Proposition 20.7.

Proposition 20.8:   *Let $f(x) \in H(\Gamma(0, r_1, r_2))$ (with $0 < r_1 < r_2$) and let $\sum_{-\infty}^{+\infty} a_n x^n$ be its Laurent series. The function $\mu \to v(f,\mu)$ is bounded in $]-\log r_2, -\log r_1[$ and equal to $\inf_{n \in \mathbb{Z}} (v(a_n) + n\mu)$. Besides we have $v(f(x)) \geq v(f, v(x))$ whenever $x \in \Gamma(0, r_1, r_2)$ and the equality holds in all of $\Gamma(0, r_1, r_2)$ except in finitely many classes of finitely many circles $C(0,r)$ $(r_1 \geq r \geq r_2)$. The left side derivative (resp. the right side derivative) of the function $v(f,.)$ at μ is equal to $N^+(f,\mu)$ (resp. to $N^-(f,\mu)$). Besides, if the function in μ $v(f,\mu)$ is not derivable at μ, then μ lies in $v(K)$.*

Moreover, the function $v(f,.)$ is concave in $]-\log r_2, -\log r_1[$. Further, the functions N^+ and N^- satisfy $N^+(fg,\mu) = N^+(f,\mu) + N^+(g,\mu)$, $N^-(fg,\mu) = N^-(f,\mu) + N^-(g,\mu)$.

Proof: Let $\Lambda = \Gamma(0, r_1, r_2)$, let $\mu_1 = -\log r_1$ and $\mu_2 = -\log r_2$. By Proposition 20.7 $v(f,\mu)$ is obviously upper bounded in $]\mu_2, \mu_1[$ hence it is piecewise linear. Now for every fixed $\mu_0 \in]\mu_2, \mu_1[$, $N^+(f,\mu_0)$ is clearly seen to be the left side derivative of the function $v(f,.)$ at μ_0 because there obviously exists $\nu < \mu_0$ such that $N^-(f,\mu) = N^+(f,\mu) = N^+(f,\mu_0)$ whenever $\mu \in]\nu, \mu_0[$. In the same way $N^-(f,\mu_0)$ is the right side derivative of $v(f,.)$ at μ_0.

If $f \in R(D)$, and if $v(f,\mu)$ is not derivable at μ, by Corollary 4.12 we know that μ lies in $v(K)$. Hence by Theorem 20.6 it is easily seen that this property holds for every $f \in H(D)$.

The inequality $v(f(x)) \geq v(f, v(x))$ comes from Proposition 20.7 and by Proposition 20.2 the equality is true in all of Λ except in finitely many classes of finitely many circles $C(0,r)$ $(r_1 < r < r_2)$ because $v(f,\mu)$ is bounded in $]\mu_2, \mu_2[$.

The functions in μ $N^+(f,\mu)$ and $N^-(f,\mu)$ are clearly decreasing because if $v(a_i) + i\nu \geq v(a_j) + j\nu$, when $i > j$, with greater reason we have $v(a_i) + i\mu \geq v(a_j) + j\mu$ whenever $\mu > \nu$. Hence the function $\mu \to v(f,\mu)$ is concave. Since $v(fg,\mu) = v(f,\mu) + v(g,\mu)$, the left side derivative $N^+(f,.)$ also satisfies $N^+(fg,\mu) = N^+(f,\mu) + N^+(g,\mu)$ and so does the right side derivative $N^-(f,.)$

Proposition 20.9   *Let $\mu \in \mathbb{R}$ and let $f, g \in H(C(0, \omega^{-\mu}))$.*
Then $N^+(fg,\mu) = N^+(f,\mu) + N^+(g,\mu)$ and $N^-(fg,\mu) = N^-(f,\mu) + N^-(g,\mu)$.
Proof: By Proposition 20.8 the relations are obvious when f and $g \in R(C(0,\omega^{-\mu}))$ because there is an annulus $\Gamma(0,r_1,r_2) \subset C(0,\omega^{-\mu})$ such that $f, g \in R(\Gamma(0,r_1,r_2))$. Now by Proposition 20.7, we may extend them to $H(C(0,\omega^{-\mu}))$ by taking h and $\ell \in R(C(0,\omega^{-\mu}))$ such that $v(f - h, \mu) > v(f,\mu)$ and $v(g - \ell, \mu) > v(g,\mu)$.

21. ELEMENTS VANISHING

ALONG A MONOTONOUS FILTER

Throughout this chapter, the set D is supposed to be infraconnected.

We remember that $\widehat{K}$ is a spherically complete algebraically closed extension of K whose residue class field is not countable and whose valuation group is equal to $\mathbb{R}$.

The question on whether an analytic element can tend to zero along a monotonous filter appears to be one of the main problems [11], [12], [14], [16], [17]. In Chapter 18 we defined elements vanishing along a filter, and elements properly vanishing along a filter. Here we need additional definitions.

Definitions: Let $f \in H(D)$ and let $\mathcal{F}$ be a monotonous filter on D.
When $\mathcal{F}$ is decreasing (resp. increasing) of center a and diameter S, f will be said to be *strictly vanishing* along $\mathcal{F}$ if $\lim_{\mathcal{F}} f(x) = 0$ and if there exists $S' > S$ (resp. $S' < S$) such that for every $r \in]S, S']$ (resp. $r \in [S', S[$) we have ${}_D\varphi_{a,r}(f) \neq 0$.
When $\mathcal{F}$ is decreasing with no center in K, it admits a canonical base $(D_n)_{n \in \mathbb{N}}$ with $D_n = d(a_n, r_n) \cap D$, and then f will be said to be *strictly vanishing* along $\mathcal{F}$ if $\lim_{\mathcal{F}} f(x) = 0$ and if there exists $S' > S$ such that ${}_D\varphi_{a_{n+1},r}(f) \neq 0$ whenever $r \in [r_n, S']$, whenever $n \in \mathbb{N}$. Actually $\mathcal{F}$ admits a center α in $\widehat{K}$ and then the definition given for decreasing filters with a center in K also applies and is obviously equivalent.
Sometimes we will also use more general definitions: f will be said to be *collapsing* (resp. *properly collapsing*, resp. *strictly collapsing*), *of limit ℓ, along $\mathcal{F}$* if $f - \ell$ is vanishing (resp. properly vanishing, resp. strictly vanishing) along $\mathcal{F}$.

Lemma 21.1 just translates these definitions into terms of valuation functions.

Lemma 21.1:   *Let $f \in H(D)$ and let $\mathcal{F}$ be a decreasing (resp. an increasing) filter of center a, of diameter S on D. Then f is strictly vanishing along $\mathcal{F}$ if and only if there exists $S' > S$ (resp. $S' \in]0, S[$) such that $v_a(f, -\log S) = +\infty, v_a(f, \mu) < +\infty$ whenever $\mu \in [-\log S', -\log S[$ (resp. $]-\log S, -\log S']$).*
Let $\mathcal{G}$ be a decreasing filter with no center, of diameter S and canonical base $(D_n)_{n \in \mathbb{N}}$, with $D_n = d(a_n, r_n) \cap D$. Then f is strictly vanishing along $\mathcal{G}$ if

and only if there exists $S' > S$ such that $\lim\limits_{n \to \infty} v_{a_n}(f, -\log r_n) = +\infty$ and $v_{a_n}(f, -\log r) < +\infty$ for $r_n \leq r < S'$.

Lemma 21.2:   Let $f \in H(D)$ and let $\mathcal{F}$ be a monotonous filter such that f is strictly vanishing along $\mathcal{F}$. Then it is properly vanishing along $\mathcal{F}$.

Proof: Let $S = diam(\mathcal{F})$. Let (D_n) be a canonical base of $\mathcal{F}$ and suppose that f is not properly vanishing along $\mathcal{F}$. Since f is vanishing along $\mathcal{F}$, there exists $q \in \mathbb{N}$ such that $f(x) = 0$ whenever $x \in D_q$. But then, in each case, it is checked that f does not satisfy the definition of an analytic element strictly vanishing along $\mathcal{F}$, because for every multiplicative semi-norm $_D\varphi_{a,r}$ whose circular filter is secant to D_q, we have $_D\varphi_{a,r}(f) = 0$. In particular, this applies to $_D\varphi_{a,r}$, for $r_q < r < S$, (resp. $S < r < r_q$) when $\mathcal{F}$ is increasing,(resp. decreasing) of center a, and $D_q = D \cap \Gamma(a, r_q, S)$, (resp. $D_q = D \cap \Gamma(a, S, r_q)$), and this applies to φ_{a_{q+1}, r_q}, when $\mathcal{F}$ has no center, whereas $D_q = D \cap d(a_{q+1}, r_q)$.

Proposition 21.3:   Let $a, b \in D$ and let $f \in H(D)$ satisfy $f(b) \neq 0$ and $_D\varphi_{a,r}(f) = 0$ for some $r \in]0, |a-b|]$. If $a = b$, then f is strictly vanishing along an increasing filter of center a and diameter $S \leq r$. If $_D\varphi_{a,|a-b|}(f) = 0$, then f is strictly vanishing along an increasing filter of center b and diameter $S \leq |a - b|$. If $_D\varphi_{a,|a-b|}(f) \neq 0$ then f is strictly vanishing along a decreasing filter of center a of diameter $S \in [r, |a-b|[$.

Proof: First we suppose $_D\varphi_{a,r}(f) = 0$, and $f(a) \neq 0$. Hence by Corollary 20.4 we know that $\lim\limits_{\mu \to +\infty} v_a(f, \mu) = v(a)$ hence there exists a unique $\gamma \geq -\log(r)$ such that $v_a(f, \gamma) = +\infty$ and $v_a(f, \mu) < +\infty$ whenever $\mu > \gamma$ hence f is strictly vanishing along the increasing filter of center a, of diameter $S = \omega^{-\gamma}$. Now we suppose $_D\varphi_{a,|a-b|}(f) = 0$, hence we have $v_b(f, -\log|a-b|) = v_a(f, -\log|a-b|) = +\infty$. Since $f(b) \neq 0$ by Corollary 20.4 we know that $\lim\limits_{\mu \to +\infty} v_b(f, \mu) = v(b)$ hence there exists a unique $\gamma \geq -\log|a-b|$ such that $v_b(f, \gamma) = +\infty$ and $v_b(f, \mu) < +\infty$ whenever $\mu > \gamma$ hence f is strictly vanishing along the increasing filter of center b and diameter $S = \omega^{-\gamma}$. Now we suppose $_D\varphi_{a,|a-b|}(f) \neq 0$. Since $_D\varphi_{a,r}(f) = 0$ we have $v_a(f, -\log r) = +\infty$ and $v_a(f, -\log|a-b|) < +\infty$ hence there exists a unique $\gamma \in]-\log|a-b|, -\log r]$ such that $v_a(f, \gamma) = +\infty$ and $v_a(f, \mu) < +\infty$ whenever $\mu < \gamma$, so f is strictly vanishing along the decreasing filter of center a, and diameter $S = \omega^{-\gamma}$.

Proposition 21.4:   Let $a, b \in D$ and let $f \in H(D)$ satisfy $f(b) \neq 0$ and $_D\varphi_{a,r}(f) = 0$ for some $r \in \mathbb{R}_+$. Then f is strictly vanishing along a monotonous filter with a center.

Proof: If $r \leq |a-b|$ the statement comes directly from Proposition 21.3. If $r > |a-b|$, then we have $v_b(f, -\log r) = v_a(f, -\log r) = +\infty$ while $\lim_{\mu \to \infty} v_b(f, \mu) = v(f(b))$ hence there exists a unique $\gamma \geq -\log r$ such that $v_b(f, \gamma) = +\infty$, and $v_b(f, \mu) < +\infty$ whenever $\mu > v_a$. Thus f is strictly vanishing along the increasing filter of center b, of diameter $S = \omega^{-\gamma}$.

Lemma 21.5:   *Let $\mathcal{F}$ be a monotonous filter on D and let $f \in H(D)$ be strictly vanishing along $\mathcal{F}$. Then $\mathcal{F}$ is pierced.*

Proof: If $\mathcal{F}$ is increasing (resp. decreasing) of center a and diameter S, and is not pierced, then there exists an annulus $\Gamma(a, S, S')$ (resp. $\Gamma(a, S', S)$) included in D such that $v_a(f, -\log S) = +\infty$ and $v_a(f, \mu) < +\infty$ whenever $\mu \in]-\log S, -\log S']$ (resp. $\mu \in [-\log S', -\log S[$). Hence by Proposition 20.9 we know that $v_a(f, \mu)$ is bounded in $]-\log S, -\log S'[$ (resp. $]\log S', -\log S[$), so this contradicts the hypothesis. When $\mathcal{F}$ has no center, we consider a center ω of $\mathcal{F}$ in $\widehat{K}$ and we consider f as an element of $\widehat{H}(\widehat{D})$. Then the disks $\widehat{d}(a_n, r_n)$ contain no hole of $\widehat{D}$ when n is big enough and therefore we have the same conclusion.

Proposition 21.6:   *Let $a \in \widetilde{D}$. Let $r, r' \in]-\log diam(D), -\log(\delta(a, D)) [$, with $r < r'$. Let $f \in H(D)$ be such that the function $v_a(f, \mu)$ is neither bounded nor identically equal to $+\infty$ in $[-\log r', -\log r]$. Then there exists a monotonous filter $\mathcal{F}$ of center a and diameter $s \in [-\log r', -\log r]$ such that f is strictly vanishing along $\mathcal{F}$.*

Proof: For convenience we assume $a = 0$. By compacity of $[-\log r', -\log r]$ there does exist $\mu \in [-\log r', -\log r]$ such that $v(f, \mu) = +\infty$. Since the function $v(f, \mu)$ is not identically $+\infty$ in $[-\log r', -\log r]$, by continuity, either there exists $\xi, \zeta \in [-\log r', -\log r]$ with $\xi < \zeta$ such that $v(f, \xi) = +\infty$, $v(f, \mu) < +\infty$ whenever $\mu \in]\xi, \zeta]$, and then f is strictly vanishing along an increasing filter of center 0 and diameter $\omega^{-\xi}$, or there exist $\xi, \zeta \in [-\log r', -\log r]$ with $\xi > \zeta$, such that $v(f, \xi) = +\infty$, $v(f, \mu) < +\infty$ whenever $\mu \in [\zeta, \xi[$ and then f is strictly vanishing along a decreasing filter of center 0 and diameter $\omega^{-\xi}$. This ends the proof.

Proposition 21.7:   *Let $f \in H(D)$ be vanishing along an increasing (resp. a decreasing) filter $\mathcal{F}$ of diameter s. Let a be a center of $\mathcal{F}$ in $\widehat{K}$, and let $E = \widehat{D} \cup (\widehat{K} \setminus \widehat{d}(a, s^-))$ (resp. $E = \widehat{D} \cup \widehat{d}(a, s)$). Then f has continuation to an element F of $\widehat{H}(E)$ such that $F(x) = 0$ whenever $x \in \widehat{K} \setminus \widehat{d}(a, s^-)$, (resp. $x \in \widehat{d}(a, s)$).*

Proof: We suppose $\mathcal{F}$ increasing. By Theorem 15.9 f has an extension $\widehat{f}$ to $\widehat{D}$. For every $r > 0$, the set of the classes of $C(a, r)$ which contain $\widehat{f}$-holes is countable. Since the residue class field of $\widehat{K}$ is not countable, in $C(a, r)$ there exist classes $\Lambda = d(b, r^-)$ which contain no $\widehat{f}$-holes. Thereby $\widehat{f}$ has continuation to an infraconnected set D' which contains $\widehat{D}$ and satisfies $\widetilde{D'} = \widehat{K}$, such that every hole is of the form $\widehat{d}(\alpha, \rho^-)$, with $d(\alpha, \rho^-)$ a hole of D. In this set, we have

$$(1) \quad _{D'}\varphi_{a,s}(\widehat{f}) = {}_D\varphi_{\mathcal{F}}(f) = 0.$$

First, we suppose that $\mathcal{F}$ is increasing. Let $V = d(a, s^-)$, and let $D'' = D' \setminus V$. Clearly V is a hole of D''. Then, as an element of $\widehat{H}(D'')$, by Theorem 15.1, (1) implies

$$(2) \quad \overline{(\widehat{f})}_V = 0.$$

Now, by Theorem 15.1, $\widehat{f}$ has a decomposition of the form $g + h$ with $g \in H_0(D' \cup (\widehat{K} \setminus \widehat{d}(a, s^-)))$ and $h \in H(D' \cup \widehat{d}(a, s^-))$. By (2), it is seen that $_{D'}\varphi_{a,s}(h) = 0$, hence $h = 0$ because h belongs to $\widehat{H}(\widehat{d}(a, s))$. As a consequence, $\widehat{f}$ belongs to $H_0(D' \cup (\widehat{K} \setminus \widehat{d}(a, s^-)))$, and therefore in K, f belongs to $H(D \cup (K \setminus d(a, s^-)))$. Besides by (2) we have $\widehat{f}(x) = 0$ whenever $x \in D''$, hence $\widehat{f}(x) = 0$ whenever $x \in K \setminus d(a, s^-)$.

If $\mathcal{F}$ is decreasing, we can easily perform a symmetric proof.

Theorem 21.8: *Let $f \in H(D)$ be vanishing along an increasing (resp. a decreasing) filter $\mathcal{F}$ of center a and diameter s. Let $E = D \cup (K \setminus d(a, s^-))$ (resp. $E = D \cup d(a, s)$). Then f has continuation to an element of $H(E)$ such that $f(x) = 0$ whenever $x \in K \setminus d(a, s^-)$, (resp. whenever $x \in d(a, s)$).*

Proof: By Proposition 21.7, f has continuation to an element $\widehat{f} \in \widehat{H}(\widehat{D} \cup (\widehat{K} \setminus \widehat{d}(a, s^-)))$, (resp. to $H(\widehat{D} \cup \widehat{d}(a, s)))$. Therefore in K, f belongs to $H(D \cup (K \setminus d(a, s^-)))$. Besides we have $\widehat{f}(x) = 0$ whenever $x \in \widehat{K} \setminus \widehat{d}(a, S^-)$, (resp. $x \in \widehat{d}(a, s)$), hence $f(x) = 0$ whenever $x \in K \setminus d(a, s^-)$, (resp. $f(x) = 0$ whenever $x \in d(a, s)$).

Theorem 21.9: *Let $f \in H(D)$ be such that $f' \in H(D)$. Let $\mathcal{F}$ be a monotonous filter on D. Then f is collapsing along $\mathcal{F}$ if and only if f' is vanishing along $\mathcal{F}$.*

Proof: Without loss of generality, we may obviously assume that $\mathcal{P}(\mathcal{F}) = \emptyset$. We may also assume that f is bounded. Indeed if f is not bounded, there does exist a bounded closed infraconnected subset A of D such that $\mathcal{F}$ is secant with D, and then f (resp. f') is collapsing (resp. vanishing) along $\mathcal{F}$ if and only if its restriction to A is collapsing (resp. vanishing) along $\mathcal{F} \cap A$. Let s be the diameter of $\mathcal{F}$. Since $\widehat{K}$ is spherically complete, $\widehat{\mathcal{F}}$ obviously admits a center a in $\widehat{K}$. Now, if $\mathcal{F}$ is decreasing, an inversion of center a maps $D \setminus \{a\}$ onto a set D', and $\mathcal{F}$ onto an increasing filter $\mathcal{F}'$. Then, putting $g(u) = f(\frac{1}{u-a})$, it seen that f (resp. f') is collapsing (resp. vanishing) along $\mathcal{F}$ if and only if so is g (resp. g') along $\mathcal{F}'$. Thus, without loss of generality, we may suppose $\mathcal{F}$ increasing.

Let $(d(a_n, s^-))_{n \in \mathbb{N}^*}$ be the set of f-holes included in $C(a, s)$, and let

$$E := \widehat{D} \cup (\widehat{d}(a, s) \setminus (\bigcup_{n=0}^{\infty} \widehat{d}(a_n, s^-))).$$ Since each f-holes of D is clearly a hole of E, then for each hole T of D, the Mittag-Leffler term $\overline{f}_T$ of f associated to T

satisfies $\|\overline{f}_T\|_D = \|\overline{f}_T\|_E$. Let $f = \overline{f}_0 + \sum_{n=1}^{\infty} \overline{f}_{T_n}$ be the Mittag-Leffler series of f

on D. f has continuation to an element $\widehat{f} \in \widehat{H}(E)$, and its Mittag-Leffler series on E is the same as its series on D. Hence all the terms of the Mittag-Leffler series of $\widehat{f}$ in $\widehat{D}$ actually belong to $H(E)$, and therefore so do their derivatives.

By Theorem 19.4, the Mittag-Leffler series of f', on D is $(\overline{f}_0)' + \sum_{n=1}^{\infty} (\overline{f}_{T_n})'$. Since

the Mittag-Leffler series of f, as an element of $H(D)$, is the same as that of $\widehat{f}$

as an element of $\widehat{H}(E)$, the series $(\overline{f}_0)' + \sum_{n=1}^{\infty} (\overline{f}_{T_n})'$ does converge in $\widehat{H}(E)$, and

is the Mittag-Leffler series of $(\widehat{f})'$, in $\widehat{H}(E)$. Hence in particular, $\widehat{f}'$ belongs to $H(E)$.

First we suppose f collapsing along $\mathcal{F}$ and will prove that f' is vanishing along $\mathcal{F}$. Let $\ell = \lim_{\mathcal{F}} f(x)$. Then $f - \ell$ is vanishing along $\mathcal{F}$, and therefore by Proposition 21.9, f has continuation to an element $h \in \widehat{H}(E)$ such that $\widehat{f}(x) - \ell = 0$ whenever $x \in E$. Then it is seen that $h(x) = \widehat{f}(x)$ whenever $x \in E \cap C(a, s)$. Indeed, given $b \in E \cap C(a, s)$, we have $d(b, s^-) \subset E$, and

$$_E\varphi_{b,s}(\widehat{f} - \ell) = {}_E\varphi_{b,s}(h - \ell) = {}_E\varphi_{a,s}(\widehat{f} - \ell) = {}_D\varphi_{a,s}(f - \ell) = 0,$$

because by definition, $f(x) = h(x)$ whenever $x \in D$. Now, since $\widehat{f}(x) = \ell$ whenever $x \in E \cap C(a, s)$, obviously we have $\widehat{f}'(x) = 0$ whenever $x \in E \cap C(a, s)$, hence $_E\varphi_{a,s}(\widehat{f}') = 0$, and therefore $_D\varphi_{a,s}(f') = 0$, hence f' is vanishing along $\mathcal{F}$.

Reciprocally, let f' be vanishing along $\mathcal{F}$. Then $\widehat{(f')}$ is vanishing along $\widehat{\mathcal{F}}$. Hence for every $b \in C(a, s) \cap E$, we have $_E\varphi_{b,s}(\widehat{f}') = {}_E\varphi_{a,s}(\widehat{f}') = 0$, hence $\widehat{f}'(x) =$

0 whenever $x \in d(b, s^-)$, because $d(b, s^-) \subset E$. Hence $f(x)$ is a constant ℓ inside $d(b, s^-)$, and therefore we have $_E\varphi_{b,s}(\widehat{f} - \ell) =_E \varphi_{a,s}(\widehat{f} - \ell) =_D \varphi_{a,s}(f - \ell) = 0$. This finishes the proof of Theorem 21.9.

Thanks to the monotonous filters we are now able to end the study of the characteristic functions.

Theorem 21.10: *Let E be a set in K. Then E is not infraconnected if and only if there exists a proper subset A of E such that the characteristic function of A belongs to $H(E)$.*

Proof: If E is not infraconnected, it admits an empty annulus $\Gamma(a, r', r'')$ and then by Lemma 10.14 the characteristic functions of $\mathcal{I}_E(\Gamma(a, r', r''))$ and $\mathcal{E}_E(\Gamma(a, r', r''))$ belong to $H(E)$.

Now we suppose E to be infraconnected and assume that there is a subset B of E whose characteristic function u belongs to $H(E)$. Let $A = E \setminus B$. Then both A, B are obviously closed and open in D. Suppose A and B are different from $\emptyset$, let $a \in A$ and let $b \in B$. Since A is open, there exists $r > 0$ such that $u(x) = 0$ whenever $x \in d(a, r) \cap E$ and then we have $_D\varphi_{a,r}(u) = 0$, hence by Proposition 21.3 there exists a monotonous filter $\mathcal{F}$ with center $\alpha \in E$ such that u is strictly vanishing along $\mathcal{F}$, hence by Lemma 21.2 f is properly vanishing along $\mathcal{F}$. But this contradicts the hypothesis "$f(x) = 0$ or 1 for all $x \in D$". This finishes proving Theorem 21.10.

Corollary 21.11: *Let $E \in \mathcal{A}$. The algebra $H(E)$ has non trivial idempotents if and only if E is not infraconnected.*

Theorem 21.12: *Let $f \in H(D)$. Then $f(D)$ is infraconnected.*

Proof: Let $D' = f(D)$ and let us suppose that D' admits an empty annulus $\Gamma(a, r', r'')$. Let $D'' = \overline{D'}$. It is seen that $\Gamma(a, r', r'')$ also is an empty annulus of D''.

Let $A'' = \mathcal{I}_{D''}(\Gamma(a, r', r''))$ and let $B'' = \mathcal{E}_{D''}(\Gamma(a, r', r''))$. Let u be the characteristic function of A''. By Proposition 10.14 we know that u belongs to $H(D'')$. Since D'' is closed and contains $f(D)$, by Corollary 11.3, $u \circ f$ belongs to $H(D)$. Let $A = f^{-1}(A'')$ and $B = f^{-1}(B'')$. Obviously we have $A \cap B = \emptyset$ and $A \cup B = D$ because $A'' \cap B'' = \emptyset, A'' \cap B'' = D''$. Besides $u \circ f$ is the characteristic function of A. But D is infraconnected hence by Theorem 21.10 $H(D)$ contains no characteristic function of any proper set . This ends the proof of Theorem 21.12.

22.   QUASI-MINORATED ELEMENTS

Throughout this chapter, the set D is supposed to be infraconnected.
The main results given here were obtained in [11], [14], [15], [16], [17].

Theorem 22.1:   *Let f be a non zero non quasi-minorated element of $H(D)$. There exists a pierced monotonous filter $\mathcal{F}$ on D such that f is strictly vanishing along $\mathcal{F}$.*

Proof: Let $(\alpha_n)_{n\in\mathbb{N}}$ be a bounded sequence in D such that $\lim_{n\to\infty} f(\alpha_n) = 0$ and such that no subsequence converges in K . For every n, let $\psi_n = {}_D\varphi_{\alpha_n}$. By Theorem 12.7 there exists a subsequence of the sequence ψ_n which converges to a limit $\psi \in Mult(H(D), \mathcal{U}_D)$. Hence, without loss of generality we may assume that $\lim_{n\to\infty} \psi_n = \psi$. Since ψ is not of the form ${}_D\varphi_\infty$, there exists a circular filter $\mathcal{F}$ on D such that $\psi = {}_D\varphi_{\mathcal{F}}$.

We first suppose $\mathcal{F}$ to be convergent, of limit $a \in \overline{D}$, and set $g(x) = x - a$. We see that $\psi_n(g) = |\alpha_n - a|$ hence $\lim_{n\to\infty} |\alpha_n - a| = \psi(g) = 0$. But this contradicts the hypothesis "the sequence (α_n) has no convergent subsequence". Hence $\mathcal{F}$ is a large circular filter, of diameter r.

We now suppose $\mathcal{F}$ to have a center $a \in K$. Hence the monotonous filter $\mathcal{G}$ on D of center a and diameter r is thinner than $\mathcal{F}$ and then f is vanishing along $\mathcal{G}$. Thus we have ${}_D\varphi_{a,r}(f) = 0$ and therefore by Proposition 21.3, f is strictly vanishing along certain monotonous filter $\mathcal{G}'$ which , by Lemma 21.5, is pierced.

Finally we suppose $\mathcal{F}$ to have no center in K. Hence it is a decreasing filter. If f is strictly vanishing along $\mathcal{F}$, then $\mathcal{F}$ is pierced by Lemma 21.5 . Hence we may suppose f not strictly vanishing along $\mathcal{F}$. We consider a cononical base $(D_n)_{n\in\mathbb{N}}$ of $\mathcal{F}$ with $D_n = D \cap d(a_n, r_n)$. Since $\lim_{\mathcal{F}} f(x) = 0$ obviously we have $\lim_{n\to\infty} v_{a_n}(f, -\log r_n) = +\infty$. But since f is not strictly vanishing along $\mathcal{F}$, by Lemma 21.1 there exists $n \in \mathbb{N}$ such that ${}_D\varphi_{a_n, r_n}(f) = 0$ hence the circular filter $\mathcal{F}'$ of center a_n, of diameter r_n satisfies $\lim_{\mathcal{F}'} f(x) = 0$ and therefore we return to the previous case. Finally in every cases, f has been proven to be strictly vanishing along a pierced monotonous filter.

Corollary 22.2:   *If D has no monotonous pierced filter, every element different from zero is quasi-minorated.*

Theorem 22.3:   *Let f_1, f_2 be quasi-minorated elements of $H(D)$. If $f_1 f_2$ belongs to $H(D)$, then it is also quasi-minorated.*

Proof: Indeed suppose that $f_1 f_2$ is not quasi-minorated. By Theorem 22.1 there exists a pierced monotonous filter $\mathcal{F}$ on D such that $_D\varphi_{\mathcal{F}}(f_1 f_2) = 0$. Hence, either $_D\varphi_{\mathcal{F}}(f_1) = 0$ or $_D\varphi_{\mathcal{F}}(f_2) = 0$. But these two options are impossible because both f_1, f_2 are quasi-minorated. Hence so is $f_1 f_2$.

Theorem 22.4:   *Let $f \in H(D)$. If f is semi-invertible then it is quasi-minorated.*

Proof: Let f be semi-invertible, of the form $P(x)g(x)$ with g invertible in $H(D)$ and let P be a polynomial whose zeros $a_1, \ldots, a_q$ belong to D. Then we suppose f not quasi-minorated. By Theorem 22.3, g is not quasi-minorated either. Hence there exists a pierced monotonous filter $\mathcal{F}$ on D such that $_D\varphi_{\mathcal{F}}(g) = 0$. But by Lemma 12.2 this contradicts the hypothesis "g invertible in $H(D)$". Hence f is quasi-minorated.

Theorem 22.5:   *Let $D \in \mathcal{A}$ be such that $\overline{D}$ is open. Then an element of $H(D)$ is quasi-minorated if and only if it is quasi-invertible.*

Proof: If f is quasi-invertible, it is semi-invertible and then by Theorem 22.4 it is quasi-minorated. Now we assume f to be quasi-minorated and will prove it to be quasi-invertible.

If D is closed and bounded, D is open and then the conclusion comes from Theorem 14.7. Now, we just suppose D to be bounded. By Corollary 10.12, $H(D)$ is of the form $S(D)H(\overline{D})$ with $S(D)$ the set of the polynomials whose zeros lie in $\overline{D} \setminus D$. There exists $Q \in S(D)$ and $h \in H(\overline{D})$ such that $f = \dfrac{h}{Q}$. If h is not quasi-minorated, there exists a monotonous filter $\mathcal{F}$ on D such that $_D\varphi_{\mathcal{F}}(h) = 0$ and then it is seen that $_D\varphi_{\mathcal{F}}(f) = 0$. This contradicts the hypothesis "f quasi-minorated" and therefore h is quasi-minorated. Then by Theorem 14.7, h is quasi-invertible, in the form Pg with P a polynomial whose zeros are interior to $\overline{D}$ and g an element invertible in $H(\overline{D})$. Since $D \in \mathcal{A}$, $\dfrac{g}{Q}$ is invertible in $H(D)$.

Let $P = P_1 P_2$ with P_1 (resp. P_2) the polynomial of the zeros of P in $\overset{\circ}{\overline{D}} \cap D$ (resp. in $\overline{D} \setminus D$). Then $\dfrac{P_2 g}{Q}$ is invertible in $H(D)$ and then f is seen to be quasi-invertible.

Now we suppose D unbounded. We may obviously assume D to have at least one hole $d(a, r)$ because if D has no hole then, $\overline{D} = K$ and then by Corollary 10.8 we have $H(D) = R(D)$, (and then every element is obviously quasi-invertible !).

Besides without loss of generality we may also assume $a = 0$. Now let $\gamma(x) = \dfrac{1}{x}$, let $D' = \gamma(D)$, and let $F(u) = f \circ \gamma(u)$. By Theorem 11.8, D' also belongs to $\mathcal{A}$. By Lemma 14.6, F is quasi-minorated in $H(D')$, hence F is quasi-invertible and factorizes in the form QG with Q a polynomial whose zeros belong to $\overline{D'}$ and G an invertible element of $H(D')$. Let $q = deg(Q)$. Let $g(x) = G(\frac{1}{x})$, and let $P(x) = x^q Q(\frac{1}{x})$. Then P is a polynomial whose zeros lie in $\overline{D}$. Thus we have $f(x) = P(x)\dfrac{g(x)}{x^q}$. Obviously $\dfrac{g(x)}{x^q}$ is an invertible element of $H(D)$ and this ends the proof of Theorem 22.5.

Theorem 22.6:   *If D belongs to $\mathcal{A}$ and has no pierced filter, every element different from zero is quasi-invertible*

Proof: Indeed, since D has no pierced filter, by Corollary 22.2 every element of $H(D)$ is quasi-minorated. But since D has no pierced filter, $\overline{D}$ is open, hence by Theorem 22.5 every element of $H(D)$ is quasi-invertible.

Corollary 22.7:   *Let D be closed bounded and let $\mathcal{T}$ be the set holes of D. If $\{\widetilde{T} | T \in \mathcal{T}\}$ is finite, then every element of $H(D)$ different from zero is quasi-invertible.*

Corollary 22.8:   *If D is closed bounded and has finitely many holes then every element of $H(D)$ different from zero is quasi-invertible.*

Corollary 22.9:   *If D is a disk $d(a,r)$, or $d(a,r^-)$, or if D is an annulus $\Gamma(a,r_1,r_2)$ (with $0 < r_1 < r_2$) or $\Delta(a,r_1,r_2)$ (with $0 < r_1 \leq r_2$) or a circle $C(a,r)$ then every element of $H(D)$ different from zero is quasi-invertible.*

Proof: Indeed D has no pierced filter hence the elements different from zero are quasi-minorated and are quasi-invertible because D is open.

We will see that when D belongs to $\mathcal{A}$, a quasi-minorated element that has no zero in D, actually is invertible in $H(D)$.

Lemma 22.10:   *Let $f \in H(\overline{D})$ be quasi-minorated in $H(D)$. Then f is quasi-minorated in $H(\overline{D})$.*

Proof: Indeed let $(a_n)_{n \in \mathbb{N}}$ be a sequence in $\overline{D}$ such that $\lim\limits_{n \to \infty} f(a_n) = 0$. There obviously exists a sequence $(b_n)_{n \in \mathbb{N}}$ such that $\lim\limits_{n \to \infty} a_n - b_n = 0$ and $\lim\limits_{n \to \infty} f(b_n) = 0$. Since f is quasi-minorated in $H(D)$, either we can extract a Cauchy sequence $(b_{n_q})_{q \in \mathbb{N}}$ from the sequence $(b_n)_{n \in \mathbb{N}}$, and then the sequence

$(a_{n_q})_{q \in \mathbb{N}}$ is a Cauchy subsequence of the sequence $(a_n)_{n \in \mathbb{N}}$, or we can extract a subsequence $(b_{n_q})_{q \in \mathbb{N}}$, such that $\lim_{n \to \infty} |b_{n_q}| = +\infty$, and then we have $\lim_{n \to \infty} |a_{n_q}| = +\infty$. Thus we have proven that f is quasi-minorated in $H(\overline{D})$.

Theorem 22.11: *Let $D \in \mathcal{A}$. Let $f \in H(D)$ be quasi-minorated and have no zero in D. Then f is invertible in $H(D)$.*

Proof: First we suppose D closed and bounded. Assume $\inf_{x \in D} |f(x)| = 0$ and let $(a_n)_{n \in \mathbb{N}}$ be a sequence in D such that $\lim_{n \to \infty} f(a_n) = 0$. Since D is bounded, and since f is quasi-minorated, we can extract a Cauchy subsequence $(a_{n_q})_{q \in \mathbb{N}}$. But since D is closed, this subsequence has a limit a in D and therefore we have $f(a) = 0$. Thus we see that $\inf_{x \in D} |f(x)| > 0$. Hence by Lemma 10.3, f is invertible in $H(D)$.

Now we just suppose D bounded. Let Q be the polynomial of the poles of f in $\overline{D} \setminus D$, and let $h(x) = Q(x)f(x)$. By theorem 22.3 h is quasi-minorated in $H(D)$. But by theorem 10.11 h belongs to $H(\overline{D})$, and therefore by Lemma 22.10, it is quasi-minorated in $H(\overline{D})$. We will prove that h has finitely many zeros in $\overline{D}$. Indeed we assume that h admits infinitely many zeros in $\overline{D}$. So we can find a sequence $(a_n)_{n \in \mathbb{N}}$ in $\overline{D} \setminus D$ such that $h(a_n) = 0$ whenever $n \neq m$, $n, m \in \mathbb{N}$. Since h is quasi-minorated in $H(\overline{D})$, and since D is bounded, we can extract a cauchy subsequence from the sequence $(a_n)_{n \in \mathbb{N}}$. This Cauchy subsequence obviously converges to a point $a \in \overline{D}$ and therefore we have $h(a) = 0$. But, as h has no zero in D, a belongs to $\overline{D} \setminus D$. And now, since D belongs to $\mathcal{A}$, then a belongs to $\overset{\circ}{\overline{D}}$. But by Corollary 14.2, a zero of h which is interior to $\overline{D}$ is isolated in $\overline{D}$, and this contradicts the definition of a. Thus we have proven that h has finitely many zeros $(b_j)_{1 \leq j \leq q}$ in $\overline{D}$, all of them in the set $\overline{D} \setminus D$ which is included in $\overset{\circ}{\overline{D}}$. Then each zero b_j has a multiplicity order n_j, $(1 \leq j \leq q)$. Let $P(x) = \prod_{j=1}^{q} (x - b_j)^{n_j}$ be the polynomial of the zeros of h in $\overset{\circ}{\overline{D}}$. By corollary 14.5 the function $g(x) = \dfrac{h(x)}{P(x)}$ belongs to $H(\overline{D})$ and obviously has no zero in $\overline{D}$. As we have already seen when D is closed, g is invertible in $H(\overline{D})$. Now, since both P, Q have all their zeros in $\overline{D} \setminus D$, they are invertible in $R(D)$ and so is $\dfrac{P}{Q}$. But then $f = \dfrac{P}{Q} g$.

23. VALUES AND ZEROS OF POWER SERIES

Most of the classical results on the zeros of the polynomial will now be extended to power series. In particular, power series convergent in a disk satisfy a Schwarz Lemma that is even simpler than in $\mathbb{C}$. We will also notice Dieudonne-Dwork's Theorem.

In this chapter, r is a strictly positive real number and r', r'' are strictly positive real numbers satisfying $r' < r''$.

Theorem 23.1: *Let $f \in H(C(0,r))$. The number of zeros of f in $C(0,r)$ is equal to $N^+(f, -\log r) - N^-(f, -\log r)$, (taking multiplicities into account).*

Proof: This equality was given for a polynomial in Theorem 4.11. First we prove it when f is an element of $H(C(0,r))$ invertible in $H(C(0,r))$. Clearly we have $N^+(\frac{1}{f}, -\log r) = -N^+(f, -\log r)$ and $N^-(\frac{1}{f}, -\log r) = -N^-(f, -\log r)$. Since any $h \in H(C(0,r))$ satisfies $N^+(h, -\log r) \geq N^-(h, -\log r)$ we see that $N^+(f, -\log r) = N^-(f, -\log r)$.

We now consider the general case. By Corollary 22.9, f has a factorization of the form Pg with P a polynomial whose zeros belong to $C(0,r)$ and g invertible in $H(C(0,r))$. Then $N^+(f, -\log r) - N^-(f, \log r) = N^+(P, -\log r) - N^-(P, -\log r) = deg(P)$ and this just ends the proof of Theorem 23.1.

Corollary 23.2: *Let $f \in H(C(0,r))$ have t zeros in $C(0,r)$.*

Let $q = N^-(f, -\log r)$. Then $r = \sqrt[t]{\left| \dfrac{a_q}{a_{q+t}} \right|}$.

Theorem 23.3: *Let $f = \displaystyle\sum_{-\infty}^{+\infty} a_n x^n \in H(C(0,r))$ have no zero in $C(0,r)$. Let $t = N^+(f, -\log r)$. Then we have $N^+(f, -\log r) = N^-(f, -\log r)$ and $v(f(x)) = v(f, -\log r) = v(a_t) - t\log r$ whenever $x \in C(0,r)$.*

Proof: Since f has no zero, by Theorem 23.1 we have $N^+(f, -\log r) = N^-(f, -\log r)$. Then $|a_t| r^t > |a_n| r^n$ whenever $n \neq t$ hence $|f(x)| = |a_t| r^t$ i.e. $v(f(x)) = v(f, -\log r)$.

Theorem 23.4: *Let $\Lambda = \Gamma(0, r', r'')$ (resp. $\Delta(0, r', r'')$) and let $f(x) =$*

$\sum_{-\infty}^{+\infty} a_n x^n$ *have no zero in* Λ. *Then we have* $N^+(f,\mu) = N^-(f,\mu) = q \in \mathbb{Z}$ *whenever* $\mu \in]-\log r'', -\log r'[$ *(resp.* $\mu \in [-\log r'', -\log r']$ *) and* $|f(x)| = |a_q x^q|$ *whenever* $x \in \Lambda$.

Proof: Let $I =]-\log r'', -\log r'[$ (resp. $I = [-\log r'', -\log r']$). By Theorem 23.1 we see that $N^+(f,\mu) = N^-(f,\mu)$ whenever $\mu \in I$ hence the functions $N^+(f,\mu)$, $N^-(f,\mu)$ are continuous, hence are constant and therefore both are equal to an integer $q \in \mathbb{Z}$ for all $\mu \in I$. Thus we have $v(f,\mu) = v(a_q) + q\mu$ whenever $\mu \in I$. Now by Theorem 23.3 we see that $v(f(x)) = v(f, v(x))$ whenever $x \in \Lambda$ hence finally $v(f(x)) = v(a_q) + qv(x)$ whenever $x \in \Lambda$.

Theorem 23.5:   *Let* $\Lambda = d(0,r)$ *and let* $f \in H(\Lambda)$. *The number of zeros of* f *in* Λ *(taking multiplicities into account) is equal to* $N^+(f, -\log r)$.

Proof: We first suppose that f is a polynomial of degree q whose zeros lie in Λ. Since P has no zeros outside Λ, by Theorem 4.11 we know that $N^+(P, -\log r) = q$. Now in the general case, by Corollary 22.9, f factorizes in the form Pg with g invertible in $H(\Lambda)$ and P a polynomial whose zeros lie in Λ. Since $g \in H(d(0,r))$ we know that $N^-(g,\mu) \geq 0$. But since g is invertible, we also have $N^-(\dfrac{1}{g}, \mu) \geq 0$. Hence by Proposition 20.8, for all $\mu \geq -\log r$ we have

$$0 \leq N^-(g,\mu) \leq N^+(g,\mu) = -N^+(\frac{1}{g},\mu) \leq -N^-(g,\mu) \leq 0$$

therefore $N^+(g,\mu) = 0$. So we have $N^+(f, -\log r) = N^+(P, -\log r)$ and this is the number of zeros of P, hence of f too.

Theorem 23.6:   *Let* Λ *be a disk of the form* $d(a,r)$ *(resp.* $d(a,r^-)$ *) and let* $f \in H(\Lambda)$ *have no zero in* Λ. *Then* $|f(x)|$ *is equal to a constant in* Λ.

Proof: We may obviously assume $a = 0$. By Theorem 23.5, we have $N^+(f,\mu) = N^-(f,\mu) = 0$ whenever $\mu \geq -\log r$ (resp. $\mu > -\log r$) hence by Proposition 20.8 $v(f,\mu)$ has a derivative equal to 0 and therefore is equal to a constant in $[-\log r, +\infty[$. Then by Proposition 20.8 we have $v(f(x)) = v(f, v(x))$ hence $v(f(x))$ is equal to a constant in Λ.

Theorem 23.7:   *Let* Λ *be a set in one of the following forms:*

    i) $\Lambda = d(0,r)$

    ii) $\Lambda = d(0,r^-)$

    iii) $\Lambda = C(0,r)$.

Let $f \in H(\Lambda)$ (f not identically 0) and let $h \in H(\Lambda)$ satisfy $\|f - h\|_\Lambda < \|f\|_\Lambda$. Then f and h have the same number of zeros in Λ (taking multiplicities into account).

Proof: Regardless of the case i), ii), iii) we know that $\|f\|_\Lambda = {}_\Lambda\varphi_{0,r}(f)$ and then we have $-\log\|f\|_\Lambda = v(f, -\log r)$. Since $\|f - h\|_\Lambda < \|f\|_\Lambda$ then $v(f - h, -\log S) > v(f, -\log S)$. Hence by Corollary 20.7 we know that $N^+(f, -\log S) = N^+(h, -\log S)$ and $N^-(f, -\log r) = N^-(h, -\log r)$, hence f has as many zeros as h in Λ, by Theorem 23.5 if $\Lambda = d(0, r)$ and by Theorem 23.1 if $\Lambda = C(0, r)$.

We now suppose $\Lambda = d(0, r^-)$. By Corollary 22.9 both f, h are quasi-invertible in $H(\Lambda)$. Let $\rho \in]0, r]$ be such that $d(0, \rho)$ contains all the zeros of h in Λ. According to the statements i) and ii) already proven, we see that f has as many zeros as h in $d(0, \rho)$ (taking multiplicities into account) and has no zero in $C(0, s)$ whenever $s \in]\rho, r[$. This ends the proof.

Corollary 23.8: *Let $a \in K$, and let $\Lambda = d(a, r)$, (resp. $d(a, r^-)$), (resp. $C(a, r)$) and let D contain Λ. Any element $f \in H(D)$ has finitely many zeros in Λ and factorizes in the form $f = Pg$ with P the polynomial of the zeros of f in Λ and g an element of $H(D)$ that has no zero in Λ.*

Proof: Indeed, by Theorem 23.7 f has finitely many zeros in Λ and then this factorization is given by Theorem 14.5.

Corollary 23.9: *Let $f(x) \in A(d(0, r^-))$ have infinitely many zeros in $d(0, r^-)$. Then the set of the zeros of f in $d(0, r^-)$ is a sequence $(\alpha_n)_{n \in \mathbb{N}}$, such that $\lim_{n \to +\infty} |\alpha_n| = r$.*

These theorems on analytic elements help obtain more accurate results on the analytic functions in an open disk.

Theorem 23.10: *Let $f(x) = \sum_{n=0}^{\infty} a_n x^n \in A(d(0, r^-))$. Then f has finitely many zeros in $d(0, r^-)$ if and only if there exists $q \in \mathbb{N}$ such that $|a_q| r^q \geq \sup_{n \in \mathbb{N}} |a_n| r^n$.*

Proof: First we suppose that there exists $q \in \mathbb{N}$ such that $|a_q| r^q \geq \sup_{n \in \mathbb{N}} |a_n| r^n$.

Let $t \in \mathbb{N}$ be the smallest of the integers q such that $|a_q| r^q \geq \sup_{n \in \mathbb{N}} |a_n| r^n$.

There does exist a unique $s \in]0, r[$ such that $|a_t| s^t \geq |a_n| s^n$ for every $n < t$. Then it is seen that for all $\rho \in]s, r[$, in $H(d(0, \rho))$ we have $N^+(f, -\log \rho) = N^-(f, -\log \rho) = t$, hence f admits exactly t zeros in $d(0, \rho)$, whenever $\rho \in]s, r[$.

Reciprocally, suppose that f admits exactly t zeros in $d(0, r^-)$. Then, there does exist $s \in]0, r[$ such that f admits exactly t zeros in $d(0, s)$, and of course in each disk $d(0, \rho)$ for every $\rho \in]s, r[$. Hence we have $N^+(f, -\log \rho) = N^-(f, -\log \rho) = t$ for every $\rho \in]s, r[$. Therefore, we have $|a_t|\rho^t > |a_n|\rho^n$ for every $n \neq t$, and for every $\rho \in]s, r[$. Finally we see that $|a_t|r^t \geq \sup |a_n|r^n$.

By Theorem 13.9 and 23.10 we obtain Corollary 23.11.

Corollary 23.11: *Let $f(x) \in A(d(0, r^-))$. If f is not bounded, then f has infinitely many zeros in $d(0, r^-)$.*

Lemma 23.12: *Let $f(t) \in A(d(0, 1^-))$ and let $r_1, r_2 \in (0, 1)$ satisfy $r_1 < r_2$. If f admits exactly q zeros in $d(0, r_1)$ (taking multiplicities into account) and has no zeros in $\Gamma(0, r_1, r_2)$, then f satisfies*

$$v(f, -\log r_2) - v(f, -\log r_1) = t(\log r_1 - \log r_2).$$

Proof: By Theorem 23.5, we have $N^+(f, -\log r_1) = q$, and by Theorem 23.3 it is seen that $v(f, \mu) - v(f, -\log r_1) = q(\mu + \log r_1)$. This is true for every $\mu \in [\mu, -\log r_[$, hence by continuity we have $v(f, -\log r_2) - v(f, -\log r_1) = q(-\log r_2 + \log r_1)$.

Theorem 23.13: *Let $f(t) \in A(d(0, 1^-))$ and let $r_1, r_2 \in]0, 1[$ satisfy $r_1 < r_2$. If f admits exactly q zeros in $d(0, r_1)$ (taking multiplicities into account) and t different zeros $\alpha_1, \ldots, \alpha_t$, of multiplicity order ζ_j $(1 \leq j \leq t)$ respectively in $\Gamma(0, r_1, r_2)$, then f satisfies*

$$v(f, -\log r_2) - v(f, -\log r_1) = -\sum_{j=1}^{t} \zeta_j(v(a_j) + \log r_2) - q(\log r_2 - \log r_1).$$

Proof: For each $j = 1, \ldots t - 1$, we denote by u_j the number of zeros of f in $d(0, |\alpha_j|)$, and then by Theorem 23.12, we have

$$v(f, v(\alpha_{j+1})) - v(f, v(\alpha_j)) = q(v(\alpha_{j+1}) - v(\alpha_j)),$$
$$v(f, v(\alpha_1)) - v(f, -\log r_1) = u_j(v(\alpha_{j+1}) + \log r_1),$$
$$v(f, -\log r_2) - v(f, v(\alpha_q)) = u_t(-\log r_2 - v(\alpha_j)).$$

Then by an immediate induction we check that

$$v(f, -\log r_2) - v(f, -\log r_1) = -\sum_{j=1}^{t} \zeta_j(v(a_j) + \log r_2) - q(\log r_2 - \log r_1).$$

By Theorems 23.11 and 23.13 we obtain Corollary 23.14

Corollary 23.14: *Let* $f(x) = \sum\limits_{n=0}^{\infty} a_n x^n \in A(d(0, r^-))$ *have a set of zeros in* $d(0, r^-)$ *that consists of a sequence* $(\alpha_n)_{n \in \mathbb{N}}$, *where each* α_n *is of order* u_n. *Then* f *is unbounded if and only if the sequence* $(\alpha_n)_{n \in \mathbb{N}}$ *satisfy* $\prod\limits_{n=0}^{\infty} \left(\dfrac{\alpha_n}{r}\right)^{u_n} = 0$.

Theorem 23.15: *Let* $f(x) = \sum\limits_{n=0}^{\infty} a_n x^n \in A(d(0, r^-))$. *The set of the zeros of* f *in* $d(0, r^-)$ *is an increasing distances sequence if and only if so is the sequence* $\dfrac{a_n}{a_{n+1}}$. *Further, if these properties are satisfied, then the sequence of the zeros of* f *in* $d(0, r^-)$ *is a sequence* $(\alpha_n)_{n \in \mathbb{N}^*}$ *such that* $\lim\limits_{n \to +\infty} |\alpha_n| = r$, *and* $|\alpha_n| = \left|\dfrac{a_n}{a_{n+1}}\right|$.

Proof: First we suppose that the set of the zeros of f in $d(0, r^-)$ is an increasing distances sequence $(\alpha_n)_{n \in \mathbb{N}^*}$. Then by Corollary 23.9, we know that $\lim\limits_{n \to +\infty} |\alpha_n| = r$. By Corollary 23.2, for each $q \in \mathbb{N}^*$, we have $N^+(f, v(\alpha_q)) - N^-(f, v(\alpha_q)) = 1$, and $|\alpha_0| = \left|\dfrac{a_0}{a_1}\right|$. Then by an immediate induction we deduce that $|\alpha_n| = \left|\dfrac{a_n}{a_{n+1}}\right|$ for every $n \in \mathbb{N}^*$.

Reciprocally, we suppose that $\dfrac{a_n}{a_{n+1}}$ is an increasing distances sequence. Hence we have $\left|\dfrac{a_n}{a_{n+1}}\right| < \left|\dfrac{a_{n+1}}{a_{n+2}}\right|$ for every $n \in \mathbb{N}$. For each $m \in \mathbb{N}^*$, we put $\nu_m = v(a_{m-1}) - v(a_m)$, and $r_m = \omega^{-\nu_m}$. Clearly, we have $N^+(f, \nu_m) - N^-(f, \nu_m) = 1$ for every $m \in \mathbb{N}^*$, and $N^+(f, \mu) = N^-(f, \mu)$, for every $\mu \in (] - \log r, +\infty[\backslash \{\nu_m | \quad m \in \mathbb{N}^*\})$. Hence by Theorem 23.1, f admits exactly one zero in each circle $C(0, r_m)$, and no other zero in $d(0, r^-)$. So, the conclusion is clear.

Remark: We can construct a strictly increasing bounded sequence $(r_n)_{n \in \mathbb{N}}$, in $|K|$, of limit r, satisfying $\dfrac{r_n}{r_{n+1}} < \dfrac{r_{n+1}}{r_{n+2}}$, for every $n \in \mathbb{N}$. So, there do exist bounded functions $f \in A(d(0, r^-))$, satisfying the hypothesis of theorem 23.15.

We take this opportunity to characterize the analytic elements in a disk that are invertible.

Theorem 23.16: *Let* $f(x) = \sum\limits_{n=0}^{\infty} a_n(x - a)^n \in H(d(a, r))$. *Then the following statements* a), b), c), d), e) *are equivalent:*

a) $|a_0| > |a_n|$ *for all* $n > 1$.
b) $\|f - f(a)\|_{d(a,r)} < |f(a)|$
c) f *has no zero in* $d(a,r)$
d) $|f(x)|$ *is constant and different from* 0 *in* $d(a,r)$
e) f *is invertible in* $H(d(a,r))$.

Proof: First a), b) are equivalent by Theorem 13.1. Second, a), c), d) are seen to be equivalent by Theorems 23.5 and 23.6. Third, by Lemma 10.3, d) implies e) and finally e) obviously implies c).

Corollory 23.17:   *Let* $f(x) = \displaystyle\sum_{n=0}^{\infty} a_n x^n \in A(d(0,r^-))$. *Then Statements* a),
b), c), d) *are equivalent:*
   a) $|a_0| \geq |a_n|$ *for all* $n > 1$.
   b) f *has no zero in* $d(0,r)$
   c) $|f(x)|$ *is constant and different from* 0 *in* $d(0,r)$
   d) f *is invertible in* $A(d(0,r^-))$.

Proof: We just have to apply Theorem 23.16 to f in $H(d(0,\rho))$ for every $\rho \in$ $]0,r[$.

Theorem 23.18:   **(Schwarz Lemma)**   *Let* $D = d(a,s)$, *let* $f \in H(D)$ *have at least* q *zeros in* $d(a,r)$ *with* $0 < r < s$. *Then we have* $\dfrac{\varphi_{a,s}(f)}{\varphi_{a,r}(f)} \geq (\dfrac{s}{r})^q$.

Proof: By Theorem 14.1, f is of the form Pg with P a q-degree polynomial that has its q zeros in $d(a,r)$, and g an element of $H(D)$. Then by Theorem 23.4 we have $\dfrac{_D\varphi_{a,s}(P)}{_D\varphi_{a,r}(P)} = (\dfrac{s}{r})^q$. Besides we have $_D\varphi_{a,s}(g) \geq {_D\varphi_{a,r}(g)}$, so the conclusion is clear.

Theorem 23.19:   *Let* $a,\ b \in K$ *and* $r,s \in \mathbb{R}_+^*$, *and let* $f \in A(d(0,r^-))$, $g \in A(d(0,s^-))$ *be such that* $f(d(a,r^-)) \subset d(b,s^-)$. *Then* $g \circ f$ *belongs to* $A(d(0,r^-))$.

Proof: Without loss of generality we can clearly assume $a = b = 0$. First, suppose that f has no zero in $d(0,r^-)$. Then $|f(x)|$ is equal to a constant c in $d(0,r^-)$, with $c < s$. Hence, of course, $f(d(0,r^-))$ is included in $d(0,c)$. Now, let $\rho \in]0,r[$. The restriction of f to $d(0,\rho)$ belongs to $H(d(0,\rho^-))$, and the restriction of g to $d(0,c)$ belongs to $H(d(0,c))$. Hence the restriction of $g \circ f$ to $d(0,\rho)$ belongs to $H(d(0,\rho))$. This is true for every $\rho \in]0,r[$ and therefore this shows that $g \circ f$ belongs to $H(d(0,r^-))$.

Now, we suppose that f admits at least one zero in $d(0, r^-)$. Hence there exists $r' \in]0, r[$ such that f has at least one zero in $d(0, r')$. Therefore by theorem 23.18, $\|f\|_{d(0,\rho)}$ is strictly increasing in ρ in the interval $[r', r[$. Now, let $\rho \in]0, r[$, and let $\sigma = \|f\|_{d(0,\rho)}$. The restriction of f to $d(0, \rho)$ belongs to $H(d(0, \rho))$, and further, $f(d(0, \rho))$ is included in $d(0, \sigma)$. Since g belongs to $H(d(0, \sigma))$, $g \circ f$ belongs to $H(d(0, \rho))$. As previously, this is true for every $\rho \in]0, r[$, hence $g \circ f$ belongs to $A(d(0, r^-))$,

Theorem 23.20: (**Dieudonné-Dwork**) *Let $f \in A(d(0, 1^-))$ satisfy $f(0) = 1$, and have no zero in $d(0, 1^-)$. There exists a sequence $(u_k)_{k \in \mathbb{N}^*}$ in $d(0, 1)$ such that $f(x) = \prod_{k=1}^{\infty}(1 - u_k x^k)$ whenever $x \in d(0, 1^-)$.*

Proof: Since $f(0) = 1$, we can write $f(x)$ in the form $1 + \sum_{n=1}^{\infty} b_n x^n$. Since f has no zero in $d(0, 1^-)$, by Corollary 23.17 we have $|b_n| \leq 1$ for every $n \in \mathbb{N}^*$. Now, suppose that we have already found $u_1, ..., u_k$ such that $f(x)$ factorizes in the form

$$(\mathcal{R}_k) \quad \prod_{j=1}^{k}(1 - u_j x^j)\left(1 + x^{k+1} \sum_{n=0}^{\infty} \theta_{n,k+1} x^n\right) \text{ with } |\theta_{n,k+1}| \leq 1 \text{ for every } n \in \mathbb{N}.$$

Actually, we have $(1 + \theta_{0,k+1} x^{k+1})\left(1 + \sum_{n=1}^{\infty}(-\theta_{0,k+1} x^{k+1})^n\right) = 1$, and therefore we can factorize $\left(1 + x^{k+1} \sum_{n=0}^{\infty} \theta_{n,k+1} x^n\right)$ in the form

$$(1 + \theta_{0,k+1} x^{k+1})\left(1 + \sum_{n=1}^{\infty}(-\theta_{0,k+1} x^{k+1})^n\right)\left(1 + x^{k+1} \sum_{n=0}^{\infty} \theta_{n,k+1} x^n\right)$$

Now, consider the function

$$g_{k+1} := \left(1 + \sum_{n=1}^{\infty}(-\theta_{0,k+1} x^{k+1})^n\right)\left(1 + x^{k+1} \sum_{n=0}^{\infty} \theta_{n,k+1} x^n\right).$$

In g_{k+1}, it is seen that the term in x^{k+1} is equal to 0, so g_{k+1} is of the form $\left(1 + x^{k+2} \sum_{n=0}^{\infty} \theta_{n,k+2} x^n\right)$, with $|\theta_{n,k+2}| \leq 1$, for every $n \in \mathbb{N}$.

Now we just put $u_{k+1} = \theta_{0,k+1}$, and then we have proven $(\mathcal{R}_{k+1})$. Since $(\mathcal{R}_0)$ is trivially satisfied, by induction we can construct a sequence $(u_k)_{k\in\mathbb{N}^*}$ in $d(0, 1^-)$, and a sequence $(g_k)_{k\in\mathbb{N}^*}$ in $A(d(0, 1^-))$, such that for each $k \in \mathbb{N}$, g_k is of the form $1 + x^{k+1} \sum_{n=0}^{\infty} \theta_{n,k+1} x^n$, and satisfies $f(x) = \prod_{j=1}^{k} (1 - u_j x^j) g_k(x)$. For each

$k \in \mathbb{N}^*$, let $f_k(x) = \prod_{j=1}^{k} (1 - u_j x^j)$.

It is seen that for each $r \in]0, 1[$, we have $\|g_k - 1\|_{d(0,r)} \leq r^{k+1}$. As a consequence, for every $r \in]0, 1[$, the sequence $(f_k)_{k\in\mathbb{N}^*}$ converges to f in $H(d(0, r))$, and

therefore we have $f(x) = \prod_{k=1}^{\infty} (1 - u_k x^k)$ for all $x \in d(0, 1^-)$. This ends the proof.

24. QUASI-INVERTIBLE ELEMENTS

Throughout this chapter D is supposed to be infraconnected.
Some of the results given here were obtained in [12], and published in [15] and
[18]. We will show that as soon as an ideal of an algebra $H(D)$ contains a
quasi-invertible element, this ideal is principal and generated by a polynomial.

Lemma 24.1:   *Let T be a hole of D and let $f \in H(D \cup T)$ be invertible in
$H(D)$. If f has no zero in T then f is invertible in $H(D \cup T)$.*

Proof: Let $(f_n)_{n \in \mathbb{N}}$ be a sequence in $R(D \cup T)$ which converges to f in
$H(D \cup T)$. Let $T = d(a, r^-)$. Since f has no zero in T, by Theorem 23.6 we have
$|f(x)| = |f(a)|$ for all $x \in T$ and therefore $|f(x)| = {}_D\varphi_{a,r}(f)$ for all $x \in T$. Now
since D is infraconnected we have ${}_D\varphi_{a,r}(g) \leq \|g\|_D$ whenever $g \in H(D)$ hence
for n big enough:

$$\left| \frac{1}{f(x)} - \frac{1}{f_n(x)} \right| = \frac{|f_n(x) - f(x)|}{|f(x)|^2} \leq \frac{\|f_n - f\|_D}{\left({}_D\varphi_{a,r}(f) \right)^2} \text{ whenever } x \in T. \text{ Hence we see}$$

that the sequence $\dfrac{1}{f_n}$ converges to $\dfrac{1}{f}$ in $H(D \cup T)$.

Theorem 24.2:   *Let $D \in \mathcal{A}$. If an ideal contains a quasi-invertible element,
then it is generated by a polynomial whose zeros belong to $\overset{\circ}{\overline{D}} \cap D$.*
Proof:
Let $\mathcal{J}$ be an ideal of $H(D)$ that contains a quasi-invertible element f, and
let $\mathcal{J}_0$ be the set of polynomials that belong to $\mathcal{J}$. By hypothesis f factorizes in
$H(D)$ in the form Pg with g invertible in $H(D)$ and $P(x) \in K[x]$, all the zeros
of P lying inside $D \cap \overset{\circ}{\overline{D}}$. Since fg^{-1} belongs to $\mathcal{J}$, obviously P belongs to $\mathcal{J}_0$.
Hence T divides P , and then all the zeros of T lie in $D \cap \overset{\circ}{\overline{D}}$. We will show that
$\mathcal{J} = TH(D)$.

First we suppose that D is bounded. It is clearly seen that $\mathcal{J}_0$ is an ideal of
$K[x]$, hence there exists $T(x) \in K[x]$ such that $\mathcal{J}_0 = T(x)K[x]$.

Let $\alpha_1, ..., \alpha_q$ be the zeros of T. Now we suppose that there exists some
$h \in \mathcal{J} \setminus TH(D)$. Since D is bounded, by Theorem 10.11 , h is of the form $\dfrac{\ell}{S}$
with $\ell \in H(\overline{D})$ and S a polynomial whose zeros belong to $\overline{D} \setminus D$. Hence ℓ belongs
to $\mathcal{J}$. Now we can find $r > 0$ such that $d(\alpha_i, r) \subset \overline{D}$, whenever $i = 1, ..., q$. Let

$\Lambda = \bigcup_{i=1}^{q} d(\alpha_i, r)$ and let $D' = D \cup \Lambda$. Since the zeros of T lie in Λ there exists $\lambda > 0$ such that $|T(x)| \geq \lambda$ whenever $x \in D \setminus \Lambda$. Now since $\overline{D}$ is closed and bounded , there exists $b \in K$ such that $\|b\ell\|_D < \lambda$. We put $\phi = T + b\ell$. Clearly outside Λ , we have $|\phi(x)| \geq \lambda$. Besides by Corollary 23.8, in each disk $d(\alpha_i, r)$, ϕ has finitely many zeros, hence in D', ϕ has finitely many zeros, all of them in Λ. Hence it factorizes in the form $Q(x)W(x)$ with $W \in H(D'), W(x) \neq 0$ whenever $x \in D'$ and Q a polynomial whose zeros belong to Λ. By Theorem 23.6 , $|W(x)|$ has a strictly positive lower bound in Λ and another non zero lower bound in $D' \setminus \Lambda$ because $|Q(x)|$ is obviously bounded in D'. Finally W has a non zero lower bound in D', therefore it is invertible in $H(D')$. Hence Q belongs to $\mathcal{J}$. But then T divides Q in $K[x]$. Since $T + b\ell$ is equal to WQ, then T divides $T + b\ell$, and ℓ, and h too. This contradicts the hypothesis $h \in \mathcal{J} \setminus TH(D)$, and finishes proving that T generates $\mathcal{J}$ when D is bounded.

Now we suppose D unbounded. We may obviously assume D to have at least one hole $d(a, r^-)$, and without loss of generality, we may assume $a = 0$. Let $\gamma(x) = \dfrac{1}{x}$, and let $D' = \gamma(D)$. Then D' is a bounded set that belongs to $\mathcal{A}$, such that $0 \notin D'$. We also have $\gamma = \gamma^{-1}$, and $\gamma(D') = D$. Let ψ be the mapping from $H(D)$ into $H(D')$ defined as $\psi(f) = f \circ \gamma$. Then ψ is a K-algebra isomorphism from $H(D)$ onto $H(D')$. Besides, $\psi(\mathcal{J})$ is an ideal $\mathcal{J}'$ of $H(D')$. Let

$$P(x) = \prod_{j=1}^{q} (x - a_i), \text{ for every } j = 1, ..., q, \text{ let } a_i' = \frac{1}{a_i}, \text{ and let } B(u) = \prod_{j=1}^{q} (u - a_i').$$

Clearly, $\psi(P) = \dfrac{B(u)}{u^q}$. As $0 \notin D'$, u is invertible in $H(D')$, and B belongs to $\mathcal{J}'$.

Hence $\mathcal{J}'$ is generated by a polynomial whose zeros lie inside $D' \cap \overset{\circ}{\overline{D}'}$. Now, let

$$B(u) = \prod_{j=1}^{t} (u - c_j). \text{ For every } j = 1, ..., t, \text{ let } e_j = \frac{1}{c_j}, \text{ and let } T(x) = \prod_{j=1}^{t} (x - e_j).$$

It is seen that for each $j = 1, ..., t$, e_j does belong to $D \cap \overset{\circ}{\overline{D}}$. Moreover, for each $h \in \mathcal{J}$, $\psi(h)$ belongs to $\mathcal{J}'$ and is of the form $B(u)G(u)$, with G invertible in $H(D')$. Putting $F = \psi(G)$, in $H(D)$ we have $h = T(x)\frac{F(x)}{x^t}$. Then $\frac{F(x)}{x^t}$ is an invertible element of $H(D)$, and then this finishes showing that T generates $\mathcal{J}$.

Notations: For any integer $n \in \mathbb{N}$ we will denote by $\mathcal{Q}_n(D)$ the set of the quasi-invertible elements $f \in H(D)$ that have exactly n zeros, taking multiplicities into account, and by $\mathcal{Q}(D)$ the set $\bigcup_{n=0}^{\infty} \mathcal{Q}_n(D)$.

Theorem 24.3:   *Let D be closed and bounded. Let $n \in \mathbb{N}$, let $f \in \mathcal{Q}_n(D)$ and let $\alpha_1, ..., \alpha_n$ be the zeros of f (taking multiplicities into account). For every $\epsilon > 0$ there exists $\eta > 0$ such that for every $h \in H(D)$ satisfying $\|f - h\|_D \leq \eta$, h belongs to $\mathcal{Q}_n(D)$ and the zeros $\beta_1, ..., \beta_n$ of h, once correctly ordered, satisfy $|\alpha_i - \beta_i| \leq \epsilon$.*

Proof: Let $f = Pg \in \mathcal{Q}_n(D)$ with g invertible in $H(D)$, and P a n-degree monic polynomial whose zeros are interior to D. Let $\gamma_1, ..., \gamma_q$ be the different zeros of P, each γ_j of order s_j (with obviously $\displaystyle\sum_{j=1}^{q} s_j = n$). Let $\theta = \inf_{j \neq \ell} |\gamma_j - \gamma_\ell|$, let $\epsilon \in]0, \theta[$, let $\Lambda_j(\epsilon) = d(\gamma_j, \epsilon)$, and let $\Lambda(\epsilon) = \displaystyle\bigcup_{j=1}^{q} d(\gamma_j, \theta)$. It is easily seen that $|P(x)|$ has a non zero lower bound in $D \setminus \Lambda(\epsilon)$. Since D is closed and bounded, $|g(x)|$ has a non zero lower bound in D. Hence $|f(x)|$ has a lower bound $\lambda > 0$ in $D \setminus \Lambda(\epsilon)$. Let $\eta = \min(\lambda, \min_{1 \leq j \leq q} \|f\|_{\Lambda_j(\epsilon)})$ and let $h \in H(D)$ satisfy

(1) $\|f - h\|_D < \eta$.   Obviously we have $|f(x)| = |h(x)| \geq \lambda$ whenever $x \in D \setminus \Lambda(\epsilon)$. But then by (1) and by Theorem 23.7, we see that h has exactly s_i zeros like f in $\Lambda_i(\epsilon)$, $(1 \leq i \leq q)$ taking multiplicities into account. Thus we have alrealy proven the statement when all the zeros of f have order 1.

Now extending this to the general case is just a question of writing. We may assume the α_i to be ordered in such a way that
$$\alpha_1 = ... = \alpha_{s_1} = \gamma_1, \quad \alpha_{s_1+1} = ... = \alpha_{s_1+s_2} = \gamma_2,$$
$$\alpha_{s_1+...s_{q-1}+1} = ... = \alpha_{s_1+...+s_q} = \gamma_q.$$

Thus, for every $j = 1, ..., q$ in $\Lambda_j(\epsilon)$, γ_j is equal to $\alpha_{s_1+...+s_{j-1}+k}$ whenever $k = 1, ... s_j$. Since f admits s_j zeros in $\Lambda_j(\epsilon)$, as h does, we may denote them by $\beta_{s_1+...s_{j-1}+1}, ..., \beta_{s_1+...+s_j}$ (some of them being eventually equal). So we obtain $|\alpha_i - \beta_i| \leq \epsilon$ whenever $i = 1, ..., n$.

Corollary 24.4:   *Let D be closed and bounded. For every $n \in \mathbb{N}$, $\mathcal{Q}_n(D)$ is open in $H(D)$ and so is $\mathcal{Q}(D)$.*

Lemma 24.5:   *Let $a \in \overline{D}$ satisfy $a \notin \overset{\circ}{D}$. There exists a quasi-minorated element $f \in H_b(D)$ which is not semi- invertible, satisfying $\lim_{\substack{x \to a \\ x \in D}} f(x) = 0$ and*
$$\limsup_{\substack{x \to 0 \\ x \in D}} \left| \frac{f(x)}{x} \right| = +\infty.$$

Proof: Without loss of generality we may obviously assume $a = 0$. Then the Cauchy filter $\mathcal{F}$ of base $\{d(0, r) \cap D \mid r > 0\}$ is pierced . Let $(T_m)_{m \in \mathbb{N}}$ be a

sequence of holes of D that runs $\mathcal{F}$ and let $D' = K \setminus (\bigcup_{m=0}^{\infty} T_m)$. By Lemma 18.4,
there exists $f \in H_b(D')$ such that $\lim_{\substack{x \to 0 \\ x \in D'}} f(x) = 0$ and such that $\limsup_{\substack{x \to 0 \\ x \in D}} \left| \frac{f(x)}{x} \right| = +\infty$. We check that D' has no monotonous pierced filter because its only holes are the T_m. Hence by Corollary 22.2, f is quasi-minorated. If 0 belongs to D it is seen that $f(0) = 0$ while $\limsup_{\substack{x \to 0 \\ x \in D}} \left| \frac{f(x)}{x} \right| = +\infty$ hence f can't factorize in the form $xg(x)$ with $g \in H(D)$ and therefore f is not semi-invertible.

Now, we suppose $0 \notin D$. Let us suppose f quasi-invertible. Then it factorizes in the form $P(x)g(x)$ with P the polynomial of the zeros of f in $\overline{D}{}^{\circ}$ and g an invertible element in $H(D)$. Since $0 \notin D$, we have $P(0) \neq 0$, hence $\lim_{\substack{x \to 0 \\ x \in D}} g(x) = 0$.

Hence, by Corollary 10.9, $\dfrac{1}{g(x)}$ admits a pole at 0. Let n be its order. Then by Corollary 10.9, $\dfrac{x^n}{g(x)}$ has a limit different from zero at 0. But since $\dfrac{f(x)}{x}$ is unbounded in any set $d(0, r) \cap (D' \setminus \{0\})$, so is $\dfrac{g(x)}{x}$ and therefore we have $\liminf_{x \to 0} \left| \dfrac{x^n}{g(x)} \right| = 0$. Hence g can't be invertible . This finally shows that f is not semi-invertible and finishes the proof of Lemma 24.5.

Lemma 24.6: *Let D be such that $\widetilde{D} \setminus \overline{D}$ is not bounded. Then there exists a quasi-minorated element $f \in H_b(D)$ satisfying*

$$(1) \quad \lim_{\substack{|x| \to \infty \\ x \in D}} f(x) = 0 \quad and \quad (2) \quad \limsup_{\substack{|x| \to \infty \\ x \in D}} |x f(x)| = +\infty.$$

Besides xf does not belong to $H(D)$.

Proof: Since D has holes, we may obviously assume that 0 belongs to a hole. Let $\gamma(x) = \dfrac{1}{x}$, and let $D' = \gamma(D)$. Then D' is bounded and 0 belongs to $\overline{D'} \setminus \overline{D'}{}^{\circ}$. By Lemma 24.5, there exists a quasi-minorated element $h \in H_b(D')$ satisfying

$$(3) \quad \lim_{\substack{x \to 0 \\ x \in D}} f(x) = 0 \quad and \quad (4) \quad \limsup_{\substack{x \to 0 \\ x \in D}} \left| \frac{f(x)}{x} \right| = +\infty.$$

Then we put $f = h \circ \gamma$. By (3), f satisfies (1), by (4), f satisfies (2). Besides, by lemma 14.6, f is quasi-minorated. Finally we check that xf does not belong to $H(D)$. Indeed suppose $xf \in H(D)$. By Theorem 10.5, xf is of the form $g(x) + P(x)$, with $g \in H_b(D)$ and $P \in K[x]$. Let $q = deg(P)$. Since xf is not bounded, we have $q > 0$. Then $x^{1-q}f$ has a limit different from 0 when $|x|$ tends to $+\infty$, and this contradicts (1). This ends the proof of Lemma 24.6.

Theorem 24.7: *If D does not belong to $\mathcal{A}$, there exist invertible elements $f, g \in H(D)$ such that fg belongs to $H(D)$ but is not semi-invertible.*

Proof: First we suppose that there exists $a \in (\overline{D} \setminus D) \setminus \overset{\circ}{\overline{D}}$. Without loss of generality, we assume $a = 0$. By Lemma 24.5, there exists $f \in H_b(D)$ such that $\lim\limits_{\substack{x \to 0 \\ x \in D}} f(x) = 0$ while $\dfrac{f(x)}{x}$ is not bounded in any set $D \cap d(0, r)$ $(r > 0)$. Since $f \in H_b(D)$, we can find $A \in K$ such that $|A| < \|f\|_D$. Let $g = A + f$. Then g is invertible in $H(D)$. Let $T = d(b, \rho^-)$ be a hole of D and let $F = \dfrac{x}{(x - b)g}$. Then both $\dfrac{x}{x - b}$, g^{-1} belong to $H_b(D)$, hence so does F. Besides, by definition, F is the product of two invertible elements of $H(D)$. We also notice that F has no zero in D.

Next, we notice that $\dfrac{f(x)(x - b)}{x}$ is not bounded in any set $D \cap d(0, r)$ $(r > 0)$, although $\lim\limits_{\substack{x \to 0 \\ x \in D}} f(x)(x - b) = 0$. Hence by Corollary 10.9 $\dfrac{f(x)(x - b)}{x}$ does not belong to $H(D)$. But then, since $\dfrac{A(x - b)}{x}$ does belong to $H(D)$, we see that F^{-1} does not belong to $H(D)$. Since F has no zero in D, it is not semi-invertible although both $\dfrac{x}{x - b}$, g^{-1} are invertible in $H(D)$.

Now we suppose that $\widetilde{D} \setminus \overline{D}$ is not bounded. Since D has holes, we may obviously assume that 0 belongs to a hole. Let $\gamma(x) = \dfrac{1}{x}$, and let $D' = \gamma(D)$.

Then D' is bounded, and 0 belongs to $\in \overline{D'} \setminus \overset{\circ}{\overline{D'}}$. Hence, as we just saw, there exist invertible elements $h, g \in H_b(D')$ such that hg belongs to $H_b(D')$ and has no zero in D' but is not invertible in $H(D')$. Then we put $\theta = h \circ \gamma$, $\psi = g \circ \gamma$, $\phi = (hg) \circ \gamma$. By Lemma 11.4, both θ, ψ are invertible in $H(D)$, ϕ belongs to $H(D)$ and has no zero in D. Since hg is not invertible in $H(D')$, by Lemma 11.4 ϕ is not invertible in $H(D)$. Since it has no zero in D, it is not semi-invertible in $H(D)$, and this finishes the proof of Theorem 24.7.

Theorem 24.8: *The following three statements are equivalent:*
   i) *D belongs to $\mathcal{A}$ and is such that $\overline{D}$ is open.*
   ii) *$\widetilde{D} \setminus \overline{D}$ is bounded and $\overline{D}$ is open.*

ii) $\widetilde{D} \setminus \overline{D}$ *is bounded and* $\overline{D}$ *is open.*

iii) *The set of the quasi-minorated elements of* $H(D)$ *is equal to the set of the quasi-invertible elements.*

Proof: By Theorem 18.8 we know that i) implies ii). Conversely, suppose ii) is satisfied. If D does not belong to $\mathcal{A}$, since $\widetilde{D} \setminus \overline{D}$ is bounded, there must exist $a \in \overline{D} \setminus D$ that does not belong to $\overset{\circ}{\overline{D}}$ and therefore this contradicts ii). Hence i) and ii) are equivalent. Now by Theorem 22.5, i) implies iii). Finally it just remains to show that if $\widetilde{D} \setminus \overline{D}$ is not bounded or if $\overline{D}$ is not open then there exist quasi-minorated elements that are not quasi-invertible.

On one hand, if $\overline{D}$ is not open, by Lemma 24.5 such an element does exist. On the other hand, if $\widetilde{D} \setminus \overline{D}$ is not bounded then D does not belong to $\mathcal{A}$, and therefore, by Theorem 24.7, there exist invertible elements f, g in $H(D)$ such that fg is not semi-invertible. But then, by Theorem 22.4, both f, g are quasi-minorated, and then by Theorem 22.3, fg is quasi-minorated. This ends the proof of Theorem 24.8.

25. ZEROS THEOREM FOR POWER SERIES

This chapter is aimed to show that, given a sequence $(a_n)_{n\in\mathbb{N}}$ such that $|a_n| < r$ for all $n \in \mathbb{N}$, and $\lim_{n\to\infty} |a_n| = r$, a sequence of integers $(q_n)_{n\in\mathbb{N}}$, and a number $B > 1$, there exists an analytic function $f \in A(d(0, r^-))$ that admits each a_n as a zero of order $t_n \geq q_n$, and such that $\|f\|_{d(0,|a_n|)} \leq B \prod_{j=0}^{n} |a_j|^{q_j}$. This result, that is very useful, was stated by J. Fresnel and B. de Mathan in [38], without any proof except a reference to a lemma of [49]. Here we will give a full proof of this.

Definition : Let $P(x) \sum_{j=0}^{q} a_j x^j \in K[x]$, and let $r \in]0, +\infty[$. P will be said to be *r-optimal* if all the zeros of P lie in $C(0, r)$.

Lemma 25.1: *A polynomial* $P(x) = \sum_{j=0}^{q} a_j x^j$ *is r-optimal if and only if*

$$|a_0| = |a_q| r^q = \sup_{a \leq j \leq q} |a_j| r^j.$$

Proof: Indeed P is r-optimal if and only if $N^+(P, -\log r) - N^-(P, -\log r) = q$, i.e. if $N^-(P, -\log r) = a_0$, $N^+(P, -\log r) = q$.

Corollary 25.2: *If a polynomial P is r-optimal, then r lies in $|K|$.*

Lemma 25.3: *Let $r \in]0, +\infty]$, let $D = C(0, r)$, let $f \in H(D)$, and let $P(x) \in K[x]$ be a r-optimal polynomial. There exist $g \in H(D)$ and $S \in K[x]$ satisfying*
(1) $f = Pg + S$
(2) $deg(S) < deg(P)$.
Besides, both g, S are unique satisfying (1) and (2). Further, g, S satisfy
(3) $\|S\|_D \leq |f\|_D$
(4) $\|g\|_D \leq \dfrac{\|f\|_D}{\|P\|_D}$
Moreover, if a set D' contains D and if f belongs to $H(D')$, then so does g.

Proof: Let $\nu = -\log r$ and let $s = deg(P)$. We know that $-\log \|f\|_D = v(f,\nu)$. Thus we have to find $g \in H(D)$ and $S \in K[x]$ satisfying (1) , (2), and

(5) $v(S,\nu) \geq v(f,\nu)$

(6) $v(g,\mu) \geq v(f,\nu) - v(P,\nu)$

We will prove that for each $n \in \mathbb{Z}$ there exist $S_n \in K[x]$ and $g_n \in H(D)$ satisfying :

(7) $x^n = Pg_n + S_n,$

(8) $deg(S_n) < s,$

(9) $v(S_n,\nu) \geq n\nu,$

(10) $v(g_n,) \geq n\nu - v(P,\nu).$

First we suppose $n \geq 0$. Since P is r-optimal there exists $\lambda \in K$ such that $|\lambda| = r$. Therefore a change of variable of the form $u = \dfrac{x}{\lambda}$ enables us to assume $r = 1$. Obviously we can also assume $\|P\|_D = 1$. But then, as P is r-optimal, P is quasi-monic. Hence by Lemma 6.3, there exist S_n and $g_n \in K[x]$ satisfying (7), (8), (9), (10).

Now, we suppose $n < 0$. We put $\check{P}(x) = x^s P\left(\dfrac{1}{x}\right)$. Clearly $\check{P}$ is a r-optimal polynomial again. Then we can perform the Euclidean division of x^{h+s-1} by $\check{P}$ and we obtain $\check{g}_n \in H(C(0,r))$ and $\check{S}_n x) \in K[x]$ satisfying

(11) $x^{n+s-1} = \check{P}\check{g}_n + \check{S}_n,$

(12) $deg(\check{S}_n) < s,$

(13) $v(\check{g}_n,-\nu) \geq -(n+s-1)\nu - v(\check{P},-\nu)$

(14) $v(\check{S}_n,-\nu) \geq -(n+s-1)\nu.$

Now we put $g_{-n}(x) = x^{-1}\check{g}_n\left(\dfrac{1}{x}\right)$ and $S_{-n}(x) = x^{s-1}\check{S}_n\left(\dfrac{1}{x}\right)$. Clearly by (11) we have :

(15) $x^{-n} = P(x)g_{-n}(x)S_{-n}(x).$

By definition and by (14) S_{-n} is a polynomial satisfying

(16) $deg(r(S_{-n})) < s$

and $v(S_{-n},\nu) = (s-1)\nu + (\check{S}_n - \nu)$, hence by (14) , we have

(17) $v(S_{-n},\nu) \geq -n\nu.$

Next, by definition, g_{-n} satisfies $v(g_{-n},\nu) = v(\check{g}_n,-\nu) - \nu$ hence by (13) we obtain $v(g_n,\nu) \geq -(n+s-1)\nu - \nu - (-s\nu + v(P,\nu)) = -n\nu - v(P,\nu).$

Thus, we have now proven the theorem when f is of the form x^n for every $n \in \mathbb{Z}$. This is easily generalized to the general case. Indeed, in the general case, f is of the form $\displaystyle\sum_{-\infty}^{+\infty} a_n x^n$ with $\displaystyle\lim_{n\to+\infty} |a_n|r^n = \lim_{n\to-\infty} |a_n|r^n = 0$. For every $n \in \mathbb{Z}$ we have $x^n = P(x)g_n(x) + S_n(x)$ with S_n, g_n satisfying (7), (8), (9), (10).

Actually we know that $v(S_n, \nu) = -\log\left(\|S\|_D\right)$ hence by (9) the series $\displaystyle\sum_{n=q}^{t} a_n S_n$ does converge in $H(D)$ to a polynomial S satisfyng (2) (3). In the same way by (10) the series $\displaystyle\sum_{n=q}^{t} a_n g_n$ converges in $H(D)$ to a limit g satisfying (1) and (2).

Finally we check that both g, S are unique. Indeed, supose we can find $h \in H(D)$ and $Q \in K[x]$ satisfying $deg(Q) < s$ and $f = Ph + Q$. Then in $H(D), Q - S$ can be factorized in the form $P(g - h)$. Since P is r-optimal, by Lemma 25.1 its s zeros lie in D. Hence $Q - S$ admits at least s zeros in D, but this contradicts the fact that $deg(Q - S) < s$.

Now let D' be a set containing D, and suppose that f belongs to $H(D')$. Then so does $f - S$. Clearly, P divides the polynomial of the zeros of $f - S$. Then by Corollary 14.5 it is seen that P divides $f - S$ in $H(D')$. Therefore $f - S$ factorizes in the form $P\ell$ with $\ell \in H(D')$, hence g is just the restriction of ℓ to D.

Lemma 25.4:   *Let $(u_n)_{n\in\mathbb{N}}$ be a sequence in $\mathbb{R}_+$ such that $\displaystyle\sum_{n=0}^{\infty} u_n < +\infty$. Let*

$$A = \sum_{n=0}^{\infty} u_n \text{ and let } B > A. \text{ There exists an increasing sequence } (q_n)_{n\in\mathbb{N}} \text{ such}$$

that $\displaystyle\lim_{n\to\infty} q_n = +\infty$ and such that $\displaystyle\sum_{n=0}^{\infty} q_n u_n \leq B$.

Proof: Let $E = B - A$. For every $n \in \mathbb{N}$ we denote by s_n the smallest integer such that $\displaystyle\sum_{j=s_n}^{\infty} u_j \leq 4^{-n} E$. Then for every $j \in \mathbb{N}$ such that $s_n \leq j < s_{n+1}$ we put $q_j = 2^n$. We have $\displaystyle\sum_{j+s_n}^{s_{n+1}-1} q_j u_j \leq 2^{-n} E$, hence $\displaystyle\sum_{j=0}^{s_{n+1}-1} q_j u_j \leq A + E \sum_{k=1}^{n} 2^{-k}$ and finally $\displaystyle\sum_{j=0}^{\infty} q_j u_j \leq B$.

Theorem 25.5:   *Let $(a_j)_{j\in\mathbb{N}}$ be a sequence in $d(0, r^-)$ such that $|a_n| \leq |a_{n+1}|$ for every $n \in \mathbb{N}$ and $\displaystyle\lim_{n\to\infty} |a_n| = r$. Let $(q_n)_{n\in\mathbb{N}}$ be a sequence in $\mathbb{N}^*$ and let $B \in]1, +\infty[$. There exists $f \in A_b(d(0, r^-))$ satisfying*

    i) $f(0) = 1$

ii) $\|f\|_{d(0,|a_n|)} \leq B \prod_{j=0}^{n} |\frac{a_n}{a_j}|^{q_j}$ *whenever* $n \in \mathbb{N}$

iii) *for each* $n \in \mathbb{N}$ a_n *is a zero of* f *of order* $z_n \geq q_n$.

Proof. We put $\rho = -\log r$. The set $\{|a_j| \mid j \in \mathbb{N}\}$ is obviously equal to the image of a strictly increasing sequence of limit r that we will denote by $(r_m)_{m \in \mathbb{N}}$. For each $m \in \mathbb{N}$, the set S_m of the a_j that lie in $C(0, r_m)$ is of the form $\{a_{h_m}, a_{h_m+1}, ..., a_{k_m}\}$. We put $P_m = \prod_{j=h_m}^{k_m} (1 - \frac{x}{a_j})^{q_j}$. Without loss of generality we can assume

(1) $|a_n| > \sqrt{\dfrac{r^2}{B}}$ for every $n \in \mathbb{N}$.

Indeed suppose we can construct f satisfying i), ii), iii) when (1) is satisfied and consider the general case. Let $t \in \mathbb{N}$ be such that $|a_n| > \sqrt{\dfrac{r^2}{B}}$ whenever $n \in \mathbb{N}$. Then we can construct $g \in A(d(0, r^-))$ satisfying $g(0) = 1$, $\|g\|_{d(0,|a_n|)} \leq B \prod_{j=t}^{n} |\frac{a_n}{a_j}|^{q_j}$ whenever $n \geq t$, and such that for each $n \geq t$, a_n is a zero of g of order $z_n \geq q_n$. Then we just put $f(x) = \prod_{j=0}^{t-1} (1 - \frac{x}{a_j})^{q_j} g(x)$ and then f is easily seen to satisfy i), ii), iii). Thus we will assume (1). For every $m \in \mathbb{N}$ we put $\mu_m = -\log r_m$, and $\lambda = -\log B$. By (1) we have $\mu_0 \leq \dfrac{\lambda}{2}$. By Lemma 25.4 there exists an increasing sequence $(t_n)_{n \in \mathbb{N}}$ in $\mathbb{N}$ such that $\lim_{n \to \infty} t_n = +\infty$ and such that

(2) $\sum_{j=0}^{\infty} t_j(\mu_{j+1}) \leq \lambda.$

For every $s \in \mathbb{N}$ we put $\sigma(s) = \sum_{j=0}^{s} t_j(\mu_j - \mu_{j+1})$, and $g_s = \prod_{m=0}^{s} P_m$. We notice that $v(g_s, \mu) = v(g_m, \mu)$ whenever $\mu \geq \mu_m$ and $s \geq m$. So, we can define the function ℓ from $]\rho, +\infty[$ into $\mathbb{R}$ as $\ell(\mu) = \lim_{s \to \infty} v(g_s, \mu)$. Then ℓ is obviously increasing and satisfies

(3) $\ell(\mu) = \sum_{j=0}^{k(m)} q_j(\mu - v(a_j))$ whenever $\mu \in [\mu_{m+1}, \mu_m]$.

We will construct a sequence $(f_s)_{s \in \mathbb{N}}$ in $K[x]$ satisfying $(\alpha_s), (\beta_s), (\gamma_s), (\delta_s)$ for every $s \in \mathbb{N}$ and $(\epsilon_s), (\varphi_s)$ for every $s \in \mathbb{N}^*$.

(α_s) $f_s(0) = 1$,

(β_s) P_j divides f_s for every $j \leq s$,

(γ_s) $v(f_s, \mu_{s+1}) \geq \ell(\mu_{s+1}) - \sigma(s)$,

(δ_s) $v(f_s, \mu) \geq \ell(\mu) - \lambda$ whenever $\mu > \rho$,

(ϵ_s) $v(f_s - f_{s-1}, \mu) \geq \ell(\mu) + t_s(\mu - \mu_s) - \lambda$ whenever $\mu \geq \mu_s$,

(φ_s) $v(f_s - f_{s-1}, \mu) \geq \ell(\mu_s) + t_s(\mu - \mu_s) - \sigma(s)$ whenever $\mu \in]\rho, \mu_s]$.

By taking $f_0 = P_0$, we check that $(\alpha_0), (\beta_0), (\gamma_0), (\delta_0)$ are obviously satisfied. We now suppose already constructed f_m satisfying $(\alpha_m), (\beta_m), (\gamma_m), (\delta_m)$ for every $m = 0, ..., s$ and $(\epsilon_m), (\varphi_m)$ for every $m = 1, ..., s$. We will define f_{s+1} satisfying $(\alpha_{s+1}), (\beta_{s+1}), (\gamma_{s+1}), (\delta_{s+1}), (\epsilon_{s+1}), (\varphi_{s+1})$. It is seen that each polynomial P_s is r_s - optimal. Let R_{s+1} be the rest of the Euclidean division of f_s by P_{s+1} in $H(C(0, r_{s+1}))$. Let $Q_{s+1} = x^{t_{s+1}} g_s$. We have

(4) $v(Q_{s+1}, \mu) = \mu t_{s+1} + v(g_s, \mu)$ whenever $\mu \in \mathbb{R}$.

We notice that Q_{s+1} admits no zero in $C(0, r_{s+1})$ and then is invertible in $H(C(0, r_{s+1}))$. As a consequence we can perform the Euclidean division of R_{s+1} by P_{s+1} in $H(C(0, r_{s+1}))$. Let V_{s+1} be the rest of this division. Thus, $\dfrac{R_{s+1}}{Q_{s+1}}$ is of the form $T_{s+1}P_{s+1} + V_{s+1}$ with $T_{s+1} \in H(C(0, r_{s+1}))$, and we have $R_{s+1} = Q_{s+1}(T_{s+1}P_{s+1} + V_{s+1})$. Now we put $f_{s+1} = f_s - Q_{s+1}V_{s+1}$. Of course, f_{s+1} satisfies (α_{s+1}). We will check that f_{s+1} satisfies β_{s+1}). By definition each P_j divides f_s for every $j = 0, ..., s$. Next, P_{s+1} divides both $f_s - R_{s+1}$ and $Q_{s+1}T_{s+1}P_{s+1}$. Hence it also divides f_{s+1} and thereby (β_{s+1}) is satisfied.

Now we will prove (φ_{s+1}). By Theorem 25.3 R_{s+1} satisfies $v(R_{s+1}, \mu_{s+1}) \geq v(f_s, \mu_{s+1})$. Hence by Relation (γ_s) we have

(5) $v(R_{s+1}, \mu_{s+1}) \geq \ell(\mu_{s+1}) - \sigma(s) = v(g_{s+1}, \mu_{s+1}) - \sigma(s)$.

Since $deg(R_{s+1}) < deg(P_{s+1}) < deg(g_{s+1})$ and since all the zeros of g_{s+1} lie in $d(0, r_{s+1})$, by Theorem 23.13 we have

$$v(R_{s+1}, \mu_{s+1}) - v(R_{s+1}, \mu) \leq v(g_{s+1}, \mu_{s+1}) - v(g_{s+1}, \mu),$$

and therefore by (5) we obtain

(6) $v(R_{s+1}, \mu) \geq v(g_{s+1}, \mu) - \sigma(s)$ whenever $\mu \in]\rho, \mu_{s+1}]$.

In the same way by Theorem 25.3 we have

(7) $v(V_{s+1}, \mu_{s+1}) \geq v(R_{s+1}, \mu_{s+1}) - v(g_s, \mu_{s+1}) - t_{s+1}\mu_{s+1}$.

Hence by (4) and (7) we obtain $v(Q_{s+1}V_{s+1}, \mu_{s+1}) \geq v(R_{s+1}, \mu_{s+1})$ and then by (6) we have

(8) $v(Q_{s+1}V_{s+1}, \mu_{s+1}) \geq v(g_{s+1}, \mu_{s+1}) - \sigma(s)$.

We notice that $deg(Q_{s+1}V_{s+1}) < deg(g_{s+1} + t_{s+1}$ and that all the zeros of $g_{s=1}$ lie in $d(0, r_{s+1})$. Hence by Theorem 23.13 and by (8) we have

(9) $v(Q_{s+1}V_{s+1}, \mu) \geq v(g_{s+1}, \mu) - t_{s+1}(\mu_{s+1} - \mu) - \sigma(s)$ for every $\mu \in]\rho, \mu_{s+1}]$.

Actually, $v(Q_{s+1}V_{s+1}, \mu) = v(f_{s+1} - f_s, \mu)$ and $v(g_{s+1}, \mu) \geq \ell(\mu)$ for every $\mu \in \,]\rho, \mu_{s+1}]$, hence by (9) we have proven φ_{s+1}. We will deduce (ϵ_{s+1}).
In particular when $\mu = \mu_{s+2}$, we obtain

$$v(Q_{s+1}V_{s+1}, \mu) \geq \ell(\mu_{s+2}) - t_{s+1}(\mu_{s+1} - \mu_{s+2}) - \sigma(s).$$

But we notice that $t_{s+1}(\mu_{s+1} - \mu_{s+2}) + \sigma(s) = \sigma(s+1)$, hence
(10) $v(Q_{s+1}V_{s+1}, \mu_{s+2}) \geq \ell(\mu_{s+2}) - \sigma(s+1).$
Anyway, by (9) we obtain
(11) $v(f_{s+1} - f_s, \mu) \geq \ell(\mu) - \lambda$ whenever $\mu \in\,]\rho, \mu_{s+1}]$
Now we take $\mu \geq \mu_{s+1}$. It is seen that

$$v(Q_{s+1}, \mu) - v(Q_{s+1}, \mu_{s+1}) = \ell(\mu) - \ell(\mu_{s+1}) + t_{s+1}(\mu - \mu_{s+1}).$$

Therefore we have

$$(Q_{s+1}V_{s+1}, \mu) \geq v(Q_{s+1}V_{s+1}, \mu_{s+1}) + \ell(\mu) - \ell(\mu_{s+1}) + t_{s+1}(\mu - \mu_{s+1}).$$

But by (9) we have $v(Q_{s+1}V_{s+1}, \mu_{s+1}) \geq \ell(\mu_{s+1}) - \lambda$, hence we obtain
$v(Q_{s+1}V_{s+1}, \mu) \geq \ell(\mu) - \lambda + t_{s+1}(\mu - \mu_{s+1})$ whenever $\mu \geq \mu_{s+1}$, and this is
(ϵ_{s+1}). In particular we have $v(f_{s+1} - f_s, \mu) \geq \ell(\mu) - \lambda$ whenever $\mu \geq \mu_{s+1}$ and
therefore by (δ_s) we obtain (δ_{s+1}).
Now we will show (γ_{s+1}). Obviously we have
(12) $v(f_0, \mu_{s+2}) \geq \ell(\mu_{s+2}).$
Next, by Relations $(\varphi_m)_{1 \leq m \leq s+1}$ for every $m \in \mathbb{N}^*$ we have

$$v(f_m - f_{m-1}, \mu_{s+2}) \geq \ell(\mu_m) - [t_m(\mu_m - \mu_{s+2}) + \sigma(m)].$$

But as the sequence $(t_m)_{m \in \mathbb{N}}$ is increasing, it is seen that

$$\sigma(m) + t_m(\mu_m - \mu_{s+2}) \leq \sum_{j=0}^{s+1} t_j(\mu_j - \mu_{j+1}) = \sigma(s+1).$$

Obviously $\ell(\mu_{s+2}) \leq \ell(\mu_m)$, hence we obtain $v(f_m - f_{m-1}, \mu_{s+2}) \geq \ell(\mu_{s+2}) - \sigma(s+1)$ whenever $m = 1, ..., s+1$. Finally by (12), f_{s+1} satisfies $v(f_{s+1}, \mu_{s+2}) \geq \ell(\mu_{s+2}) - \sigma(s+1)$ and this is (γ_{s+1}).

The sequence $(f_s)_{s \in \mathbb{N}}$ is now constructed satisfying (α_s), (β_s), (γ_s), (δ_s), (ϵ_s), (φ_s). By Relations (ϵ_s) the sequence is easily seen to converge in each algebra $H(d(0, u))$ whenever $u \in\,]0, r[$. Indeed, given $u \in\,]0, r[$ and $N \in \mathbb{N}$ such that $\mu_N < \log u$, by (ϵ_{s+1}) we have $-\log(\|f_{s+1} - f_s\|_{d(0,u)}) = v(f_{s+1} - f_s, -\log u) \geq$

$\ell(-\log u) - t_s(\log u + \mu_{s+1}) - \lambda$ hence $-\log(\|f_{s+1} - f_s\|_{d(0,u)}) \geq \ell(-\log u) - t_s(\log u + \mu_N) - \lambda$, whenever $s > N$. As $\lim\limits_{s \to +\infty} t_s = +\infty$, it is seen that

$\lim\limits_{s \to \infty} \|f_{s+1} - f_s\|_{d(0,u)} = 0$. Let f be the function defined in $d(0, r^-)$ as the limit of the sequence $(f_s)_{s \in \mathbb{N}}$ in each disk $d(0, u)$. Obviously, as an element of $H(d(0, u))$ for every $u \in]0, r[$, f belongs to $A(d(0, r^-))$. By Relations (α_s), f obviously satisfies i).

We will check ii). Let $u \in]0, r[$ be such that $\mu_N \leq -\log u \leq \mu_{N+1}$. We have

(13) $-\log \|f\|_{d(0,u)} = v(f, -\log u)$.

When s is big enough, $v(f_s - \log u)$ is clearly equal to $v(f, -\log u)$, hence f satisfies $-\log \|f\|_{d(0,u)} = v(f_s - \log u) \geq \ell(-\log u) - \lambda$. Hence by (3) we obtain ii). Now we just have to check that every a_j is a zero of f of order $z_j \geq q_j$. Let m be such that $h_m \geq j$. For every $s \geq m, (1 - \dfrac{x}{a_j})^{q_j}$ divides f_s in $H(d(0, u))$ (for every $u \in]0, r]$). Therefore, by Corollary 14.3 $(1 - \dfrac{x}{a_j})^{q_j}$ divides f in $H(d(0, u))$ and this finishes the proof of theorem 25.5.

Remark: It is known that when K is not spherically complete, one can't hope to find an analytic function that admits each a_n as a zero of order q_n exactly, and no other zero [49]. But at least, by Theorem 25.5 we can reduce the discrepency caused by the additional zeros as much as we want.

26.   IMAGE OF A DISK

Theorem 26.1: $Let\ f(x) = \sum_{n=0}^{\infty} a_n(x-a)^n \in H(d(a,r))\ and\ let\ s = \sup_{n\geq 1} |a_n| r^n.$
$Then\ f(d(a,r)) = d(a_0,s)\ and\ v_a(f - a_0, -\log r) = -\log s.$

Proof: On one hand $f(d(a,r))$ is clearly included in $d(a_0,s)$. On the other hand, we take $b \in d(a_0,s)$ and consider the element $g(x) = f(x) - b = a_0 - b + \sum_{n=1}^{\infty} a_n(x - a)^n$. Without loss of generality we may obviously assume $a = 0$. Besides, Theorem 26.1 is trivial when $s = 0$ so we may assume $s > 0$. Hence by hypothesis we have $N^+(g, -\log r) \geq 1$ because of the inequality $\sup_{n\geq 1} |a_n| r^n \geq |a_0 - b|$. But then, by Theorem 23.5 g admits at least one zero in $d(0,r)$ and therefore b belongs to $f(d(0,r))$. Now, by Proposition 20.7 we see that $v(f - a_0, -\log r) = \inf_{n\geq 1} v(a_n) - n\log r$ and therefore $v(f - a_0, -\log r) = -\log s$.

Corollary 26.2: $Let\ f(x) = \sum_{n=0}^{\infty} a_n(x - a)^n \in A_b(d(a,r^-))\ and\ let\ s = \sup_{n\geq 1} |a_n| r^n.\ If\ s > 0\ then\ f(d(a,r^-)) = d(a_0,s^-).$

Proof: On one hand $f(d(a,r^-))$ is obviously included in $d(a_0,s^-)$. On the other hand, let $b \in d(a_0,s^-)$ and let $\rho \in]0,r[$ satisfy $\sup_{n\geq 1} |a_n|\rho^n \geq |b - a_0|$. Then by Theorem 26.1, b does belong to $f(d(a,\rho))$ because $f \in H(d(a,\rho))$.

Remark: In particular Corollary 26.2 applies to any element of $H(d(a,r^-))$.

Proposition 26.3: $Let\ D\ be\ a\ set\ of\ the\ form\ \bigcup_{i\in I} d(\alpha_i,r^-)\ with\ |\alpha_i - \alpha_j| = r$ $whenever\ i \neq j.\ Let\ f \in H(D).\ Let\ \ell\ be\ fixed\ in\ I\ ,\ let\ f(x) = \sum_{n=0}^{\infty} a_n(x - \alpha_\ell)^n$ $whenever\ x \in d(\alpha_\ell,r^-)\ and\ let\ s = \sup_{n\geq 1} |a_n| r^n.\ For\ every\ i \in I,\ let\ \beta_i = f(\alpha_i).$ $Then\ we\ have\ f(d(\alpha_i,r^-)) = d(\beta_i,s^-)\ for\ every\ i \in I\ and\ |\beta_i - \beta_j| \leq s\ whenever$

$i, j \in I$. Besides if I is not finite, the equality $|\beta_i - \beta_\ell| = s$ holds for every $i \in I$ but finitely many.

Proof: We put $a = \alpha_\ell$ and $g = f - \beta_\ell$. By Theorem 26.1, for every $\rho \in]0, r[$ we have $v_a(g, -\log \rho) = \inf_{n \geq 1} v(a_n) - n \log \rho$ hence by continuity:

$$v_a(g, -\log r) = \inf_{n \geq 1} v(a_n) - n \log r = -\log s.$$

Now let $j \in I$ and let $b = \alpha_j$. Since $|b - a| = r$, by Proposition 20.2 we have $v_a(g, -\log r) = v_b(g, -\log r)$ hence $v(g(x)) \geq v_a(g, -\log r) = -\log s$ for all $x \in d(b, r^-)$. Thus we see that $|f(x) - \beta_\ell| \leq s$ for all $x \in d(b, r^-)$. Hence, $f(d(b, r^-))$ is a disk $d(\beta_j, t)$ with $t \leq s$. Since α_ℓ and α_j play the same role, in the same way we show that $s \leq t$ and therefore $s = t$. Thus we have proven that $f(d(\alpha_i, r^-)) = d(\beta_i, s^-)$ for every $i \in I$ and $|\beta_i - \beta_j| \leq s$ whenever $i, j \in I$.

Now we suppose I infinite. The equality $v(g(x)) = v_a(g, -\log r) = -\log s$ is true in all the classes of $C(a, r)$ but finitely many ones, hence we have $|\beta_j - \beta_\ell| = s$ for every $j \in I$ but finitely many and this ends the proof of Proposition 26.3.

Theorem 26.4: *Let D be an open set, let $f \in H(D)$ and let $D' = f(D)$. Then D' is open and satisfies $\delta(D', (K \setminus D')) \geq \delta(D, (K \setminus D)) \inf_{x \in D} |f'(x)|$.*

Proof: By Theorem 26.1 D' is clearly open. Let $a' \in D'$, let $a \in D$ be such that $f(a) = a'$ and let $r = \delta(a, K \setminus D)$. In $d(a, r^-)$, $f(x)$ is equal to a power series $\sum_{n=0}^{\infty} a_n(x - a)^n$. Let $s = \sum_{n \geq 1} |a_n| r^n$. By Corollary 26.2 we have $d(a', s) \subset D'$ hence (1) $\delta(a', (K \setminus D')) \geq s$. We put $t = \inf_{x \in D} |f'(x)|$. In $d(a, r^-)$ we have $|f'(x)| \leq \sup_{n \geq 1} |na_n| r^{n-1}$ hence $|h'(x)| \leq \dfrac{s}{r}$ and then $s \geq rt$. Finally by (1) we have $\delta(a', (K \setminus D')) \geq rt \geq t\delta(D, (K \setminus D))$. Since this is true for every $a' \in D'$ we see that $\delta(D', (K \setminus D')) \geq t\delta(D, K \setminus D)$.

Lemma 26.5: *Let $f \in H(d(0, r))$ satisfy $N^+(f, -\log r) \geq 1$, let $b \in f(d(0, r))$ and let $g(x) = f(x) - b$. Then we have $N^+(g, -\log r) = N^+(f, -\log r)$.*

Proof: Let $f(x) = \sum_{n=0}^{\infty} a_n x^n$ and let $t = N^+(f, -\log r)$. By hypothesis we have $|a_t| r^t \geq |a_n| r^n$ whenever $n < t$ and $|a_t| r^t > |a_n| r^n$ whenever $n > t$. Now let $g(x) = \sum_{n=0}^{\infty} b_n x^n$. Hence $b_0 = a_0 - b, b_n = a_n$ whenever $n \geq 1$. By hypothesis and by Theorem 26.1 we have $|b_0| \leq \sup\limits_{n \geq 1} |a_n| r^n$ hence $|b_0| \leq |b_t| r^t$ and finally $N^+(g, -\log r) = t$.

Lemma 26.6: *Let $f \in H(d(a, r))$ have t zeros in $d(a, r)$ with $t \geq 1$ (taking multiplicities into account) and let $b \in f(d(a, r))$. Then $f - b$ also admits t zeros in $d(a, r)$ (taking multiplicities into account).*

Proof: We assume $a = 0$. By Lemma 26.5 we know that $N^+(f, -\log r) = t$. Hence we have $N^+(f, -\log r) = t$ and $v(f, -\log r) = v(a_t) - t \log r$. Besides $v(f(x)) \geq v(f, -\log r)$ for all $x \in d(0, r)$ hence $v(b) \geq v(f, -\log r)$ and therefore $v(b) \geq v(a_t) - t \log r$. Hence we see that $N^+(f - b, -\log r) = t$ and this ends the proof.

Lemma 26.7: *K is supposed to have characteristic zero. Let $f \in H(d(0, r))$, let $t = N^+(f, -\log r)$ and assume $t \geq 1$. Let $\alpha_1, ..., \alpha_q$ be the zeros of f' in $d(0, r)$. For every $b \in f(d(0, r)) \setminus f(\{\alpha_1, ..., \alpha_q\})$, $f - b$ admits exactly t zeros of order 1 in $d(0, r)$.*

Proof: Let $b \in f(d(0, r)) \setminus f(\{\alpha_1, ...\alpha_q\})$. By Lemma 26.6 , $f - b$ admits t zeros in $d(0, r)$ (taking multiplicities into account). But for each zero α of $f - b$, (as $\alpha \neq \alpha_j$ whenever $j = 1, ..., q$) we have $f'(\alpha) \neq 0$ hence the t zeros of $f - b$ have order one.

27.   STRICTLY INJECTIVE ANALYTIC ELEMENTS

In this chapter D is just an open set in K.

Definition: Let D be an open set and let $f \in H(D)$. Then f will be said to be *strictly injective* in D if f is injective and if $f'(x) \neq 0$ whenever $x \in D$.

In the same way, given $a \in k$, and $r > 0$, an analytic function $f(x) \in \mathcal{A}(d(a,r^-))$ will be said to be *strictly injective* in $d(a,r^-)$ if f is injective and if $f'(x) \neq 0$ whenever $x \in D$.

Theorem 27.1: *Let K have characteristic zero. Let $f \in H(D)$ be injective in D. Then f is strictly injective.*

Proof: Suppose that f is not strictly injective, and let $\alpha \in D$ be a zero of f'. Let $d(\alpha, r)$ be a disk included in D. Without loss of generality we may assume $\alpha = 0$. Hence in $d(0,r)$, $f(x)$ is equal to a series of the form $a_0 + \sum\limits_{n=q}^{\infty} a_n x^n$ with $q \geq 2$ and $a_q \neq 0$. Therefore we have $N^+(f, -\log r) \geq q \geq 2$. Let $t = N^+(f, -\log r)$. Since K has characteristic zero, f' is not identically zero, and therefore admits finitely many zeros $\alpha_1 = 0, \alpha_2, ..., \alpha_s$ in $d(0,r)$. Let $b \in f(d(0,r)) \setminus f(\{\alpha_1, ... \alpha_s\})$. By Lemma 26.7 $f - b$ admits t simple zeros in $d(0,r)$ and this contradicts the hypothesis "f injective in $d(0,r)$".

Theorem 27.2: *Let $a \in K$, $r \in \mathbb{R}_+$, let $f(x) = \sum\limits_{n=0}^{\infty} a_n(x-a)^n \in H(d(a,r))$, and let $s = \sup\limits_{n \geq 1} |a_n| r^n$ be > 0. Then the following statements are equivalent:*

   $\alpha)$ $|a_1| > |a_n| r^{n-1}$ *whenever* $n > 1$
   $\beta)$ $|f(x) - f(y)| = |x - y||a_1|$ *whenever* $x, y \in d(a,r)$
   $\gamma)$ f *is strictly injective in* $d(a,r)$.

Besides when conditions $\alpha), \beta), \gamma)$ *are satisfied, then we have* $s = |a_1| r$ *and* $|f'(x)| = |a_1|$ *whenever* $x \in d(a,r)$.

Proof: Without loss of generality we may obviously assume $a = 0$.

First, we suppose $\alpha)$ is satisfied and consider

$$f(x) - f(y) = (x - y)\left(a_1 + \sum\limits_{n=2}^{\infty} a_n \left(\sum\limits_{j=0}^{n-1} x^j y^{n-1-j} \right) \right).$$

For every $n \geq 2$ it is seen that $|x^j y^{n-1-j}| \leq r^{n-1}$. Hence we have

$$|a_1| > \left| \sum_{n=2}^{\infty} a_n \left(\sum_{j=0}^{n-1} x^j y^{n-1-j} \right) \right|$$ and thereby $|f(x) - f(y)| = |a_1| \, |x - y|$. At the same time we notice that $\alpha)$ implies $|f'(x)| = |a_1|$ whenever $x \in d(a, r)$ while by Theorem 26.1 we have $s = |a_1| r$.

Second, we suppose $\beta)$ is satisfied. Since $s > 0$ by Theorem 26.1 f is not a constant hence $|a_1| \neq 0$. Then by $\beta)$ we have $f(x) \neq f(y)$ whenever $x \neq y$.

Third we suppose $\gamma)$ is satisfied. Let $b \in f(d(a, r))$ let $g = f - b$, and let $t = N^+(g, -\log r)$. If $t \geq 2$ either g admits several different zeros or g admits a zero α of order t. In both cases we see that g is not strictly injective, hence neither is f. Finally we have $t = 1$ and then $\alpha)$ is satisfied. This ends the proof of Theorem 27.2.

Theorem 27.2 is easily applied to analytic functions in a disk.

Corollary 27.3: *Let* $f(x) = \displaystyle\sum_{n=0}^{\infty} a_n(x - a)^n \in A_b(d(a, r^-))$*, and suppose that*

$s := \displaystyle\sup_{n \geq 1} |a_n| r^n$ *is strictly positive. Then conditions* $\alpha), \beta), \gamma), \delta)$ *are equivalent.*

    $\alpha)$ $|a_1| \geq |a_n| r^{n-1}$ *whenever* $n > 1$
    $\beta)$ $|f(x) - f(y)| = |x - y| \, |a_1|$ *whenever* $x, y \in d(a, r^-)$
    $\gamma)$ *f is strictly injective in* $d(a, r^-)$*.*
    $\delta)$ $s = |a_1| r$*.*

Besides when conditions $\alpha), \beta), \gamma), \delta)$ *are satisfied, we have* $|f'(x)| = |a_1|$ *whenever* $x \in d(a, r^-)$*.*

Proof: For every $\rho \in]0, r[$ we apply Theorem 27.2 to $f \in H(d(a, \rho))$.

Lemma 27.4: *Let* $f \in d(\alpha, r^-)$ *be injective and such that* f' *is not identically zero. Then* f *is strictly injective.*

Proof: We may obviously assume that $f'(\alpha) \neq 0$ and $\alpha = 0$. Hence, f is of the form $\displaystyle\sum_{n=0}^{\infty} a_n x^n$ with $a_1 \neq 0$. If f' has a zero β, there exists an integer $q > 1$ such that
(1) $|q a_q| \, |\beta|^{q-1} = |a_1|$. Let $g(x) = f(x) - f(0)$. Then g is also injective and has a simple zero at 0. But by (1) we have $|a_q| \, |\beta|^q \geq |a_1| \, |\beta|$ hence we know that g has at least q zeros in $d(0, r^-)$ and then admits another zero $\gamma \neq 0$, which contradicts the fact that it is injective. Hence, we see that f' has no zero in $d(\alpha, r^-)$ i.e., f is strictly injective.

We are now able to study the reciprocal of an analytic element in a disk $d(a, r)$.

Theorem 27.5:   *Let $a \in K$, $r \in \mathbb{R}_+$, let $f(x) \in H(d(a,r))$ be strictly injective in $d(a,r)$, and let $s = |f'(a)|r$. Let $b = f(a)$, and $s = r|f(a)|$. The homomorphism ψ from $H(d(b,s))$ into $H(d(a,r))$ defined as $\psi(h) = h \circ f$ is an isometric isomorphism from $H(d(b,s))$ onto $H(d(a,r))$.*

Proof: Without loss of generality we may obviously suppose $a = b = f(a) = 0$, and $f'(0) = 1$. We put $E = d(0,r)$. Hence, by Theorem 27.2 in $d(0,r)$ $f(x)$ is equal to a series of the form $x + \sum_{j=2}^{\infty} a_j x^j$ with $\sup_{j \geq 2} |a_j| r^j < r$. Besides, by Theorem 26.1 r is equal to s. It is obviously seen that $\|\psi(h)\|_E = \|h\|_E$ for every $h \in H(E)$. So, we only have to prove that ψ is surjective. Let $\lambda = \inf_{P \in K[x]} \|\psi(P) - x\|_E$, and suppose $\lambda > 0$. We put $\theta = \sum_{j=2}^{\infty} a_j x^j$ and $\rho = \frac{1}{r}\|\theta\|_E$. Hence ρ belongs to $]0,1[$. For every $n \in \mathbb{N}^*$, we put $\theta_n = f^n - x^n$. It is seen that we have (1) $\|\theta_n\|_E \leq \|\theta\|_E r^{n-1} = \rho r^n$.   Since $\lambda > 0$ and since $\rho \in]0,1[$, we can find $P \in K[x]$ satisfying (2) $\|P \circ f - x\|_E < \dfrac{\lambda}{\sqrt{\rho}}$. Let $P \circ f = \sum_{n=0}^{\infty} b_n x^n$. By (2) we have $|b_n| r^n \leq \dfrac{\lambda}{\sqrt{\rho}}$ whenever $n \in \mathbb{N}$. But on the other hand we have

$$\sum_{n=0}^{\infty} b_n x^n = \sum_{n=0}^{\infty} b_n f^n - \sum_{n=0}^{\infty} b_n \theta_n.$$ Finally by (1) we obtain $\|\sum_{n=0}^{\infty} b_n \theta_n\|_E \leq \lambda\sqrt{\rho}$.

But since $\rho < 1$, this contradicts the hypothesis and shows that $\lambda = 0$. Hence x does belong to the closure of $\psi(H(E))$. But as ψ is isometric, $\psi(H(E))$ is obviously closed in $H(E)$, and therefore x belongs to $\psi(H(E))$. As a consequence $\psi(H(E)) = H(E)$.

Corollary 27.6:   *Let $f \in H(d(a,r))$ be strictly injective and let $d(b,s) = f(d(a,r))$. Then $\overset{-1}{f}$ belongs to $H(d(b,s))$.*

Corollary 27.7:   *Let K have characteristic 0, let $f \in H(a,r))$ be injective and let $d(b,s) = f(d(a,r))$. Then $\overset{-1}{f}$ belongs to $H(d(b,s))$.*

Corollary 27.8:   *Let $f \in A(d(0,r^-))$ be strictly injective in $d(0,r^-)$ and let $s = r|f'(0)|$. Then $\overset{-1}{f}$ belongs to $A(d(0,s^-))$.*

Proof: Indeed, by Corollary 27.6 $\overset{-1}{f}$ belongs to $H(0, u)$ for every $u \in [0, s]$.

Corollary 27.9: *Let K have characteristic 0, let $f \in A(d(0, r^-))$ be injective in $d(0, r^-)$ and let $s = r|f'(0)|$. Then $\overset{-1}{f}$ belongs to $A(d(0, s^-))$.*

Remark: *Let E be a topological space and let F and G be subsets dense in E. If F is open then $F \cap G$ is dense in E.*

Indeed let $a \in E$, let V be an open neighbourhood of a and let $u \in V \cap F$. Since F is open, $V \cap F$ is a neighbourhood of u. Hence as G is dense in E, there exists $x \in (V \cap F) \cap G$. Therefore, $V \cap (F \cap G) \neq \emptyset$.

We remember that Condition B) was defined in Chapter 18.

Lemma 27.10: *Let $\overline{D}$ be open. Let a be a point of $\overline{D} \backslash D$ and let $f \in H(D \cup \{a\})$ be strictly injective in D. There exists an open set D' sayisfying:*

$\overline{D'}$ is open,

$D \cup \{a\} \subset D' \subset \overline{D}$

f belongs to $H(D')$ and is strictly injective in D'.

Proof: By Theorem 10.5 we know that f is of the form $g + h$ with $g \in H(\overline{D})$ and $h \in R(K \setminus (\overline{D} \setminus (D \cup \{a\})))$. Since $\overline{D}$ is open, there does exist $\sigma > 0$ such that $d(a, \sigma) \subset \overline{D}$. Now, as $h \in R(D \cup \{a\})$ there exists $\tau > 0$ such that $h \in R(D \cup d(a, \tau))$. Let $\rho = \min(\sigma, \tau)$ and let $D' = D \cup d(a, \rho)$. It is seen that $\overline{D'} = \overline{D} \cup d(a, \rho)$, and therefore $\overline{D'}$ is open. Next, both g, h belongs to $H(D')$, hence so does f.

We suppose that f is not injective in D'. Let b and $c \in D'$ be such that $f(b) = f(c)$ and let $\omega = f(b)$. Let $r \in \mathbb{R}_+^*$ be such that $d(b, r) \cup d(c, r) \subset D'$ whereas $d(b, r) \cap d(c, r) = \emptyset$. Then $f(d(b, r))$ is a disk $d(\omega, s)$ whereas $f(d(c, r))$ is a disk $d(\omega, t)$. We may obviously assume $s \leq t$. Let $\Sigma = d(b, r) \cap D$ and let $\Lambda = d(c, r) \cap D$. Obviously, Σ is dense in $d(b, r)$ whereas Λ is dense in $d(c, r)$ hence $f(\Sigma)$ is dense in $d(\omega, s)$ whereas $f(\Lambda)$ is dense in $d(\omega, t)$. Hence $f(\Lambda) \cap d(\omega, s)$ is dense in $d(\omega, s)$. Since Σ and Λ are open sets in K, by Corollary 26.2, both $f(\Sigma)$ and $f(\Lambda)$ are open sets in K because f is not a constant in $d(b, r)$ or in $d(c, r)$. Hence both $f(\Lambda) \cap d(\omega, s)$, $f(\Sigma)$ are dense open subsets of $d(\omega, s)$. Therefore, by the Remark above we see that $f(\Sigma) \cap f(\Lambda)$ is dense in $d(\omega, s)$ and certainly is not empty. Let $y \in f(\Sigma) \cap f(\Lambda)$, and $\alpha \in \Sigma$, $\beta \in \Lambda$ satisfy $f(\alpha) = f(\beta) = y$. By definition of Δ and Λ we see that α and $\beta \in D$ whereas $\alpha \neq \beta$. This contradicts the hypothesis "f is injective in D" and finally shows f to be injective in D'.

Now since f is strictly injective in $d(a, r^-) \cap D$ we have $f'(x) \neq 0$ whenever $x \in d(a, r^-) \cap D$, hence by Lemma 27.4 f is strictly injective in $d(a, r^-)$. Finally, we have $f'(x) \neq 0$ whenever $x \in D'$ and this ends the proof of Lemma 27.10.

Definition: A set D is said to be *analytic* if for every $f \in H(D)$ for every $a \in D$ and $r > 0$, the property $f(x) = 0$ whenever $x \in d(a, r) \cap D$ implies $f(x) = 0$ whenever $x \in D$.

Theorem 27.11:   *Let $\overline{D}$ be open, let D' satisfy $D \subset D' \subset \overline{D}$ and let $f \in H(D')$ be strictly injective in D. There exists an open set D'' satisfying $D' \subset D'' \subset \overline{D}$ such that f belongs to $H(D'')$ and is strictly injective in D''.*

Proof: By Lemma 27.10, for every $a \in D'$ there exists an open set D_a such that $\overline{D_a}$ is open, satisfying Condition B), such that $D \cup \{a\} \subset D_a \subset \overline{D}$ and such that f belongs to $H(D_a)$ and is strictly injective in D_a. Let $D'' = \bigcup_{a \in D'} D_a$.

Then D'' is open and such that $D' \subset D'' \subset \overline{D}$. By Theorem 10.5 f has a unique decomposition in the form $g + h$ with $g \in H(\overline{D})$ and $h \in R(K \setminus (\overline{D} \setminus D'))$. Since $f \in H(D_a)$ obviously $h \in H(D_a)$. Hence h has no pole in D_a whenever $a \in D'$, therefore $h \in R(D'')$. Hence f belongs to $H(D'')$. Besides, by Lemma 27.10 we have $f'(x) \neq 0$ whenever $x \in D_a$, hence whenever $x \in D''$.

Now we just have to check that f is injective in D''. Let $a, b \in D'$ satisfy $f(a) = f(b)$. Since D_a satisfies Condition B), we may apply Lemma 27.10 to D_a and b and then we have an open set E such that $D_a \cup \{b\} \subset E \subset \overline{D_a} = \overline{D}$ and such that f belongs to $H(E)$ and is strictly injective in E. Hence the hypothesis $f(a) = f(b)$ is impossible, and therefore f is injective in D'. But actually D'' does satisfy the hypothesis on D'. Hence f is injective in D'', and this finishes proving Theorem 27.11.

Proposition 27.12:   *Let $(d(\alpha_i, r_i^-))_{i \in I}$ be a partition of D. Let $h \in H(D)$ be injective in D and let $f \in H(D)$ satisfy*

    i) $|f'(\alpha_i) - h'(\alpha_i)| < |h'(\alpha_i)|$ *whenever $i \in I$*

    ii) $\|f - h\|_{d(\alpha_i, r_i^-)} < |h'(\alpha_i)| r_i$ *whenever $i \in I$*

Then f is strictly injective. Furthermore, for all $i \in I$ we have $f(d(\alpha_i, r_i^-)) = h(d(\alpha_i, r_i^-))$.

Proof: By i) h' is not identically zero in $d(\alpha_i, r_i^-)$ hence by Lemma 27.4 h is strictly injective in $d(\alpha_i, r_i^-)$. For every $i \in I$ we put $s_i = |h'(\alpha_i)| r_i$. By Corollary 26.2 we have $h(d(\alpha_i, r_i^-)) = d(h(\alpha_i), s_i^-)$ hence by ii) it is seen that

$f(d(\alpha_i, r_i^-)) \subset d(h(\alpha_i), s_i^-)$. Let $f(d(\alpha_i, r_i^-)) = d(f(\alpha_i), t_i^-)$. Therefore, we have
(1) $t_i \leq s_i$ while obviously
(2) $t_i \geq |f'(\alpha_i)| r_i$.
But by i) we have $|f'(\alpha_i)| = |h'(\alpha_i)|$, hence by (1) and (2) we see that we have $t_i = s_i$ hence $t_i = |f'(\alpha_i)| r_i$, and therefore by Theorem 27.2 f is strictly injective in $d(\alpha_i, r_i^-)$. Suppose that f is not injective in all of D. Then there exists a and $b \in D$ such that $f(a) = f(b)$. Since f is injective in each disk $d(\alpha_i, r_i^-)$ we see that there exist j and $m \in I$ with $j \neq m$ such that $a \in d(\alpha_j, r_j^-)$ and $b \in d(\alpha_m, r_m^-)$. Since we have just proven that $f(d(\alpha_i, r_i^-)) = h(d(\alpha_i, r_i^-))$ for all $i \in I$ we see that there exist $a' \in d(\alpha_j, r_j^-)$ and $b' \in d(\alpha_m, r_m^-)$ such that $h(a') = f(a), h(b') = f(b)$. This clearly contradicts the hypothesis "h is injective in D" and shows that f is injective in all of D. Therefore by Lemma 27.4 f is strictly injective.

Theorem 27.13 shows the set of the strictly injective elements to be open in $H(D)$.

Theorem 27.13:   *Let D be such that $\delta(D, K \setminus D) = \rho > 0$. Let $\lambda \in]0, +\infty[$. Let $h \in H(D)$ be injective and satisfy $|h'(x)| \geq \lambda$ whenever $x \in D$. For every $f \in H(D)$ such that $\|f - h\|_D < \lambda\rho$, f is strictly injective in D and satisfies $f(d(\alpha, \rho^-)) = h(d(\alpha, \rho^-))$ for all $\alpha \in D$.*

Proof: Let $h \in H(D)$ satisfy $\|f - h\|_D < \lambda\rho$. Since the distance from D to $K \setminus D$ is $\rho > 0$, there exists a partition of D in the form $(d(\alpha_i, \rho^-))_{i \in I}$ and then we have $\|f - h\|_{d(\alpha_i, \rho^-)} < \lambda\rho \leq |h'(\alpha_i)|\rho$. Besides, given any $g \in H(d(\alpha, \rho^-))$, by Theorem 19.1 we have $\|g'\|_{d(\alpha, \rho^-)} \leq \dfrac{1}{\rho}\|g\|_{d(\alpha, \rho^-)}$ hence

$$|f'(\alpha_i) - h'(\alpha_i)| \leq \frac{1}{\rho}\|f - h\|_{d(\alpha_i, \rho^-)} < \lambda \leq |h'(\alpha_i)|.$$ So Conditions i) and ii) of Proposition 27.12 are clearly satisfied.

Remark: As we know when D has holes , ρ is just the lower bound of the diameters of the holes.

Theorem 27.14:   *Let $d(\alpha_i, r_i^-)_{i \in I}$ be a partition of D, let $h \in H(D)$ and let $A \in]0, 1[$. Let $\phi \in H(D)$ satisfy $\|\phi\|_D < 1$ and $\|h\phi\|_{d(\alpha_i, r_i^-)} \leq A r_i \|h'\|_{d(\alpha_i, r_i^-)}$ for all $i \in I$. Then for every $B \in]A, 1[$ there exists a family $(\beta_i)_{i \in I}$ with $\beta_i \in d(\alpha_i, r_i^-)$ such that*

    i)   *$|h(\beta_i)\phi'(\beta_i)| \leq B|h'(\beta_i)|$ whenever $i \in I$*

    ii)   *$\|h\phi\|_{d(\alpha_i, r_i^-)} \leq B|h'(\beta_i)| r_i$ whenever $i \in I$.*

Let $f = h(1 + \phi)$ and let h be strictly injective. Then f is strictly injective and satisfies $f(d(\alpha_i, r_i^-)) = h(d(\alpha_i, r_i^-))$ for all $i \in I$.

Proof: Let $B \in]A, 1[$, let us fix $i \in I$ and put $\alpha = \alpha_i, r = r_i$. For every $g \in H(d(\alpha, r^-))$ we have $\|g\|_{d(\alpha, r-)} = \lim\limits_{|x - \alpha| \to r^-} |g(x)|$ hence there clearly exists

$\beta \in d(\alpha, r^-)$ such that

(1) $|h'(\beta)| \geq \dfrac{A}{B} \, \|h'\|_{d(\alpha, r-)}$, hence the hypothesis implies

(2) $\|h\phi\|_{d(\beta, r-)} \leq B \dfrac{A}{B} \, |h'(\beta)|r$

Besides we know that $\|\phi'\|_{d(\alpha, r-)} \leq \dfrac{1}{r} \, \|\phi\|_{d(\alpha, r-)}$ hence we have

$$|h(\beta)\phi'(\beta)| \leq |h(\beta)|\frac{1}{r}\|\phi\|_{d(\alpha, r-)} \leq \frac{1}{r}\|h\|_{d(\alpha, r-)}\|\phi\|_{d(\alpha, r-)}.$$

As the norm $\| \cdot \|_{d(a, r-)}$ is multiplicative, we have $|h(\beta)\phi'(\beta)| \leq \dfrac{1}{r}\|h\phi\|_{d(\alpha, r-)}$ hence by the above hypothesis $|h(\beta)\phi'(\beta)| \leq A\|h'\|_{d(\alpha, r-)}$ and finally by (1) we obtain

(3) $|h(\beta)\phi'(\beta)| \leq B|h'(\beta)|$

Hence, we just have to put $\beta_i = \beta$ and do this for every $i \in I$ in order to obtain i) and ii) from (2) and (3). Since $\|\phi\|_D \leq 1$ we may take $B \geq \|\phi\|_D$. We see that $f' - h' = h'\phi + h\phi'$ hence by Condition i) we obtain $|f'(\beta_i) - h'(\beta_i)| \leq B|h'(\beta_i)|$. Then , as $B < 1$, and as $h'(x) \neq 0$ whenever $x \in D$, we see that $|f'(\beta_i) - h'(\beta_i)| < |h'(\beta_i)|$ whenever $i \in I$. This is just Condition i) in Proposition 27.12. Besides, Condition ii) is seen to imply Condition ii) in Proposition 27.12. Hence by Proposition 27.12, f is strictly injective and satisfies $f(d(\alpha_i, r_i^-) = h(d(\alpha_i, r_i^-))$ for all $i \in I$.

Theorem 27.15: *Let D be analytic, and let F be the set of the injective elements of $H(D)$. The closure of F in $H(D)$ is equal to $F \cup K$.*

Proof: Let $\overline{F}$ be the closure of F and let $f \in \overline{F} \setminus K$. Suppose that f is not injective and let $a, b \in D$ be such that $f(a) = f(b)$. Without loss of generality we may obviously assume $f(a) = 0$. Now let $r \in]0, |a - b|[$ be such that $d(a, r) \cup d(b, r)$ is included in D. Since f is not identically zero in D, and since D is an analytic set, the restriction of f to $d(a, r)$, (resp. $d(b, r)$), is not identically zero. Let $h \in F$ satisfy $\|f - h\|_D < \min(\|f\|_{d(a, r)}, \|h\|_{d(b, r)})$. By Theorem 22.7 h admits a zero in $d(a, r)$ and another in $d(b, r)$. But since $r < |a - b|$, it is seen that these two zeros are different, and therefore this contradicts the hypothesis "$h \in F$". So this ends the proof.

28. LOGARITHM AND EXPONENTIAL

In this chapter the field K is supposed to have characteristic zero.

We will define the p-adic logarithm and the p-adic exponential, and will shortly study them, in connection with the study of the roots of 1 made in Chapter 8.

Lemma 28.1: K *is supposed to have residue characteristic $p \neq 0$. Let $r \in \,]0,1[$, and for each $n \in \mathbb{N}$, let $h_n(x) = (1+x)^{p^n}$. The sequence h_n converges to 1 with respect to the uniform convergence on $d(0,r)$.*

Proof: Without loss of generality, we may assume $|p| = \dfrac{1}{p}$. Let $E = d(0,r)$, let $\| \, . \, \|_E$ denote the norm of uniform convergence on E, and for each $n \in \mathbb{N}$, let $u_n = p^n$ and let q_n be the integral part of $\dfrac{n}{2}$. Now we put $t_n = p^{q_n}$, and we denote by h the identical function on E. Then $h_n - 1 = \displaystyle\sum_{j=1}^{u_n} \binom{u_n}{j} h^j$. By Lemma 8.1 we have $\left| \binom{u_n}{j} \right| \leq \dfrac{p^{-n}}{|j|}$ hence

$$(1) \qquad \left| \binom{u_n}{j} \right| \leq j \, p^{-n} \leq t_n \, p^{-n} \leq p^{-\frac{n}{2}} \text{ whenever } j = 1, ..., t_n. \text{ Now we have}$$

$$h_n - 1 = \sum_{j=1}^{t_n} \binom{u_n}{j} h^j + \sum_{j=t_n+1}^{u_n} \binom{u_n}{j} h^j. \text{ By (1) it is seen that}$$

$$(2) \qquad \left\| \sum_{j=1}^{t_n} \binom{u_n}{j} h^j \right\|_E \leq p^{-\frac{n}{2}} \text{ while}$$

$$(3) \qquad \left\| \sum_{j=t_n+1}^{u_n} \binom{u_n}{j} h^j \right\|_E \leq \|h\|_E^{t_n+1}.$$

Now by (2) and (3) it is seen that $\displaystyle\lim_{n \to \infty} \|h_n - 1\|_E = 0$.

Notation: For convenience we denote by r_q the positive number such that $\log(r_q) = -\dfrac{1}{p^{q-1}(p-1)}$. We denote by $f(x)$ the series $\displaystyle\sum_{n=1}^{\infty} (-1)^{n-1} \frac{x^n}{n}$.

Theorem 28.2: f *has a radius of convergence equal to* 1. *If the residue characteristic of K is $p \neq 0$, then f is unbounded in $d(0, 1^-)$. If the residue characteristic is zero, then $|f(x)|$ is bounded by 1 in $d(0, 1^-)$. The function defined in $d(1, 1^-)$ as $Log(x) = f(x-1)$ has a derivative equal to $\dfrac{1}{x}$, and satisfies $Log(ab) = Log(a) + Log(b)$ whenever a, $b \in d(1, 1^-)$.*

Proof: It is clearly seen that that the radius of f is 1, because $|n| \geq \dfrac{1}{n}$. As in the archimedean case the property $Log(ab) = Log(a) + Log(b)$ comes from the fact that both Log and the fuction h_a defined as $h_a(x) = Log(ax)$ have the same derivative. The other statements are immediate.

In Chapter 8, when K has residue characteristic $p \neq 0$, we have introduced the group A of the p^s-th roots of 1, i.e. the set of the $u \in K$ satisfying $u^{p^s} = 1$ for some $s \in \mathbb{N}$.

Theorem 28.3: K *is supposed to have residue characteristic $p \neq 0$ (resp. 0). All the zeros of Log are of order 1. The set of the zeros of the function Log is equal to* A, *(resp. 1 is the only zero of Log). The restriction of Log to the disk $d(1, (r_1)^-)$ (resp. $d(1, 1^-)$ is injective and is a bijection from $d(1, (r_1)^-)$ onto $d(0, (r_1)^-)$ (resp. from $d(1, 1^-)$ onto $d(0, 1^-)$).*

Proof: It is seen that the zeros of Log are of order 1 because the derivative of Log has no zero. First, we suppose K to have residue characteristic $p \neq 0$. It is obviously seen that each root of 1 in $d(1, 1^-)$ is a zero of Log. Besides, by Theorem 8.11 we know that the only roots of 1 in $d(1, 1^-)$ are the p^n-th roots. Now we will check that Log admits no zero other than the roots of 1. Indeed, suppose that a is a zero of Log but is not a root of 1, and for each $n \in \mathbb{N}$, let $y_n = x^{p^n}$. Since y belongs to $d(1, 1^-)$, by Lemma 28.1 we have $\lim_{n \to \infty} y_n = 1$. But obviously we have $Log(y_n) = 0$ for every $n \in \mathbb{N}$, hence this contradicts the fact that 1 is an isolated zero of Log.

Thus, Log has no zero in the disk $d(1, (r_1)^-)$, except 1, and therefore by Theorem 22.5 the series $f(x) = \sum_{n=1}^{\infty} (-1)^{n-1} \dfrac{x^n}{n}$ satisfies $N^+(f, -\log r) = 1$ for every $r \in \,]0, r_1[$, hence $r > \dfrac{r^n}{|n|}$ for all $r \in \,]0, r_1[$, for every $n \in \mathbb{N}^*$. Therefore, by Corollary 27.3 it is injective in $d(0, r_1^-)$. Then by Corollary 26.2 we see that $Log(d(1, r_1^-)) = d(0, r_1^-)$.

Now we suppose that K has residue characteristic zero. Then, the function

$$f(x) = \sum_{n=1}^{\infty} (-1)^{n-1} \frac{x^n}{n} \quad \text{satisfies} \quad N^+(f, -\log r) = 1 \quad \text{for every} \quad r \in]0,1[, \text{ hence}$$

$r > \dfrac{r^n}{n}$ for all $r \in]0,1[$, for every $n \in \mathbb{N}^*$. Therefore, f has no zero different from 1 in $d(0,1^-)$, and, by Corollary 27.3, is injective in $d(0,1^-)$. Then by Theorem 26.2 we see that $Log(d(1,1^-)) = d(0,1^-)$. This ends the proof.

Corollary 28.4:   *K is supposed to have residue characteristic 0. There is no root of 1 in $d(1,1^-)$, except 1.*

Proof: Indeed any root of 1 should be a zero of Log in $d(1,1^-)$.

Notation: If K has residue characteristic $p \neq 0$, we denote by exp the reciprocal function of the restriction of Log to $d(1, r_1^-)$ which obviously is a function defined in $d(0, r_1^-)$, with values in $d(1, r_1^-)$.

If K has residue characteristic 0 we denote by exp the reciprocal function of Log which is obviously defined in $d(0,1^-)$, and takes values in $d(1,1^-)$.

Theorem 28.5:   *K is supposed have residue characteristic $p \neq 0$ (resp. 0). The function exp belongs to $A_b(d(0, r_1^-))$, (resp. $A_b(d(0,1^-))$), is a bijection from $d(0, r_1^-)$ onto $d(1, r_1^-)$ (resp. from $d(0,1^-)$ onto $d(1,1^-)$), and satisfies*

$$exp(x) = exp'(x) = \sum_{n=0}^{\infty} \frac{x^n}{n!} \quad \text{whenever } x \in d(0, r_1^-), \; (\text{resp. } x \in d(0,1^-)). \text{ Besides}$$

the disk of convergence of its series is equal to $d(0, r_1^-)$ (resp. $d(0,1^-)$).

Proof: By Corollary 27.8 we know that exp belongs to $A_b(d(0, r_1^-))$ (resp. $A_b(d(0,1^-))$), and is obviously a bijection from $d(0, r_1^-)$ onto $d(1, r_1^-)$ (resp. from $d(0,1^-)$ onto $d(1,1^-)$). As it is the reciprocal of Log, it must satisfy $exp(x) = exp'(x)$ for all $x \in d(0, r_1^-)$, (resp. $x \in d(0,1^-)$), and therefore $exp(x) = \sum_{n=0}^{\infty} \frac{x^n}{n!}$ whenever $x \in d(0, r_1^-)$, (resp. $x \in d(0,1^-)$). Thus the radius of convergence r is at least r_1 (resp. 1). If the residue characteristic is 0, it is obviously seen that the series cannot converge for $|x| = 1$, hence the disk of convergence is $d(0,1^-)$.

Now we suppose that the residue characteristic is $p \neq 0$. Suppose that the power series of exp converges in $d(0, r_1)$. Then exp has continuation to an element of $H(d(0, r_1))$. On the other hand, since $N(f, -\log r) = 1$ for all $r \in]0, r_1[$, we have $N^-(f, -\log r_1) = 1$, and then by Theorem 26.1 $Log(d(1, r_1))$ is equal to $d(0, r_1)$. Hence we can consider $exp(Log(x))$ in all the disk $d(0, r_1)$, and then by Corollary 11.3 this is an element of $H(d(1, r_1))$. But this element is equal to x in all of $d(1, r_1^-)$, and therefore in all of $d(1, r_1)$. Of course this contradicts the fact that Log is not injective in the circle $C(1, r_1)$. This finishes proving that the disk of convergence of exp is just $d(0, r_1^-)$.

The exponential function admits an extension to a continuous group isomorphism defined in $\mathbb{C}_p$ onto a subgroup E of $d(0, 1^-)$. First a concrete construction was made in [53] ,[58]. One finds more abstract proofs in [60]. Here is a summary of this definition, with the help of the purely algebraic Lemma 28.6 below.

Notations: Given two topological abelian groups $\mathcal{E}$ and $\mathcal{F}$, we will denote by $\mathcal{H}om(\mathcal{E}, \mathcal{F})$ the abelian group of the continuous group homomorphisms from $\mathcal{E}$ into $\mathcal{F}$.

This lemma is classical in algebra.

Lemma 28.6:   *Let $(\mathcal{E}, +)$ and $(\mathcal{F}, +)$ be two topological abelian groups with $\mathcal{F}$ divisible. Let $\mathcal{E}_o$ be a subgroup of $\mathcal{E}$ which is a neighbourhood of 0 in $\mathcal{E}$. Let $\varphi_o \in \mathcal{H}om(\mathcal{E}_o, \mathcal{F})$. There exists $\varphi \in \mathcal{H}om(\mathcal{E}, \mathcal{F})$ such that $\varphi(x) = \varphi_o(x)$ whenever $x \in \mathcal{E}_o$.*

Theorem 28.7:   *The analytic function exp has continuation to a continuous injective homomorphism Exp from $(\mathbb{C}_p, +)$ to $d(1, 1^-)$ satisfying $Log(Exp(x)) = x$ whenever $x \in \mathbb{C}_p$ and $\dfrac{Exp(Log(u))}{u} \in A$ whenever $u \in d(1, 1^-)$.*

Proof: Since $d(0, 1^-)$ is a divisible multiplicative group , the exponential has continuation to a continuous homomorphism from $\mathbb{C}_p$ to $d(0, 1^-)$, denoted by Exp. One checks (1) $Log(Exp(x)) = x$ whenever $x \in \mathbb{C}_p$ and (2) $\dfrac{Exp(Log(u))}{u} \in A$ whenever $u \in B$ [53],[58]. In particular, by (1) Exp is clearly injective in $\mathbb{C}_p$. For all $u \in d(0, 1^-)$, we put $\tau(u) = \dfrac{u}{Exp(Log(u))}$. Then τ is a continuous homomorphism from $d(0, 1^-)$ into A such that $\tau(u) = u$ for all $u \in A$. Let $B = Ker(\tau)$. Then $d(0, 1^-)$ is equal to the direct product of B by A and we have $B = Exp(\mathbb{C}_p)$ and $d(1, r_1^-) \subset B$. Thus Exp is a continuous isomorphism from $\mathbb{C}_p$ onto B. Besides, we check that $Log(Exp(Log(u))) = Log(u)$, hence $\dfrac{u}{Exp(Log(u))}$ belongs to A.

29. A FINITE INCREASINGS PROPERTY

In this chapter D is just an open set in K.

When a homographic function $h \in R(D)$ satisfies certain property of finite increasings, this property still holds for elements of the form $h\phi$ when ϕ is a product of homographic functions satisfying certain conditions.

Definition: Let f, ϕ be functions from D to K. Then ϕ will be said to be *f-dominated* if

i) $\qquad |\phi(x)| = 1$ whenever $x \in D$

ii) $\qquad |f(y) - f(x)|^2 > |f(x)f(y)|\, |\phi(y) - \phi(x)|^2$ whenever $x, y \in D$ such that $x \neq y$.

Lemma 29.1: (L. Haddad) *Let f, ϕ be functions from D to K. If ϕ is f-dominated then the function $g = f\phi$ satisfies $|g(y) - g(x)| = |f(y) - f(x)|$ whenever $x, y \in D$. Besides for every f-dominated function ψ from D to K, ψ is g-dominated.*

Proof: Let x, y be two different points of D and let ϕ be f-dominated. By ii) it is seen that at least one of these two relations is satisfied

(1) $\qquad |f(y) - f(x)| > |f(x)|\, |\phi(y) - \phi(x)|$

(2) $\qquad |f(y) - f(x)| > |f(y)|\, |\phi(y) - \phi(x)|$

Due to the symmetry we may obviously assume (1) without loss of generality

We have $g(y) - g(x) = \phi(y)(f(y) - f(x)) + f(x)(\phi(y) - \phi(x))$.

By i) and (1) we see that $|\phi(y)(f(y) - f(x))| > |f(x)(\phi(y) - \phi(x))|$

hence finally $|g(y) - g(x)| = |\phi(y)(f(y) - f(x))| = |f(y) - f(x)|$.

Now let ψ be f-dominated. We see that

$$|g(y) - g(x)|^2 = |f(y) - f(x)|^2 > |f(x)f(y)|\, |\psi(y) - \psi(x)|^2.$$

But $|f(x)| = |g(x)|$ in all of D because $|\phi(x)| = 1$ hence

$$|g(y) - g(x)|^2 > |g(y)\, g(x)|\, |\psi(y) - \psi(x)|^2$$

hence ψ is g-dominated. This ends the proof of Lemma 29.1.

Corollary 29.2: *Let f be a function from D to K and let $(\phi_n)_{n \in \mathbb{N}^*}$ be a sequence of f-dominated functions such that the sequence $\left(f \prod_{i=1}^{n} \phi_i \right)_{n \in \mathbb{N}^*}$ converges to a limit g for the simple convergence in D. Then we have $|g(y) - g(x)| = |f(y) - f(x)|$ whenever $x, y \in D$.*

Lemma 29.3: (L. Haddad) *Let ϕ be a homographic function. For every x and y which are not poles of ϕ we have $(\phi(y) - \phi(x))^2 = \phi'(x)\phi'(y)(y - x)^2$.*

Proof: If $\phi(x)$ is in the form $A + \dfrac{\lambda}{(x-b)}$ then we have $\phi'(x)\phi'(y) =$

$\dfrac{\lambda^2}{(x-b)^2(y-b)^2}$ while $\phi(x) - \phi(y) = \dfrac{\lambda(y-x)}{(x-b)(y-b)}$ hence $(\phi(y) - \phi(x))^2 =$ $\phi'(x)\phi'(y)(y-x)^2$. Otherwise $\phi(x)$ is of the form $Ax + B$ and then the equality is even more obvious.

Lemma 29.4: (L. Haddad) *Let f be a derivable function from D to K satisfying $|f(x) - f(y)|^2 \geq |f'(x)f'(y)|\,|y - x|^2$ whenever $x, y \in D$ and let $\phi \in R(D)$ be a homographic function satisfying $|\phi(x)| = 1$ whenever $x \in D$ and $|f(x)\phi'(x)| < |f'(x)|$ whenever $x \in D$. Then ϕ is f-dominated.*

Proof: We see that for every $x, y \in D$ with $x \neq y$ we have
(1) $|f(y) - f(x)|^2 > |f(x)|\,|\phi'(x)|\,|f(y)|\,|\phi'(y)|\,|y - x|^2$.
But by Lemma 29.3 we have $|\phi'(x)\phi'(y)|\,|y - x|^2 = |\phi(y) - \phi(x)|^2$.

Now we have to introduce the strongly copiercing sequences.

Definitions: Let E be a set of diameter $r \in\,]0 + \infty[$. A couple $(\alpha, \beta) \in (K \setminus \overline{E})^2$ will be said to be a *strongly copiercing couple associated to E* if $|\alpha - \beta| < \delta(\alpha, E)$. Hence it is seen that either both α, β belong to a same hole of E or both α, β belong to $K \setminus \widetilde{E}$ and more precisely to the same class of a circle whose center lies in $\widetilde{E}$.

A finite sequence $(a_n, b_n)_{1 \leq n \leq q}$ will be said to be a *strongly copiercing sequence associated to E* if for every $n = 1, ..., q$, (a_n, b_n) is a strongly copiercing couple associated to E.

Finally, an infinite sequence $(a_n, b_n)_{n \in \mathbb{N}}$ will be said to be *a strongly copiercing sequence associated to E* if for every $n \in \mathbb{N}, (a_n, b_n)$ is a strongly copiercing couple associated to E and if $\lim_{n \to \infty} \dfrac{|a_n - b_n|}{\delta(a_n, E)} = 0$.

Now a homographic function $\phi \in R(E)$ will be said to be *strongly copiercing with respect to E* if ϕ is of the form $\dfrac{x - a}{x - b}$ with (a, b) a strongly copiercing couple associated to E. As usual by *homographic function* we mean a function in the form $\dfrac{ax + b}{cx + d}$ with $ad - bc \neq 0$. A sequence $(\phi_n)_{n \in I}$ of homographic functions will be said to be *strongly copiercing with respect to E* if each ϕ_n is of the form $\dfrac{x - a_n}{x - b_n}$ with $(a_n, b_n)_{n \in I}$ a strongly copiercing finite or infinite sequence associated to D.

As usual , a function f defined in a set E is said to be *isometric* if it satisfies $|f(x) - f(y)| = |x - y|$ for all x and y in E.

Lemma 29.5 is clearly seen.

Lemma 29.5:   *Let (a, b) be a strongly copiercing couple associated to D, and let $\phi(x) = \dfrac{x - a}{x - b}$. Then $\|\phi - 1\|_D = \dfrac{|a - b|}{\delta(a, D)} < 1$. If $(\phi_n)_{n \in \mathbb{N}}$ is an infinite sequence of homographic functions strongly copiercing with respect to D, the product $\prod\limits_{n=0}^{\infty} \phi_n$ converges in $H(D)$.*

Now we may show that the finite increasing property is conserved when multiplying by the product of the terms of a copiercing sequence.

Theorem 29.6:   *Let f be a derivable function from D to K satisfying $|f(x) - f(y)|^2 = |x - y|^2 |f'(x) \, f'(y))|$. Let I be a subset of $\mathbb{N}$, let $(\phi_n)_{n \in I}$ be a sequence of homographic functions strongly copiercing with respect to D, satisfying $|f(x)\phi'_n(x)| < |f'(x)|$ whenever $x \in D$, and $n \in I$. Let $g = f\left(\prod\limits_{n \in I} \phi_n\right)$.*

Then g satisfies

$\quad$ (1) $\quad$ $|g(x)| = |f(x)|$ $\qquad$ *whenever $x \in D$*

$\quad$ (2) $\quad$ $|g'(x)| = |f'(x)|$ $\qquad$ *whenever $x \in D$*

$\quad$ (3) $\quad$ $|g(y) - g(x)| = |f(y) - f(x)|$ *whenever $x, y \in D$*

$\quad$ (4) $\quad$ $|g(y) - g(x)|^2 = |y - x|^2 |g'(x)g'(y)|$ *whenever $x, y \in D$.*

Proof: Without loss of generality we may obviously assume $I = \mathbb{N}^*$. Since the sequence (a_n, b_n) is copiercing by Lemma 29.5 we know that $|\phi_n(x)| = 1$ whenever $x \in D$, and $n \in \mathbb{N}^*$ hence Statement (1) is obvious. Besides if we put

$$\psi_n = \prod_{j=1}^{n} \phi_j \quad (n \in \mathbb{N}^*),$$

the sequence ψ_n converges in $H(D)$ to a limit ψ. Then, as D is open, the sequence $\psi'_n(x)$ converges to $\psi'(x)$ for all $x \in D$. But since $|f(x)\phi'_n(x)| < |f'(x)|$ we see that $|(f\psi_n)'(x)| = |f'(x)|$ for every $n \in \mathbb{N}^*$ hence finally $|g'(x)| = |f'(x)|$ whenever $x \in D$. This shows (2). Now by Lemma 29.4 we see that ϕ_n is f-dominated for every $n \in \mathbb{N}^*$ hence by Corollary 29.2 we just have (3) $|g(y) - g(x)| = |f(y) - f(x)|$ hence Statement (4) follows from Statement (2) and (1) and from the hypothesis $|f(x) - f(y)|^2 = |f'(x) \, f'(y)| \, |x - y|^2$.

Corollary 29.7:   *Let $f(x) = \sum\limits_{n=0}^{\infty} a_n(x - \alpha)^n$ be a power series convergent and strictly injective in a disk $d(\alpha, r)$ (resp. $d(\alpha, r^-)$) which contains D. Let I*

be a subset of $\mathbb{N}$, let $(\phi_n)_{n \in I}$ be a sequence of homographic functions strongly copiercing with respect to D , satisfying $|f(x)\phi'_n(x)| < |f'(x)|$ whenever $x \in D$, and $n \in I$. Let $g = \left(\prod\limits_{n \in I} \phi_n\right) f$. Then g satisfies Relations (1), (2), (3), (4) in Theorem 29.6.

(Indeed Theorem 27.2 (resp. Corollary 27.3) shows f to satisfy the hypothesis of Theorem 29.6.)

Example: We suppose $K = \mathbb{C}_p$. Let F be the function defined in $d(0, 1^-)$ as

$$F(x) = Log(1 + x) = \sum_{n=1}^{\infty} \frac{(-1)^{n-1} x^n}{n}.$$

Let $r_1 = p^{-\frac{1}{(p-1)}}$, let $r_2 = p^{-\frac{1}{p(p-1)}}$, and let $E = d(0, r_2{}^-)$. Let $u_1, ..., u_{p-1}$ be the pth roots of 1 other than 1, let $r \in \left]0, p^{-\frac{1}{p-1}}\right[$ and let $D = E \setminus \left(\bigcup_{i=1}^{p-1} d(u_i - 1, r^-)\right)$.

Let $b_1, ..., b_{p-1} \in E \setminus D$ satisfy

(1) $|b_i - u_i| < r^2 p^{\frac{1}{p-1}}$

and let $G(x)$ be the function defined in E as

$$G(x) = \frac{Log(1 + x)}{\prod\limits_{i=1}^{p-1}(1 + x - b_i)}.$$

We will show that G is a strictly injective element of $H(D)$ satisfying

(2) $|G(y) - G(x)|^2 = |G'(x)G'(y)| \, |x - y|^2$ whenever $x, y \in D$.

Let $P(x) = \prod\limits_{i=1}^{p-1}\left(1 - \frac{1+x}{u_i}\right)$. Obviously F belongs to $H(E)$ and its zeros in E are 0 and the $u_i - 1$ $(1 \leq i \leq p - 1)$, all of them of order 1. Hence the function $f(x) = \dfrac{F(x)}{P(x)}$ extends to an element of $H(E)$ which admits 0 as a zero of order 1 and has no other zero in E. Hence the power series $\sum\limits_{n=1}^{\infty} a_n x^n$ of f in E satisfies $|a_1| r_2 > |a_n|(r_2)^n$ whenever $n > 1$, while we have

(3) $|f'(x)| = |a_1| = |f'(0)| = \dfrac{|F'(0)|}{|P(0)|} = 1$ whenever $x \in E$.

(4) $|f(x)| = |x| \, |a_1| = |x|$ whenever $x \in E$.

Now we see that each homographic function $\phi_i(x) = \dfrac{(1 + x - u_i)}{(1 + x - b_i)}$ obviously belongs to $H(D)$. Hence with the help of (4) we have

(5) $|f(x)\,\phi_i'(x)| = |b_i - u_i|\dfrac{|x|}{|1 + x - b_i|^2}$ whenever $x \in D$.

But we have $\dfrac{|x|}{|1 + x - b_i|^2} \leq \max\left(\dfrac{1}{|1 + x - b_i|}, \dfrac{|b_i|}{|1 + x - b_i|^2}\right)$. Hence, as

$|1 + x - b| \geq r$, we obtain $\dfrac{|x|}{|1 + x - b_i|^2} \leq \dfrac{1}{r}\max\left(1, \dfrac{r_1}{r}\right) = \dfrac{r_1}{r^2}$ and therefore by (3), (4), (5) we have

(6) $\qquad |f(x)|\,|\phi_i'(x)| \leq |b_i - u_i|\left(\dfrac{r_1}{r^2}\right) < 1 = |f'(x)|$ whenever $x \in D$.

Now let $g(x) = f(x)\Big(\displaystyle\prod_{i=1}^{p-1}\phi_i\Big)$. Obviously g belongs to $H(D)$. By (6) and by Corollary 29.7 g is strictly injective and satisfies

$$|g(y) - g(x)|^2 = |g'(x)|\,|g'(y)|\,|y - x|^2.$$

But G is clearly equal to $\Big(\displaystyle\prod_{i=1}^{p-1}u_i\Big)^{-1} g$ hence G also is a strictly injective element of $H(D)$ satisfying (2).

Now we will consider the case when h is a homographic function .

Lemma 29.8: *Let $h \in R(D)$ be a homographic function. Then h is isometric if and only if $|h'(x)| = 1$ whenever $x \in D$.*

Proof: First we suppose h isometric in D. Since D is open, each point $a \in D$ lies in a disk $d(a, r) \subset D$ and then by Theorem 27.2 it is seen that $|h'(x)| = 1$ whenever $x \in D$. Now let us suppose $|h'(x)| = 1$ whenever $x \in D$. If h has no pole, it is just a function $ax + b$ with $|a| = 1$ hence h is clearly isometric. Therefore we assume h to be of the form $\dfrac{\lambda}{x - \beta} + b$. Hence we have $|h'(x)| = \left|-\dfrac{\lambda}{(x - \beta)^2}\right| = 1$ whenever $x \in D$. Let $r = \delta(\beta, D)$. Then D is clearly included in $C(\beta, r)$ and we have   (1) $|\lambda| = r^2$.

Now consider $h(x) - h(y) = \dfrac{\lambda(x - y)}{(x - \beta)(y - \beta)}$ when x and y belong to D. Since $D \subset C(\beta, r)$ we see that $|x - \beta| = |y - \beta| = r$, hence by (1) we have $|h(x) - h(y)| = |x - y|$.

Theorem 29.9: *Let $h \in R(D)$ be a homographic function, let I be a subset of $\mathbb{N}$, let $(\phi_n)_{n \in I}$ be a sequence of homographic functions strongly copiercing with respect to D, satisfying $|h(x)\phi_n'(x)| < |h'(x)|$ whenever $x \in D$, whenever $n \in I$ and let $g = h \prod_{n \in I} \phi_n$. Then g is strictly injective in D, and satisfies*

$$(1) \qquad |g(x)| = |h(x)| \quad \text{whenever } x \in D$$
$$(2) \qquad |g'(x)| = |h'(x)| \quad \text{whenever } x \in D$$
$$(3) \qquad |g(y) - g(x)| = |h(y) - h(x)| \quad \text{whenever } x, y, \in D$$
$$(4) \qquad |g(y) - g(x)|^2 = |y - x|^2 |g'(x)g'(y)| \quad \text{whenever } x, y \in D$$

Moreover, g is isometric if and only if $|h'(x)| = 1$ whenever $x \in D$.

Proof: As in Theorem 29.6 we assume I to be equal to $\mathbb{N}^*$. By Lemma 29.3 applied to the homographic function $h(x)$ we see that h satisfies the hypothesis of f in Theorem 29.6. Hence g satisfies (1), (2), (3), (4) and is strictly injective in D. By Lemma 29.8, g is isometric if and only if $|g'(x)| = 1$ whenever $x \in D$. But by Theorem 29.6 we have $|g'(x)| = |h'(x)|$ whenever $x \in D$ hence g is isometric if and only if $|h'(x)| = 1$ whenever $x \in D$.

Theorem 29.10: *Let $h \in H(D)$ be a homographic function, let $(\phi_n)_{n \in I}$ be a sequence of homographic functions strongly copiercing with respect to D, satisfying $|h(x)\phi_n'(x)| < |h'(x)|$ whenever $x \in D$, and $n \in I$.*

Let $c_1, ..., c_m \in K \setminus \tilde{D}$ satisfy $\delta(c_j, D)|h'(x)| > |h(x)|$ whenever $x \in D$ and $j = 1, ..., m$. Let $u_1, ..., u_m \in \mathbb{Z}$ with $u_j = \pm 1 (1 \leq j \leq m)$.

Let $g = h \Big(\prod_{n \in I} \phi_n \Big) \Big(\prod_{j=1}^{m} (x - c_j) \Big)^{u_j}$ and let $A = \prod_{j=1}^{m} \big(\delta(c_j, D) \big)^{u_j}$. Then g is strictly injective in D and satisfies

$$(1) \qquad |g(x)| = A|h(x)| \quad \text{whenever } x \in D$$
$$(2) \qquad |g'(x)| = A|h'(x)| \quad \text{whenever } x \in D$$
$$(3) \qquad |g(y) - g(x)| = A|h(y) - h(y)| \quad \text{whenever } x, y \in D$$
$$(4) \qquad |g(y) - g(x)| = |y - x|^2 |g'(x)g'(y)| \quad \text{whenever } x, y \in D.$$

Proof: Let $f(x) = h(x) \Big(\prod_{i \in I} \phi_i(x) \Big)$. By Theorem 29.9 we have Relations (5), (6), (7), (8)

$$(5) \qquad |f(x)| = |h(x)|$$
$$(6) \qquad |f'(x)| = |h'(x)|$$
$$(7) \qquad |f(y) - f(x)| = |h(y) - h(x)|$$
$$(8) \qquad |f(y) - f(x)|^2 = |y - x|^2 |f'(x)f'(y)|$$

Let $a \in D$ and for every $j = 1, ..., m$ let $\psi_j(x) = \left(\dfrac{x - c_j}{a - c_j}\right)^{u_j}$. Since $c_j \in K \setminus \tilde{D}$

it is seen that $|x - c_j| = |a - c_j| > |a - x|$ and therefore we have

(9) $\qquad |\psi_j(x) - 1| < 1$ whenever $x \in D$

(10) $\qquad \psi_j'(x) = u_j(a - c_j)^{-u_j}(x - c_j)^{u_j - 1}.$

Since $|a - c_j| = |x - c_j| = \delta(c_j, D)$ whenever $x \in D$, Relation (10) gives us

$|\psi_j'(x)| = \dfrac{1}{\delta(c_j, D)}$ hence by the hypothesis $|h'(x)|\delta(c_j, D) > |h(x)|$. Then by (5)

and (6) we see that each ψ_j satisfies

(11) $\qquad |f(x)\psi_j'(x)| < |f'(x)|$

Hence, thanks to (8) and (9) we may apply Lemma 29.4 and Corollary 29.2 to

the product $F(x) = f(x) \displaystyle\prod_{j=1}^{m} \psi_j(x)$. By Lemma 29.4, each ψ_j is f-dominated and

therefore by Corollary 29.2 we have

(12) $\qquad |F(y) - F(x)| = |f(y) - f(x)|$ and by (9), (11), (12) we obtain

(13) $\qquad |F(x)| = |f(x)|$

(14) $\qquad |F'(x)| = |f'(x)|$

(15) $\qquad |F(y) - F(x)|^2 = |y - x|^2 |F'(x)F'(y)|$

But we have $g(x) = F(x)\left(\displaystyle\prod_{j=1}^{m}(a - c_j)^{u_j}\right)$ hence $|g(x)| = A|F(x)|$ and finally

Relations (13), (14), (15), (12) just give Statements (1), (2), (3), (4) announced
above.

Example: $\qquad$ Let $\alpha, \beta \in d(0, 1)$ and let $h(x) = \dfrac{x - \alpha}{x - \beta}$. Let $(a_n)_{n \in \mathbb{N}}$ be a sequence

in $d(0, 1^-)$ such that $0 < |a_n| < |a_{n+1}| < 1$, $\displaystyle\lim_{n \to \infty} |a_n| = 1$. Let $(r_n)_{n \in \mathbb{N}}$

be a sequence in $]0, 1[$ such that $r_n \leq |a_n|$ whenever $n \in \mathbb{N}$, and let $D = $

$d(0, 1) \setminus \left(\left(\displaystyle\bigcup_{n \in \mathbb{N}} d(a_n, r_n^-)\right) \cup \{\beta\}\right)$. Let $(b_n)_{n \in \mathbb{N}}$ be a sequence in $d(0, 1^-)$ such

that $0 < |a_n - b_n| < r_n^2 |\alpha - \beta|$. Since $r_n < 1$ we have $|a_n - b_n| < r_n$.

$\qquad$ The holes of D are seen to be the disks $T_n = d(a_n, r_n^-)$, while $b_n \in T_n$. Hence

the homographic function $\phi_n(x) = \dfrac{x - a_n}{x - b_n}$ belongs to $R(D)$ for each $n \in \mathbb{N}$. It is

seen that $\|\phi_n'\|_D = \dfrac{|a_n - b_n|}{r_n^2}$ while $\|\dfrac{h}{h'}\|_D = \dfrac{1}{|\alpha - \beta|}$ hence we have $\|\phi_n' \dfrac{h}{h'}\|_D \leq$

$\|\phi_n'\|_D \ \|\dfrac{h}{h'}\|_D = \dfrac{|a_n - b_n|}{|\alpha - \beta| r_n^2} < 1$ whenever $n \in \mathbb{N}$ whence the sequence ϕ_n does satisfy the hypothesis of Theorem 29.9 and therefore $h\left(\prod_{n\in\mathbb{N}}\phi_n\right)$ is injective in D. Now let $c \in K$ satisfy $|c| > \dfrac{1}{|\alpha - \beta|}$. Then by Theorem 29.10 we see that $h\left(\prod_{n\in\mathbb{N}}\phi_n\right)(x - c)$ is also injective in D.

Remark 1: Given a homographic function h, the function $\dfrac{h}{h'}$ is seen to be a polynomial of degree at most 2. Hence we see that the hypothesis of Corollary 29.7 and Theorem 29.9 are satisfied:

when $\|\dfrac{f}{f'}\phi'\|_D < 1$ in Corollary 29.7 and

when $\|\dfrac{h}{h'}\phi'\|_D < 1$ in Theorem 29.9 and 29.10.

However the hypothesis $|h(x)\phi'(x)| < |h'(x)|$ whenever $x \in D$ is wider because $\dfrac{h}{h'}\phi'$ is not assured to reach its maximum when D is not strongly infraconnected. In the same way in Theorem 29.10 the hypothesis $\delta(c_i, D)|h'(x)| > |h(x)|$ whenever $x \in D$ is satisfied when $\delta(c_i, D) > \|\dfrac{h}{h'}\|_D$ although this last hypothesis is more restrictive.

Remark 2: We would be tempted to generalize Theorem 29.9 by trying to show that if any $\phi \in H(D)$ satisfies $\|\phi - 1\|_D < 1$ and $\|\dfrac{h}{h'}\phi'\|_D < 1$, then $h\phi$ is injective in D again. Actually, this is false at least when the residue characteristic p is not zero, as this counter example shows. Indeed, let $D = d(0, 1)$, let a and $\theta \in K$ satisfy $|\theta| < 1 < |a\theta| < \dfrac{1}{p}$, let $h = a + x$ and let $\phi = \theta x^p + 1$.

Since $|\theta| < 1$ we have $\|\phi - 1\|_D < 1$ and since $|a\theta| < \dfrac{1}{p}$ we have $\|\dfrac{h}{h'}\phi'\|_D < 1$.

Now $h\phi = a + x + a\theta x^p + \theta x^{p+1}$. Since $|a\theta| > 1$ it is seen that $h\phi - a$ has p zeros in $d(0, 1)$, hence $h\phi$ is not injective. Thus we think that Theorem 29.9 has to be stated in terms of copiercing homographic functions. We will see later how to translate it when D is infraconnected. In [37] Yvette Fencyrol-Perrin proved that in hypervaluated fields, the strictly injective elements are just the homographic functions. Here, Theorem 29.9 suggests that the sufficient condition to obtain injective elements might also be necessary provided D satisfies certain condition. This question is considered again in Chapter 48.

30. MAXIMUM PRINCIPLE

Throughout this chapter D is supposed to be infraconnected and closed.

We consider the question whether every element of $H_b(D)$ reaches its maximum $\|f\|_D$ at some point $\alpha \in D$ [22].

Definition: D will be said to be *calibrated* if every hole has a diameter that belongs to $|K|$ and if $diam(D)$ either belongs to $|K|$ or is infinite.

Lemma 30.1: *Let $\mathcal{F}$ be a decreasing filter with no center, of canonical base $(D_n)_{n\in\mathbb{N}}$ with $D_n = d(a_n, r_n) \cap D$. Let $f \in H_b(D)$ be not vanishing along $\mathcal{F}$. There exists $t \in \mathbb{N}$ such that $|f(x)| = ct$ whenever $x \in D_t$.*

Proof: Let $\lambda = \lim\limits_{n\to\infty} \|f\|_{D_n}$. Since f is not vanishing along $\mathcal{F}$, we have $\lambda > 0$, hence there exists $h \in R(D)$ such that (1) $\|f - h\|_D < \lambda$. Since $\mathcal{F}$ has no center, there exists $t \in \mathbb{N}$ such that h has neither any zero nor any pole inside $\widetilde{D_t}$ because otherwise, certain zero or certain pole α would belong to $\left(\bigcap\limits_{n\in\mathbb{N}} \widetilde{D_n} \right)$ and thereby would be a center of $\mathcal{F}$. Then by Theorem 23.6, $|h(x)|$ is equal to a constant C in D_t. But by hypothesis we have $\|f\|_{D_t} \geq \lambda$ hence $\|h\|_{D_t} \geq \lambda$, hence $C \geq \lambda$ and therefore by (1) we have $|f(x)| \geq \lambda$ whenever $x \in D_t$.

Lemma 30.2: *We suppose $0 \in \widetilde{D}$. Let $S = diam(D)$ and let $\rho = \delta(0, D)$. Let $r', r'' \in \mathbb{R}_+$ satisfy $\rho \leq r' < r'' \leq S$ and let $h \in H(\Gamma(0, r', r''))$ satisfy $N^-(h, \mu) = N^+(h, \mu) = t \neq 0$ whenever $\mu \in]-\log r'', -\log r'[$. Then $_D\varphi_{0,r'}(h) \in |K|$, (resp. $_D\varphi_{0,r''}(h) \in |K|$) if and only if $r' \in |K|$, (resp. $r'' \in |K|$).*

Proof: Since h belongs to $H(\Gamma(0, r', r''))$, $h(x)$ is of the form $\sum\limits_{-\infty}^{+\infty} \alpha_n x^n$ with $|\alpha_t| r^t > |\alpha_n| r^n$ whenever $n \neq t$, whenever $r \in]r', r''[$ and therefore $|h(x)| = |\alpha_t||x|^t$ whenever $x \in \Gamma(0, r', r'')$. Hence we have $_D\varphi_{0,r'}(h) = |\alpha_t| r'^t$ and $_D\varphi_{0,r''}(h) = |\alpha_t| r''^t$. Since $t \neq 0$ it is seen that $_D\varphi_{0,r'}(h) \in |K|$ if and only if $r' \in |K|$ and $_D\varphi_{0,r''}(h) \in |K|$ if and only if $r'' \in |K|$.

Lemma 30.3: *Let D be bounded and let $f \in H(D)$. Either we have $\|f\|_D = {}_D\varphi_D(f)$ or there exists a f-hole T of D such that $\|f\|_D = {}_D\varphi_T(f)$. Further, if*

$f \in R(D)$ *satisfies* $\|f\|_D > {}_D\varphi_D(f)$ *then there exists a hole* T *of* D *that contains at least one pole of* f *such that* $\|f\|_D = {}_D\varphi_T(f)$.

Proof: Let $(T_n)_{n \in \mathbb{N}^*}$ be the sequence of the f-holes. By Corollary 15.2, either $\|f\|_D = \|\overline{f}_0\|_D$ and then we have $\|f\|_D = {}_D\varphi_D(f)$ or we have $\|f\|_D = \|\overline{f}_{T_n}\|_D$ for some $n > 0$ and then we have $\|f\|_D = {}_D\varphi_{T_n}(f)$. In particular if f belongs to $R(D)$ and if $\|f\|_D > {}_D\varphi_D(f)$ then $\|f\|_D$ is of the form $\|\overline{f}_{T_n}\|_D$ for some $n > 0$ and therefore f does have at least one pole in T_n.

Theorem 30.4: *Let* D *be unbounded and* $\overline{D}$ *be different from* K *and let* $f \in H_b(D)$. *There exists a* f-*hole* T *of* D *such that* ${}_D\varphi_T(f) = \|f\|_D$.

Proof: Since $\overline{D} \neq K, D$ has at least one hole $\Lambda = d(\alpha, r^-)$ and then the inversion $\mathcal{I}_a(x) = \dfrac{1}{x-a}$ maps D onto a bounded set D' of diameter $S = \dfrac{1}{r}$. Let $g\left(\dfrac{1}{x-a}\right) = f(x)$. As $f \in H_b(D)$, we know that $g \in H(\overline{D'})$ whereas obviously $\|g\|_{\overline{D'}} = \|f\|_D$. Further, it is seen that ${}_{D'}\varphi_{D'}(g) = {}_D\varphi_\Lambda(f)$ whereas every hole T of D other than Λ is mapped onto a hole T' of D' and we have ${}_{D'}\varphi_{T'}(g) = {}_D\varphi_T(f)$ (besides D' has no hole other than the images of the holes of D).

Now by Lemma 30.3, either $\|g\|_{\overline{D'}} = {}_{\overline{D'}}\varphi_{\overline{D'}}(g)$ and then $\|f\|_D = {}_D\varphi_\Lambda(f)$ or there exists a g-hole T' of D' such that $\|g\|_D = {}_{\overline{D'}}\varphi_{T'}(g)$ and then $\|f\| = {}_D\varphi_T(f)$. Besides, since T' is a g-hole of D, f does not belong to $H(D \cup T)$, and therefore T is a f-hole of D.

The strongly infraconnected sets were defined in Chapter 2.

Proposition 30.5: *Let* D *be strongly infraconnected and closed, and let* $f \in H_b(D)$ *satisfy* $\|f\|_D \in |K|$. *Then there exists* $\alpha \in D$ *such that* $\|f\|_D = |f(\alpha)|$.

Proof: We first assume $f \in R(D)$, and we suppose the conclusion to be false. By Lemmas 30.3 and 30.4, either we have $\|f\|_D = {}_D\varphi_D(f)$ with D bounded, or there exists a hole T of D such that $\|f\|_D = {}_D\varphi_T(f)$. Hence, anyway, there do exist $a \in \widetilde{D}$ and $r \in \mathbb{R}_+$ such that ${}_D\varphi_{a,r}(f) = \|f\|_D$. We will show that $r \in |K|$.

Indeed let us suppose that $r \notin |K|$. Then D is not included in $C(a, r)$. Hence , either the increasing filter $\mathcal{G}'$ of center a and diameter r is secant with D, or the decreasing filter $\mathcal{G}''$ of center a and diameter r is secant with D.

Without loss of generality we may obviously assume $a = 0$. For example let us suppose $\mathcal{G}'$ to be secant with D. Then there is $r' < r$ such that $f \in R(\Gamma(0, r', r))$, and then $f(x)$ is of the form $\displaystyle\sum_{-\infty}^{+\infty} a_n(x)^n$ with $N^-(h, \mu) =$

$N^+(h,\mu) = t \in \mathbb{Z}$ whenever $\mu \in]-\log r, -\log r'[$. If $t = 0$, we know that $|f(x)| = |a_0|$ whenever $x \in \Gamma(0, r', R)$, hence $_D\varphi_{0,r}(f) = |a_0| = |f(x)|$ whenever $x \in \Gamma(o, r', r) \cap D$. This contradicts the hypothesis. Hence we have $t \neq 0$ and then by Lemma 30.2, we see that $r \in |K|$.

In the same way when $\mathcal{G}''$ is secant with D, there exists $r'' > r$ such that $f \in R(\Gamma(0, r, r''))$, hence $f(x)$ is of the form $\sum_{-\infty}^{+\infty} a_n x^n$ inside $\Gamma(0, r, r'')$, with $N^-(h,\mu) = N^+(h,\mu) = t \in \mathbb{Z}$ and then we see that $t \neq 0$, hence by Lemma 30.2 we have $r \in |K|$.

Now, since D is strongly infraconnected there exists a sequence $(x_n)_{n\in\mathbb{N}}$ in D such that $|x_n| = |x_n - x_m|$ whenever $n \neq m$. Hence by Corollary 20.3 there exists $n \in \mathbb{N}$ such that $v(f(x_n)) = v(f, -\log r)$ whenever $n \geq \mathbb{N}$ hence $|f(x_n)| = {}_D\varphi_{0,r}(f)$ whenever $n \geq \mathbb{N}$. Finally we have a contradiction to the hypothesis when $f \in R(D)$.

The proof is easily extended to the case when $f \in H(D)$. Indeed let $h \in R(D)$ satisfy $\|f - h\|_D < \|f\|_D$. Then we have $\|h\|_D = \|f\|_D$ while there exists $a \in D$ such that $|h(a)| = \|h\|_D$. Hence $|f(a)| = |h(a)|$ and therefore $|f(a)| = \|f\|_D$.

Theorem 30.6: *D is calibrated and strongly infraconnected if and only if for every $f \in H_b(D)$ there exists $\alpha \in D$ such that $|f(\alpha)| = \|f\|_D$.*

Proof: We first suppose D calibrated and strongly infraconnected. By Proposition 30.5 we just have to show that $\|f\|_D \in |K|$. Let $(T_n)_{n\in\mathbb{N}^*}$ be the sequence of the f-holes.

First we suppose that $\|f\|_D = \|\overline{f}_{T_q}\|_D$ for some $q \geq 1$, and we put $T_q = (a_q, r_q^-)$. By Theorem 15.1 we have $\|\overline{f}_{T_q}\|_D = {}_D\varphi_{a_q, r_q}(\overline{f}_{T_q})$. Since $\overline{f}_{T_q}$ belongs to $H_0(K \setminus T_q)$ it is of the form $\sum_{m=1}^{\infty} b_m(x - a_q)^{-m}$ and we have ${}_D\varphi_{a_q, r_q}(\overline{f}_{T_q}) = \sup_{m\geq 1} |b_m| r_q^{-m}$. In particular, there exists $l \in \mathbb{N}^*$ such that ${}_D\varphi_{a_q, r_q}(\overline{f}_{T_q}) = |b_l| r_q^{-l}$. Since D is calibrated, r_q belongs to $|K|$ hence so does ${}_D\varphi_{a_q, r_q}(\overline{f}_{T_q})$ and then so does $\|f\|_D$.

Now, we suppose $\|f\|_D = \|\overline{f}_0\|_D$. If D is not bounded, $\overline{f}_0$ belongs to K hence we have nothing to prove. We suppose D bounded, of diameter $S \in |K|$. Then $\|\overline{f}_0\|_D$ is of the form $_D\varphi_{a,S}(\overline{f}_0)$. But $\overline{f}_0$ is of the form $\sum_{m=0}^{\infty} b_m(x - a)^m$ and $_D\varphi_{a,S}(f) = \sup_{m\in\mathbb{N}} |b_m| S^m \in |K|$, with $_D\varphi_{a,S}(f) = |b_l| S^l \in |K|$ for certain $l \in \mathbb{N}^*$.

Hence $\|f\|_D \in |K|$. This finishes proving that if D is calibrated and strongly infraconnected, there exists $\alpha \in D$ such that $|f(\alpha)| = \|f\|_D$.

Now we suppose that D is not at the same time calibrated and strongly infraconnected.

First we suppose D is not calibrated. For example we suppose that D admits a hole $T = d(a, r^-)$ with $r \notin K$. Let $f(x) = \dfrac{1}{x-a}$. By definition of the holes we have $r = \inf_{\xi \in D} |\xi - a|$ hence clearly $\|f\|_D = \dfrac{1}{r} \notin |K|$ and then there does not exist any $\alpha \in D$ such that $|f(\alpha)| = \|f\|_D$. In the same way if D is bounded, of diameter $S \notin K$, given $a \in D$, it is clear that $\|x - a\|_D = S \notin |K|$.

Now we suppose that D is not strongly infraconnected and let $\Gamma = d(a, r^-)$ be a hole such that all the classes of $C(a, r)$ have empty intersection with D except finitely many ones, denoted by $\Lambda_i = d(a_i, r^-)$ $(1 \le i \le q)$. Let $b \in C(a, r) \setminus$

$(\bigcup_{i=1}^{q} \Lambda_i)$. Let $f(x) = \dfrac{\prod_{j=1}^{q}(x - a_j)}{(x - b)^{q+1}}$. It is easily seen that when $x \in K \cap d(a, r)$

we have $|f(x)| < \dfrac{1}{r}$ because when $x \in \Lambda_i$, we have $\left| \dfrac{\prod_{1 \le j \le q, j \ne i}(x - a_j)}{(x - b)^{q+1}} \right| < \dfrac{1}{r^2}$. On

the other hand, when $x \in D \setminus d(a, r)$ we have $|f(x)| = \dfrac{1}{|x - a|} < \dfrac{1}{r}$ hence finally

$|f(x)| < \dfrac{1}{r}$ whenever $x \in D$. Now, since D is infraconnected, by Proposition 3.14 the circular filter $\mathcal{F}$ of center a and diameter r is secant with D and then we have $_D\varphi_{a,r}(f) = \dfrac{1}{r}$, hence $\|f\|_D = \dfrac{1}{r}$. That finishes proving the existence of an element which does not reach its maximum on D and that ends the proof of Theorem 30.6.

31. ANALYTIC ELEMENTS

MEROMORPHIC IN A HOLE

Throughout this chapter D is infraconnected, T is a hole of D and V is a disk of the form $d(a, r)$ or $d(a, r^-)$, included in $\widetilde{D}$, such that $\widetilde{V} \cap D \neq \emptyset$. f denotes an element of $H(D)$.

We will carefully study the analytic elements whose singularities in V just consist of a few poles .

Definitions: f will be said to be *meromorphic in V* if there exist finitely many points $(a_i)_{(1 \leq i \leq n)}$ in V such that f has continuation to an element of $H((D \cup V) \setminus \{a_i \mid 1 \leq i \leq n\})$.

Let f be meromorphic in V, and belong to $H((D \cup V) \setminus \{a_i \mid 1 \leq i \leq n\})$.

For each $i = 1, ..., n$ if $f \notin H((D \cup V) \setminus \{a_h \mid h \neq i\})$ then by Corollary 10.9 and Theorem 10.10, a_i is a pole of f as an element of $H((D \cup V) \setminus \{a_j \mid 1 \leq j \leq n\})$. Let q_i be its order. a_i will be called *a pole of f of order q_i in V.* The polynomial

$$P(x) = \prod_{i=1}^{n} (x - a_i)^{q_i}$$ will be called *the polynomial of the poles of f in V.*

Let $a \in \widetilde{D}$. Then f will be said to be *meromorphic at a* if:

either a belongs to $\overline{D}$,

or a belongs to a hole T of D, and f is meromorphic in T.

Further, f will be said to be *holomorphic at a* if f is meromorphic at a and does not admit a as a pole.

Remark: By definition the polynomial of the poles P of f in V satisfies: $Pf \in H(V)$ and $(Pf)(a_i) \neq 0$ whenever $i = 1, ...n$.

This Lemma is immediate:

Lemma 31.1: *Let $a_1, ..., a_n$ be points in T , let $q_1, ..., q_n \in \mathbb{N}^*$ and let $P(x) = \prod_{i=1}^{n} (x - a_i)^{q_i}$. The set of the $h \in H(D)$, meromorphic in T, whose polynomials of poles in T divides P, is a K-subvector space of $H(D)$ which is closed with respect to $\mathcal{U}_D$.*

Lemma 31.2: *Let D be bounded. If f is meromorphic in T, the polynomial of its poles P in T satisfies $Pf \in H(D \cup T)$.*

Proof: Indeed , let $D' = (D \cup T) \setminus \{a_1, ..., a_n\}$. Then D' is bounded and therefore by Theorem 10.4, Pf belongs to $H(D')$. But by construction Pf is bounded at each point a_i and therefore by Corollary 10.6, Pf belongs to $H(D \cup T)$.

Lemma 31.3: *Let D belong to $\mathcal{A}$. If f is meromorphic in T, the polynomial of its poles P in T satisfies: $Pf \in H(D \cup T)$.*

Proof: Indeed , let $D' = (D \cup T) \setminus \{a_1, ..., a_n\}$. Then by Theorem 18.8 , D' belongs to $\mathcal{A}$ and therefore Pf belongs to $H(D')$. But by construction Pf is bounded at each point a_i and therefore by Corollary 10.6, Pf belongs to $H(D \cup T)$.

Lemma 31.4: *Let D be bounded, (resp. let $D \in \mathcal{A}$), and let f be invertible in $H(D)$. Then f is meromorphic in T if and only if so is $\dfrac{1}{f}$.*

Proof: First we suppose f meromorphic in T. Let P be the polynomial of its poles in T and let $g(x) = f(x)P(x)$. Since D is bounded, (resp. belongs to $\mathcal{A}$) by Lemma 31.2 (resp. by Lemma 31.3), g belongs to $H(D \cup T)$. Let Q be the polynomial of the zeros of g in T. Since f has no zero in D, Q actually is the polynomial of the zeros of g in $D \cup T$ and then g is of the form $Q(x)h(x)$ with h an element of $H(D \cup T)$ that has no zero in T. Hence we have $\dfrac{1}{h} = \dfrac{1}{f}\dfrac{Q}{P}$. If $D \in \mathcal{A}$, $\dfrac{1}{h}$ obviously belongs to $H(D)$. If D is bounded, we have $\dfrac{Q}{P} \in R_b(D)$ and then by Lemma 10.4, $\dfrac{1}{h}$ belongs to $H(D)$. Thus $\dfrac{1}{h}$ belongs to $H(D)$ anyway. Now, by Lemma 24.1, h is invertible in $H(D \cup T)$. Hence f factorizes in the form $\dfrac{P}{Q}h$ and then $\dfrac{1}{f} = \dfrac{1}{Q}\dfrac{P}{h}$. But $\dfrac{P}{h}$ belongs to $H(D \cup T)$ and therefore $\dfrac{1}{f}$ is meromorphic in T and admits Q as the polynomial of its poles in T. We may obviously apply the same reasonning to $\dfrac{1}{f}$ and this shows the reciprocal.

Theorem 31.5: *Let $F \in H(D)$ be meromorphic in T and satisfy $\|F - 1\|_D < 1$. Then F has as many poles as many zeros in T.*

Proof: Without loss of generality we may obviously assume that D is bounded because the hypothesis remains true in any set $D \cap d(0, R)$. We may also assume that $T = d(0, r^-)$. Let P (resp. Q) be the polynomial of the zeros (resp.

the poles) of F in T . Then by Lemma 31.2 F factorizes in the form $\dfrac{f}{Q}$ with $f \in H(D \cup T)$. Now, the zeros of f inside T are just those of F, hence f factorizes in the form Pg with $g \in H(D \cup T)$. Let $h = \dfrac{P}{Q}$. Since g has no zero in T by Theorem 23.6 $|g(x)|$ is equal to a constant inside T. Let $s \in]0, r[$ be such that all the zeros of P and of Q lie in $d(0, s)$. Then obviously F belongs to $H(\Gamma(0, s, r))$. Now by hypothesis there exists $\lambda > 0$ such that $v(F, \mu) \geq \lambda$ for all $\mu \leq -\log r$. Hence by continuity , there exists $\lambda' > 0$ and s' in $]s, r[$ such that $v(F, \mu) \geq \lambda'$ for all $\mu \leq -\log s'$. Thus there exists $b \in]0, 1[$ and t in $]s', r[$ such that $|F(x) - 1| \leq b$ for all $x \in \Gamma(0, t, r)$ and then, it is seen that $|h(x)|$ is constant in $\Gamma(0, t, r)$. Hence we have $\dfrac{d}{d\mu} v(h, \mu) = 0$ for all $\mu \in [-\log r, -\log t]$. Since h has neither any zero nor any pole in $C(0, t)$, by Corollary 4.11, h has as many zeros as many poles in $d(0, t)$ and therefore in $d(0, r)$ and this ends the proof.

In connection with the Mittag-Leffler series , this theorem is quite useful.

Theorem 31.6:   *Let $f \in H_b(D)$ be meromorphic in T, and admit only one pole b inside T. Let q be the multiplicity order of b. Then the Mittag-Leffler term of f associated to T is of the form* $\displaystyle\sum_{j=1}^{q} \frac{a_j}{(x - b)^j}$, *with $a_q \neq 0$.*

Proof: First, we may clearly assume that D is bounded, because given a suitable disk $d(0, S)$ that contains T, the Mittag-Leffler term $\overline{f_T}$ of f associated to T, with respect to the infraconnected set D is the same as the one with respect to the infraconnected set $D \cap d(0, S)$.

We can write $\overline{f_T}$ in the form $\displaystyle\sum_{j=1}^{\infty} \frac{a_j}{(x - b)^j}$. But since D is bounded, $(x - b)^q f$ belongs to $H(D \cup T)$, hence so does $(x - b)^q \overline{f_T}$. Therefore we have $\displaystyle\sum_{j=q+1}^{\infty} \frac{a_j}{(x - b)^j} = 0$, hence finally $a_j = 0$ whenever $j \geq q + 1$. Besides we check that $a_q \neq 0$, because if $a_q = 0$, then $(x - b)^{q-1} \overline{f_T}$ clearly belongs to $H(D \cup T)$ and then, so does $(x - b)^{q-1} f$, but this contradicts the hypothesis: "b is a pole of order q".

Before closing this Chapter, it is useful to consider again elements meromorphic at a point.

Lemma 31.7:   *Let $a \in \overline{D}$ and let f be f meromorphic but not holomorphic at a. Then a is a pole of f.*

Proof: By hypothesis, there exists $r > 0$ such that f belongs to $H\big(D \cup (d(a,r) \setminus \{a\})\big)$. Suppose that a is not a pole of f. Then by Theorem 10.10 f belongs to $H(D \cup d(a,r))$ and then f is holomorphic at a.

Corollary 31.8:   *Let $a \in \overline{D}$. Then f is meromorphic at a, and admits a as a pole of order q if and only if there exists a disk $d(a,r)$ included in $\overline{D}$ and an element $h \in H(d(a,r))$ such that $f(x)(x-a)^q = h(x)$ whenever $x \in d(a,r) \setminus \{a\}$, and $h(a) \neq 0$.*

Corollary 31.9:   *Let D satisfy Condition B). For every $a \in \overline{D}$ f is meromorphic at a. For every $a \in D$, f is holomorphic at a.*

Remark: Let $a \in \overline{D}$ and let f admit a as a pole of order q. This does not imply that f is meromorphic at a. Indeed, by Lemma 18.4 we know that there exist infraconnected sets E with a point $a \in E \setminus \overset{\circ}{E}$ and elements $h \in H(E)$ such that $\displaystyle\lim_{|x-a| \to 0} h(x) = 0$, and such that $\displaystyle\limsup_{|x-a| \to 0} \left| \frac{h(x)}{x-a} \right| = \infty$. Let $E' = E \setminus \{a\}$ and let $g = \dfrac{1+h}{x-a}$. It is easily seen that a is a pole of order 1 for g. But h does not belong to any space $H(d(a,r))$, whenever $r > 0$ (because if it did, it should factorize in $H(d(a,r))$ in the form $(x-a)\ell(x)$, with $\ell \in H(d(a,r))$). Thus it is seen that $(x-a)g$ does not belong to $H(E)$, and therefore g is not meromorphic at a.

32.   MOTZKIN FACTORIZATION

Throughout this chapter, D is a closed infraconnected set and f belongs to $H(D)$.

The idea of factorizing semi-invertible analytic elements into a product of singular factors is due to E. Motzkin [51]. This factorization has tight links with the Mittag-Leffler series, as it was shown in [34] and more precisely in [5]. However, this factorization requires that the set D belongs to $\mathcal{A}$, as shown in [5].

Lemma 32.1:   *Let $T = d(a, r^-)$, with $a \in K$, and $r > 0$, let $E = K \setminus T$ and let $b \in T$. Let $g \in H(E)$ be invertible in $H(E)$. Then there exist $\lambda \in K$, $q \in \mathbb{Z}$, and $h \in H(E)$ invertible in $H(E)$, satisfying $\|h - 1\|_E < 1$, $\lim\limits_{|x| \to +\infty} h(x) = 1$ and $g(x) = \lambda(x - b)^q h(x)$. Besides λ, q, h, are respectively unique, satisfying those relations. Further, both λ, q do not depend on b in T, $\dfrac{g'}{g}$ belongs to $H_0(E)$.*

Proof: Without loss of generality we may obviously assume $a = 0$. Besides, as g is invertible, if g belongs to $H_0(E)$, then $\dfrac{1}{g}$ does not. So, we may clearly assume g does not belong to $H_0(E)$. By Theorem 10.5. g is of the form $\widetilde{g} + \widehat{g}$, with $\widetilde{g} \in K[x]$, $\widetilde{g} \neq 0$, and $\widehat{g} \in H_0(E)$. Let $q = deg(\widetilde{g})$ and let λ be its coefficient of degree q. Now we put $h(x) = \dfrac{g(x)}{\lambda(x - b)^q}$. By definition it is seen that both λ, q do not depend on b in T. Hence we may also assume $b = 0$. Clearly, as $H(E)$ is a K-algebra, h is invertible in $H(E)$ and satisfies $\lim\limits_{|x| \to +\infty} h(x) = 1$. In particular we notice that h is bounded in E. Now we check that $\|h - 1\|_E < 1$.

Let $s = \dfrac{1}{r}$, let $A = d(0, s)$ and let $F(u) = h(\frac{1}{u})$ whenever $u \in d(0, s)$, $u \neq 0$. Then F belongs to $H(d(0, s) \setminus \{0\})$. But since h is bounded in E, F is bounded in $d(0, s) \setminus \{0\}$ and therefore F belongs to $H(d(0, s))$. Besides, the condition $\lim\limits_{|x| \to +\infty} h(x) = 1$ shows that $F(0) = 1$, hence F has no zero in $d(0, s)$. Thus $F(u)$ is of the form $\sum\limits_{n=0}^{\infty} a_n u^n$ with $a_0 = 1$ and by Theorem 23.16, we have $|a_n|s^n < 1$ whenever $n > 0$. Let $\epsilon = \sup\{|a_n|s^n \,|\, n > 0\}$. Then we have $\|F - 1\|_{d(0,s)} = \|h - 1\|_E = \epsilon$. Now, h, q, λ are easily seen to be unique. Indeed let $g(x) = \alpha x^t l(x)$ with l invertible in $H(E)$, satisfying $\lim\limits_{|x|\to+\infty} l(x) = 1$. Then we have $1 = x^{q-t}\dfrac{\lambda h(x)}{\alpha l(x)}$. Thus we see that $q = t$, $\lambda = \alpha$ and therefore $h = l$.

Finally, we check that $\dfrac{g'}{g}$ belongs to $H_0(E)$. Indeed $\dfrac{g'}{g} = \dfrac{q}{x-b} + \dfrac{h'}{h}$. Obviously, $\dfrac{q}{x-b}$ belongs to $H_0(E)$. Since $\lim\limits_{|x|\to\infty} |h(x)| = 1$, it is seen that $h(x)$ is of the form $1 + \sum\limits_{n=1}^{\infty} \dfrac{a_n}{x^n}$ with $\lim\limits_{n\to\infty} |\dfrac{a_n}{s^n}| = 0$, and therefore h' is an element of $H(E)$ such that $\lim\limits_{|x|\to\infty} |h'(x)| = 0$. As a consequence $\dfrac{h'}{h}$ belongs to $H_0(E)$. Hence so does $\dfrac{g'}{g}$ and this ends the proof of Lemma 32.1.

Definitions: Let $E = K \setminus d(a, r^-)$ with $a \in K$, and $r > 0$. Let $f \in H(E)$ be invertible in $H(E)$ and let $\lambda(x-a)^q h(x)$ be the factorization given in Lemma 32.1. The integer q will be named *the index of f* and will be denoted by $m(f, d(a, r^-))$. The element f will be called *a pure factor associated to* $d(a, r^-)$ if $\lambda = 1$.

Lemma 32.2:   *Let $T = d(a, r^-)$, let $E = K \setminus T$ with $a \in K$, and let f be a pure factor associated to T such that $\|f - 1\|_E < 1$. Then $m(f, T) = 0$.*

Proof: Indeed, let $q = m(f, T)$ and let $f = (x-a)^q h$. If $q \neq 0$, this contradicts the uniqueness of q and h shown in Lemma 32.1.

Definitions: Let f belong to $H(D)$. Let T be a hole of D and let h be a pure factor associated to T. Then f will be said *to admit h as a Motzkin factor in the hole T* if $\dfrac{f}{h}$ belongs to $H(D \cup T)$ and has no zero inside T.

Denoting by $\mathcal{G}^T$ the group of the invertible elements of $H(K \setminus T)$, by Lemma 32.1 we have Corollary 32.3.

Corollary 32.3:   *Let $T = d(a, r^-)$. The set of the pure factors associated to T is a sub-multiplicative group of the group $\mathcal{G}^T$. Further, every element of $\mathcal{G}^T$ is of the form λh with h a pure factor associated to T and $\lambda \in K^*$.*

Theorem 32.4:   *Let T be a hole of D and let f have a Motzkin factor h in T. Then h is unique. Further, if T is not a f-hole, h is the polynomial of the zeros of f inside T.*
Besides, if E is another infraconnected set included in D admitting T as a hole, and if g denotes the restriction of f to E, then g admits a Motzkin factor in the hole T as an element of $H(E)$, and this Motzkin factor is equal to h.

Proof: Let f have another Motzkin factor l in T, let $F = \dfrac{f}{h}$ and let $G = \dfrac{f}{l}$. Since G has no zeros inside T, by Proposition 24.9 there exists a closed bounded infraconnected set D' satisfying $T \subset D' \subset (D \cup T)$, $T \neq D'$, such that G is invertible in $H(D')$. Hence in $H(D')$ we have $\dfrac{F}{G} = \dfrac{l}{h}$. Since $T \neq D'$ it is seen that $D' \cap (K \setminus T)$ is an infraconnected closed bounded set included in D that admits T as a hole. Besides we have $D' \cup (K \setminus T) = K$, and therefore by Theorem 15.10, we see that $\dfrac{l}{h}$ belongs to $H(K)$ and therefore is a polynomial P. Since $\dfrac{F}{G}$ belongs to $H(D')$ and has no zeros in T, it is seen that $m(h, T) = m(l, T)$, so we have $\displaystyle\lim_{|x| \to \infty} \dfrac{h(x)}{l(x)} = 1$. Hence $P = 1$ and this proves that h is unique.

Now we assume that T is not a f-hole. Hence f belongs to $H(D \cup T)$. Let Q be the polynomial of the zeros of f inside T. Then by Theorem 14.5, $\dfrac{f}{Q}$ belongs to $H(D \cup T)$ and has no zeros inside T. Since its Motzkin factor h is unique, we have $h = Q$. The last statement about g is obvious because $\dfrac{g}{h}$ clearly belongs to $H(E \cup T)$ and has no zero inside T. This ends the proof of Theorem 32.4.

Definitions: We will call *the f-supersequence of D* the sequence of the holes $(T_n)_{n \in I}$ such that either T_n is a f-hole or f belongs to $H(D \cup T)$ and has at least one zero inside T_n. If f admits a Motzkin factor h in a hole T, it will be denoted by f^T, and $m(h, T)$ will be called *the Motzkin index of f in T*. For every hole which does not belong to the f-supersequence, we put $f^T = 1$.

Lemma 32.5 is immediate.

Lemma 32.5:   *Let $D \in \mathcal{A}$, let T be a hole of D, and let $f, g \in H(D)$ be such that each admits a Motzkin factor in the hole T. Then fg admits a Motzkin factor in T, and we have $(fg)^T = f^T g^T$ and $m(fg, T) = m(f, T) + m(g, T)$. Besides, if f is invertible in $H(D)$, then f^{-1} admits a Motzkin factor in T which is $(f^T)^{-1}$, and satisfies $m(f^{-1}, T) = -m(f, T)$.*

Lemma 32.6:   *Let the f-supersequence $(T_n)_{n \in \mathbb{N}}$ be such that for every $n \in \mathbb{N}$, f^{T_n} exists and such that $\displaystyle\lim_{n \to \infty} f^{T_n} - 1 = 0$. Then there exists $N \in \mathbb{N}$ such that $m(f^{T_n}, T_n) = 0$ whenever $n > N$. Besides, if $D \in \mathcal{A}$, the product $\left(\displaystyle\prod_{n=1}^{t} f^{T_n}\right)\left(\displaystyle\prod_{n=t+1}^{\infty} f^{T_n}\right)$ does not depend on t whenever $t \geq N$.*

Proof: Indeed, there exists $N \in \mathbb{N}$ such that we have $\|f^{T_n} - 1\|_D < 1$ whenever $n \in \mathbb{N}$, and therefore , by Lemma 32.2, $m(f^{T_n}, T_n) = 0$. Now in $H_b(D)$, we have $\left(\prod\limits_{n=N+1}^{\infty} f^{T_n} \right) = \left(\prod\limits_{n=N+1}^{t} f^{T_n} \right) \left(\prod\limits_{n=t+1}^{\infty} f^{T_n} \right)$.

But then, if D belongs to $\mathcal{A}$, we have $\left(\prod\limits_{n=1}^{t} f^{T_n} \right) \left(\prod\limits_{n=t+1}^{\infty} f^{T_n} \right) =$

$$\left(\prod\limits_{n=1}^{N} f^{T_n} \right) \left(\prod\limits_{n=N+1}^{t} f^{T_n} \right) \left(\prod\limits_{n=t+1}^{\infty} f^{T_n} \right) = \left(\prod\limits_{n=1}^{N} f^{T_n} \right) \left(\prod\limits_{n=N+1}^{\infty} f^{T_n} \right).$$

Definitions: Let $(T_n)_{n \in I}$ be the f-supersequence of D with I a subset of $\mathbb{N}$ which is either finite or equal to $\mathbb{N}$.

If I is finite, f will be said to have *a finite Motzkin factorization* if it factorizes in $H(D)$ in the form $\left(f^0 \prod\limits_{n \in I} f^{T_n} \right)$ with f^0 an element of $H(\widetilde{D})$ whose zeros belong to D and for each $n \in \mathbb{N}$, f^{T_n} a Motzkin factor in T_n.

If I is infinite and equal to $\mathbb{N}$, f will be said to have *an infinite Motzkin factorization* if it admits a sequence of Motzkin factors f^{T_n} satisfying $\lim\limits_{n \to \infty} f^{T_n} - 1 = 0$

such that f factorizes in $H(D)$ in the form $\left(f^0 \prod\limits_{n=1}^{t} f^{T_n} \right) \left(\prod\limits_{n=t+1}^{\infty} f^{T_n} \right)$, with f^0 an element of $H(\widetilde{D})$ whose zeros belong to D.

In both cases, f^0 will be called *the principal factor of f*.

Corollary 32.7:   *Let D be bounded and let f have an infinite Motzkin factorization with a f-supersequence $(T_n)_{n \in \mathbb{N}}$. Then we have $f = f^0 \left(\prod\limits_{n=1}^{\infty} f^{T_n} \right)$.*

Corollary 32.8:   *Let f have an infinite Motzkin factorization with a f-supersequence $(T_n)_{n \in \mathbb{N}}$ such that $m(f^{T_n}, T_n) = 0$ for all $n > 0$. Then we have* $f = f^0 \left(\prod\limits_{n=1}^{\infty} f^{T_n} \right)$.

Remark: Let $f \in H(D)$ be unbounded and have Motzkin factorization of the form $\left(f^0 \prod\limits_{n=1}^{N} f^{T_n} \right) \left(\prod\limits_{n=N+1}^{\infty} f^{T_n} \right)$. One cannot claim that the product $\left(f^0 \prod\limits_{n=1}^{\infty} f^{T_n} \right)$ converges in $H(D)$, even if D is closed and belongs to $\mathcal{A}$. Indeed, let $r \in]0, 1[$, let $(a_n)_{n \in \mathbb{N}}$ be a sequence in $d(0,1)$ such that $|a_n - a_m| = 1$ whenever $n \neq m$, and $a_1 = 0$. For every $n \in \mathbb{N}^*$, we put $T_n = d(a_n, r^-)$ and $E = K \setminus (\bigcup_{n=1}^{\infty} T_n)$. Clearly the holes of E are the T_n. Let $(\lambda_n)_{n \geq 2}$ be a sequence in $d(0, r^-)$ such that $\lim_{n \to \infty} \lambda_n = 0$. For every $n \geq 2$, we put $g_n = 1 + \dfrac{\lambda_n}{x - a_n}$. The sequence $(g_n)_{n \geq 2}$ is seen to satisfy $\|g_n - 1\|_E \leq \dfrac{|\lambda_n|}{\rho} < 1$, and therefore we have $\lim_{n \to \infty} \|g_n - 1\|_E = 0$. Hence the product $h = \prod_{n=2}^{\infty} g_n$ obviously converges in $H(E)$.

Since E clearly belongs to $\mathcal{A}$, we see that $x^2 h$ belongs to $H(E)$ and is invertible in $H(E)$. Besides f clearly has Motzkin factorization with $f^{T_n} = g_n$ for every $n \geq 2$, $f^{T_1} = x^2$, and $f^0 = 1$. However, we will check that the sequence $(f_n)_{n \in \mathbb{N}^*}$ defined by $f_n = x^2 \prod_{j=2}^{n} g_j$ does not converge in $H(E)$. Indeed we have

$$f_{n+1}(x) - f_n(x) = x^2 \Big(\prod_{j=2}^{n} g_j(x) \Big)(g_{n+1}(x) - 1).$$ For every $x \in K \setminus d(0,1)$, we have

$$|x^2 \prod_{j=2}^{n} g_j(x)\| = |x^2|, \text{ and } |g_{n+1}(x) - 1| = \Big|\frac{\lambda_{n+1}}{x}\Big| \text{ hence } |f_{n+1}(x) - f_n(x)| =$$ $|x||\lambda_{n+1}|$. Thus $f_{n+1} - f_n$ is not bounded in $H(E)$, and therefore the sequence $(f_n)_{n \in \mathbb{N}^*}$ does not converge in $H(E)$. According to Theorem 4 in [51] the product $\prod_{n=1}^{\infty} f_n$ should converge to $x^2 h$ in $H(E)$. Here we see that this is not true in the general case. Actually the proof given in [51] only shows the simple convergence of the sequence (f_n), and the uniform convergence on bounded subsets of D.

By Lemma 32.5, Lemma 32.9 is immediate.

Lemma 32.9: *Let $D \in \mathcal{A}$, let f, $g \in H(D)$ have Motzkin factorization. Then so does fg. Besides, we have $(fg)^0 = f^0 g^0$. Further, if f is invertible, f^{-1} also has Motzkin factorization, and it satisfies $(f^{-1})^0 = (f^0)^{-1}$.*

Corollary 32.10: (K.Boussaf) *Let $D \in \mathcal{A}$, let f have an infinite Motzkin factorization of the form $(f^0 \prod_{n=1}^{t} f^{T_n})(\prod_{n=t+1}^{\infty} f^{T_n})$. Let $N \in \mathbb{N}$ be such that*

$m(f^{T_n}, T_n) = 0$ *for all* $n > N$. *Then we have*

$$f = f^0 \Big(\prod_{n=1}^{N} f^{T_n} \Big) \Big(\prod_{n=N+1}^{\infty} f^{T_n} \Big).$$

Proposition 32.11 : *Let* $f \in H(D)$ *satisfy* $\|f - 1\|_D < 1$ *and have Motzkin factorization of the form* $f^0 \Big(\prod_{n=1}^{\infty} f^{T_n} \Big)$ *with* $(T_n)_{n \in \mathbb{N}^*}$ *the* f*-supersequence of* D. *Then for each* $n \geq 1$ *we have* $m(f^{T_n}, T_n) = 0$.

Proof: For every $n \in \mathbb{N}^*$, we put $q_n = m(f^{T_n}, T_n)$. By Lemma 32.6, we may assume the $(T_n)_{n \in \mathbb{N}^*}$ ranged in such a way that $q_n \neq 0$ for $n \leq N$ while $q_n = 0$ whenever $n > N$. When $n \leq N$, f^{T_n} is of the form $(x - \alpha_n)^{q_n}(1 + \omega_n)$ with $\omega_n \in H_0(K \setminus T_n)$, $\| \omega_n \|_{K \setminus T_n} < 1$ and $\alpha_n \in T_n$. When $n > N$, f^{T_n} is just in the form $(1 + \omega_n)$ with $\omega_n \in H_0(K \setminus T_n)$ and $\|\omega_n\|_{K \setminus T_n} < 1$. Besides, as f has no zero in D, obviously f^0 has no zero in D and therefore it has no zero in $\widetilde{D}$. Hence by Theorem 23.16, f^0 is of the form $A(1 + \omega_0(x))$ with $\omega_0 \in H(\widetilde{D})$, $\|\omega_0\|_D < 1$. Let $h(x) = A \prod_{n=1}^{N} (x - \alpha_n)^{q_n}$. We see that f factorizes in the form $h \prod_{n=0}^{\infty} (1 + \omega_n)$. Since $\|\omega_n\|_D < 1$ for every $n \in \mathbb{N}$ and since $\lim_{n \to \infty} \omega_n = 0$ it is seen that h satisfies

(1) $\|h - 1\|_D < 1$ as f does.

Let us suppose $q_1 \neq 0$. We may obviously assume $\alpha_1 = 0$. Let $T_1 = d(0, r^-)$. Thus, in T_1, h admits 0 as a zero of order q_1 if $q_1 > 0$ (resp. a pole of order $-q_1$ if $q_1 < 0$) and has neither any zero nor any pole different from 0. Anyway, when $x \in T_1$, we have

(2) $|h(x)| = B|x^{q_1}|$ with $B = |A| \prod_{n=2}^{N} |\alpha_n|^{q_n}$.

First we suppose $r \notin |K|$. There does exist $r' > r$ such that h has neither any zero nor any pole in $\Delta(0, r, r')$, hence h has neither any zero nor any pole in $d(0, r') \setminus \{0\}$. Therefore $|h(x)|$ is of the form $B|x|^{q_1}$ in all of $d(0, r') \setminus \{0\}$. In particular, this is true in $\Delta(0, r, r') \cap D$ and shows that $_D\varphi_{0,r}(h) = Br^{q_1}$ while $_D\varphi_{0,r'}(h) = Br'^{q_1}$. Hence we see that $|h(x)|$ is not constant in $\Delta(0, r, r') \cap D$ and therefore this contradicts Relation (1).

Now we suppose that r belongs to $|K|$ and then by a classical linear change of variable, we may restrict ourselves to the case where $r = 1$. Hence we have

$T_1 = d(0, 1^-)$. By (2) we have $_D\varphi_{0,1}(h) = B$. But by definition B belongs to $|K|$ and therefore we may clearly assume $B = 1$ without loss of generality. Hence h is of the form $\dfrac{P(x)}{Q(x)}$ with

(3) $\|P\|_{d(0,1)} = \|Q\|_{d(0,1)} = 1$

and P prime to Q.

Let us suppose $q_1 > 0$. By definition of h, P has no zero different from 0 in T_1 while Q has no zero in T_1. Hence Q satisfies

(4) $|Q(x)| = |Q(0)|$ whenever $x \in T_1$

and then by (3) we have $|Q(0)| = 1$, while obviously $P(0) = 0$. Hence by (3) it is seen that $\|P - Q\|_{d(0,1)} = 1$ and therefore by (4) we have $\|h - 1\|_{d(0,1)} = 1$. But we know that for every $g \in R(D \cup d(0,1))$ we have $\|g\|_{d(0,1^-)} \leq \|g\|_{D \cup d(0,1^-)} = \|g\|_D$ hence we see that $\|h - 1\|_D \geq 1$ and this contradicts (1).

We now suppose $q_1 < 0$. By definition h is obviously invertible in $R(D)$. Hence we put $F = \dfrac{1}{h}$ and we see that F satisfies $\|F - 1\|_D < 1$ and admits 0 as a unique zero in T_1 while it has no pole in T_1. Hence the same process lets us get to the same contradiction and finishes showing that $q_n = 0$ for every $n \geq 1$.

Lemma 32.12: *Let $f \in H(D)$ be invertible in $H(D)$ and have Motzkin factorization, and let $a \in D$. Then f satisfies $\|\dfrac{f}{f(a)} - 1\|_D < 1$ if and only if for every hole T of the f-supersequence of D we have $m(f, T) = 0$.*

Proof: Without loss of generality, we may obviously assume $f(a) = 1$. By Lemma 32.11, we already know that if f satisfies $\|f - 1\|_D < 1$, then for every hole of the f-supersequence, we have $m(f, T) = 0$. Now we suppose that for every hole T of the f-supersequence we have $m(f, T) = 0$ and will prove that $\|f - 1\|_D < 1$. Indeed, by Lemma 32.1, for each hole of the f-supersequence, we have $\|f^T - 1\|_D < 1$. Besides, since f is invertible, f^0 must also be invertible, hence by Theorem 23.16 it is of the form $(1 + \psi(x))$, with $\|\psi\|_D < 1$. Then it is seen that $\|f - 1\|_D < 1$.

We will show that all the semi-invertible elements have Motzkin factorization, step after step, and first we consider rational functions.

Proposition 32.13: *Let $f \in R(D)$. Then f admits Motzkin factorization.*

Proof: The f-supersequence is obviously finite. Let $T_1, ..., T_q$ be this f-supersequence. For each $n = 1, ..., q$ we denote by f^{T_n} the rational function whose numerator (resp. denominator) is the polynomial of the zeros (resp. of the poles) of f inside T_n. Finally we denote by f^0 the rational function whose numerator (resp. denominator) is the polynomial whose zeros are the zeros (resp. the poles) of f in $D \cup (K \setminus \widetilde{D})$, (resp. in $K \setminus \widetilde{D}$). It is seen that f^0 is the principal factor of f, and that for each $n = 1, ...q$, f^{T_n} is the Motzkin factor of f associated to the hole T_n.

Proposition 32.14:   *Let $\phi \in H(D)$ satisfy $\|\phi - 1\|_D < 1$. Then ϕ admits Moztkin factorization $\phi^0 \left(\prod_{n=1}^{\infty} \phi^{T_n} \right)$ with $(T_n)_{n \in \mathbb{N}^*}$ the f-supersequence. For every ϕ -hole T we have $\|\phi^T - 1\|_D = \|\overline{\phi}_T\|_D$. Besides ϕ^0 satisfies $\|\phi^0 - 1\|_D = \|\overline{\phi}_0 - 1\|_D$.*

Proof :   First we suppose $\phi \in R(D)$. Then by Proposition 32.13, ϕ admits Motzkin factorization. Now by Lemma 32.12, for each $n > 0$ we have $m(\phi, T_n) = 0$, and therefore, ϕ^{T_n} is of the form $1 + \omega_n$ with $\|\omega_n\|_D < 1$ whenever $n > 0$ while $\phi^0 = 1 + \omega_0$ with $\|\omega_0\|_D < 1$. Hence we see that $\overline{\phi}_{T_n} = \overline{\left(\omega_n \prod_{\substack{j \neq n \\ j \in \mathbb{N}}} (1 + \omega_j) \right)}_{T_n}$.

Clearly, $\prod_{\substack{j \neq n \\ j \in \mathbb{N}}} (1 + \omega_j)$ is of the form $1 + \psi_n$ with $\|\psi_n\|_D < 1$, hence $\|(\omega_n \psi_n)\|_D < \|\omega_n\|_D$ and we obtain

(1) $\|(\overline{\omega_n \psi_n})_{T_n}\|_D \leq \|\omega_n \psi_n\|_D < \|\omega_n\|_D$.

But ω_n is clearly equal to $\overline{(\phi^{T_n})}_{T_n}$ and then we have

(2) $\|\overline{(\omega_n)}_{T_n}\|_D = \|\omega_n\|_D > \|\omega_n \psi_n\|_D \geq \|\overline{(\omega_n \psi_n)}_{T_n}\|_D$.

Besides $\overline{(\omega_n + \omega_n \psi_n)}_{T_n} = \overline{(\omega_n)}_{T_n} + \overline{(\omega_n \psi_n)}_{T_n}$ hence by (1) and (2) we have $\|\overline{(\omega_n(1 + \psi_n))}_{T_n}\|_D = \|\omega_n\|_D$ and finally $\|\overline{\phi}_{T_n}\|_D = \|\overline{(\omega_n(1 + \psi_n))}_{T_n}\|_D = \|\omega_n\|_D = \|\phi^{T_n} - 1\|_D$. In the same way we put $\prod_{n=1}^{\infty}(1 + \omega_n) = 1 + \psi$ with

(3) $\|\psi\|_D < 1$.

It is seen that ψ belongs to $H_0(K \setminus (\bigcup_{n=1}^{\infty} T_n))$. Hence Theorem 15.1, when applied to ψ, shows that

(4) $\overline{\psi}_0 = 0$.

Next, we have $\phi = (1 + \omega_0)(1 + \psi) = 1 + \omega_0 + \psi + \omega_0\psi$ hence
$\overline{\phi}_0 = 1 + \overline{(\omega_0)}_0 + \overline{\psi}_0 + \overline{(\omega_0\psi)}_0$. By defintion $\omega_0 \in H(\widetilde{D})$ hence $\omega_0 = \overline{(\omega_0)}_0$ and then
by (4) we have $\overline{\phi}_0 = 1 + \omega_0 + \overline{(\omega_0\psi)}_0$. But by (3) it is seen that $\|\overline{(\omega_0\psi)}_0\|_D <$
$\|\omega_0\|_D$ hence finally we obtain $\|\overline{\phi}_0 - 1\|_D = \|\omega_0\|_D = \|\phi^0 - 1\|_D$. Thus we have
proven the inequalities satisfied by the ϕ^{T_n} and by ϕ^0 when ϕ belongs to $R(D)$.

Now we consider the general case when $\phi \in H(D)$. Let $(f_m)_{m \in \mathbb{N}}$ be a
sequence in $R(D)$ such that $\lim_{m \to \infty} \|\phi - f_m\|_D = 0$. Let $\varepsilon \in]0, 1[$ and let $N \in \mathbb{N}$
be such that $\|f_m - \phi\|_D \leq \varepsilon$ whenever $m \geq N$. Let T be a hole of the ϕ-
supersequence. We will show that the sequence $((f_m)^T)_{m \in \mathbb{N}}$ converges in $H(D)$
and that this convergence is uniform with respect to the ϕ-supersequence. We
fix $m \geq N$. It is seen that $\|f_m - 1\|_D < 1$ and then by Lemmas 32.1 and 32.11 we
have $\|(f_m)^T - 1\|_D < 1$ and in particular $\|(f_m)^T\|_D = 1$. Besides we remember
that in $H(K \setminus T)$, the norm $\| \, . \, \|_D$ is multiplicative and actually equal to ${}_D\varphi_T$.
Now let $s \geq N$. We have
$$\|(f_m)^T - (f_s)^T\|_D = \|\frac{(f_m)^T}{(f_s)^T} - 1\|_D \leq \max(\|\frac{(f_m)^T}{(f_s)^T} - (f_m)^T\|_D, \|(f_m)^T - 1\|_D).$$
Therefore we obtain

(5) $\|(f_m)^T - (f_s)^T\|_D \leq \max(\|(f_m)^T - 1\|_D), \|(f_s)^T - 1\|_D).$

But, as we have just proved about elements of $R(D)$, we have
$\|(f_m)^T - 1\|_D = \|\overline{(f_m)}_T\|_D$ and $\|(f_s)^T - 1\|_D = \|\overline{(f_s)}_T\|_D$. Hence by (5) and by
Theorem 15.1, we obtain

(6) $\|(f_m)^T - (f_s)^T\|_D \leq \varepsilon.$

Relation (6) does not depend on the hole T and it shows that, for each fixed
$n \in \mathbb{N}^*$, the sequence $((f_m)^{T_n})_{m \in \mathbb{N}}$ is a Cauchy sequence which converges in
$H(K \setminus T_n)$, to an element whose index is equal to 0, and this convergence is
uniform with respect to n. For each $n \in \mathbb{N}^*$, we put $\phi_n = \lim_{m \to \infty} (f_m)^{T_n}$. Then
it is seen that $\prod_{n=1}^{\infty} \phi_n = \lim_{m \to \infty} \prod_{n=1}^{\infty} (f_m)^{T_n}$. As a consequence, the sequence $(f_m)^0$
is also convergent in $H(D)$, and actually in $H_b(\widetilde{D})$. Let ϕ_0 be its limit. Then
we have this factorization: $\phi = \prod_{n=0}^{\infty} \phi_n$. It is seen that this is the Motzkin
factorization for ϕ. Obviously, for each fixed $n > 0$, the equality satisfied by the
$(f_m)^{T_n}$ holds for ϕ^{T_n} and shows that $\|\overline{(\phi)}_{T_n}\|_D = \|(\phi)^{T_n} - 1\|_D$. In the same way,
the equality satisfied by the $(f_m)^0$ shows that $\|\overline{(\phi)}_0 - 1\|_D = \|(\phi)^0 - 1\|_D$. This
ends the proof of Proposition 32.14.

Theorem 32.15 is given in [34].

Theorem 32.15:   *Let $a \in D$. Let $\phi \in H_b(D)$ be such that $|\phi(a)| \neq 0$. The following statements* i) , ii) , iii) *are equivalent*

     i) $\|\phi - \phi(a)\|_D < |\phi(a)|$.

     ii)   *For every hole T we have* $\|\overline{\phi}_T\|_D < |\phi(a)|$ *and* $\|\overline{\phi}_0 - \overline{\phi}_0(a)\|_D < |\phi(a)|$.

     iii) ϕ *is invertible, admits a Motzkin factorization, and for every hole* T , ϕ^T *satisfies* $\|\phi^T - 1\|_D < 1$ *and* ϕ^0 *satisfies* $\|\phi^0 - \phi^0(a)\|_D < |\phi(a)|$.

Besides if statements i) , ii) , iii) *are satisfied then we have*

     (u) $m(\phi, T) = 0$ *for every hole T.*

     (v) $\|\overline{\phi}_T\|_D = \|\phi^T - 1\|_D |\phi(a)|$ *for every hole T.*

     (w) $\|\phi^0 - \phi^0(a)\|_D = \|\overline{\phi}_0 - \overline{\phi}_0(a)\|_D$.

Proof: Without loss of generality we may obviously assume $|\phi(a)| = 1$ and

(1) $|\phi(a) - 1| < 1$.

Let $(T_m)_{m \in I}$ be the ϕ-supersequence of D. We notice that when i) is satisfied, ϕ is obviously invertible.

First we suppose i) is satisfied and will show that so is ii). By Theorem 15.1 we have

(2) $\|\overline{(\phi - \phi(a))}_{T_m}\|_D \leq \|\phi - \phi(a)\|_D$. But it is seen that $\overline{(\phi - \phi(a))_{T_m}} = \overline{\phi}_{T_m}$.
Hence by (2) we see that

(3) $\|\overline{\phi}_{T_m}\|_D \leq \|\phi - \phi(a)\|_D < 1$, whenever $m \in I$.

In the same way $\overline{(\phi - \phi(a))_0} = \overline{\phi}_0 - \phi(a)$ and then by Theorem 15.1 we have

(4) $\|\overline{\phi}_0 - \phi(a)\|_D < 1$.

Besides by (3) we see that $\| \sum_{m=1}^{\infty} \overline{\phi}_{T_m} \|_D < 1$ hence $\|\phi - \overline{\phi}_0\|_D = \| \sum_{m=1}^{\infty} \overline{\phi}_{T_m} \|_D < 1$

and therefore $|\phi(a) - \overline{\phi}_0(a)| < 1$, hence by (4) we see that

(5) $\|\overline{\phi}_0 - \overline{\phi}_0(a)\|_D < 1$.

Finally by (3) and (5), Statement ii) is clearly proven.

Now we will show that each of the statements ii) and iii) separately implies i). We suppose ii) is satisfied. Hence we have

(6) $\| \sum_{m \in I} \overline{\phi}_{T_m} \|_D < 1$.

If D is bounded, by Statement ii) and by (6) we obtain i). Now let D be not bounded. Then $\overline{\phi}_0$ is a constant λ. Hence ϕ is in the form $\lambda + \sum_{m=1}^{\infty} \overline{\phi}_{T_m}$ with $\|\overline{\phi}_{T_m}\|_D < 1$ whenever $m \geq 1$ hence

(7) $\left\| \sum_{m=1}^{\infty} \overline{\phi}_{T_m} \right\|_D < 1.$

Now we have $\phi - \phi(a) = \sum_{m=0}^{\infty} \overline{\phi}_{T_m} - \overline{\phi}_{T_m}(a) = \sum_{m=1}^{\infty} (\overline{\phi}_{T_m} - \overline{\phi}_{T_m}(a)).$ By (7) we

see that $\left\| \sum_{m=1}^{\infty} (\overline{\phi}_{T_m} - \overline{\phi}_{T_m}(a)) \right\|_D < 1$ hence finally i) $\|\phi - \phi(a)\|_D < 1.$

We now suppose iii) is satisfied. Hence we have

(8) $\left\| \phi^{T_m} - 1 \right\|_D < 1$ for all $m \in I.$

If D is bounded we have $\|\phi^0 - \phi^0(a)\|_D < 1$ hence by (8) we directly have i). If D is not bounded then ϕ^0 is a constant ν such that $\phi(a) = \nu \prod_{m \in I} \phi^{T_m}(a)$ hence by (8) and (1) we see that $|\nu - 1| < 1$ hence by (8) we obtain i) again.

Thus, i) is implied as well by ii) as by iii). Obviously by (1), i) implies $\|\phi - 1\|_D < 1$ and therefore we may apply Proposition 32.14. Next, we suppose that either ii) or iii) is satisfied. Hence so is i) and so are (u) and (v) by Proposition 32.14.

Finally, we will show w) and at the same time we will finish proving the equivalence between ii) and iii). Let $\psi = (\overline{\phi}_0(a))^{-1}\phi.$ We may apply Proposition 32.14 to ψ and we have

(9) $\|\overline{\psi}_0 - 1\|_D = \|\psi^0 - 1\|_D.$

But we have

(10) $\|\psi^0 - 1\|_D \geq \|\psi^0 - \psi^0(a)\|_D = \|\phi^0 - \phi^0(a)\|_D$

(11) $\|\overline{\phi}_0 - \overline{\phi}_0(a)\|_D = \|\overline{\psi}_0 - \overline{\psi}_0(a)\|_D = \|\overline{\psi}_0 - 1\|_D.$

Hence by (9), (10), (11) we obtain

(12) $\|\phi^0 - \phi^0(a)\|_D \leq \|\overline{\phi}_0 - \overline{\phi}_0(a)\|_D.$

Now let $\gamma = \phi^0(a)$ and let $\chi = \gamma^{-1}\phi.$ By (1) and (7) we see that $|\gamma - 1| < 1,$ hence we may apply Proposition 32.14 to χ and we have

(13) $\|\chi^0 - 1\|_D = \|\overline{\phi}_0 - \overline{\phi}_0(a)\|_D$

while $\|\overline{\phi}_0 - \overline{\phi}_0(a)\|_D = \|\overline{\chi}_0 - \overline{\chi}_0(a)\|_D \leq \|\overline{\chi}_0 - 1\|_D$ and $\|\chi^0 - 1\|_D = \|\chi^0 - \chi^0(a)\|_D = \|\phi^0 - \phi^0(a)\|_D.$ Hence by (13) we see that $\|\overline{\phi}_0 - \overline{\phi}_0(a)\|_D \leq \|\phi^0 - \phi^0(a)\|_D$ and therefore by (12) we obtain w). This finishes proving the equivalence between ii) and iii) and ends the proof of Theorem 32.15.

Remark: If D is not bounded, as ϕ is bounded, both ϕ^0 , $\overline{\phi}_0$ are constant and therefore, the statements $\|\overline{\phi}_0 - \overline{\phi}_0(a)\|_D < |\phi(a)|$ and $\|\phi^0 - \phi^0(a)\|_D < |\phi(a)|$ are automatically satisfied. Statement ii) is then equivalent to:

ii')   *For every hole T we have $\|\overline{\phi}_T\|_D < |\phi(a)|$*

and Statement iii) is equivalent to:

iii')   *ϕ is invertible and for every hole T , ϕ^T satisfies $\|\phi^T - 1\|_D < 1$.*

When D is closed and bounded Theorem 32.16 is an easy consequence of Theorem 32.15.

Theorem 32.16:   *Let $D \in \mathcal{A}$. Then f has Motzkin factorization if and only if it is semi-invertible .*

Proof: Without loss of generality we may assume the f-supersequence to be infinite. We denote it by $(T_n)_{n \in \mathbb{N}^*}$. Let f have Motzkin factorization

$$f^0 \left(\prod_{n=1}^{t} f^{T_n} \right) \left(\prod_{n=t+1}^{\infty} f^{T_n} \right).$$ By definition, f^0 is semi-invertible in $H(\widetilde{D})$ hence in

$H(D)$. Besides $\left(\prod_{n=1}^{t} f^{T_n} \right) \left(\prod_{n=t+1}^{\infty} f^{T_n} \right)$ is clearly invertible in $H(D)$. So f is semi-invertible.

Now, we suppose f to be semi-invertible and will show it to have Motzkin factorization. By Lemma 32.9 we may clearly suppose that f is invertible without loss of generality.

First, we suppose that there exists M in $\mathbb{R}_+^*$ satisfying

(1) $M \leq |f(x)|$, whenever $x \in D$.

Let $h \in R(D)$ satisfy

(2) $\|f - h\|_D < \dfrac{M}{2}$,

and let $h^0 \prod_{n=1}^{N} h^{T_n}$ be the Motzkin factorization of h. For every $n = 1, ... N$, let

$q_n = m(h, T_n)$, let $a_n \in T_n$, let $h_n = (x - a_n)^{-q_n} h^{T_n}$ and let $h_0 = h^0$. Let $u(x) = \prod_{n=1}^{N} (x - a_n)^{q_n}$ and let $l(x) = h^0 \prod_{n=1}^{N} h_n$. By (1), (2) it is seen that h has no zero

in D. Let $a \in D$. Then, by Theorem 23.16, h^0 satisfies $\|h^0 - h^0(a)\|_D < |h^0(a)|$, and of course for every $n > 0$, h_n satisfy $\|h_n - 1\|_D < 1$. Hence , we have

(3) $\|l - l(a)\|_D < |l(a)|$. Let $b = |l(a)|$.

In particular, we have $|l(x)| = b$ whenever $x \in D$. Besides, we notice that we have

(4) $\dfrac{M}{|l(a)|} \leq |u(x)|.$

Let $F = \dfrac{f}{u}$. Then F does belong to $H_b(D)$. By (3) and (4) we check that

$|F(x) - l(x)| < \dfrac{b}{2}$ and therefore by (3) again, we have $|F(x)| = b$ and

$\|F - F(a)\|_D \leq \dfrac{|F(a)|}{2}$. Now we can apply Theorem 32.15 to F and then F has

Motzkin factorization $\quad F^0 \displaystyle\prod_{n=1}^{\infty} F^{T_n}$, with $m(F, T_n) = 0$ whenever $n > 0$. As

a consequence f also has Motzkin factorization $(f^0 \displaystyle\prod_{n=1}^{N} f^{T_n})(\displaystyle\prod_{n=N+1}^{\infty} f^{T_n})$ with

$f^0 = F^0$, and for each $n = 1, ..., N$, $f^{T_n} = (x - a_n)^{q_n} F^{T_n}$, and finally for each $n > N$, $f^{T_n} = F^{T_n}$.

Now we suppose that $\inf\{|f(x)| \,|x \in D\} = 0$. Since D is closed, and since f is invertible, we see that D is unbounded and that the element $G = \dfrac{1}{f}$ is not

bounded in D. Hence by Corollary 10.7 there exists $q \in \mathbb{N}^*$ such that $x^{-q}\dfrac{1}{f}$ has

a non zero limit when $|x|$ tends to $+\infty$, $(x \in D)$. Then it is easily seen that there exists $m > 0$ such that $|G(x)| \geq m$ for all $x \in D$. Indeed, on one hand there exists r such that $|G(x)| \geq 1$ for all $x \in D \setminus d(0, r)$ and on the other hand f is bounded in $D \cap d(0, r)$, hence there does exist $m \in]0, 1[$ such that $|G(x)| \geq m$ whenever $x \in D \setminus d(0, r)$. Thus G admits Motzkin factorization and then by Lemma 32.9 so does f. This ends the proof of the Theorem.

Remark: If a closed set B does not belong to $\mathcal{A}$, there are counter-examples

of invertible elements F which admit certain Motzkin factor F^T such that $\dfrac{F}{F^T}$

does not belong to $H(B)$ (and obviously does not belong to $H(B \cup T)$). Indeed, don't let B belong to $\mathcal{A}$. Since by hypothesis B is closed we know that $\widetilde{B} \setminus \overline{B}$ is not bounded. Hence by Lemma 24.6, there exists a quasi-minorated element $f \in H_b(B)$ satisfying (1) $\displaystyle\lim_{\substack{|x| \to \infty \\ x \in B}} f(x) = 0$ and such that xf does not belong to

$H(B)$. Since $f \in H_b(B)$ we can take it such that $\|f\|_B < 1$. Without loss of generality, we may assume that 0 belongs to a hole of B. Let $T = d(a, r^-)$ be

another hole of B, and let $F = \dfrac{x(1 + f)}{(x - a)}$. Then it is seen that F belongs to

$H_b(B)$ and is invertible in $H_b(B)$ because both $\dfrac{x}{(x - a)}$, $1 + f$ are invertible

in $H_b(B)$. Hence F admits Motzkin factorization. In particular we see that

$F^T = \dfrac{1}{(x-a)}$. However we check that $(x-a)F$ does not belong to $H(B)$ because $(x-a)F = x(1+f)$ and by hypothesis, xf does not belong to $H(B)$.

In the same way, let $G = \dfrac{1}{F}$. Since F is invertible in $H_b(B)$, so is G. But then we see that $\dfrac{1}{x-a}G$ does belong to $H_b(B)$ and has no zero in B, but obviously its inverse does not belong to $H(B)$. Therefore $\dfrac{1}{x-a}G$ is not semi-invertible in $H(B)$. Thus, there exist invertible elements h, g in $H(B)$ such that hg is not semi-invertible, although it does belong to $H(B)$. This contradicts Theorem 1 in [51] which states that $\dfrac{f}{f^T}$ extends to an element of $H(D \cup T)$.

Theorem 32.17: (K. Boussaf) *Let D belong to $\mathcal{A}$ and let $T = d(a, r^-)$ be a hole of D. Then f admits a Motzkin factor in the hole T if and only if $_D\varphi_{a,r}(f) \neq 0$.*

Proof: On the first hand, we suppose that f admits a Motzkin factor in the hole T. Let $f = gf^T$. Since g belongs to $H(D \cup T)$, and has no zero in T, of course we have $_D\varphi_{a,r}(g) \neq 0$. Next, as an invertible element of $H(D)$, it is seen that $_D\varphi_{a,r}(f^T) \neq 0$. Hence $_D\varphi_{a,r}(f) \neq 0$.

On the second hand, we suppose $_D\varphi_{a,r}(f) \neq 0$. Let $\mathcal{F}$ be the circular filter of center a, of diameter r, and let $M = {_D\varphi_{a,r}(f)}$. There do exist $a_1, ..., a_q \in d(a, r)$ and s, t satisfying $s < r < t$, such that $|f(x)| \geq M$ whenever $x \in D \bigcap (\bigcap_{j=1}^{q} \Gamma(a_j, s, t))$. Let $A = D \bigcap (\bigcap_{j=1}^{q} \Gamma(a_j, s, t))$. Then T is clearly a hole of A. Besides, the restriction g of f to the infraconnected set $A = D \bigcap (\bigcap_{j=1}^{q} \Gamma(a_j, s, t))$ is invertible in $H(A)$, and therefore, by Theorem 32.16, it admits a Motzkin factor g^T in the hole T. But then, $\dfrac{f}{g^T}$ belongs to $H(D)$ and to $H(A \cup T)$. Let $E = A \cup T$. Clearly, a hole of $D \cap E$ is either a hole of D included in E, or a hole of E. Hence D and E are infraconnected sets that satisfy the hypothesis of Theorem 15.10, and then $\dfrac{f}{g^T}$ belongs to $H(D \cup A) = H(D \cup T)$. Besides, as g^T is the Motzkin factor of g in T, $\dfrac{g}{g^T}$ has no zero inside T. This ends the proof.

Theorem 32.18: *Let $D \in \mathcal{A}$ and let $\mathcal{G}$ be the multiplicative group of the invert-*

ible elements in $H(D)$. Let $\mathcal{T}$ be the set of the holes of D. Let $\mathcal{G}^0$ be the subgroup of the elements invertible in $H(\widetilde{D})$. Let $\mathcal{H} = \mathcal{G}^0 \prod_{T \in \mathcal{T}} \mathcal{G}^T$. The product $\mathcal{H}$ is a direct product and is dense in $\mathcal{G}$.

Proof: The product is direct because for each element, Motzkin factorization is unique. Thus $\mathcal{H}$ is the set of the invertible elements whose Motzkin factorization is finite. Since every element of $\mathcal{G}$ has Motzkin factorization, it obviously belongs to the closure of $\mathcal{H}$.

33. APPLICATIONS OF

THE MOTZKIN FACTORIZATION

Thanks to the Motzkin factorization, the question on whether the n-th root of an analytic element is an analytic element appears to be linked to the number of zeros of each Motzkin factor. Several of these results were given in [32].

Theorem 33.1: *Let D be a closed bounded infraconnected set and let $f \in H(D)$ be semi-invertible. Let T be a hole of D . We assume f^s to have continuation to an element of $H(D \cup T)$ for some $s \in \mathbb{N}^*$. If f does not belong to $H(D \cup T)$, then the number of the zeros of f^s inside T is a multiple of s different from 0 (taking mutiplicities into account).*

Proof: By Theorem 32.9 we have $(f^s)^T = (f^T)^s$. But as f^s belongs to $H(D \cup T)$, by Theorem 32.4 $(f^s)^T$ is the polynomial of the zeros of f inside T. Let $Q = (f^s)^T$. Then we have $deg(Q) = m((f^s), T) = sm(f, T)$. So s divides q. Now assume $q = 0$. We have $Q = 1$, $m(f, T) = 0$ and therefore $(f^T)^s = 1$, and $\lim_{|x| \to \infty} f^T(x) = 1$. Thus f^T is just the constant 1, and therefore f does not belong to $H(D \cup T)$. So this ends the proof of Theorem 33.1.

In particular Theorem 33.1 applies to open disks.

Theorem 33.2: *Let $r \in |K|$ and let $f \in H(d(0, r^-)) \setminus H(d(0, r))$ satisfy $f^s \in H(d(0, r))$. Then f has continuation to an analytic element in a set D of the form $d(0, r) \setminus (\bigcup_{i=1}^{t} d(a_i, r^-))$ with $|a_i| = r = |a_i - a_j|$ whenever $i \neq j$, such that for each $i = 1, ..., t$ the number of zeros of f^s in $d(a_i, r_i)$ is a multiple of s different from 0.*

Proof: The Mittag- Leffler series of f in $d(0, r^-)$ is of the form $\sum_{n=0}^{\infty} f_n$ with $f_0 = \overline{f}_0 \in H(d(0, r))$, and for every $n > 0$, $f_n = \overline{f}_{T_n}$, with $T_n = d(a_n, r^-)$, and $|a_n - a_j| = |a_n| = r$ whenever $n \neq j$. Since $f \notin H(d(0, r))$, at least one of the f_n is different from 0. Let l be an integer such that $f_l \neq 0$. Then f has continuation to an element of $H(d(0, r^-) \cup T_l)$. Now since f^s belongs to $H(d(0, r))$, by Theorem

33.1, f^s has a number of zeros inside T_l which is different from 0 and a multiple of s. Since any element of $H(d(0,r))$ has finitely many zeros in $d(0,r)$, we see that there are finitely many integers l such that $f_l \neq 0$. Let I be the finite set of the $l \in \mathbb{N}^*$ such that $f_l \neq 0$ and let $D = d(0,r) \setminus \left(\bigcup_{l \in I} d(a_n, r^-) \right)$. Then by definition f belongs to $H(D)$ and for every $l \in I$, the number of zeros of f^s in T_n is different from 0 and a multiple of s. This ends the proof Theorem 33.2.

Theorem 33.3 is just an application of Theorem 33.2.

Theorem 33.3: *Let f be a power series whose radius of convergence is r though f does not belong to $H(d(0,r))$. If for some $s \in \mathbb{N}^*$, f^s has a radius of convergence r' strictly superior to r, and if f^s has strictly less than s zeros inside $C(0,r)$ (taking multiplicities into account), then f does not belong to $H(d(0,r^-))$.*

Proof: Since $r' > r$ obviously we have $s > 1$. We assume that f belongs to $H(d(0,r^-))$, and therefore r must belong to $|K|$. Since $r' > r$, f^s belongs to $H(d(0,r))$, and then by Theorem 33.2, its number of zeros inside $C(0,r)$ is different from 0 and a multiple of s, which contradicts the hypothesis. Hence finally f does not belong to $H(d(0,r^-))$.

We have now got to recall the definition of the function $\sqrt[q]{u}$ when $u \in d(1,1^-)$.

In the following, q will denote an integer superior or equal to 2, prime to p if $p \neq 0$.

Let $(1+x)^q = 1 + qx + \sum_{j=2}^{q} b_j x^j$. It is seen that $|q| = 1$, while $|b_j| \leq 1$ whenever $j = 2, ..., q$. By Corollary 26.2 and Corollary 27.3 the mapping g defined in $d(0,1^-)$ by $g(x) = (1+x)^q$ is a bijection from $d(0,1^-)$ onto $d(1,1^-)$ and therefore the mapping θ_q defined in $d(1,1^-)$ by $\theta_q(u) = u^q$ is a bijection from $d(1,1^-)$ onto $d(1,1^-)$. We denote by $\sqrt[q]{u}$ the reciprocal mapping from $d(1,1^-)$ onto $d(1,1^-)$ and we put $\psi_q(x) = \sqrt[q]{1+x}$ whenever $x \in d(0,1^-)$.

Let K have characteristic different from 0. Then this characteristic is equal to the residue characteristic p, and K is an extension of $\mathbb{F}_p$. For each $t \in \mathbb{N}$, let $\phi_t(x)$ be the t-degree monic polynomial $\binom{x}{t}$.

Lemma 33.4: *For every $a \in \mathbb{Z}_p$, $\phi_t(a)$ belongs to $\mathbb{Z}_p$.*

Proof: Since $\phi_t(y)$ is a polynomial function in y, it is obviously continuous in $\mathbb{Q}_p$. But when $y \in \mathbb{N}$, $\phi_t(y)$ belongs to $\mathbb{N}$. Let $a \in \mathbb{Z}_p$, and let a_n be a sequence in $\mathbb{N}$ such that $\lim\limits_{n \to +\infty} a_n = a$. Then we have $\phi_t(a) = \lim\limits_{n \to +\infty} \phi_t(a_n)$, and therefore $\phi_t(a) \in \mathbb{Z}_p$.

Notations: We have to recall the definition of the function $(1 + x)^a$. $\mathbb{F}_p$ is isomorphic to the field quotient of $\mathbb{Z}_p$ by its maximal ideal $p\mathbb{Z}_p$. For every $x \in \mathbb{Z}_p$, $\overline{x}$ denotes the class of x in $\mathbb{F}_p$. For $x \in d(0,1^-)$ and $a \in \mathbb{Z}_p$, we put

$$g_a(x) = \sum_{n=0}^{\infty} \phi_n(a)x^n \text{ if } K \text{ has characteristic } 0, \text{ and } g_a(x) = \sum_{n=0}^{\infty} \overline{\phi_n(a)}x^n \text{ if } K \text{ has}$$

characteristic different from 0. .

Lemma 33.5:   *For all $a \in \mathbb{Z}_p$, g_a belongs to $A_b(d(0,1^-))$ and satisfies $g_a(x) \in d(1,1^-)$ whenever $x \in d(0,1^-)$. Further, for every $b \in \mathbb{Z}_p$, they satisfy $g_{a+b}(x) = g_a(x)g_b(x)$.*

Proof: By definition, the $\phi_n(a)$ belong to $\mathbb{Z}_p$ if K has characteristic 0, and to $\mathbb{F}_p$ if K has characteristic $p \neq 0$, and therefore satisfy $|\phi_n(a)| \leq 1$. Hence g_a belongs to $A_b(d(0,1^-))$ and satisfies $g_a(x) \in d(1,1^-)$ whenever $x \in d(0,1^-)$. The last statement is just a consequence of the obvious identities $\phi_n(a + b) = \sum\limits_{i+j=n} \phi_i(a)\phi_j(b)$ in $\mathbb{Z}_p$ when K has characteristic 0, and

$$\overline{\phi}_n(a + b) = \sum_{i+j=n} \overline{\phi}_i(a)\overline{\phi}_j(b) \text{ in } \mathbb{F}_p \text{ when } K \text{ has characteristic } p \neq 0. \ .$$

Lemma 33.6: $\psi_q = g_{\frac{1}{q}}$.

Proof: By Lemma 33.5, we have $(g_{\frac{1}{q}}(x))^q = 1 + x = \psi_q(x)^q$. Since both $g_{\frac{1}{q}}(x)$, $\psi_q(x)$ belong to $d(1,1^-)$, they are equal.

Theorem 33.7:   *ψ_q belongs to $A_b(d(0,1))$ but ψ_q does not belong to $H(d(0,1^-))$.*

Proof: First, assume that ψ_q belongs to $H(d(0,1))$. Then it must satisfy $(\psi_q(-1))^q = 0$, hence $\psi_q(-1) = 0$ and therefore $\psi_q(x)^q$ admits a zero of order q at -1. But this contradicts the identity $(\psi_q(x))^q = 1 + x$ and therefore $\psi_q(x)$ does not belong to $H(d(0,1))$. Finally, since $(\psi_q(x))^q$ has only one zero in $C(0,1)$, by Theorem 33.3, we see that ψ_q does not belong to $H(d(0,1^-))$.

Remark 1:   Since ψ_q belongs to $A_b(d(0,1^-))$, obviously ψ_q belongs to $H(d(0,r))$, whenever $r \in]0,1[$. Now, let E be a closed bounded set in K and let $h \in H(E)$ satisfy $\|h\|_E < 1$. Then by Corollary 11.3, $\psi_q \circ h$ belongs to $H(E)$. In other words, if $g \in H(E)$, and if $\|g - 1\|_E < 1$, then $\sqrt[q]{g}$ also belongs to $H(E)$.

Remark 2:   Theorems 33.1 and 33.2 couldn't be significantly improved as this example shows. Let q be an integer prime to p, let $a, b \in d(1, 1^-)$, with $a \neq b$, and let $P(x) = (a-x)^{q-1}(b-x)$. It is easily seen that $|P(x) - 1| < 1$ whenever $x \in d(0, 1^-)$ and then we can consider $f(x) = \sqrt[q]{P(x)}$. We will show that $f \in H(d(0, 1^-)) \setminus H(d(0, 1))$. Indeed we have

$$f(x) = \sqrt[q]{(a-x)^q \left(\frac{b-x}{a-x}\right)} = (a-x)\sqrt[q]{1 + \left(\frac{b-a}{a-x}\right)}$$

$$= (a-x)\sum_{n=0}^{\infty} \binom{\frac{1}{q}}{n} \left(\frac{b-a}{a-x}\right)^n = a - x + \frac{1}{q}(b-a) + \sum_{j=1}^{\infty} \binom{\frac{1}{q}}{j+1} \frac{(b-a)^{j+1}}{(a-x)^j}.$$

This is just a Mittag-Leffler series of the form $f_0 + f_1 \in H(d(0, 1^-))$, with

$$f_0 = -x + a + \frac{1}{q}(b-a) \in H(d(0, 1)),$$

and

$$f_1 = -\sum_{j=1}^{\infty} \binom{\frac{1}{q}}{j+1} \frac{(a-b)^{j+1}}{(x-a)^j} \in H_0(K \setminus d(1, 1^-)).$$

Thus we see that f belongs to $H(d(0, 1^-))$, and more precisely $f \in H(K \setminus d(1, 1^-))$, but $f \notin H(d(0, 1))$. Actually f^q has exactly q zeros in $d(1, 1^-)$.

34. MAXIMUM IN A CIRCLE WITH HOLES

Throughout this chapter D denotes a closed infraconnected set of diameter $S \in \overline{\mathbb{R}}_+$.

Here we will give technical results useful to the next chapters.

Notations: Let D be bounded, of diameter r, and included in a disk $d(a, r^-)$. For every $q \in \mathbb{N}$ we will denote by $\mathcal{V}(D, q)$ the set of the monic polynomials P of degree q whose zeros lie in $d(a, r^-) \setminus D$. We denote by $\vartheta(D, q)$ the number

$$r^q \inf_{P \in \mathcal{V}(D,q)} \left\| \frac{1}{P} \right\|_D.$$

Proposition 34.1: *Let D be bounded, of diameter r, and included in a disk $d(a, r^-)$. Let ψ be defined as $\psi(x) = \alpha x + \beta$, and let $D' = \psi(D)$. Then for every $q \in \mathbb{N}$ we have $\vartheta(D', q) = \vartheta(D, q)$.*

Proof: Let $u = \psi(x)$. Hence we have $x = \psi^{-1}(u) = \dfrac{u - \beta}{\alpha}$. Now, for every

$$P(x) = \prod_{i=1}^{q} (x - a_i) \in \mathcal{V}(D, q),$$

we put $\overline{\psi}(P) = \alpha^q P \circ \psi^{-1}$. It is easily checked that $\overline{\psi}$ defines bijection from $\mathcal{V}(D, q)$ onto $\mathcal{V}(D', q)$. Besides, by Proposition 11.6 we have $\left\| \dfrac{1}{P \circ \psi^{-1}} \right\|_{D'} = \left\| \dfrac{1}{P} \right\|_D$ hence $\left\| \dfrac{1}{\overline{\psi}(P)} \right\|_{D'} = \dfrac{1}{|\alpha|^q} \left\| \dfrac{1}{P} \right\|_D$. But the diameter r' of D' is obviously equal to $|\alpha|r$, hence finally we have $r'^q \left\| \dfrac{1}{\overline{\psi}(P)} \right\|_{D'} = r^q \left\| \dfrac{1}{P} \right\|_D$, and this ends the proof.

Notations: Let $a \in \widetilde{D}$, let $r \in\,]\delta(a, D), S] \cap |K|$ and let $(T_i)_{i \in I}$ be the set of the holes of D included in $C(a, r)$. We will denote by $\mathcal{T}(D, a, r)$ the set $\bigcup_{i \in I} T_i$.

For every $q \in \mathbb{N}$ we will denote by $\mathcal{S}(D, a, r, q)$ the set of the monic polynomials P of degree q whose zeros lie in $\mathcal{T}(D, a, r)$.

Lemma 34.2: *Let $a \in \widetilde{D}$, let $r \in [\delta(a, D), diam(D)] \cap |K|$, let $q \in \mathbb{N}$ and let $P \in \mathcal{S}(D, a, r, q)$. Let $(T_j)_{1 \leq j \leq \ell}$ be the holes of D included in $C(a, r)$ which contain at least one zero of P. For every $j = 1, ..., \ell$ let $T_j = d(a_j, r_j^-)$, and let*

t_j be the number of zeros of P in T_j (taking multiplicities into account). Then we have

$$\left\|\frac{1}{P}\right\|_D = \left\|\frac{1}{P}\right\|_{D\cap C(a,r)} = \frac{1}{\displaystyle\inf_{1\le j\le \ell}\Big(r_j^{t_j}\prod_{\substack{1\le m\le \ell\\ m\ne j}}|a_m-a_j|^{t_m}\Big)}$$

Proof: Let $D' = K\setminus\Big(\bigcup_{j=1}^{\ell} T_j\Big)$, let $\psi_j = {}_D\varphi_{T_j}\,(1\le j\le \ell)$ and let $\lambda = \min_{1\le j\le \ell}\dfrac{1}{\psi_j(P)}$. It is seen that $\dfrac{1}{P}$ belongs to $H_0(D')$ and then by Theorem 15.1 we have $\left\|\dfrac{1}{P}\right\|_{D'} = \lambda$. But then, we see that $\left\|\dfrac{1}{P}\right\|_{D'}\ge\left\|\dfrac{1}{P}\right\|_D\ge\left\|\dfrac{1}{P}\right\|_{D\cap C(a,r)}\ge\lambda$ and therefore we have (1) $\left\|\dfrac{1}{P}\right\|_D = \lambda$. Now, it is easily seen that $\psi_j(x-\alpha) = r_j$ when $\alpha\in T_j$, and that $\psi_j(x-\alpha) = |\alpha-a_j|$ when $\alpha\notin T_j$ so that we have $\psi_j(P) = r_j^{t_j}\prod_{\substack{1\le m\le \ell\\ m\ne j}}|a_m-a_j|^{t_m}$. Finally by (1) the conclusion is clear.

Notation: Let $a\in\tilde{D}$. Let $r\in[\delta(a,D),diam(D)]\cap|K|$ and let $q\in\mathbb{N}$. We put

$$\gamma_D(a,r,q) = r^q\inf_{P\in\mathcal{S}(D,a,r,q)}\left\|\frac{1}{P}\right\|_D.$$

Corollary 34.3: Let $a\in D$, let $r\in[\delta(a,D),diam(D)]\cap|K|$ and let $q\in\mathbb{N}$. Let $(T_i)_{i\in I}$ be the set of the holes of D included in $C(a,r)$ with $T_i = d(a_i,r_i^-)$. Then we have

$$\gamma_D(a,r,q) = \frac{r^q}{\displaystyle\sup_{1\le\ell\le q,(i_1,\dots,i_\ell)\in I^\ell,\sum_{j=1}^{\ell}t_j=q}\Big(\inf_{1\le j\le\ell}r_{i_j}^{t_j}\prod_{\substack{1\le m\le\ell\\ m\ne i_j}}|a_m-a_{i_j}|^{t_m}\Big)}.$$

Besides, for every $\epsilon > 0$, there exist classes Λ_j, $(1\le j\le t)$ of $C(a,r)$, and integers q_j, $(1\le j\le t)$ such that $\sum_{j=1}^{t}q_j = q$, and $\max_{1\le j\le t}\vartheta(\Lambda_j\cap D,q_j)\le\gamma_D(a,r,q)+\epsilon$.

Proposition 34.4: Let $a\in D$ and $r\in[\delta(a,D)],diam(D)]\cap|K|$. Let $f\in R(D)$ and let s (resp. t) be the number of zeros (resp. of the poles) of f inside $C(a,r)$ (taking multiplicities into account). We suppose $s\le t$ and put $q = t-s$. If $C(a,r)\cap D\ne\emptyset$ then we have $\|f\|_{D\cap C(a,r)}\ge{}_D\varphi_{a,r}(f)\gamma_D(a,r,q)$.

Proof: Let $f = \dfrac{P}{Q} g$ where P and Q are monic polynomials whose zeros lie in $C(a,r)$ while $g \in R(D)$ has neither any zero nor any pole in $C(a,r)$. By Lemma 4.1 we know that $|g(x)| = {}_D\varphi_{a,r}(g)$ for all $x \in C(a,r)$. Hence without loss of generality we may assume $g = 1$. If $P = 1$ the inequality we want to prove is obvious. Thus the inequality is already proven when $s = 0$ (whenever t).

Now, given $m, n \in \mathbb{N}$, with $m \leq n$, we assume the inequality proven when $s \leq m$, $t \leq n$, and $s \leq t$ and will prove it when $s = m+1$, $t = n+1$. Indeed, suppose $s = m+1$, $t = n+1$. Let E be the set of the couples (ξ, η) such that ξ is a zero of P, and η is a zero of Q. Now let $(\alpha, \beta) \in E$ be such that $|\alpha - \beta| = \inf\{|\xi - \eta| \mid (\xi, \eta) \in E\}$. Let $u(x) = \dfrac{x - \alpha}{x - \beta}$ and let $f(x) = h(x) u(x)$. We will show

(1) $\|f\|_{D \cap C(a,r)} \geq \|h\|_{D \cap C(a,r)}$.

We notice

(2) ${}_D\varphi_{a,r}(u) = 1$.

First we suppose that $\|h\|_{D \cap C(a,r)} = {}_D\varphi_{a,r}(h)$. We have ${}_D\varphi_{a,r}(f) = {}_D\varphi_{a,r}(h)$ hence $\|f\|_{D \cap C(a,r)} \geq {}_D\varphi_{a,r}(f) = \|h\|_{D \cap C(a,r)}$.

Now we suppose that ${}_D\varphi_{a,r}(h) < \|h\|_{D \cap C(a,r)}$. Let $D' = D \cap C(a,r)$. It is seen that ${}_D\varphi_{a,r}(h) = {}_{D'}\varphi_{D'}(h)$ and then by Lemma 29.3, there exists a hole $T = d(b, \rho^-)$ of $D \cap C(a,r)$ which contains at least one zero of Q, such that ${}_{D \cap C(a,r)}\varphi_{b,\rho}(h) = \|h\|_{D \cap C(a,r)}$. Actually T is a hole of D included in $C(a,r)$ because it contains a pole of h. Hence we may assume that b is just a pole of h. We will show

(3) ${}_D\varphi_{b,\rho}(u) \geq 1$.

By definition of (α, β) we have

(4) $|\alpha - b| \geq |\alpha - \beta|$.

If $\alpha \in d(b, \rho^-)$ then β also belongs to $d(b, \rho^-)$ and we have ${}_D\varphi_{b,\rho}(u) = 1$.
If $\alpha \notin d(b, \rho^-)$:
either $|b - \beta| > |\alpha - \beta|$ while $\rho \leq |b - \beta|$ and then we have ${}_D\varphi_{b,\rho}(u) = 1$,
or $|b - \beta| = |\alpha - \beta|$ while $|b - \alpha| = |\beta - \alpha|$ and then we have ${}_D\varphi_{b,\rho}(u) = 1$ again,
or $|b - \beta| < |\alpha - \beta|$ and then we have ${}_D\varphi_{b,\rho}(u) > 1$.
Thus (3) is now proven in all the cases. Hence we have

$$\|f\|_{D \cap C(a,r)} \geq {}_D\varphi_{b,\rho}(h) = \|h\|_{D \cap C(a,r)}$$

and this finishes showing (1). But now h clearly has m zeros and n poles in $C(a,r)$. Hence it satisfies the inequality

$$\|f\|_{D \cap C(a,r)} \geq {}_D\varphi_{a,r}(h) \gamma_D(a,r,q)$$

and therefore by (1) and (2) we have

$$\|f\|_{D \cap C(a,r)} \geq \, _D\varphi_{a,r}(h) \, _D\varphi_{a,r}(u) \gamma_D(a,r,q).$$

We have now proven the inequality when $s = m+1$, $t = n+1$. Therefore, we can check that the inequality announced in Proposition 34.4 is proven for every couple (s,t). Since the inequality is true for $(0, t-s)$, it is true for $(1, t-s+1), ...(s,t)$. This ends the proof of Proposition 34.4.

Lemma 34.5: Let $a \in \tilde{D}$, let $r \in [\delta(a,D), diam(D)] \cap |K|$. Let $h \in R(D)$ have at least as many poles as many zeros in $C(a,r)$. There exists a hole T of D included in $C(a,r)$ such that

$$\|h\|_{D \cap C(a,r)} \geq \, _D\varphi_{a,r}(h)\left(\frac{r}{diam(T)}\right).$$

Proof: Let s (resp. t) be the number of zeros (resp. of the poles) of h in $C(a,r)$ and let $q = t - s$. Let $\varepsilon \in]0,1[$. By Proposition 34.4 we have

(1) $\quad \|h\|_{D \cap C(a,r)} \geq \, _D\varphi_{a,r}(h)\gamma_D(a,r,q).$

Now by definition of $\gamma_D(a,r,q)$ there exists a monic polynomial P of degree q whose zeros belong to $C(a,r) \setminus D$, such that

(2) $\quad \gamma_D(a,r,q) \geq r^q \|\frac{1}{P}\|_D (1 - \varepsilon).$

Let $D' = D \cup d(a, r^-) \cup \left(K \setminus d(a,r)\right)$. Clearly $\frac{1}{P}$ belongs to $R_b(D')$ and every hole of D' is a hole of D. By Theorem 29.4, there exists a $\frac{1}{P}$-hole T of D' such that

(3) $\quad _D\varphi_T\left(\frac{1}{P}\right) = \|\frac{1}{P}\|_{D'}.$

Let $\rho = diam(T)$. Since T is a hole of D we see that $\|\frac{1}{P}\|_D \geq \, _D\varphi_T\left(\frac{1}{P}\right) = \, _{D'}\varphi_T\left(\frac{1}{P}\right)$. Hence by (3) and by the inclusion $D \subset D'$, we see that

(4) $\quad \|\frac{1}{P}\|_D = \|\frac{1}{P}\|_{D'} = \, _D\varphi_T\left(\frac{1}{P}\right) = \frac{1}{_D\varphi_T(P)}.$

Now since T is a $\frac{1}{P}$- hole of D', P has at least one zero inside T and therefore by Theorem 23.18 it is seen that

(5) $\quad \dfrac{_D\varphi_T(P)}{_D\varphi_{a,r}(P)} \leq \dfrac{\rho}{r}.$

But since P is monic and has all its zeros in $C(a,r)$, it is seen that $_D\varphi_{a,r}(P) = r^q$, hence by (5) we have $\dfrac{1}{_D\varphi_T} \geq \dfrac{1}{\rho r^q}$. Finally, by (1), (2), (4), we obtain

$$\|h\|_{D\cap C(a,r)} \geq \; _D\varphi_{a,r}(h)\left(\frac{r^q}{_D\varphi_T(P)}\right)(1-\varepsilon) \geq \; _D\varphi_{a,r}(h)\left(\frac{r}{\rho}\right)(1-\varepsilon).$$

This ends the proof of Lemma 34.5.

Proposition 34.6: Let $a \in \tilde{D}$, let $t \in]\delta(a,D), diam(D)[$. Let $f \in H(D)$ satisfy $v_a'^r(f, -\log t) < v_a''^l(f, -\log t)$. Then $C(a,t)$ contains at least one hole T of D such that

$$\|f\|_{C(a,t)\cap D} \geq \frac{t}{diam(T)} \; _D\varphi_{a,t}(f).$$

Proof: Without loss of generality we may assume $a = 0$. Let $m = \; _D\varphi_{0,t}(f)$ and let $h \in R(D)$ satisfy (E) $\|f - h\|_D < m$. Hence there exists $s, u \in \mathbb{R}_+$ satisfying $s < t < u$ and $_D\varphi_{0,t}(f) = \; _D\varphi_{0,t}(h)$ whenever $t \in [s, u]$. In particular, we have

$$v_a'^r(f, -\log t) = v_a'^r(h, -\log t), \quad v_a''^l(h, -\log t) = v_a''^l(h, -\log t).$$

Let q_1 (resp. q_2) be the number of the zeros (resp. the poles) of h inside $C(0,t)$ and let $q = q_2 - q_1$. By Corollary 4.12 we have $v_a'^r(h, -\log t) - v_a''^l(h, -\log t) = q$. Hence by Lemma 34.5, there exists a hole T included in $C(0,t)$ such that

$$(F) \quad \|h\|_{C(0,t)\cap D} \geq \frac{t}{diam(T)} \; _D\varphi_{0,t}(f).$$

But by (E), it is seen that $_D\varphi_{0,t}(f) = \; _D\varphi_{0,t}(h)$, and since $\|f\|_{C(0,t)\cap D} \geq \|f\|_{C(0,t)\cap D}$, we have $\|f\|_{C(0,t)\cap D} = \|h\|_{C(0,t)\cap D}$. As a consequence, the inequality (F) shown for h also holds for f, and this ends the proof.

35. *T*-FILTERS AND *T*-SEQUENCES

Many questions are linked to the problem of the analytic elements strictly vanishing along a monotonous filter. This problem just involves a particular kind of monotonous pierced filters called T-filters, [11], [14], [17]. We will introduce them.

In all of this chapter D denotes an infraconnected set of diameter $S \in \overline{\mathbb{R}}_+$.

Definitions: Let $\mathcal{F}$ be an increasing filter on D of center $a \in \tilde{D}$ and diameter S. The filter $\mathcal{F}$ will be called *an increasing T- filter* if there exists a sequence of circles $(\Sigma_m)_{m \in \mathbb{N}}$ with $\Sigma_m = C(a, d_m)$ $(m \in \mathbb{N})$ such that $\lim_{m \to \infty} d_m = S$ and $d_m < d_{m+1}$ together with a sequence of natural integers $(q_m)_{m \in \mathbb{N}}$ satisfying

$$(1) \quad \lim_{m \to \infty} \gamma_D(a, d_m, q_m) \prod_{j=1}^{m-1} \left(\frac{d_j}{d_m} \right)^{q_j} = 0.$$

A decreasing filter $\mathcal{F}$ will be called *a decreasing T-filter* if it admits a base $(D_m)_{m \in \mathbb{N}}$ with $D_m = d(a_m, d_m) \cap D \setminus \left(\bigcap_{m \in \mathbb{N}} d(a_m, r_m) \right)$ together with a sequence of natural integers $(q_m)_{m \in \mathbb{N}}$ such that, putting $\Sigma_m = C(a_m, d_m)$, the sequences $(a_m)_{m \in \mathbb{N}}, (d_m)_{m \in \mathbb{N}}, (q_m)_{m \in \mathbb{N}}$ satisfy

$$(2) \quad \lim_{m \to \infty} \gamma_D(a, d_m, q_m) \prod_{j=1}^{m-1} \left(\frac{d_m}{d_j} \right)^{q_j} = 0 \, .$$

In particular, this definition holds when $\mathcal{F}$ is a decreasing filter of center a and diameter S, and then we can take $a_m = a$ for every $m \in \mathbb{N}$.

Another way to define the T-filters consists of introducing the T-sequences. We will call *a weighted sequence* a sequence $(T_{m,i}, q_{m,i})_{\substack{1 \le i \le s(m) \\ m \in \mathbb{N}}}$ with $(T_{m,i})_{\substack{1 \le i \le s(m) \\ m \in \mathbb{N}}}$ a monotonous distances holes sequence and $(q_{m,i})_{\substack{1 \le i \le s(m) \\ m \in \mathbb{N}}}$ a sequence of integers.

A weighted sequence $(T_{m,i}, q_{m,i})_{\substack{1 \le i \le s(m) \\ m \in \mathbb{N}}}$ will be said to be *idempotent* if $q_{m,i} = 0$ or 1 for all (m, i), $(1 \le i \le s(m), \ m \in \mathbb{N})$.

All the definitions given about monotonous distances holes sequences will apply in the same way to weighted sequences: a weighted sequence $(T_{m,i}, q_{m,i})_{\substack{1 \le i \le s(m) \\ m \in \mathbb{N}}}$

will be said "to be something" (or "to satisfy a certain property") if "so is" (or "so does") the monotonous distances holes sequence $(T_{m,i})_{\substack{1 \le i \le s(m) \\ m \in \mathbb{N}}}$. In particular, the diameter, the monotony, the piercing of the monotonous distances holes sequence $(T_{m,i})_{\substack{1 \le i \le s(m) \\ m \in \mathbb{N}}}$ will be called, respectively, *the diameter, the monotony, the piercing* of the weighted sequence $(T_{m,i}, q_{m,i})_{\substack{1 \le i \le s(m) \\ m \in \mathbb{N}}}$.

Now let $(T_{m,i}, q_{m,i})_{\substack{1 \le i \le s(m) \\ m \in \mathbb{N}}}$ be an increasing (resp. decreasing) weighted sequence of monotony $(d_m)_{m \in \mathbb{N}}$, associated to $\mathcal{F}$ and for every (i, m), $(1 \le i \le h_m, m \in \mathbb{N})$, let $T_{m,i} = d(a_{m,i}, \rho_{m,i}^-)$ and let $q_m = \sum_{i=1}^{h_m} q_{m,i}$.

The weighted sequence will be said to be a *T-sequence* if it satisfies

$$(3) \quad \lim_{m \to \infty} \left(\sup_{1 \le j \le s(m)} \left[\left(\frac{d_m}{\rho_{m,j}} \right)^{q_{m,j}} \prod_{\substack{i \ne j \\ 1 \le i \le s(m)}} \left(\frac{d_m}{|a_{m,i} - a_{m,j}|} \right)^{q_{m,i}} \right] \prod_{n=1}^{m-1} \left(\frac{d_n}{d_m} \right)^{q_n} \right) = 0$$

(resp.

$$(4) \quad \lim_{m \to \infty} \left(\sup_{1 \le j \le s(m)} \left[\left(\frac{d_m}{\rho_{m,j}} \right)^{q_{m,j}} \prod_{\substack{i \ne j \\ 1 \le i \le s(m)}} \left(\frac{d_m}{|a_{m,i} - a_{m,j}|} \right)^{q_{m,i}} \right] \prod_{n=1}^{m-1} \left(\frac{d_m}{d_n} \right)^{q_n} \right) = 0).$$

In both cases, if this weighted sequence is a T-sequence, we will call *subsidence* of the T-sequence the number

$$\sup_{m \in \mathbb{N}} \left(\sup_{1 \le j \le s(m)} \left[\log \left(\frac{d_m}{\rho_{m,j}} \right)^{q_{m,j}} + \sum_{\substack{i \ne j \\ 1 \le i \le s(m)}} \log \left(\frac{d_m}{|a_{m,i} - a_{m,j}|} \right)^{q_{m,i}} \right] - \right.$$

$$\left. - \sum_{n=1}^{m-1} \left| \log \left(\frac{d_n}{d_m} \right)^{q_n} \right|_{\infty} \right)$$

Given a T-sequence $(T_{m,i}, q_{m,i})_{\substack{1 \le i \le s(m) \\ m \in \mathbb{N}}}$,, a monotonous filter $\mathcal{F}$ will be said *to admit the T-sequence* $(T_{m,i}, q_{m,i})_{\substack{1 \le i \le s(m) \\ m \in \mathbb{N}}}$ if it is associated to the monotonous distances holes sequence $(T_{m,i})_{\substack{1 \le i \le s(m) \\ m \in \mathbb{N}}}$.

Remark 1:   Given a weighted sequence $(T_{m,i}, q_{m,i})_{\substack{1 \le i \le s(m) \\ m \in \mathbb{N}}}$, defined as above, it is a T-sequence if and only if it satisfies

$$(5) \quad \lim_{m \to \infty} \left(- \sup_{1 \le j \le s(m)} \left[q_{m,j}(\log d_m - \log \rho_{m,j}) + \right. \right.$$

$$\left. \left. + \sum_{\substack{i \ne j \\ 1 \le i \le s(m)}} q_{m,i}(\log d_m - \log |a_{m,i} - a_{m,j}|) \right] + \sum_{n=1}^{m-1} q_n |\log d_m - \log d_n|_\infty \right) = +\infty.$$

Remark 2: Given a T-sequence $(T_{m,i}, q_{m,i})_{\substack{1 \le i \le s(m), \\ m \in \mathbb{N}}}$, for every $t \in \mathbb{N}$, the weighted sequence $(T_{m,i}, q_{m,i})_{\substack{1 \le i \le s(m) \\ m \ge t}}$ is a T-sequence again.

Remark 3: Let $(T_{m,i}, q_{m,i})_{\substack{1 \le i \le s(m), \\ m \in \mathbb{N}}}$ be a weighted sequence of D with $T_{m,i}, q_{m,i} = d(a_{m,i}, r^-_{m,i})$. For each couple (m,i) in $\widehat{K}$ we put $\widehat{T}_{m,i} = \widehat{d}(a_{m,i}, r^-_{m,i})$, $(1 \le i \le s(m), m \in \mathbb{N})$. Now let D' be a set in $\widehat{K}$ such that $D \subset D' \subset (\widehat{K} \setminus \bigcup_{\substack{1 \le i \le s(m) \\ m \in \mathbb{N}}} \widehat{T}_{m,i})$. Then the weighted sequence $(T_{m,i}, q_{m,i})_{\substack{1 \le i \le s(m) \\ m \in \mathbb{N}}}$ of D is a T-sequence in K if and only if so is the weighted sequence $(\widehat{T}_{m,i}, q_{m,i})_{\substack{1 \le i \le s(m) \\ m \in \mathbb{N}}}$ of $\widehat{D}$ in $\widehat{K}$.

By Lemma 34.1, this lemma is almost immediate.

Lemma 35.1 *Let $(T_{m,i}, q_{m,i})_{\substack{1 \le i \le s(m) \\ m \in \mathbb{N}}}$ be an increasing (resp. decreasing) weighted sequence, let $q_m = \sum_{i=1}^{s(m)} q_{m,i}$ and let $(C(a_m, d_m))_{m \in \mathbb{N}}$ be a sequence of circles that runs the weighted sequence. The weighted sequence is a T-sequence if and only if there exists a sequence of monic polynomials $(Q_m)_{m \in \mathbb{N}}$ such that for each $(m,i)_{1 \le i \le s(m)}$, Q_m admits exactly a zero of order $q_{m,i}$ in $T_{m,i}$ and has no other zero in K, satisfying further*

$$\lim_{m \to \infty} \left(\left({}_D\varphi_{a_m, d_m}(Q_m) \left\| \frac{1}{Q_m} \right\|_{C(a_m, d_m) \cap D} \right) \prod_{n=1}^{m-1} \left(\frac{d_n}{d_m} \right)^{q_n} \right) = 0$$

$$\left(resp. \quad \lim_{m \to \infty} \left(\left({}_D\varphi_{a_m, d_m}(Q_m) \left\| \frac{1}{Q_m} \right\|_{C(a_m, d_m) \cap D} \right) \prod_{n=1}^{m-1} \left(\frac{d_m}{d_n} \right)^{q_n} \right) = 0. \right)$$

Lemma 35.2: *Let $\mathcal{F}$ be a monotonous filter on D. Then $\mathcal{F}$ is a T-filter if and only if there exists a T-sequence associated to $\mathcal{F}$.*

Proof: Let $(T_{m,i}, q_{m,i})_{\substack{1 \leq i \leq s(m), \\ m \in \mathbb{N}}}$ be a T-sequence associated to $\mathcal{F}$, and for every $m \in \mathbb{N}$ let Σ_m be a circle $C(a_m, d_m)$ which contains all the $(T_{m,i})_{1 \leq i \leq s(m)}$. (Obviously when $\mathcal{F}$ has a center a we may take $a_m = a$ for all $m \in \mathbb{N}$.) For every $i = 1, ..., h_m$ we put $T_{m,i} = d(a_{m,i}, \rho_{m,i}^-)$. Clearly the polynomial

$$P(x) = \prod_{i=1}^{s(m)} (x - a_{m,i})^{q_{m,i}} \text{ satisfies}$$

$$\left\| \frac{1}{P} \right\|_{\Sigma_m \cap D} = \sup_{1 \leq j \leq s(m)} \left(\frac{1}{\rho_{m,j}} \right)^{q_{m,j}} \prod_{\substack{i \neq j, \\ 1 \leq i \leq s(m)}} \left(\frac{1}{|a_{m,i} - a_{m,j}|} \right)^{q_{m,i}}$$

hence

$$\gamma_D(a_m, d_m, q_m) \leq d_m^{q_m} \left\| \frac{1}{P} \right\|_D \leq \sup_{1 \leq j \leq s(m)} \left(\frac{d_m}{\rho_{m,j}} \right)^{q_{m,j}} \prod_{\substack{i \neq j, \\ 1 \leq i \leq s(m)}} \left(\frac{d_m}{|a_{m,i} - a_{m,j}|} \right)^{q_{m,i}}$$

and therefore if the T-sequence is increasing (resp. decreasing) we have

$$\lim_{m \to \infty} \gamma_D(a_m, d_m, q_m) \prod_{n=1}^{m-1} \left(\frac{d_n}{d_m} \right)^{q_n} = 0$$

$$\left(\text{resp. } \lim_{m \to \infty} \gamma_D(a_m, d_m, q_m) \prod_{n=1}^{m-1} \left(\frac{d_m}{d_n} \right)^{q_n} = 0 \right).$$

Thus we see that $\mathcal{F}$ is a T-filter.

Reciprocally, we assume $\mathcal{F}$ to be a T-filter. Let $\mathcal{F}$ be an increasing (resp. decreasing) filter. Let $(\Sigma_m)_{m \in \mathbb{N}}$ be a sequence of circles, with $\Sigma_m = C(a_m, d_m)$, and let $(q_m)_{m \in \mathbb{N}}$ be a sequence of integers such that Relation (1) (resp. (2)) is satisfied. For every $m \in \mathbb{N}$ we may find holes $T_{m,i} = d(a_{m,i}, \rho_{m,i})$ $(1 \leq i \leq h(m))$ in Σ_m, and integers $q_{m,i}$ such that $\sum_{i=1}^{s(m)} q_{m,i} = q_m$ and such that the polynomial $P(x) = \prod_{i=1}^{s(m)} (x - a_{m,i})^{q_{m,i}}$ satisfies $\left\| \frac{1}{P} \right\| d_m{}^{q_m} \leq \gamma_D(a_m, d_m, q_m) + 1$ and then, by Relation (1) (resp. (2)) it is seen that the weighted sequence $(T_{m,i}, q_{m,i})_{\substack{1 \leq i \leq s(m) \\ m \in \mathbb{N}}}$ is a T-sequence. This ends the proof.

Proposition 35.3:   *Let $\mathcal{F}$ be an increasing (resp. decreasing) T-filter of center α and diameter r. Let $h(x) = \dfrac{1}{x - a}$, let $D' = h(D)$ and $\mathcal{F}' = h(\mathcal{F})$. If $|a - \alpha| <$*

r, $\mathcal{F}'$ is a decreasing (resp increasing) T-filter of center α and diameter $\dfrac{1}{r}$. If $|a - \alpha| \geq r$, $\mathcal{F}'$ is an increasing (resp. decreasing) T-filter of center $h(\alpha)$ and diameter $\dfrac{r}{|a - \alpha|^2}$.

Proof: All the statements are already known by Corollaries 3.6 and 3.9, except the fact that $\mathcal{F}'$ is a T-filter. Actually, by Propositions 3.4 and 3.7, if $|a-\alpha| < r$, it appears that Statement (1) in $\mathcal{F}$ becomes (2) in $\mathcal{F}'$ and reciprocally. If $|a-\alpha| \geq r$, then both (1) and (2) are conserved.

Proposition 35.4: Let D admit a T-sequence $(T_{m,i}, q_{m,i})_{1 \leq i \leq k(m)}$, $m \in \mathbb{N}$, and let $(V_{m,j})_{1 \leq j \leq l(m)}$, $m \in \mathbb{N}$ be a monotonous distance holes sequence of an infra-connected set $E \subset D$ such that for every (m,j), $1 \leq j \leq l(m)$, $m \in \mathbb{N}$, the set $\mathcal{T}_{m,j}$ of the (m,i) such that $T_{m,i}$ is included in $V_{m,j}$ is not empty. For every (m,j) $(1 \leq j \leq l(m))$, we put $s_{m,j} = \displaystyle\sum_{(m,i) \in \mathcal{T}_{m,j}} q_{m,i}$. Then the weighted sequence $(V_{m,j}, s_{m,j})_{1 \leq j \leq l(m)}$, $m \in \mathbb{N}$ is a T-sequence of E.

Proof: Since the sequence $(T_{m,i}, q_{m,i})_{1 \leq i \leq k(m)}$, $m \in \mathbb{N}$, is a T-sequence, Relation (3) (resp. (4)) is satisfied. Let $T_{m,i} = d(b_{m,i}, \rho_{m,j}^-)$ $(1 \leq i \leq k(m))$, and $V_{m,j} = d(c_{m,j}, \sigma_{m,j}^-)$, $(1 \leq j \leq l(m))$. Then we check that

$$\sum_{i=1}^{k(m)} q_{m,i} = \sum_{j=1}^{l(m)} s_{m,j} = q_m$$

and that

$$\sup_{1 \leq i \leq k(m)} \left[\left(\frac{d_m}{\rho_{m,i}} \right)^{q_{m,i}} \prod_{\substack{e \neq i \\ 1 \leq e \leq k(m)}} \left(\frac{d_m}{|b_{m,i} - b_{m,e}|} \right)^{q_{m,e}} \right] \geq$$

$$\geq \sup_{1 \leq j \leq l(m)} \left[\left(\frac{d_m}{\sigma_{m,j}} \right)^{s_{m,j}} \prod_{\substack{h \neq j \\ 1 \leq h \leq l(m)}} \left(\frac{d_m}{|c_{m,j} - c_{m,h}|} \right)^{s_{m,h}} \right].$$

36.   EXAMPLES AND COUNTER-EXAMPLES

ABOUT *T*-FILTERS

We give examples of T-filters and T-sequences. We construct a closed infra-connected set whose interior is empty, which admits no T-filter.

We remember that a monotonous distances holes sequence $(T_{m,i})_{\substack{1 \leq i \leq h(m) \\ m \in \mathbb{N}}}$ is said to be *simple* if $h(m) = 1$ whenever $m \in \mathbb{N}$. A simple monotonous distances holes sequence will be denoted by $(T_m)_{m \in \mathbb{N}}$.

When a monotonous distances holes sequence is simple, it is much easier to determine whether there exists a T-sequence $(T_m, q_m)_{m \in \mathbb{N}}$. Indeed if $(\sigma_m)_{m \in \mathbb{N}}$ is a sequence of circles that run the sequence $(T_m)_{m \in \mathbb{N}}$, then putting $\sigma_m = C(a_m, d_m)$ and $\rho_m = diam(T_m)$ we have $\gamma_D(a_m, d_m, q) = \left(\dfrac{d_m}{\rho_m}\right)^q$ for every $q \in \mathbb{N}$. Hence a weighted sequence $(T_m, q_m)_{m \in \mathbb{N}}$ is a T-sequence if and only if

$$\lim_{m \to \infty}\left(-q_m |\log d_m - \log \rho_m|_\infty + \sum_{j=0}^{m-1} q_j |\log d_m - \log d_j|_\infty\right) = +\infty.$$

In Theorems 36.1, 36.6, and Corollaries 36.4, 36.5 $(T_m)_{m \in \mathbb{N}}$ is a simple monotonous distances holes sequence of monotony $(d_m)_{m \in \mathbb{N}}$ and diameter r with $\rho_m = diam(T_m)$. Besides, denoting by σ_m the circle $C(a_m, d_m), (m \in \mathbb{N})$, the sequence $(\sigma_m)_{m \in \mathbb{N}}$ is a sequence of circles that runs the sequence $(T_m)_{m \in \mathbb{N}}$.

Theorem 36.1:   *If $d_m = \rho_m$ is satisfied for infinitely many m, then there exists a sequence $(q_m)_{m \in \mathbb{N}}$ such that $(T_m, q_m)_{m \in \mathbb{N}}$ is a T-sequence.*

Proof: Indeed without loss of generality we may assume $d_m = \rho_m$ for every $m \in \mathbb{N}$, just by considering the subsequence of the T_m such that $d_m = \rho_m$. Then we have $\gamma_D(a_m, d_m, q_m) = 1$ whenever q_m, whence we can take a sequence q_m such that

$$\sum_{j=0}^{\infty} q_j |\log d_m - \log d_j|_\infty = +\infty.$$

Lemma 36.2: *Let $(u_n)_{n\in\mathbb{N}}$ be a strictly decreasing sequence of limit 0 in $]0,\infty[$, and let $(q_n)_{n\in\mathbb{N}}$ be sequence of natural integers such that $\sum\limits_{n\in\mathbb{N}} q_n u_n = +\infty$. Then*

we have $\lim\limits_{n\to\infty} \sum\limits_{j=0}^{n} q_j(u_j - u_n) = +\infty$.

Proof: Let $B \in \mathbb{R}_+$, and let $N \in \mathbb{N}$ be such that $\sum\limits_{j=0}^{N} q_j u_j \geq 2B$. Let $q = \sum\limits_{j=0}^{N} q_j$, and let $t \in \mathbb{N}$ be such that $u_t < \dfrac{B}{q}$. Now let $m \in \mathbb{N}$ satisfy $m \geq t$. We have

$$\sum_{j=0}^{m} q_j(u_j - u_m) \geq \sum_{j=0}^{m} q_j(u_j - u_t) \geq \sum_{j=0}^{N} q_j(u_j - u_t) = \sum_{j=0}^{N} q_j u_j - q u_t.$$

But we have $q u_t \leq B$, hence $\sum\limits_{j=0}^{m} q_j(u_j - u_m) \geq B$. This ends the proof.

In Lemma 36.2, the implication "(1) implies (2)" is due to E. Motzkin and Ph. Robba [50].

Lemma 36.3: *Let $(u_n)_{n\in\mathbb{N}}$ be a strictly decreasing sequence of limit 0 in $]0,\infty[$, and let $(\theta_n)_{n\in\mathbb{N}}$ be a sequence in $]0,+\infty[$. If there exists a sequence of natural integers $(q_n)_{n\in\mathbb{N}}$ such that*

$$(1)\quad \lim_{n\to\infty} \sum_{j=0}^{n} q_j(u_j - u_n) - q_n\theta_n = +\infty,$$

then we have

$$(2)\quad \sum_{n=0}^{\infty} \frac{u_n}{\theta_n} = +\infty.$$

Conversely, if (2) is satisfied and if the sequences $(u_n)_{n\in\mathbb{N}}$, $(\theta_n)_{n\in\mathbb{N}}$ satisfy at least one of these 2 additional conditions:

α)   the sequence $(\theta_n)_{n\in\mathbb{N}}$ is bounded,

β) $\sum\limits_{n=0}^{\infty} u_n < +\infty$,

then there exists a sequence of natural integers $(q_n)_{n\in\mathbb{N}}$ satisfying (1).

Proof: First we suppose the existence of a sequence of integers $(q_n)_{n\in\mathbb{N}}$ satisfying (1) and will deduce that $\displaystyle\sum_{m=0}^{\infty}\frac{u_m}{\theta_m} = +\infty$. For each $m \in \mathbb{N}$, we put

$$\phi_m = \sum_{j=1}^{m} q_j(u_j - u_m) - q_m\theta_m.$$ By hypothesis we have $\displaystyle\lim_{m\to+\infty}\phi_m = +\infty$. Since

$\displaystyle\lim_{m\to\infty}\phi_m = +\infty$ there exists $N \in \mathbb{N}$ such that $\displaystyle q_m\theta_m < \sum_{j=1}^{m-1} q_j(u_j - u_m)$ whenever

$m > N$ and then with greater reason we have

$$(3) \quad q_m\theta_m < \sum_{j=1}^{m-1} q_j u_j.$$

Now since $\displaystyle q_m\theta_m = \sum_{j=1}^{m} q_j(u_j - u_m) - \phi_m,$ we have

$$q_{m+1}\theta_{m+1} = \sum_{j=1}^{m} q_j(u_j - u_{m+1}) - \phi_{m+1} =$$

$$\sum_{j=1}^{m-1} q_j(u_j - u_m) + q_m(u_m - u_{m+1}) - \phi_{m+1} \quad \text{and therefore}$$

$$q_{m+1}\theta_{m+1} < \sum_{j=1}^{m-1} q_j u_j + q_m u_m - \phi_{m+1}.$$

Now we can write

$$\sum_{j=1}^{m} q_j u_j = \Big(\sum_{j=1}^{m-1} q_j u_j\Big)\Big[1 + \frac{q_m u_m}{\displaystyle\sum_{j=1}^{m-1} q_j u_j}\Big]$$

and then by (3) we have

$$(4) \quad \sum_{j=1}^{m} q_j u_j < \Big(\sum_{j=1}^{m-1} q_j u_j\Big)\Big[1 + \frac{u_m}{\theta_m}\Big]$$

Since $\displaystyle\lim_{m\to\infty}\phi_m = +\infty$ there exists $N \in \mathbb{N}$ such that $\phi_m > 0$ whenever $m > N$.
By induction on m, from (4) we have

$$\sum_{j=1}^{m} q_j u_j < \Big(\sum_{j=1}^{N} q_j u_j\Big)\Big(\prod_{j=N}^{m}\Big(1 + \frac{u_j}{\theta_j}\Big)\Big)$$

and therefore we obtain

(5) $\quad q_m \theta_m < \left(\sum_{j=1}^{N} q_j u_j \right) \left(\prod_{j=N}^{m} \left(1 + \frac{u_j}{\theta_j} \right) \right) - \phi_m$ whenever $m > N$.

Since $\theta_m \geq 0$ whenever $m \in \mathbb{N}$ and since $\lim_{m \to \infty} \phi_m = +\infty$ one sees that

$$\lim_{m \to \infty} \prod_{j=N}^{\infty} \left(1 + \frac{u_j}{\theta_j} \right) = +\infty \text{ and therefore the series } \sum_{m=0}^{\infty} \frac{u_m}{\theta_m} \text{ diverges.}$$

Conversely, now we suppose that (2) is satisfied together with one of the hypothesis $\alpha)$, $\beta)$, and we will show that there exists a sequence of integers $(q_n)_{n \in \mathbb{N}}$ satisfying (1).

First we assume

(6) $\quad \theta_n \leq \lambda$ for every $n \in \mathbb{N}$.

Then we put $q_n = int(\frac{1}{\theta_n} + 1)$. By (2), we see

(7) $\quad \displaystyle\sum_{n=0}^{\infty} q_n u_n = +\infty$,

and by (6) we have (8) $\quad q_n \theta_n \leq \lambda + 1$ for all $n \in \mathbb{N}$.

But by Lemma 36.2, (7) shows that $\lim_{n \to +\infty} \sum_{j=0}^{n} q_j (u_j - u_n) = +\infty$, hence by (8) we finally obtain

(9) $\quad \displaystyle\lim_{n \to +\infty} \sum_{j=0}^{n} q_j (u_j - u_n) - q_n \theta_n = +\infty$.

Now, we stop assuming the sequence $(\theta_n)_{n \in \mathbb{N}}$ to be bounded, but we suppose

(10) $\quad \displaystyle\sum_{n=0}^{\infty} u_n < +\infty$.

Let J be the subset of the $n \in \mathbb{N}$ such that $\theta_n > 1$. Clearly by (10) we have $\sum_{n \in J} \frac{u_n}{\theta_n} < +\infty$, hence $\sum_{n \in \mathbb{N} \setminus J} \frac{u_n}{\theta_n} = +\infty$. Therefore we can consider an increasing bijection ϕ from $\mathbb{N}$ onto $\mathbb{N} \setminus J$ such that $(u_{\phi(s)})_{s \in \mathbb{N}}$, $(\theta_{\phi(s)})_{s \in \mathbb{N}}$ satisfy $\sum_{s=0}^{\infty} \frac{u_{\phi(s)}}{\theta_{\phi(s)}} = +\infty$. So, we define a sequence of integers $(q_{\phi(s)})_{s \in \mathbb{N}}$ satisfying

$$\lim_{s \to +\infty} \sum_{j=0}^{s} q_{n_j} (u_{n_j} - u_{\phi(s)}) - q_{\phi(s)} \theta_{\phi(s)} = +\infty. \qquad \text{Finally we put } q_n = 0 \text{ for every}$$

$n \in J$, and then the sequence $(q_n)_{n \in \mathbb{N}}$ is defined for every $n \in \mathbb{N}$, and clearly satisfies (9) again. This finishes the proof of Lemma 36.3.

Corollary 36.4:   *If the sequence* $(T_m, q_m)_{m \in \mathbb{N}}$ *is a T-sequence, then we have*

$$\sum_{m=0}^{+\infty} \frac{|\log r - \log d_m|_\infty}{\log d_m - \log \rho_m} = +\infty.$$

Conversely, if

$$\sum_{m=0}^{+\infty} \frac{|\log r - \log d_m|_\infty}{\log d_m - \log \rho_m} = +\infty$$

and if $\sum_{m=0}^{\infty} |\log r - \log d_m|_\infty < +\infty$ *and* $\rho_m < d_m$ *for all* m, *then there exists a sequence* $(q_m)_{m \in \mathbb{N}}$ *such that the weighted sequence* $(T_m, q_m)_{m \in \mathbb{N}}$ *is a T-sequence.*

Proof: Indeed, we put $u_n = |\log r - \log d_n|_\infty$, and $\theta_n = \log r - \log \rho_n$, $(n \in \mathbb{N})$, and we apply Lemma 36.3 to these sequences.

Proposition 36.5:   *If* $\sum_{m=0}^{\infty} |\log r - \log d_m|_\infty < +\infty$ *and if* $\liminf_{n \to \infty}(\log d_m - \log \rho_m) > 0$, *then there exists no sequence of integers* $(q_n)_{n \in \mathbb{N}}$ *such that* $(T_m, q_m)_{m \in \mathbb{N}}$ *is a T-sequence.*

Proof: Indeed, as above, we put $u_n = |\log r - \log d_n|_\infty$, and $\theta_n = \log r - \log \rho_n$, $(n \in \mathbb{N})$. It is easily seen that $\sum_{m=0}^{+\infty} \frac{|\log r - \log d_m|_\infty}{\log d_m - \log \rho_m} < \infty$, hence by Corollary 36.4, there exists no T sequence of the form $(T_m, q_m)_{m \in \mathbb{N}}$.

Remark 1:   In Lemma 36.3, we would like to have a genuine reciprocal to the first claim, without assuming one of the additional conditions α), or β). Actually, if the sequence $(\theta_n)_{n \in \mathbb{N}}$ is not bounded, and if $\sum_{n=0}^{\infty} u_n = +\infty$, it is far from clear whether (2) implies the existence of a sequence $(q_n)_{n \in \mathbb{N}}$ satisfying (1). However, this example shows that there is little hope to generalize the reciprocal.

For every $n \in \mathbb{N}$, we put $u_n = \dfrac{1}{\sqrt{n}}$ and $\theta_n = 3\sqrt{n+1}$. Then (2) is clearly satisfied. Now, we can check that the sequence $(q_n)_{n \in \mathbb{N}}$ defined as $q_n = 1$ whenever $n \in \mathbb{N}^*$ does not satisfy (1). Indeed we have $\sum_{j=1}^{n} u_j \leq 2\sqrt{n+1}$, hence

$$\sum_{j=1}^{n} u_j - u_n - \theta_n < \sum_{j=1}^{n} u_j - \theta_n \leq -\sqrt{n+1}.$$

Remark 2: In particular, if $\displaystyle\sum_{m=0}^{\infty} |\log r - \log d_m|_\infty < +\infty$ and if $\displaystyle\limsup_{m\to\infty}\left(\frac{\rho_m}{r}\right) <$ 1 then there is no T-sequence $(T_m, q_m)_{m\in\mathbb{N}}$.

Proposition 36.6: *We have* $\displaystyle\lim_{m\to\infty}\sum_{j=0}^{m-1} |\log d_m - \log d_j|_\infty = +\infty$ *if and only if* $\displaystyle\sum_{j=0}^{\infty} |\log r - \log d_j|_\infty = +\infty.$

Proof: We just have to show that $\displaystyle\sum_{j=0}^{\infty} |\log r - \log d_j|_\infty = +\infty$ implies $\displaystyle\lim_{m\to\infty}\sum_{j=0}^{m-1} |\log d_m - \log d_j|_\infty = +\infty$. We put $u_m = |\log r - \log d_m|_\infty$ and then, by Lemma 36.2 we have $|\log d_m - \log d_j|_\infty = u_m - u_j$ because both $\log d_m - \log d_j$, $\log r - \log d_m$ have the same sign for $j < m$. Hence the conclusion is clear.

Remark: The condition $\displaystyle\liminf_{m\to\infty} \rho_m > 0$ is not required provided the sequence $(d_m)_{m\in\mathbb{N}}$ satisfies good conditions.

Proposition 36.7: *There exist simple T-sequences $(T_m, q_m)_{m\in\mathbb{N}}$ with*

$$\lim_{m\to\infty} diam(T_m) = 0.$$

Proof: Let d_m satisfy $\displaystyle\sum_{j=0}^{\infty} |\log r - \log d_j|_\infty = +\infty$ and let ρ_m satisfy

$$\log \rho_m = -\frac{1}{2}\sum_{j=0}^{m} |\log d_m - \log d_j|_\infty.$$

By Theorem 36.6 we know that $\displaystyle\lim_{m\to\infty}\sum_{j=0}^{m-1} |\log d_m - \log d_j|_\infty = +\infty$ hence $\displaystyle\lim_{m\to\infty} \rho_m = 0$. However it is seen that

$$\lim_{m\to\infty}\left(-(\log d_m - \log \rho_m) + \sum_{j=0}^{m-1} |\log d_m - \log d_j|_\infty\right) = +\infty$$

hence the weighted sequence $(T_m, 1)_{m \in \mathbb{N}}$ is a T-sequence.

Theorem 36.8:   *If K is separable, there exist closed infraconnected sets whose interiors are empty, which admit no T-filter.*

Proof: We construct a set $D \subset d(0,1)$, by defining its holes first. Since K is separable we can take an injective sequence $S = (a_s)_{s \in \mathbb{N}}$ dense in $d(0,1)$. Let $\rho_s = p^{(2^s)}$ and let $\tau_s = d(a_s, \rho_s^-)$ $(s \in \mathbb{N})$. We can easily construct a sequence of natural integers $(s(n))_{n \in \mathbb{N}}$ satisfying

(1) $s(n) > s(n-1)$

(2) $\tau_{s(n)} \cap \tau_{s(j)} = \emptyset$ whenever $j \neq n$

(3) $a_i \in \displaystyle\bigcup_{j=0}^{n} \tau_{s(j)}$ whenever $i \neq 0, \ldots s(n)$.

Indeed let $s_0 = 0$, assume the $s(j)$ already obtained up to $j = n$, satisfying (1), (2), (3) and let us define $s(n+1)$. We take for $s(n+1)$ the lowest integer m such that $\tau_m \cap \left(\displaystyle\bigcup_{j=0}^{n} \tau_{s(j)} \right) = \emptyset$. Such an integer $s(n+1)$ is easily seen to exist because we can find $m \in \mathbb{N}$ such that the distance from a_m to $\displaystyle\bigcup_{j=0}^{n} \tau_{s(j)}$ is strictly superior to ρ_m. Then (1) and (2) are now satisfied up to the rank $n+1$. Let $i \in \mathbb{N} \cap [s(n)+1, s(n+1)-1]$. Since τ_i has a not empty intersection with one $\tau_{s(j)}$ for a certain $j \leq n$, and since the sequence $(\rho_s)_{s \in \mathbb{N}}$ is strictly decreasing, obviously we have $\tau_i \subset \tau_{s(j)}$. Then (3) is clearly satisfied for $i \leq s(n+1) - 1$, and then also for $i \leq s(n+1)$ because $a_{s(n+1)} \in \tau_{s(n+1)}$.

We put $a_n = \alpha_{s(n)}$, $T_n = \tau_{s(n)}$, $r_n = \rho_{s(n)}$. Clearly $\displaystyle\bigcup_{n=0}^{\infty} T_n$ is dense in $d(0,1)$ because by (3) it contains all the a_s . We fix $N \in \mathbb{N}$. We will first show that for every $\epsilon > 0$, there exists $N' > N$ such that $r_N < \epsilon$ and $\tau_{N'} \subset C(a_N, r_N)$. Indeed $C(a_N, r_N)$ is not included in any T_n because otherwise, we should have $T_N \subset T_n$. Since $r_N \in |K|$, there exist infinitely many T_m included in $C(a_N, r_N)$, one of them has a radius $r_m < \epsilon$, and we may call N' this m. By induction we shortly define a convergent subsequence of the sequence $(a_n)_{n \in \mathbb{N}}$ whose terms belong to $C(a_N, r_N)$. Indeed $C(a_{N'}, r_{N'})$ contains a $T_{N''}$ such that $r_{N''} < \epsilon^2$ whereas $T_{N''} \subset C(a_{N'}, r_{N'}) \subset C(a_N, r_N)$, and so on.

Let D be the set of the limits of the convergent subsequences of the sequence $(a_n)_{n \in \mathbb{N}}$. By definition D is closed and included in $d(0,1)$. Moreover the T_n appear to be the holes of D. Indeed D obviously has an empty intersection with each T_n because $T_n \cap T_m = \emptyset$ for every $n \neq m$. Next , each T_n is included in $d(0,1)$

and $C(a_n, r_n)$ contains points of D, hence $\tilde{D} = d(0,1)$. But then the distance from T_n to D is just r_n and therefore T_n is a hole of D. Finally we check that D has no hole other than the T_n. Indeed let $\alpha \in d(0,1) \setminus \bigcup_{n=1}^{\infty} T_n$. Since $\bigcup_{n \in \mathbb{N}^*} T_n$ is dense in $d(0,1)$, there exists a subsequence of the sequence $(T_n)_{n \in \mathbb{N}^*}$ which converges to α, hence α is the limit of a subsequence of the sequence $(a_n)_{n \in \mathbb{N}}$, hence $\alpha \in D$. Thus if $\alpha \in d(0,1) \setminus D$, α belongs to a certain T_N and then this finishes proving that every hole of D is a T_n.

Now it is easily seen that D is infraconnected. Indeed let $\alpha \in D$ and let $r \in |K|$. Since $\bigcup_{n \in \mathbb{N}^*} T_n$ is dense in $d(0,1)$, there exists $N \in \mathbb{N}$ such that $T_N \cap C(\alpha, r) \neq \emptyset$. Since α belongs to T_N , T_N is obviously included in $C(\alpha, r)$, hence we have $r_N < r$, and we know that $C(a_N, r_N)$ contains points of D which obviously belong to $C(\alpha, r)$, hence $C(\alpha, r) \cap D \neq \emptyset$. We check that D has empty interior because by definition every point $\alpha \in D$ is the limit of a subsequence of the sequence $(a_n)_{n \in N}$ no term of which belong to D.

Now we will show that D has no T-filter. Indeed assume D to have a T-filter $\mathcal{F}$, of diameter r. We know that even if $\mathcal{F}$ is decreasing with no center in K, it admits a center α in $\widehat{K}$. With no loss of generality we may assume $\mathcal{F}$ to be decreasing. Then there exists a T-sequence $(U_m, u_m)_{m \in \mathbb{N}^*}$ associated to $\mathcal{F}$ where each U_m is a hole of D, whose diameter is denoted by σ_m, and where u_m belongs to $\mathbb{N}^*$, whereas the distance d_m from U_m to α satisfies $d_{m+1} \leq d_m$, $\lim_{m \to \infty} d_m = r$ and $d_m < r$ whenever $m \in \mathbb{N}^*$. By relation (4) in the definition of the T-sequences (chapter 35) it is easily checked that our T-sequence $(U_m, u_m)_{m \in \mathbb{N}^*}$ must satisfy

$$(4) \qquad \lim_{n \to \infty} \left[\left(\frac{d_n}{\sigma_n} \right)^{u_n} \prod_{j=1}^{n-1} \left(\frac{d_n}{d_j} \right)^{u_j} \right] = 0.$$

We put $\lambda_n = \log \sigma_n$. Then by (4) we have

$$\lim_{n \to \infty} \left(u_n \left(\log r + \lambda_n \right) + \left(\sum_{j=1}^{n-1} u_j \right) \left(\log r - \log d_1 \right) \right) = +\infty$$

hence there obviously exists a positive constant C such that

$$(5) \qquad u_n \left(\log r + \lambda_n \right) \leq C \sum_{j=1}^{n-1} u_j.$$

Since $\lim_{m \to \infty} r_m = 0$ and since the sequence $(U_n)_{n \in \mathbb{N}}$ appears to be a reordered subsequence of the sequence T_n it is clearly seen that $\lim_{n \to \infty} \lambda_n = +\infty$. Then there exists a positive constant B such that $u_n(\log r + \lambda_n) \leq B\lambda_n$ whenever $n \in \mathbb{N}^*$

and then by (5) there exists a positive constant A such that

$$(6) \qquad u_n \lambda_n \le A \sum_{j=1}^{n-1} u_j.$$

By applying (6) at the rank $n-1$ we have

$$u_{n-1} \le \frac{A}{\lambda_{n-1}} \sum_{j=1}^{n-2} u_j \text{ hence by (6) at the rank } n \text{ we obtain}$$

$$u_n \le \frac{A}{\lambda_n} \Big(\sum_{j=1}^{n-2} u_j \Big) \Big(1 + \frac{A}{\lambda_{n-1}} \Big)$$

and then by an immediate downing induction we have

$$u_n \le \frac{Au_1}{\lambda_n} \Big(\prod_{j=1}^{n-1} \Big(1 + \frac{A}{\lambda_j} \Big) \Big)$$

We now consider the sequence $(\omega_n)_{n \in \mathbb{N}}$ defined as

$$\omega_n = \log\Big(\frac{1}{\lambda_n} \prod_{j=1}^{n-1} \Big(1 + \frac{A}{\lambda_j} \Big) \Big) = -\log \lambda_n + \sum_{j=1}^{n-1} \log\Big(1 + \frac{A}{\lambda_j} \Big).$$

It is seen that

$$(7) \qquad \omega_n \le -\log \lambda_n + A \sum_{j=1}^{n-1} \frac{1}{\lambda_j}.$$

Now, by construction of D, every U_n is a certain T_m, hence every λ_j is of the form 2^m while the sequence $n \to \lambda_n$ is injective, hence

$$\sum_{j=1}^{n-1} \frac{1}{\lambda_j} \le \sum_{m \in \mathbb{N}^*} \frac{1}{2^m} = 1.$$

Therefore by (7) we see that $\lim_{n \to \infty} \omega_n = -\infty$, hence $u_n = 0$ when n is big enough.

This contradicts the hypothesis $u_n \in \mathbb{N}^*$ and finishes proving that D has no T-filter, and this ends the proof of Theorem 36.8.

Lemma 36.9:   *Let E be a set which is not countable and let f be a function from D into $\mathbb{R}_+$. There exists a sequence $(x_m)_{m \in \mathbb{N}}$ in E and $\lambda > 0$ such that $f(x_m) \ge \lambda$ for all $m \in \mathbb{N}$.*

Proof: We assume that such a λ and such a sequence $(x_m)_{m \in \mathbb{N}}$ do not exist. Hence for every $q \in \mathbb{N}^*$, the set A_q of the $x \in E$ such that $f(x) \ge \frac{1}{q}$ is finite.

But E is clearly equal to $\bigcup_{q=1}^{\infty} A_q$ and therefore E is countable.

Definition: The field K will be said to be *weakly valued* if both these two sets are countable: $|K|$, $\mathcal{K}$. Else, K will be said to be *strongly valued*.

Remark: $\mathbb{C}_p$ is weakly valued. Indeed, in Chapter 8, we saw that the valuation group of $\mathbb{C}_p$ is isomorphic to $\mathbb{Q}$, and that the the residue class field of $\mathbb{C}_p$ is an algebraic closure of $\mathbb{F}_p$, and therefore is countable.

Proposition 36.10:   *Let K be strongly valued. Let $a \in \widetilde{D}$ be such that $\delta(a, D) < diam(D)$. We assume that for every $r \in]\delta(a, D), \ diam(D)[\cap |K|$, each class of $C(a, r)$ contains at least one hole of D.*
Then for each $r \in]\delta(a, D), diam(D)]$ there exists on D an increasing T-filter and a decreasing T-filter of center a and diameter r.

Proof: Let $r \in]\delta(a, D), diam(D)]$. We will construct an increasing idempotent T-sequence $(T_n)_{n \in \mathbb{N}}$ of D, of center a and diameter r.
Let $J =]\delta(a, D), diam(D)[\cap |K|$, and let $(r_n)_{n \in \mathbb{N}}$, $(r'_n)_{n \in \mathbb{N}}$ be sequences of J satisfying $r_n < r'_n < r_{n+1}$, $\lim_{n \to \infty} r_n = r$. By hypothesis, either the residue class field $\mathcal{K}$ of K is not countable, or $|K|$ is not.

First we suppose that $|K|$ is not countable. Let n be fixed. By Lemma 36.9 there exists $\rho_n > 0$ together with an infinite family of circles $C(a, r)$, with $r_n < r < r'_n$, every one of them contains at least one hole of diameter bigger than ρ_n. Let $q_n \in \mathbb{N}$ satisfy

$$(1) \qquad \left(\frac{r_n}{r'_n}\right)^{q_n} \frac{r'_{n+1}}{\rho_{n+1}} < \frac{1}{n+1}.$$

Thus, we have defined the sequences $(\rho_n)_{n \in \mathbb{N}}$, $(q_n)_{n \in \mathbb{N}}$. Now, for each $n \in \mathbb{N}$, we put $t_n = \sum_{h=1}^{n} q_h$. From above, we can clearly find q_n circles $C(a, d_m)_{t_n \le m < t_{n+1}}$, with $r_n < d_m < d_{m+1} < r'_n$, which all contain at least one hole T_m of diameter bigger than ρ_n.
Then by definition T_m is included in $C(a, d_m)$ while $T_h \cap C(a, d_m) = \emptyset$ whenever $h \neq m$. Besides when $t_n \le m < t_{n+1}$, by hypothesis we have $diam(T_m) \ge \rho_n$. Thus by (1) the weighted sequence $(T_m, 1)$ is seen to satisfy

$$\frac{d_m}{diam(T_m)} \prod_{j=1}^{m} \frac{d_j}{d_m} \le \frac{1}{n} \qquad \text{whenever } t_n \le m < t_{n+1}.$$

This shows that we have $\lim_{m \to \infty} \dfrac{d_m}{diam(T_m)} \prod \dfrac{d_j}{d_m} = 0$ and therefore the weighted sequence $(T_m, 1)$ is an idempotent T-sequence.

Now we suppose that the residue class field $\mathcal{K}$ is not countable. By Lemma 36.9, for each $n \in \mathbb{N}$, there exist $\rho_n > 0$ together with an infinite family of classes of $C(a, r_n)$, such that each contains at least one hole of diameter bigger than ρ_n. In the same way as the previous case, for each $n \in \mathbb{N}$, we take $q_n \in \mathbb{N}$ satisfying (1) again and then, we have defined the sequences $(\rho_n)_{n \in \mathbb{N}}$, $(q_n)_{n \in \mathbb{N}}$. For each $n \in \mathbb{N}$, we put $t_n = \sum_{h=1}^{n} q_h$. Then in $C(a, r_n)$, we can clearly find q_n different classes such that each contains at least one hole of diameter bigger than ρ_n. Let $(T_m)_{t_n \leq m < t_{n+1}}$ be these holes. Thus by definition, we have

(2) $\delta(T_j, T_h) = r_n$ whenever h, j such that $t_n \leq h < t_{n+1}, t_n \leq j < t_{n+1}, h \neq j$.

By (2), it is easily checked that $\gamma_D(a, r_n, q_n) \leq \dfrac{r_n}{\rho_n}$. Therefore the weighted sequence $(T_m, 1)_{m \in \mathbb{N}}$ is an idempotent T-sequence because by (1) we have

$$\gamma_D(a, r_n, q_n) \prod_{j=1}^{n} \left(\frac{r_j}{r_n}\right)^{q_j} \leq \frac{1}{n}.$$

Symmetrically, we can prove the existence of a decreasing idempotent T-sequence of center a and diameter r: in the proof, we just have to replace (1) by

(1)' $\left(\dfrac{r'_n}{r_n}\right)^{q_n} \dfrac{r_{n+1}}{\rho_{n+1}} < \dfrac{1}{n+1}.$

This finishes the proof of Theorem 36.10.

37. CHARACTERISTIC PROPERTY

OF THE T-FILTERS

Throughout this chapter D is infraconnected.

We will prove that there exist elements strictly vanishing along a monotonous filter if and only if this filter is a T-filter [11], [14], [17]. It is quite easy to show that a pierced filter admitting strictly vanishing elements is a T-filter (Proposition 37.1). The big problem consists of proving that given any T-filter, there do exist analytic elements strictly vanishing along it (Theorem 37.2, [11], [14]).

Proposition 37.1, roughly, was proven separately and simultaneously by E. Motzkin and Ph. Robba [50], and by the author in March 1969 [11], [14] (however, Motzkin-Robba's claim was not stated in terms of T-filters, but in terms of sequences of holes that look like T-sequences).

Proposition 37.1:   *Let $\mathcal{F}$ be a monotonous filter on D. Let $f \in H(D)$ be strictly vanishing along $\mathcal{F}$. Then $\mathcal{F}$ is a T-filter.*

Proof :   Let us suppose $\mathcal{F}$ to be increasing, of center a and diameter S. With no loss of generality we may assume $a = 0$. Let $\lambda = -\log S$. There exists $\nu > \lambda$ such that $v(f, \mu) < +\infty$ whenever $\mu \in]\lambda, \nu]$ while $v(f, \lambda) = +\infty$. Let $r = \omega^{-\nu}$. Hence the function $v(f, .)$ is bounded in every interval $[\xi, \nu]$ with $\xi \in]\lambda, \nu[$ and therefore by Proposition 20.2 the equality $v(f(x)) = v(f, v(x))$ is true in all of $D \cap \Gamma(0, r, S)$ but inside finitely many classes of circles $C_m = C(0, r_m)$ with $r_m < r_{m+1}$, $\lim\limits_{m \to \infty} r_m = S$.

We fix $m \in \mathbb{N}$ and take r', r'' satisfying $r \le r' < r_m < r'' < S$. If $D \cap C_m \ne \emptyset$ we put $\theta_m = \|f\|_{D \cap C_m}$ and if $D \cap C_m = \emptyset$ we put $\theta_m = {}_D\varphi_{0, r_m}(f)$. Since $v(f, \mu)$ is bounded in $[-\log r'', -\log r']$ by a constant M we may find $h \in R(D)$ such that $\|h - f\|_D < \omega^{-M}$. Hence we have

$$(1) \qquad {}_D\varphi_{0, r_m}(f) = {}_D\varphi_{0, r_m}(h)$$

and if $D \cap C_m \ne \emptyset$ we have

$$(2) \qquad \|h\|_{C_m \cap D_m} = \|f\|_{C_m \cap D_m} = \theta_m.$$

Let $(T_{m,i})_{1 \le i \le s(m)}$ be the holes of D inside C_m which contain at least as many poles as many zeros and, for each one, let $q_{m,i}$ be the difference between the number of the poles and the number of the zeros (taking multiplicities into account).

Let $q_m = \displaystyle\sum_{i=1}^{s(m)} q_{m,i}$. Then we know that

(3) $v'^r(h, -\log r_m) - v'^l(h, -\log r_m) \leq q_m$.

By Proposition 34.4 and by Relation (1) we have

(4) $\theta_m \geq \gamma_D(0, r_m, q_m)_D \varphi_{0,r_m}(h)$

when $D \cap C_m \neq \emptyset$ and $\theta_m = {}_D\varphi_{0,r_m}(h)$ when $D \cap C_m = \emptyset$.

Hence Relation (4) is true anyway. In terms of valuations (4) is equivalent to

$$-\log \theta_m \leq -\log \gamma_D(0, r_m, q_m) + v(f, -\log r_m).$$

Now by (3) we see that

$$v(f, -\log r_m) \leq v(f, -\log r_{m-1}) + q_{m-1}(\log r_m - \log r_{m-1}).$$

Hence by induction we easily obtain

$$v(f, -\log r_m) \leq v(f, -\log r_1) + \sum_{j=1}^{m-1} q_j(\log r_m - \log r_j)$$

and finally $-\log \theta_m \leq \log \gamma_D(0, r_m, q_m) + \displaystyle\sum_{j=1}^{m-1} q_j(\log r_m - \log r_j)$. Since f is

vanishing along $\mathcal{F}$,   we have $\displaystyle\lim_{m\to\infty}(-\log \theta_m) = +\infty$ hence

$$\lim_{m\to\infty} -\log \gamma_D(0, r_m, q_m) + \sum_{j=1}^{m-1} q_j(\log r_m - \log r_j) = +\infty.$$

This just shows $\mathcal{F}$ to be an increasing T-filter. A symmetric reasonning is made when $\mathcal{F}$ is a decreasing filter provided with a center.

  Now let $\mathcal{F}$ be decreasing with no center. Then in $\widehat{K}$ we denote by $(d(\alpha_j, \rho_j))_{j\in J}$ the family of the holes of D and we put $d(a, r) = \widetilde{D}$ if D is bounded. If D is bounded (resp. unbounded), in $\widehat{K}$ we put

$$\widehat{D} = \widehat{d}(a, r) \setminus \left(\bigcup_{j\in J} \widehat{d}(\alpha_j, \rho_j^-)\right) \;(\text{ resp. } \widehat{D} = \widehat{K} \setminus \left(\bigcup_{j\in J} \widehat{d}(\alpha_j, \rho_j^-)\right)).$$

In $\widehat{K}$, $\mathcal{F}$ has a center a. Then the filter $\widehat{\mathcal{F}}$ of center a and diameter S on $\widehat{D}$ is a T-filter because f belongs to $H_{\widehat{K}}(\widehat{D})$ and is strictly vanishing along $\widehat{\mathcal{F}}$. Hence there exists a decreasing T-sequence $(\widehat{T}_{m,i}, \; m_{q,i})_{\substack{1 \leq i \leq s(m) \\ m \in \mathbb{N}}}$ of center a and diameter R that runs $\widehat{\mathcal{F}}$. But for each (m, i), $\widehat{T}_{m,i} \cap K$ is a hole $T_{m,i}$ of D, and then, by Remark 3 in Chapter 35. the weighted sequence $(T_{m,i}, \ m_{q,i})_{\substack{1 \leq i \leq s(m) \\ m \in \mathbb{N}}}$ is a T-sequence of D.

Theorem 37.2: *Let $\mathcal{F}$ be a T-filter on D, let $(T_{m,i}, \ m_{q,i})_{\substack{1 \leq i \leq s(m) \\ m \in \mathbb{N}}}$ be a T-sequence associated to $\mathcal{F}$, and $(\sigma_m)_{m \in \mathbb{N}}$ be the sequence of circles carrying this T-sequence. Let $D' = K \backslash (\bigcup_{\substack{1 \leq i \leq s(m) \\ m \in \mathbb{N}}} T_{m,i})$, and for every (m, i), let $a_{m,i} \in T_{m,i}$. Let $\mathcal{F}'$ be the T-filter on D' associated to the T-sequence $(T_{m,i}, q_{m,i})$. Let $\alpha \in C(F')$. There exists $g \in H(D')$ satisfying these properties:*

i) g is meromorphic in $T_{m,i}$, admits $a_{m,i}$ as a pole of order at most $q_{m,i}$ and has no other pole in $T_{m,i}$,

ii) g is strictly vanishing along $\mathcal{F}'$ and equal to zero in $\mathcal{P}_{D'}(\mathcal{F}')$,

iii) for every circular filter $\mathcal{G}$ different from $\mathcal{F}$ and secant with $C(\mathcal{F}')$, $_{D'}\varphi_{\mathcal{G}}(g)$ is different from 0,

iv) $g(\alpha) \neq 0$.

Proof: Without loss of generality we may obviously assume $D' = D$ hence $\mathcal{F}' = \mathcal{F}$. Besides we may assume $\mathcal{F}$ to be decreasing because an increasing filter of center a can be obtained by an inversion of center a of a decreasing filter of center a.

Actually the biggest difficulty happens when $\mathcal{F}$ has no center. Hence we do not suppose $\mathcal{F}$ to have a center (although it might have one). Indeed, here we have to construct the element f as a limit of rational functions with coefficients in K. If we take an origin in $\widehat{K}$, we work with a variable in L so that it is not clear how to obtain rational functions with coefficients in K. For this reason we will perform a sequence of change of origin. (However the proof becomes much easier to understand by assuming that $\mathcal{F}$ has center 0 and avoiding these changes of variable.)

Let s be the diameter of $\mathcal{F}$, and let (d_m) be the monotony of the T-sequence $(T_{m,i}, q_{m,i})$. For every $m \in \mathbb{N}$, let $q_m = \sum_{i=1}^{s(m)} q_{m,i}$, and let $C_m = C(b_m, d_m)$. Let $\Lambda_m = \widetilde{C_m} \cap D$, let $D_m = \Lambda_m \setminus \mathcal{P}(\mathcal{F})$ and let $b_m = a_{m+1,1}$. The sequence (D_m) is a canonical base of $\mathcal{F}$. We will construct f satisfying

(1) $\quad v_{b_m}(f, \mu) < +\infty \quad$ whenever $\quad \mu < -\log s$

(2) $\quad \|f\|_{\Lambda_m} < \dfrac{1}{m}$

Then f will clearly be strictly vanishing along $\mathcal{F}$ and equal to zero in all of $\mathcal{P}(\mathcal{F})$. Let us suppose we have defined increasing sequences of integers $(h(n))_{n \in \mathbb{N}}$ and $(\ell(n))_{n \in \mathbb{N}}$ satisfying

$h(n) < \ell(n) < \ell(n) + 1 = h(n+1)$.

Now, for every $m \in \mathbb{N}$, let $Q_m = \displaystyle\prod_{i=1}^{s(m)} (x - a_{m,i})^{q_{m,i}}$. Let $\lambda_m = (d_m)^{q_m} \left\| \dfrac{1}{Q_m} \right\|_{C_m \cap D}$

and for every $n \in \mathbb{N}$ let $E_n = \displaystyle\prod_{m=h(n)}^{\ell(n)} Q_m$. Since the weighted sequence $(T_{m,i}, q_{m,i})$

is a T-sequence we have

$$(3) \qquad \lim_{m \to \infty} \lambda_m \prod_{j=1}^{m-1} \left(\frac{d_m}{d_j} \right)^{q_j} = 0.$$

Let $a_n \in \Lambda_{h(n+2)}$ and let $X_n = x - a_n$. Clearly we have $|X_n| \le d_{h(n)}$ if and only if $x \in \widetilde{\Lambda}_{h(n)}$. We notice that as $a_n \in \Delta_{h(n+2)}, (x \in D_{h(n+1)} \setminus D_{h(n+2)})$ is equivalent to

$$(d_{h(n+2)} < |X_n| \le d_{h(n+1)}, \ X_n \in A_n)$$

In particular $(x \in D_m \setminus D_{m+1})$ is equivalent to

$$(d_{m+1} < |X_n| \le d_m, \ X_n \in A_n)$$

whenever $m \le \ell(n+1)$. We set $E_n(x) = U_n(X_n)$ and develop $U_n : U_n(X_n) = A_{0,n} + \ldots + A_{\tau(n),n} X_n^{\tau(n)}$ with

$$(4) \quad \tau(n) = deg(U_n) = deg(E_n) = \sum_{m=h(n)}^{\ell(n)} q_m.$$

Before going on, we notice this property. For every integer $e \le \tau(n)$, we denote by $j(e)$ the unique integer such that

$$q_{h(n)} + \ldots + q_{j(e)-1} \le \tau(n) - e < q_{h(n)} + \ldots + q_{j(e)}.$$

Now putting $\beta(e) = \tau(n) - e - (q_{h(n)} + \ldots + q_{j(e)-1})$, we have $\beta(e) \ge 0$ and

$$(5) \quad |A_{e,n}| \le \left(d_{h(n)} \right)^{q_{h(n)}} \ldots \left(d_{j(e)-1} \right)^{q_{j(e)-1}} \left(d_{j(e)} \right)^{\beta(e)}.$$

Indeed this comes from the fact that $A_{e,n}$ is a sum of products of $\tau(n) - e$ zeros of U_n (taking multiplicities into account). Hence $|A_{e,n}|$ is bounded by the product of the $\tau(n) - e$ "biggest" zeros of U_n. If ζ_m is a zero of Q_m, then $\zeta_m - a_m$ is a zero of U_m and then satisfies $|\zeta_m - a_m| = d_m$. Thus we obtain (5).

We are now going to introduce a sequence of rational functions F_n which will be involved in the construction of f. Let $(\sigma(n))_{n \in \mathbb{N}}$ and $(\tau(n))_{n \in \mathbb{N}}$ be sequences of integers satisfying $0 < \sigma(n) < \tau(n)$

$$(6) \quad q_{h(n)} + \ldots + q_{t(n)-1} < \tau(n) - \sigma(n) \le q_{h(n)} + \ldots + q_{t(n)} ,$$

and let $\alpha_n = \tau(n) - \sigma(n) - (q_{h(n)} + ... + q_{t(n)-1})$.

Let $S_n(X_n) = A_{\sigma(n)+1,n}X_n^{\sigma(n)+1} + ... + A_{\tau(n),n}X_n^{\tau(n)}$, $\quad G_n(X_n) = \dfrac{S_n(X_n)}{U_n(X_n)}$,

$F_n(x) = G_n(X_n)$ and $f_n = \displaystyle\prod_{i=1}^{n} F_i$.

We will prove that the sequence f_n converges in $H(D)$. The biggest problem consists of finding for $|1 - F_n(x)|$ a good upper bound on $D \setminus D_{h(n)}$ while $|F_n(x)|$ is equal to 1. We will show

$$(7) \quad |1 - F_n(x)| \leq \Big(\frac{d_{h(n)}}{|x - a_n|}\Big)^{\psi(n)} \leq \Big(\frac{d_{h(n)}}{d_{h(n-1)}}\Big)^{\psi(n)} \quad \text{whenever} \ \ x \in D \setminus D_{h(n)}.$$

In order to prove (7) we will notice this relation. Let e_1, e_2 be such that $0 \leq e_1 < e_2 < \tau(n)$. When $|X_n| > d_{h(n)}$ we have

$$(8) \quad |X_n|^{e_1}(d_{h(n)})^{q_{h(n)}}...(d_{j(e_1)-1})^{q_{j(e_1)-1}}.(d_{j(e_1)})^{\beta(e_1)} <$$
$$< |X_n|^{e_2}.(d_{h(n)})^{q_{h(n)}}...(d_{j(e_2)-1})^{q_{j(e_2)-1}}(d_{j(e_2)-1})^{\beta(e_2)}.$$

(Indeed by hypothesis the d_m are less than $|X_n|$. Since the sum of the powers of $|X_n|$ and that of the d_m are equal in the two members, the bigger member is the right one, which has the bigger power in X_n). Hence by (8) and (5) we see that when $|X_n| > d_{h(n)}$, for every $e \leq \sigma(n)$ we have

$$|X_n|^{e}|A_{e,n}| < |X_n|^{e}(d_{\ell(n)})^{q_{h(n)}}...(d_{j(e)-1})^{q_{j(e)-1}}(d_{j(e)})^{\beta(e)} \leq$$

$$\leq |X_n|^{\sigma(n)}(d_{h(n)})^{q_{h(n)}}...(d_{t(n)-1})^{q_{t(n)-1}}(d_{t(n)})^{\alpha_n}.$$

This shows that $\quad |(U_n - S_n)(X_n)| \leq$

$$\sup_{0 \leq e \leq \sigma(n)} |A_{e,n}||X_n|^{\ell} \leq |X_n|^{\sigma(n)}(d_{h(n)})^{q_{h(n)}}...(d_{t(n)-1})^{q_{t(n)-1}}(d_{t(n)})^{\alpha_n}$$

whenever $|X_n| > d_{h(n)}$. But as $d_m < d_{h(n)}$ whenever $m = h(n) + 1, ..., t(n)$, we see that

$$|(U_n - S_n)(X_n)| \leq |X_n|^{\sigma(n)}(d_{h(n)})^{\tau(n)-\sigma(n)}.$$

On the other hand, we have $|U_n(X_n)| = |X_n|^{\tau(n)}$ whenever $|X_n| > d_{h(n)}$ because the zeros of U_n do belong to $d(0, d_{h(n)})$.

So, we have finally proven $|1 - G_n(X_n)| \leq \Big(\dfrac{d_{h(n)}}{|X_n|}\Big)^{\psi(n)}$ whenever $|X_n| > d_{h(n)}$

and this complete the proof of Relation (7).

Now we will give $|S_n(X_n)|$ an upper bound when $|X_n| \leq d_{h(n)}$. By (3) we know that

$|X_n|^e|A_{e,n}| \leq |X_n|^e (d_{h(n)})^{q_{h(n)}}...(d_{j(e)-1})^{q_{j(e)-1}}.(d_{j(e)})^{\beta(e)}.$

But all the terms in S_n have an index $e \geq \sigma(n) + 1$. Hence we have

$|X_n|^e|A_{e,n}| \geq |X_n|^{\sigma(n)+1}|X_n|^{e-\sigma(n)-1} d_{h(n)}^{q_{h(n)}}...(d_{j(e)-1})^{q_{j(e)-1}}.(d_{j(e)})^{\beta(e)}.$

Now, as $|X_n| \leq d_{h(n)}$ and $d_m < d_{h(n)}$ for all $m = h(n) + 1, ..., j(e)$ we see that

$|X_n|^e|A_{e,n}| \leq |X_n|^{\sigma(n)+1}(d_{h(n)})^{\tau(n)-\sigma(n)-1}$

for every $e \geq \sigma(n) + 1$.

Finally when $|X_n| \leq d_{h(n)}$ we have

$|S_n(X_n)| \leq |X_n|^{\sigma(n)+1}(d_{h(n)})^{\tau(n)-\sigma(n)-1}.$

With greater reason we have Relation (9)

(9) $|S_n(X_n)| \leq |X_n|^{\sigma(n)}(d_{h(n)})^{\tau(n)-\sigma(n)}$ whenever $|X_n| \leq d_{h(n)}.$

We are now going to give $|U_n(X_n)|$ a lower bound when $|X_n| \leq d_{h(n)}$, with $X_n + a_n \in D$. Let $A_n = \{X_n \mid X_n + a_n \in D\}$. For $m = h(n), ..., \ell(n)$, let

$$V_m(X_n) = Q_m(x). \text{ Then } U_n(X_n) = \prod_{m=h(n)}^{\ell(n)} V_m(X_m). \text{ We consider the } V_j(X_n)$$

when $d_{m+1} < |X_n| \leq d_m$ with $h(n) \leq m \leq \ell(n)$ and $h(n) \leq j \leq \ell(n)$. Obviously we have

(10) $|V_j(X_n)| = |X_n|^{q_j}$ when $j > m.$ and

(11) $|V_j(X_n)| = (d_j)^{q_j}$ when $j < m.$

Besides we have

(12) $|V_m(X_n)| \geq \dfrac{(d_m)^{q_m}}{\lambda_m}$ because

$|V_m(X_n)| = (d_m)^{q_m}$ when $|X_n| < d_m$ and

$|V_m(X_n)| \geq \dfrac{(d_m)^{q_m}}{\lambda_m}$ when $|X_n| = d_m$ $(X_n \in A_n).$

We will study $|F_n(x)|$ when $x \in D_{h(n)} \setminus D_{h(n+1)}$ i.e., $d_{h(n+1)} < |X_n| \leq d_{h(n)}$, $X_n \in A_n$. By (10), (11), (12) we obtain

(13) $|U_n(X_n)| \geq \dfrac{|X_n|^{q_{\ell(n)}+...+q_{m+1}}(d_{h(n)})^{q_{h(n)}}...(d_m)^{q_m}}{\lambda_m}$

whenever $d_{m+1} < |X_n| \leq d_m$ (with $h(n) \leq m \leq \ell(n)$). Then by (9) and (12) we see that

$$|G_n(X_n)| \leq \frac{|X_n|^{\sigma(n)}(d_{h(n)})^{\tau(n)-\sigma(n)}\lambda_m}{|X_n|^{q_{\ell(n)}+...+q_{m+1}}(d_{h(n)})^{q_{h(n)}}..(d_m)^{q_m}}.$$

This may also be written

$$|G_n(X_n)| \leq \frac{\lambda_m|X_n|^{\tau(n)}|X_n|^{q_{h(n)}+...+q_m}d_{h(n)}^{\tau(n)-\sigma(n)}}{|X_n|^{q_{h(n)}+...+q_{\ell(n)}}|X_n|^{\tau(n)-\sigma(n)}(d_{h(n)})^{q_{h(n)}}...(d_m)^{q_m}}.$$

                    221

Since the diameter of $calF$ is equal to s, and since Since $\tau(n) = \sum\limits_{j=h(n)}^{\ell(n)} q_j$ we

have $|G_n(X_n)| \leq$

$$\leq \lambda_m \prod_{j=h(n)}^{m} \Big(\frac{|X_n|}{d_j}\Big)^{q_j} \Big(\frac{d_{h(n)}}{|X_n|}\Big)^{\tau(n)-\sigma(n)} \leq \lambda_m \prod_{j=h(n)}^{m} \Big(\frac{|X_n|}{d_j}\Big)^{q_j} \Big(\frac{d_{h(n)}}{s}\Big)^{\tau(n)-\sigma(n)}.$$

In particular when $d_{m+1} < |X_n| \leq d_m$ we obtain

$$|G_n(X_n)| \leq \lambda_m \prod_{j=h(n)}^{m} \Big(\frac{d_m}{d_j}\Big)^{q_j} \Big(\frac{d_{h(n)}}{s}\Big)^{\tau(n)-\sigma(n)}.$$

Finally we have proven Relation (14)

$$(14) \quad |F_n(x)| \leq \lambda_m \prod_{j=h(n)}^{m} \Big(\frac{d_m}{d_j}\Big)^{q_j} \Big(\frac{d_{h(n)}}{s}\Big)^{\tau(n)-\sigma(n)}$$

whenever $x \in D_m \setminus D_{m+1}$ with $h(n) \leq m \leq \ell(n)$.

Now we consider $|F_n(x)|$ when $x \in D_{h(n+1)} \setminus D_{h(n+2)}$ i.e., $d_{h(n+2)} < |X_n| \leq d_{h(n+1)}$, $X_n \in A_n$. It is seen that for every $j = h(n),...,\ell(n)$ we have $|V_j(X_n)| = (d_j)^{q_j}$ and then we have

$$(15) \quad |U_n(X_n)| = \prod_{j=h(n)}^{\ell(n)} (d_j)^{q_j}.$$

Now let us suppose $d_{m+1} < |X_n| \leq d_m$ with $h(n) \leq m \leq \ell(n)$. By (9) we have again $|S_n(X_n)| \leq |X_n|^{\sigma(n)}(d_{h(n)})^{\tau(n)-\sigma(n)}$, hence by (15) we see that $|G_n(X_n)| \leq 1$ whenever $|X_n| \leq d_{h(n+1)}$ hence

$(16) \quad |F_n(X)| \leq 1$ whenever $x \in \Lambda_{h(n+1)}.$

Besides, since $|X_n| \leq d_m$ we have $|S_n(X_n)| \leq d_m^{\sigma(n)} d_{h(n)}^{\tau(n)-\sigma(n)}$. Hence by (15) we have

$$|G_n(X_n)| \leq \frac{d_m^{\sigma(n)} d_{h(n)}^{\tau(n)-\sigma(n)}}{\prod\limits_{j=h(n)}^{\ell(n)} d_j^{q_j}} = \prod_{j=h(n)}^{\ell(n)} \Big(\frac{d_m}{d_j}\Big)^{q_j} \Big(\frac{d_{h(n)}}{d_m}\Big)^{\tau(n)-\sigma(n)} \leq$$

$$\leq \prod_{j=h(n)}^{\ell(n)} \Big(\frac{d_m}{d_j}\Big)^{q_j} \Big(\frac{d_{h(n)}}{s}\Big)^{\tau(n)-\sigma(n)}.$$

Finally we have proven (17)

(17) $\displaystyle |F_n(x)| \leq \Big(\prod_{j=h(n)}^{\ell(n)} \big(\frac{d_m}{d_j}\big)^{q_j} \Big)\big(\frac{d_{h(n)}}{s}\big)^{\tau(n)-\sigma(n)}$ whenever $x \in D_m \setminus D_{m+1}$, for every $m = h(n+1),...,\ell(n+1)$.

We are now able to construct the F_n just by defining the sequences $h(n)$ and $\sigma(n)$. For convenience we will use $\psi(n) = \tau(n)-\sigma(n)$, and then it is equivalent to define the sequences $h(n)$ and $\psi(n)$. Let us suppose these sequences $(h(n))_{n\in\mathbb{N}}$ and $(\psi(n))_{n\in\mathbb{N}}$ are already defined up to the rank N, satisfying for every $n \leq N$ the following relations (18), (19), (20)

(18) $\displaystyle \big(\frac{d_{h(n-1)+1}}{d_{h(n-1)}}\big)^{\psi(n)} < \frac{1}{n}$

(19) $\displaystyle \big(\frac{d_{h(n)}}{s}\big)^{\psi(n)} < 2$

(20) $\displaystyle \lambda_m \prod_{j=h(n-1)}^{m} \big(\frac{d_m}{d_j}\big)^{q_j} < \frac{1}{4n}$ whenever $m \geq h(n)$.

Thus we know that $\ell(N-1) = h(N) - 1$, and we want to choose $\psi(N+1)$ and $\ell(N)$ satisfying (18), (19), (20). Of course, we can find $\psi(N+1)$ big enough to satisfy

(21) $\displaystyle \big(\frac{d_{h(N)+1}}{d_{h(N)}}\big)^{\psi(N+1)} < \frac{1}{N+1}.$

Next, as $\displaystyle \lim_{m\to\infty} d_m = s$ there does exist a rank $M > N$ such that

(22) $\displaystyle \big(\frac{d_m}{s}\big)^{\psi(N+1)} < 2$ whenever $m \geq M$.

Finally by hypothesis we have $\displaystyle \lim_{m\to\infty} \lambda_m \prod_{j=h(N)}^{m} \big(\frac{d_m}{d_j}\big)^{q_j} = 0$, hence there exists a rank $M' \geq M$ such that

(23) $\displaystyle \lambda_m \prod_{j=h(N)}^{m} \big(\frac{d_m}{d_j}\big)^{q_j} < \frac{1}{8(N+1)}$ whenever $m \geq M'$.

We take $h(N+1) = M'$ and then by (21), (22), (23) we see that (18), (19), (20) are satisfied at the rank $N+1$. In order to begin the recurrence we can take $h(0) = 1, \sigma(0) = 0, \psi(1) = 1$ and then the sequences are defined for all $n \in \mathbb{N}$.

As it was announced above, we put $\displaystyle f_n = \prod_{i=1}^{n} F_i$ and we will show that the sequence $(f_n)_{n\in\mathbb{N}}$ converges in $H(D)$ to the element f announced in the Theorem.

Let $n > 0$. We will study $|f_n(x)|$ in these 3 cases:

$\alpha)\ x \in D \setminus \Lambda_{h(n)}$, $\beta)\ x \in \Lambda_{h(n+1)}$, $\gamma)\ x \in \Lambda_{h(n)} \setminus \Lambda_{h(n+1)}.$

   $\alpha)$ By (17) we have $|F_n(x)| \leq 1$ hence $|f_n(x)| \leq |f_{n-1}(x)| \leq 1$.

β)   By (16) we notice that for every $j \leq n$, we have $|F_j(x)| \leq 1$. Next, by (17) we have

$$|F_n(x)| \leq \prod_{j=h(n)}^{\ell(n)} \left(\frac{d_{h(n+1)}}{d_j}\right)^{q_j} \left(\frac{d_{h(n)}}{s}\right)^{\psi(n)}$$

and then by (19) and (20) we see that $|F_n(x)| \leq \dfrac{1}{2(n+1)}$. So, by α) applied to f_{n-1}, we obtain

(24) $|f_n(x)| \leq \dfrac{1}{2(n+1)}.$

γ)   Let $m \in \mathbb{N}$ be such that $x \in D_m \setminus D_{m+1}$. (Obviously $m \in [h(n), \ell(n)]$.) We know that $|f_n(x)| \leq |F_{n-1}(x)F_n(x)|$, and by (17) we have

$$|F_{n-1}(x)F_n(x)| \leq \lambda_m \prod_{j=h(n-1)}^{m} \left(\frac{d_m}{d_j}\right)^{q_j} \left(\frac{d_{h(n-1)}}{s}\right)^{\psi(n-1)} \left(\frac{d_{h(n)}}{s}\right)^{\psi(n)}$$

hence by (19) and (20) we see that $|F_{n-1}(x)F_n(x)| \leq \dfrac{1}{n}$, hence $|f_n(x)| \leq \dfrac{1}{n}$.

Thus, after studying the cases α), β), γ), we have proven that $\|f_n\|_D \leq \|f_0\|_D$ and that

(25) $\|f_n\|_{D_{h(n)}} \leq \dfrac{1}{n}.$

Hence we have obtained

(26) $\|f_{n+1} - f_n\|_{D_{h(n)}} \leq \dfrac{1}{n}.$

Now when $x \in D \setminus D_{h(n)}$ by (7) we have $|1 - F_n(x)| \leq \left(\dfrac{d_{h(n)}}{d_{h(n-1)}}\right)^{\psi(n)}$ and therefore

we have $|1 - F_n(x)| \leq \left(\dfrac{d_{h(n-1)+1}}{d_{h(n-1)}}\right)^{\psi(n)}$. Hence by Relation (18) we have

(27) $|1 - F_n(x)| < \dfrac{1}{n}$ whenever $x \in D \setminus D_{h(n)}.$

Finally by (26) and (27) we have $\|f_{n+1} - f_n\|_D \leq \dfrac{1}{n} \sup(\|f_1\|_D, 1)$ and this shows that the sequence converges to a limit f in $H(D)$.

By construction it is clear that for each hole $T_{m,i}$, the sequence $\left((x - a_{m,i})^{q_{m,i}} f_n\right)_{n \in \mathbb{N}}$ converges in $H(D \cup T_{m,i})$, hence f admits $a_{m,i}$ as a unique pole of order $\leq q_{m,i}$. Next, by (16) and (25) it is seen that $\|f_j\|_{D_{h(n)}} \leq \dfrac{1}{n}$ whenever $j \geq n$, hence $\|f\|_{D_{h(n)}} \leq \dfrac{1}{n}$ and therefore f is vanishing along $\mathcal{F}$ whereas $f(x)$ is equal to zero in all of $\mathcal{P}(\mathcal{F})$. Thus i) is satisfied.

We check that ii) and iii) are satisfied. Indeed by (7) we have $|F_j(x)| = 1$ whenever $j > n$ and $x \in K \setminus D_{h(n)}$, hence we have $|f_n(x)| = |f_j(x)| = |f(x)|$ whenever $j > n$ and $x \in K \setminus D_{h(n)}$.
This shows that Statements ii) and iii) are satisfied.

Finally, if $f(\alpha) = 0$, since D' is open, f factorizes in $H(D')$ in the form $(x - \alpha)^q g(x)$ with $g(\alpha) \neq 0$, and then it is immediately seen that g also satisfies i), ii), ii), and further, satisfies iv). This finishes the proof of Theorem 37.2.

Corollary 37.3:   *Let $\mathcal{F}$ be a monotonous filter on D. There exist elements of $H(D)$ strictly vanishing along $\mathcal{F}$ if and only if $\mathcal{F}$ is a T-filter.*

38. APPLICATIONS OF T-FILTERS

Throughout this chapter, D is infraconnected.

We will apply the T-filters for characterizing the main algebraic properties of the algebras $H(D)$: noetherian algebras $H(D)$, existence of divisors of zeros. we will also characterize the property for a set E in K to be analytic, (i.e. if an analytic element is equal to zero in a disk, it is identically zero everywhere). [12], [14], [17], [50].

By Propositions 21.1, 21.2 21.3, 37.1 and 37.2, Lemma 38.1 is immediate:

Lemma 38.1: *Let $b \in D$, $l > 0$ and let $f \in H(D)$ satisfy $f(b) \neq 0$ and $_D\varphi_{b,l} = 0$. There exists an increasing T-filter $\mathcal{F}$ of center b and diameter $t \in]0, l[$ such that f is strictly vanishing along $\mathcal{F}$ and satisfies $_D\varphi_{b,s}(f) > 0$ for every $s \in]0, t[$.*

Let $a \in \tilde{D}$ and let $r, s \in \mathbb{R}$ satisfy $\delta(a, D) \leq r \leq s \leq diam(D)$. There exists $f \in H(D)$ satisfying $_D\varphi_{a,s}(f) > 0$, $_D\varphi_{a,r}(f) = 0$, (resp. $_D\varphi_{a,r}(f) > 0$, $_D\varphi_{a,s}(f) = 0$) if and only if there exists an increasing (resp. a decreasing) T-filter of center a and diameter $t \in]r, s]$, (resp. $t \in [r, s[$). Moreover, if f satisfies $_D\varphi_{a,s}(f) > 0$, $_D\varphi_{a,r}(f) = 0$, (resp. $_D\varphi_{a,s}(f) = 0$, $_D\varphi_{a,r}(f) > 0$) then there exists an increasing (resp. a decreasing) T-filter $\mathcal{F}$ of center a and diameter $t \in]r, s]$, (resp. $t \in [r, s[$) such that f is strictly vanishing along $\mathcal{F}$.

Theorem 38.2: *Let $f \in H(D) \setminus \{0\}$. Then f is not quasi-minorated if and only if there exists a T-filter $\mathcal{F}$ on D such that f is strictly vanishing along $\mathcal{F}$.*

Proof: If D admits a T-filter $\mathcal{F}$, by Theorem 37.2 there exists an element f strictly vanishing along $\mathcal{F}$ hence f is not quasi-minorated. Now let $f \in H(D)$ be not quasi-minorated. By Theorem 22.1, f is strictly vanishing along a filter $\mathcal{F}$ and then by Theorem 37.1 $\mathcal{F}$ is a T-filter.

Corollary 38.3: *Every non zero element of $H(D)$ is quasi-minorated if and only if D has no T-filter.*

Theorem 38.4: *Every non zero element of $H(D)$ is quasi-invertible if and only if D satisfies these three properties.*

α) D has no T-filter.

β) $\overline{D}$ is open.

$\gamma)$ $\widetilde{D} \setminus \overline{D}$ *is bounded.*

Proof: Let D satisfy $\alpha)$, $\beta)$, $\gamma)$ and let $f \in H(D) \setminus \{0\}$. By $\gamma)$ and Corollary 38.3 f is quasi-minorated. But then, by $\alpha)$, $\beta)$ and by Theorem 24.8 we see that f is quasi-invertible. Reciprocally, we assume all the $f \in H(D) \setminus \{0\}$ to be quasi-invertible. Hence by Theorem 22.5 every element different from 0 is quasi-minorated and therefore the set of the quasi-invertible elements is equal to the set of the quasi-minorated ones. Then by Theorem 24.8, D satisfies $\beta)$, $\gamma)$. Further, by Corollary 38.3 D has no T-filter, hence it satisfies $\alpha)$.

Theorem 38.5:   *Let A be an infraconnected subset of D not reduced to a single point, and let $b \in D \setminus A$. Let $f \in H(D)$ satisfy $f(u) = 0$ for all $u \in A$ and $f(b) \neq 0$. Then, f is strictly vanishing along a T-filter $\mathcal{F}$ on D such that $A \subset \mathcal{P}(\mathcal{F})$ and $b \in \mathcal{C}(\mathcal{F})$.*

Proof: Let $a \in A$, let $r = diam(A)$, and let $\ell = |a - b|$. Since A is not reduced to a single point, we have $r > 0$.

First suppose $\ell \leq r$. Since A is infraconnected, we have $_D\varphi_{b,\ell}(f) = {}_A\varphi_{b,\ell}(f) = 0$. But since $f(b) \neq 0$, there does exists $t \in]0,\ell]$ such that $_D\varphi_{b,s}(f) > 0$ for all $s \in]0,t[$, and $_D\varphi_{b,t}(f) = 0$. Hence, f is strictly vanishing along the increasing filter $\mathcal{F}$ of center b and diameter t, and therefore A is included in $\mathcal{P}(\mathcal{F})$.

Now, suppose $\ell > r$. If $_D\varphi_{a,\ell}(f) = 0$ then $_D\varphi_{b,\ell}(f) = 0$ hence there exists an increasing T-filter $\mathcal{F}$ of center b and diameter $t \in]0,\ell]$, so we have $A \subset \mathcal{P}(\mathcal{F})$. Hence it only remains to consider the case $_D\varphi_{a,\ell}(f) > 0$. But then, there exists $t \in [r,\ell[$ such that $_D\varphi_{a,s}(f) > 0$ for all $s \in]t,\ell[$ and $_D\varphi_{a,t}(f) = 0$. Hence f is strictly vanishing along a decreasing T-filter $\mathcal{F}$ of center a and diameter t, and therefore A is included in $\mathcal{P}(\mathcal{F})$.

Corollary 38.6:   *Let D have no T-filter. Let $a \in D$ and let $r \in \mathbb{R}_+^*$. Let $f \in H(D)$ satisfy $f(x) = 0$ whenever $x \in d(a,r) \cap D$. Then f is identically 0.*

Theorem 38.7:   *Let D have a unique T-filter $\mathcal{F}$ and let $f \in H(D)$, $f \neq 0$, be vanishing along $\mathcal{F}$. Then f is strictly vanishing along $\mathcal{F}$ and satisfies $f(x) = 0$ whenever $x \in \mathcal{P}(\mathcal{F})$.*

Proof: Since f is not quasi-minorated, by Theorems 24.1 and 37.1 there exists a T-filter $\mathcal{G}$ such that f is strictly vanishing along $\mathcal{G}$. But $\mathcal{F}$ is the unique T-filter of D, hence $\mathcal{F} = \mathcal{G}$. Now let $D' = \mathcal{P}(\mathcal{F})$ and let $\widetilde{f}$ be the restriction of f to D'. Let $\mathcal{G}$ be the circular filter on D less thin than $\mathcal{F}$. Then $\mathcal{G}$ is secant with D', and we have $_{D'}\phi_{\mathcal{G}}(\widetilde{f}) = 0$. Hence $\widetilde{f}$ is not quasi-minorated in $H(D')$ and then $\widetilde{f}$ is either strictly vanishing along a T-filter of D', or identically zero. But since D has no T-filter other than $\mathcal{F}, D'$ has no T-filter hence by Corollary 38.6, we have $f(x) = 0$ whenever $x \in D'$.

Analytic sets were introduced in Ch.27. This notion was used first by M. Krasner and E. Motzkin, and Ph. Robba. The analytic sets were characterized in [12].

Theorem 38.8:   *Let E be a set in K. Then E is analytic if and only if E is infraconnected such that any T-filter $\mathcal{F}$ on E satisfies $\mathcal{P}(\mathcal{F}) = \emptyset$.*

Proof: If E is not infraconnected, it admits an empty annulus $\Gamma(a, r', r")$ and then by Theorem 10.14 $H(E)$ contains the characteristic function of $\mathcal{I}(\Gamma(a, r', r"))$ $= d(a, r') \cap E$, hence clearly D is not analytic. Now let D have a T-filter $\mathcal{F}$ with a not empty beach.

We first suppose $\mathcal{F}$ increasing, of center a and diameter S. By Theorem 37.2 there exists $f \in H(E)$ strictly vanishing along $\mathcal{F}$ such that $f(x) = 0$ whenever $x \in \mathcal{P}(\mathcal{F})$. Now let $b \in D$ such that $|a - b| \geq S$ and let $s \in]0, S[$. We have $d(b, s) \cap D \subset \mathcal{P}(\mathcal{F})$ and therefore $f(x) = 0$ whenever $x \in d(b, s) \cap D$, while f is not identically zero in $d(a, S^-)$. Hence E is not analytic.

We now suppose $\mathcal{F}$ is decreasing and has diameter S. Since $\mathcal{P}(\mathcal{F}) \neq \emptyset$, $\mathcal{F}$ admits a center $a \in D$ and then by Theorem 37.2 there exists $f \in H(E)$ strictly vanishing along $\mathcal{F}$, such that $f(x) = 0$ whenever $x \in d(a, S) \cap D$.

Now reciprocally, we suppose E not to be analytic. There exist $f \in H(D)$, $a, b \in E$ and $r > 0$ such that $f(b) \neq 0$, and $f(x) = 0$ whenever $x \in d(a, r) \cap D$. Hence by Proposition 38.5 there exists a T-filter $\mathcal{F}$ such that $d(a, r) \subset \mathcal{P}(\mathcal{F})$.

T-filters make it easy to characterize the algebras $H(E)$ which are principal ideal rings.

Theorem 38.9:   *Let $E \in \mathcal{A}$. Properties* a), b), c), d) *are equivalent.*

  a) *Every ideal of $H(E)$ is generated by a polynomial whose zeros belong to $\overline{\overset{\circ}{E}} \cap E$*
  b) *$H(E)$ is a principal ideal algebra with no idempotent different from 0 and 1.*
  c) *$H(E)$ is a Noetherian algebra with no idempotent different from 0 and 1.*
  d) *E is infraconnected, with no T-filter and $\overline{E}$ is open.*
  e) *Every element of $H(E)$ is quasi-invertible.*

Proof: Obviously a) implies b) and b) implies c). Next, by Theorem 24.2, e) implies a). We check that d) implies e). Since $D \in \mathcal{A}$, by Property A) $\widetilde{D} \setminus \overline{D}$ is bounded. Hence by Theorem 38.4 d) implies e).

Hence we only have to show that c) implies d). For this, we suppose that d) is not satisfied.

First, if E is not infraconnected with greater reason E is not and therefore by Theorem 10.14 $H(E)$ admits idempotents other than 0 and 1 whence obviously c) is not satisfied.

Henceforth we assume E to be infraconnected.

If $\overline{E}$ is not open, E admits a pierced Cauchy filter $\mathcal{F}$. Let a be its limit. Since E belongs to $\mathcal{A}$, by Theorem 18.8 $\overline{E} \setminus E$ is included in $\overset{\circ}{\overline{E}}$ and therefore a belongs to E. Hence by Theorem 18.10, the maximal ideal $\mathcal{M}(a)$ of the $f \in H(E)$ such that $f(a) = 0$ is not of finite type, and therefore c) is not true.

Now we suppose that E admits a T-filter $\mathcal{F}$. By Lemma 12.2 for every $h \in H(E)$ there exists $A \in \mathcal{F}$ such that $h(x)$ is bounded in A. Thus E admits a pierced filter $\mathcal{F}$ with elements $f \in H(E)$ properly vanishing along $\mathcal{F}$ and then for every $h \in H(E)$, there exists $A \in \mathcal{F}$ such that $h(x)$ is bounded in A. Let $\mathcal{J}$ be the ideal of the $f \in H(E)$ such that $\lim_{\mathcal{F}} f(x) = 0$. We will prove that $\mathcal{J}$ is not of finite type. Indeed, suppose that $\mathcal{J}$ is of finite type and let $g_1, ..., g_q$ be a system of generators. By Lemma 18.4, there exists $f \in \mathcal{J}$ and a sequence z_n in E thinner than $\mathcal{F}$ satisfying

$\quad\alpha)$ $g_1(z_n) \neq 0$ whenever $n \in \mathbb{N}$ and

$\quad\beta)$ $\displaystyle \lim_{n \to \infty} \Big(\frac{|f(z_n)|}{\max_{1 \leq i \leq q} |g_i(z_n)|} \Big) = +\infty.$

For every $n \in \mathbb{N}$ let $t_n = \max_{1 \leq i \leq q} |g_i(z_n)|$. Since $\mathcal{J}$ admits $g_1, ..., g_q$ as a system of generators, there exist $h_1, ..., h_q \in H(E)$ such that $f = g_1 h_1 + ... + g_q h_q$. But as we noticed above there exists $A \in \mathcal{F}$ such that each $h_i(x)$ is bounded in A. Then it is seen that there exists $M > 0$ such that $\dfrac{|h_i(z_n)|}{t_n} \leq M$ whenever $n \in \mathbb{N}$ and $i = 1, ..., q$. This clearly contradicts the property $\beta)$ and therefore we have proven that $\mathcal{J}$ is not of finite type. This finishes showing that c) implies d). This ends the proof of Theorem 38.9.

Theorem 38.10:   *Let $A \in \mathcal{A}$. The following statements are equivalent.*

$\quad$a) *Every ideal of $H(E)$ is principal.*

$\quad$b) *$H(E)$ is noetherian*

$\quad$c) *$\overline{E}$ is open, has finitely many infraconnected components, and each one has no T-filter.*

Besides, when Statements a), b), c) are satisfied, the algebra $H(E)$ is isomorphic to the direct product $H(E_1) \times ... \times H(E_q)$ with $E_1, ..., E_q$ the infraconnected components of E.

Proof: As a) implies b), we first suppose b) to be satisfied and will show c). Obviously $H(E)$ has finitely many idempotents, hence by Proposition 10.14 E has finitely many empty annuli. Therefore, by Lemma 2.13, E has finitely many infraconnected components, $E_1, ..., E_q$. Then by Theorem 10.15, for each $i = 1, ..., q$ the characteristic function u_i of E_i belongs to $H(E)$. But for each

$i = 1, ... q$, $H(E_i)$ is isometrically isomorphic to $u_i H(E)$. Besides by definition the $(u_i)_{1 \le i \le q}$ consists of a system of idempotents such that $u_i u_j = 0$ whenever $i \ne j$ and $\sum_{i=1}^{n} u_i = 1$. Hence $H(E)$ is isometrically isomorphic to the direct product $u_1 H(E) \times ... \times u_q H(E)$. Thus finally, we see that $H(E)$ is isometrically isomorphic to $H(E_1) \times ... \times H(E_q)$. Obviously, each algebra $H(E_i)$ must be noetherian. But then, since E_i is infraconnected, $H(E_i)$ has no idempotent other than 0 and 1, and then, by Theorem 38.9, $\overline{E_i}$ is open and E_i has no T-filter. Hence we have shown that $\overline{E}$ is open, that E has finitely many infraconnected components and that each one has no T-filter. Thus we have shown c) is satisfied. Further we have seen that Statements a), b), c), do imply that $H(E)$ is isomorphic to $H(E_1) \times ... \times H(E_q)$.

Now we suppose c) satisfied, and denote by $E_i, (1 \le i \le q)$, the infraconnected components of E. Hence every algebra $H(E_i)$ is a principal ideal algebra because E_i has no T-filter and $\overline{E_i}$ is open. Besides, by Theorem 10.15, for each $i = 1, ..., q$, the characteristic function u_i of E_i belongs to $H(E)$. As previously, the $(u_i)_{1 \le i \le q}$ consists of a system of idempotents such that $u_i u_j = 0$ whenever $i \ne j$ and $\sum_{i=1}^{n} u_i = 1$. Hence $H(E)$ is isometrically isomorphic to the direct product $H(E_1) \times ... \times H(E_q)$. Therefore it is seen that every ideal of $H(E)$ is principal and therefore a) is satisfied. This ends the proof of Theorem 38.10.

Now we will characterize the sets E such that $H(E)$ is an algebra with no divisor of zero.

Definition: Two monotonous filters $\mathcal{F}_1, \mathcal{F}_2$ will be said to be *complementary* if $\mathcal{P}(\mathcal{F}_1) \cup \mathcal{P}(\mathcal{F}_2) = D$.

Theorem 38.11: *Let E be a set in K. There exist f and $g \in H(E)$ not identically equal to zero such that fg is identically zero if and only if:*
either E is not infraconnected. or E has two complementary T-filters.

Proof: If E is not infraconnected, it admits an empty annulus $\Gamma(a, r', r'')$ and then the characteristic functions u of $\mathcal{I}(\Gamma(a, r', r''))$ (resp. $w = 1 - u$ of $\mathcal{E}(\Gamma(a, r', r''s)))$ belong to $H(E)$ and satisfy $uw = 0$ though they both are not identically zero.

If E is infraconnected with two complementary T-filters $\mathcal{F}_1, \mathcal{F}_2$, there exists f_1 (resp. f_2) strictly vanishing along $\mathcal{F}_1$ (resp. $\mathcal{F}_2$) and identically equal to zero in $\mathcal{P}(\mathcal{F}_1)$ (resp. $\mathcal{P}(\mathcal{F}_2)$) hence we have $f_1(x) f_2(x) = 0$ whenever $x \in \mathcal{P}(\mathcal{F}_1) \cup \mathcal{P}(\mathcal{F}_2)$ hence whenever $x \in E$.

Now we suppose E infraconnected and we suppose that there exist f_1 and f_2 not identically equal to zero such that $f_1(x)f_2(x) = 0$ whenever $x \in D$. Let $a_1 \in E$ be such that $f_1(a_1) \neq 0$ and let $a_1 \in E$ be such that $f_2(a_2) \neq 0$. Let $r = |a_1 - a_2|$.

We first suppose that ${}_E\varphi_{a_1,r}(f_1) = 0$. Then by Proposition 21.2 there exists an increasing T-filter $\mathcal{F}_1$ of center a_1 and diameter $s \in]0,r]$ such that f_1 is strictly vanishing along $\mathcal{F}_1$. Hence there exists $l \in]0,s[$ such that ${}_E\varphi_{a_1,l}(f_1) \neq 0$, hence we have ${}_E\varphi_{a_1,l}(f_2) = 0$. If ${}_E\varphi_{a_2,r}(f_2) = 0$, then f_2 is strictly vanishing along an increasing T-filter $\mathcal{F}_2$ of center a_2 and diameter $t \in]0,r]$ and we have $\mathcal{P}(\mathcal{F}_1) \cup \mathcal{P}(\mathcal{F}_2) = D$.

If ${}_E\varphi_{a_2,r}(f_2) \neq 0$, as ${}_E\varphi_{a_2,r}(f_2) = {}_E\varphi_{a_1,r}(f_2)$, at the same time we have ${}_E\varphi_{a_1,l}(f_2) = 0$ and ${}_E\varphi_{a_1,r}(f_2) \neq 0$. Let t be the supremum of the set of the $\xi \leq r$ such that ${}_E\varphi_{a_1,\xi}(f_2) = 0$. Then f_2 is strictly vanishing along a decreasing T-filter $\mathcal{F}_2$ of center a_1 and diameter t. Hence we have $\mathcal{P}(\mathcal{F}_1) \cup \mathcal{P}(\mathcal{F}_2) = E$.

We now suppose that ${}_E\varphi_{a_1,r}(f_1) \neq 0$, hence ${}_E\varphi_{a_2,r}(f_1) \neq 0$ and then we have ${}_E\varphi_{a_2,r}(f_2) = 0$. Since $f_2(a_2) \neq 0$ there exists $l \in]0,r[$ such that ${}_E\varphi_{a_2,l}(f_2) \neq 0$. Let s be the infimum of the set of the l such that ${}_E\varphi_{a,l}(f_2) = 0$. We see that f_2 is strictly vanishing along an increasing T-filter $\mathcal{F}_2$ of center a_2 and diameter s and then we have ${}_E\varphi_{a_2,l}(f_1) = 0$ whenever $l \in [0,s[$ hence f is strictly vanishing along a decreasing T-filter $\mathcal{F}_1$ of center a_2 and diameter $t \in [s,r]$. Finally we have $\mathcal{P}(\mathcal{F}_1) \cup \mathcal{P}(\mathcal{F}_2) = D$. This ends the proof.

Corollary 38.12:   *Let $E \in \mathcal{A}$. The algebra $H(E)$ has no divisor of zero if and only if E is infraconnected with no couple of complementary T-filters.*

Theorem 38.13:   *Let $\mathcal{F}$ be a T-filter on D that admits no T-filter complementary to $\mathcal{F}$. Then $\mathcal{I}(\mathcal{F}) = \mathcal{I}_0(\mathcal{F})$.*

Proof: Of course we suppose $\mathcal{P}(\mathcal{F}) \neq \emptyset$. For instance, suppose $\mathcal{F}$ decreasing, and let $r = diam(\mathcal{F})$. Suppose that there exists $f \in \mathcal{I}(\mathcal{F}) \setminus \mathcal{I}_0(\mathcal{F})$, and let $a \in \mathcal{P}(\mathcal{F})$ be such that $f(a) \neq 0$. Since we have ${}_D\varphi_{a,r}(f) = 0$, by Lemma 38.1 D admits an increasing T-filter of center a and diameter $s \in]0,r]$, which is obviously complementary to $\mathcal{F}$. In the same way, if $\mathcal{F}$ is increasing, we perform a symmetric proof. This ends the proof of Theorem 38.13.

Now, about composition, we have sufficient conditions on D to assure that $g \circ f$ belongs to $H(D)$.

Theorem 38.14: (K. Boussaf, A. Escassut)   *Let $D \in \mathcal{A}$, and let $f \in H(D)$ be not collapsing along any T-filter of D. Then for every $g \in H(f(D))$, $g \circ f$ belongs to $H(D)$.*

Proof: By Lemma 11.1 we may assume that g belongs to $R(f(D))$ with no loss of generality. Let $g(u) = \dfrac{P(u)}{Q(u)}$. Since $D \in \mathcal{A}$, both $P \circ f$, $Q \circ f$ belong to $H(D)$.

Actually, since $\dfrac{P(u)}{Q(u)}$ belongs to $R(f(D))$, Q has no zero in $f(D)$, and therefore $Q \circ f$ has no zero in D. But by hypothesis, for every $\lambda \in K \setminus D$, $f - \lambda$ is not vanishing along any T-filters, hence is quasi-minorated, whereas it has no zero in D, hence by Theorem 22.11, it is invertible in $H(D)$, and therefore so is $Q \circ f$. So, as D belongs to $\mathcal{A}$, it is seen that $g \circ f$ belongs to $H(D)$.

Corollary 38.15: *Let $D \in \mathcal{A}$ have no T-filter and let $f \in H(D)$. Then for every $g \in H(f(D))$, $g \circ f$ belongs to $H(D)$.*

However, at the end of Chapter 50 we will see that when the set D admits certain T-filters, the composition of analytic elements, sometimes, is not an analytic element.

Corollary 38.16: *Let $D \in \mathcal{A}$ and let $D' \in \mathcal{A}$ have no T-filter. Then $\Omega(D, D')$ is equal to the set of the $f \in H(D')$ such that $f(D') \subset D$.*

As a consequence, according to the hypothesis of Corollary 38.16, by Theorem 11.7 the $f \in H(D')$ such that $f(D') \subset D$ characterize the homomorphisms from $H(D)$ into $H(D')$. Besides, if both D, D' have no T-filter, then the bijections f from D' onto D' such that $f \in H(D')$ and $f^{-1} \in H(D)$ characterize the isomorphims from $H(D)$ onto $H(D')$.

Theorem 38.17: (K. Boussaf, A. Escassut) *Let $D, D' \in \mathcal{A}$ be infraconnected. For each $\lambda \in \Omega(D', D)$, ϕ_λ is not injective if and only if it satisfies at least one of the follwing two conditions:*

1) λ is a constant.

2) There exists a T-filter $\mathcal{F}$ on D such that $\lambda(D') \subset \mathcal{P}(\mathcal{F})$.

Proof: If one of these conditions is satisfied, it is seen that there does exists $f \in H(D)$, not identically zero in D, such that $f(u) = 0$ whenever $u \in \lambda(D')$. Therefore, ϕ_λ is not injective.

Now, we suppose that λ is not a constant and that ϕ_λ is not injective. Let $A = \lambda(D')$ and let $f \in Ker(\phi_\lambda)$ be not identically zero. By Theorem 21.12 A is infraconnected. Besides, since λ is not a constant, A is not reduced to a point. Hence by Theorem 38.5 there exists a T-filter $\mathcal{F}$ on D such that $A \subset \mathcal{P}(\mathcal{F})$. This ends the proof of Theorem 38.17.

Corollary 38.18: *Let $D, D' \in \mathcal{A}$ be infraconnected. Then, ϕ_λ is injective for every non constant $\lambda \in \Omega(D', D)$ if and only if D is an analytic set.*

39. INTEGRALLY CLOSED ALGEBRAS $H(D)$

Throughout this chapter, D is infraconnected and belongs to $\mathcal{A}$.

By Theorem 38.8, we know that Conditions b),c),d) are equivalent:
   b) *$H(D)$ is a principal ideal ring,*
   c) *$H(D)$ is noetherian,*
   d) *$\overline{D}$ is open with no T-filter.*

Here we will show that if D just has no T-filter, then $H(D)$ is algebraically closed. But we will show that "most of the time", when an algebra $H(D)$ admits a T-filter, then $H(D)$ is not algebraically closed.

Notations :   Let D have no T-filter. By Corollary 38.11, $H(D)$ has no divisors of zero.

Throughout the chapter, u, w will denote two elements of $H(D)$ which are not identically zero. J will denote the set of the $x \in D$ such that $w(x) = 0$.

Let $a \in \tilde{D}$ and let $r = \delta(a, D)$. If D is bounded, of diameter S (resp. unbounded), for every $\mu \in [-\log S, -\log r]$, (resp. for every $\mu \leq -\log r$), we put $v_a(\frac{u}{w}, \mu) = v_a(u, \mu) - v_a(w, \mu)$.

Lemma 39.1:   *Let D be bounded, of diameter S, let $a \in \tilde{D}$ and let $r = \delta(a, D)$. As a function in μ, $v_a(\frac{u}{w}, \mu)$ takes values in $\mathbb{R}$, is continuous and piecewise linear in $[-\log S, -\log r]$. Further, the equality $v(\frac{u(x)}{w(x)}) = v_a(\frac{u}{w}, v(x - a))$ holds in all of $(D \backslash J) \cap \Delta(a, r, S)$ except maybe in finitely many classes of finitely many circles $C(a, s_i)_{1 \leq i \leq q}$, with $r \leq s_i \leq S$.*

Proof: The statement is clearly true when $w = 1$, and when $u = w$, because D has no T-filter and therefore every element of $H(D)$ is quasi-minorated. As a consequence, it is seen that this statement remains true for $\frac{u}{w}$ in the general case.

Lemma 39.2:   *Let D be closed and bounded, with no T-filter. Let $f := \frac{u}{w}$ belong to the field of fractions of $H(D)$ and assume f to be integral over $H(D)$, ($u, w \in$*

$H(D)$). *There exists a function $\widetilde{f}$ continuous in D, with values in K, such that*
$$\widetilde{f}(x) = \frac{u(x)}{w(x)} \quad whenever \ x \in D \ such \ that \ w(x) \neq 0.$$

Proof: Obviously, $\dfrac{u}{w}$ defines a continuous function in $D \setminus J$. Since D has no T-filter, we know that the interior of J is empty and then we just have to show that for every $\alpha \in J$, $\dfrac{u}{w}$ has a limit at α when x approaches α, whiles it lies in in $D \setminus J$. Since f is integral over $H(D)$, f satisfies an equation of the form
$$f^n + a_{n-1}f^{n-1} + ... + a_0 = 0 \text{ with } a_i \in H(D) \quad (0 \leq i \leq n-1).$$
For each $x \in D$, let $P_x(Z) := Z^n + a_{n-1}(x)Z^{n-1} + ... + a_0(x)$. We can consider the n zeros $l_1, ... l_n$, of the polynomial $P_\alpha(Z)$, taking multiplicities into account. For every $\lambda \in K$, $f + \lambda$ clearly satisfies an equation of the form $y^n + b_{n-1}y^{n-1} + ... + b_0 = 0$ with $b_i \in H(D)$, $(0 \leq i \leq n-1)$, so it is equivalent to show that f or y admits a limit at α. Then the zeros of the polynomial $Q_\alpha(Z) := x^n + b_{n-1}(\alpha)Z^{n-1} + ... + b_0(\alpha)$ are the $l_j + \lambda \ (1 \leq j \leq n)$. Like in Chapter 4, we denote by $\| \, . \, \|$ the Gauss norm defined on $K[x]$ by $\left\| \sum_{j=0}^{q} c_j x^j \right\| = \max_{0 \leq j \leq q} |c_j|$. Let ε be > 0. Thanks to the continuity of the functions a_j in D, we see that there exists $\eta(\varepsilon)$ such that $\|P_x - P_\alpha\| \leq \varepsilon^n$ whenever $x \in d(\alpha, \eta)$. Then by Theorem 6.9 , given a zero β of P_x, at least one of the zeros l_j of P_α satisfies $\beta - l_j \leq \varepsilon$. In particular, it is seen that $\dfrac{u(x)}{w(x)}$ is a zero of $P_x(Z)$ and therefore, we have

$$(1) \quad \min_{1 \leq j \leq n} \left| \frac{u(x)}{w(x)} - l_j \right| \leq \varepsilon \text{ for every } x \in d(\alpha, \eta(\varepsilon)) \setminus J.$$

Now we suppose that $f(x)$ does not converge to a limit at α. Then there exists two subsequences $(\xi_m)_{m \in \mathbb{N}}$ and $(\zeta_m)_{m \in \mathbb{N}}$ in $D \setminus J$, both of limit α, such that the sequence $(f(\xi_m))_{m \in \mathbb{N}}$ converges to certain l_i, while the sequence $(f(\zeta_m))_{m \in \mathbb{N}}$ converges to another l_j (with $l_i \neq l_j$). For convenience, we will assume $i = 1$, $j = 2$. We can obviously choose $\lambda \in K$ such that $|l_1 + \lambda| < |l_2 + \lambda|$ and such that $l_1 + \lambda \neq 0$ whenever $j = 1, ..., n$. Let $l'_j = l_j + \lambda \ (1 \leq j \leq n)$. Let $s = \dfrac{1}{2} \min_{1 \leq j \leq n} |l'_j|$ and $g = f + \lambda$. There does exist $\rho > 0$ such that for every $x \in d(\alpha, \rho) \cap (D \setminus J)$, we have $\min_{1 \leq j \leq n} |g(x) - l'_j| \leq s$, and therefore, for every $x \in d(\alpha, \rho) \cap (D \setminus J)$, there exists $j(x) \in \{1, 2, ..., n\}$ such that $|g(x)| = |l'_{j(x)}|$. Thus we have
$$(2) \quad |g(x)| \in \{|l'_1|, ..., |l'_n|\} \text{ whenever } x \in d(\alpha, \rho) \cap (D \setminus J).$$
We will deduce that for every $a \in D \cap d(\alpha, \rho)$, the function in $\mu \ v_a(g, \mu)$ is constant

in the interval $[-\log\rho, +\infty[$. Let $S = diam(D)$ and let $\rho' \in]0, \rho]$. Let $h \in H(D)$. Since D has no T-filter, by Corollary 38.2 every element is quasi-minorated and by Proposition 20.5 the function $v_a(h, \mu)$ is bounded in $[-\log\rho, -\log\rho']$. So the equality $v(h(x)) = v_a(h, v(x - a))$ holds in all of $D \cap \Delta(a, \rho', \rho)$ except maybe in finitely many classes of finitely many circles $C(a, r)$, with $\rho' \leq r \leq \rho$. Since g is of the form $\dfrac{\phi}{\psi}$, with ϕ, $\psi \in H(D)$, it is seen that the equality $v(g(x)) = v_a(g, v(x - a))$ is true in all of $(D \setminus J) \cap \Delta(a, \rho', \rho)$ except in finitely many classes of finitely many circles $C(a, r)$ with $\rho' \leq r \leq \rho$. Now by (2) we see that $v_a(g, \mu)$ takes all its values in the finite set $\{v(l'_1), ..., v(l'_n)\}$ when $\mu \in [-\log\rho, -\log\rho']$. Since the function $v_a(f, .)$ is continuous, it is constant in $[-\log\rho, -\log\rho']$ and this is true for every $\rho' \in]0, \rho]$. Hence this function actually is constant in $[-\log\rho, +\infty[$.

Now, there obviously exists $t \in \mathbb{N}$ such that $\xi_t \in d(\alpha, \rho) \cap D$ and $|g(\xi_t)| = |l'_1|$. Let $\sigma = \xi_t$. We see that $v_\sigma(g, \mu) = v(l'_1)$ when μ tends to $+\infty$. But since the function $v_\sigma(g, .)$ is constant in $[-\log\rho, +\infty[$, we obtain

(3) $v_\sigma(g, \mu) = v(l'_1)$ for every $m \geq -\log\rho$.

In the same way, there exists $s \in \mathbb{N}$ such that $\zeta_s \in d(\alpha, \rho) \cap D$ and $|g(\zeta_s)| = |l'_2|$. We put $\nu = \zeta_s$ and we have

(4) $v_\nu(g, \mu) = v_\sigma(l'_2)$ when $\mu \in [-\log\rho, v(\sigma - \nu)]$.

But when $\mu \in [-\log\rho, v(\sigma - \nu)]$, we have $v_\sigma(g, \mu) = v_\nu(g, \mu)$ and then (3) and (4) are contradictory. Thus this contradiction shows the hypothesis we made on the sequences $(\xi_m)_{m \in \mathbb{N}}$ and $(\zeta_m)_{m \in \mathbb{N}}$ is false and therefore f has a limit at α. This ends the proof of Lemma 39.2.

Theorem 39.3:   *If D has no T-filter, $H(D)$ is integrally closed.*

Proof: Let $f = \dfrac{u}{w}$ be integral over $H(D)$. Then f satisfies an equation of the form

(1) $f^n + c_{n-1}f^{n-1} + ... + c_1 f + c_0 = 0$ with $c_j \in H(D)$, $(0 \leq j \leq n - 1)$.

We first suppose D closed and bounded. By Lemma 39.2, f extends to a function $\tilde{f}$ that is continuous in D. We will prove that $\tilde{f}$ actually belongs to $H(D)$. Let $\mathcal{H}$ be the set of the holes of D that are either u-holes, or w-holes, or c_i-holes for any $i = 0, ...n - 1$. Hence $\mathcal{H}$ is a countable family that we can denote by $(T_m)_{m \in \mathbb{N}^*}$. For each $m \in \mathbb{N}^*$, we put $T_m = d(a_m, r_m^-)$. Let $\tilde{D} := d(b, \ell)$. Without loss of generality we may assume that D is equal to the set $D' = d(b, \ell) \setminus \left(\bigcup_{m \in \mathbb{N}^*} T_m \right)$. Indeed, let $T = d(a, r^-)$ be a hole of D that is not one of the T_m. By Theorem 15.1, u, w, and all the c_j, belong to $H(D')$, and then we can consider $\sum_{j=0}^{n} u^j c_j w^{n-j}$. This is an element of $H(D')$ that is identically zero in D, and therefore is identically zero in T, hence finally in all of D'. It is seen that D' has no T-filter, and that u, w extend respectively to elements $\overline{u}$, $\overline{w}$ of $H(D')$, such that $\dfrac{\overline{u}}{\overline{w}}$ is integral over $H(D')$. So we will assume that the only holes of D are the T_m, $(m \in \mathbb{N}^*)$.

In $\widehat{K}$, the notations $\widehat{d}(a,r)$, $\widehat{d}(a,r^-)$, $\widehat{C}(a,r)$, are the ones given at the bottom of chapter 2. Then for each $m \in \mathbb{N}$, we put $\widehat{T}_m = \widehat{d}(a_m, r_m)$ and as in chapter 2, we put $\widehat{D} = \widehat{d}(b, \ell) \setminus \left(\bigcup_{m=0}^{\infty} \widehat{d}(a_m, r_m^-) \right)$. Since the residue class field of $\widehat{K}$ is not countable, for every $a \in \widehat{D}$, and for every $r \in |\widehat{K}| \cap \,]0, \ell]$, all the classes of the circle $\widehat{C}(b,r)$ are included in $\widehat{D}$ except a countable family of them. Therefore $\widehat{D}$ is obviously infraconnected. Also, $\widehat{D}$ is clearly closed and bounded. Next, since D has no T-filter, and since for every hole of $\widehat{D}$ its diameter is that of a certain hole of D, it is immediately seen that $\widehat{D}$ has no T-filter too.

For every set A in $\widehat{K}$, we will denote by $\widehat{H}(A)$ the set of the analytic elements on A, with respect to the field $\widehat{K}$. Since $\widehat{D}$ has no T-filter, by Corollary 38.11, $\widehat{H}(\widehat{D})$ has no divisor of zero. Since $\widehat{D} \cap \widehat{T}_m = \emptyset$, the Mittag-Leffler series of u (resp. w, resp c_i $(0 \le i \le n-1)$) converges in $\widehat{H}(\widehat{D})$ to an element $\widehat{u}$ (resp. $\widehat{w}$, resp. $\widehat{c}_i$ $(0 \le i \le n-1)$). Since $\widehat{H}(\widehat{D})$ has no divisor of zero, we can consider its field of fractions and we put $\widehat{f} = \dfrac{\widehat{u}}{\widehat{w}}$ which obviously belongs to this field of fractions. We will check that $\widehat{f}$ satisfies the equation $\sum_{j=0}^{n} \widehat{c}_j \widehat{f}^j = 0$, with $\widetilde{c}_n = 1$.

Indeed the element $\sum_{j=0}^{n} \widehat{c}_j \widehat{u}^j \widehat{w}^{n-j}(x)$ is equal to 0 in all D, hence also in every disk $\widehat{d}(a,r)$ such that $d(a,r) \subset D$. Since $\widehat{D}$ has no T-filter, by Theorem 38.7 it is an analytic set and therefore $\widehat{f}$ is identically zero in all $\widehat{D}$, i.e. $\sum_{j=0}^{n} \widehat{c}_j \left(\dfrac{\widehat{u}}{\widehat{w}} \right)^j (x) = 0$.

Now by Lemma 39.2, $\widehat{f}$ defines a continuous function $\widetilde{\widehat{f}}$ in $\widehat{D}$ with values in $\widehat{K}$, such that $\widetilde{\widehat{f}}(x) = \dfrac{\widehat{u}(x)}{\widehat{w}(x)}$ whenever $x \in \widehat{D}$ such that $\widehat{w}(x) \ne 0$. We see that $\widetilde{f}$ is induced by $\widetilde{\widehat{f}}$ in D. We put $F = \widetilde{\widehat{f}}$ and we will show that F

is the limit of a sequence (h_m) in $R(D)$ and then, we will have proven that $\widetilde{f}$ belongs to $H(D)$. Let $S := \widehat{D} \setminus \overset{\circ}{\widehat{D}}$. We fix $\epsilon > 0$. For every $\alpha \in S$, we denote by $\mathcal{N}_\alpha$ the set of the $r > 0$ such that $|F(x) - F(\alpha)| \leq \epsilon$ whenever $x \in \widehat{D} \cap d(\alpha, r^-)$. Let $\sigma(\alpha) := \sup(\mathcal{N}_\alpha)$, and let $\tau(\alpha) := d(\alpha, \sigma(\alpha^-))$. It is seen that $\sigma(\alpha)$ belongs to $\mathcal{N}_\alpha$. Let $\beta \in \tau(\alpha)$ and let $x \in \widehat{D} \cap d(\beta, \sigma(\alpha)^-)$. Then we have $|F(x) - F(\beta)| \leq \max(|F(x) - F(\alpha)|, |F(\alpha) - F(\beta)|)$. Clearly $|F(\alpha) - F(\beta)| \leq \epsilon$. Since $|x - \beta| < \sigma(\alpha)$, we have $|x - \alpha| < \sigma(\alpha)$ hence $|F(x) - F(\alpha)| \leq \epsilon$ and $|F(x) - F(\beta)| \leq \epsilon$, and therefore $\sigma(\alpha) \leq \sigma(\beta)$. This just shows that $\tau(\alpha) \subset \tau(\beta)$. But then, α belongs to $\tau(\beta)$ and then $\sigma(\alpha) \geq \sigma(\beta)$, hence finally $\sigma(\alpha) = \sigma(\beta)$.

Now we see that there exists a subfamily E of S such that $\tau(\alpha) \cap \tau(\beta) = \emptyset$ whenever $\alpha, \beta \in E$, with $\alpha \neq \beta$, and such that $\{\tau(\alpha) \mid \alpha \in E\} = \{\tau(\alpha) \mid \alpha \in S\}$. We will check that E is countable.

Indeed, as the T_m ($m \in \mathbb{N}^*$) are the only holes of D, the $\widehat{T}_m$ are the only holes of $\widehat{D}$, which is a closed set. Therefore, every point α of S is the limit of a subsequence of the sequence $(T_m)_{m \in \mathbb{N}}$. Hence for every $\alpha \in E$, $\tau(\alpha)$ does contain terms of the sequence $(\widehat{T}_m)_{m \in \mathbb{N}^*}$. As a consequence we can define an injection from E into $\mathbb{N}$ hence E is countable.

We put $D_\epsilon = \widehat{D} \setminus \left(\bigcup_{\alpha \in E} \tau(\alpha) \right)$. Let $r \in |\widehat{K}| \cap [0, \ell]$. Since the holes of $\widehat{D}$ are the $\widehat{T}_m$, any hole of D_ϵ is either a $\widehat{T}_m$, or a disk $\tau(\alpha)$, with $\alpha \in E$. Thus the set of the holes of D_ϵ is clearly countable. Now, since $\widehat{K}$ has an uncountable residue class field, $\widehat{C}(b, r)$ has a not countable set of classes which are included in D_ϵ and this finshes showing that D_ϵ is infraconnected. Of course, D_ϵ is closed and bounded. We will check that D_ϵ is also open. Let $a \in D_\epsilon$ be a point of $D_\epsilon \setminus \overset{\circ}{D}_\epsilon$. We have $\delta(a, (\bigcup_{m \in \mathbb{N}^*} T_m)) > 0$ because every point of D in the closure of $\bigcup_{m \in \mathbb{N}} T_m$ does belong to S, even though $S \cap D_\epsilon = \emptyset$. Thus there exists a sequence $(\alpha_q)_{q \in \mathbb{N}}$ in E, such that $\lim_{q \to \infty} \tau(\alpha_q) = a$. But we saw that every disk $\tau(\alpha_q)$ contains at least one disk $\widehat{T}_{m(q)}$. Then there exists a subsequence $(\widehat{T}_{m(q)})_{q \in \mathbb{N}}$ which converges to a, and therefore we have got to a contradiction again. This finishes showing that D_ϵ is open.

Let f_ϵ be the restriction of F to D_ϵ. Since $\widehat{D}$ has no T-filter , $\widehat{u}$ (resp. $\widehat{w}$) is quasi-minorated in $\widehat{H}(\widehat{D})$, hence with greater reason its restrictions to D_ϵ that we denote by $\widehat{u}_\epsilon$ (resp. $\widehat{w}_\epsilon$), is quasi-minorated in $\widehat{H}(D_\epsilon)$. Since D_ϵ is open, both $\widehat{u}_\epsilon$, $\widehat{w}_\epsilon$ are quasi-invertible in $\widehat{H}(D_\epsilon)$. Thus, $\widehat{u}_\epsilon$ (resp. $\widehat{w}_\epsilon$) factorizes in $\widehat{H}(D_\epsilon)$

in the form $P(x)g(x)$, (resp. $Q(x)h(x)$), with g (resp. h) invertible in $\widehat{H}(D_\epsilon)$ and P (resp. Q) a polynomial whose zeros lie in D_ϵ. Since $\dfrac{\widehat{u}_\epsilon}{\widehat{w}_\epsilon}$ defines a function that is continuous in $\widehat{D}$, hence in in D_ϵ, Q must divide P in $\widehat{K}[x]$ and thereby f_ϵ belongs to $\widehat{H}(D_\epsilon)$. We put $D'' := \widehat{D} \cup (\bigcup_{\alpha \in E} \tau(\alpha))$. We now consider the Mittag-Leffler series of f_ϵ in $\widehat{H}(D_\epsilon)$. It is of the form $g_\epsilon + \sum_{\alpha \in E} h_\alpha$, with $g_\epsilon \in \widehat{H}(D'')$, and

$h_\alpha \in \widehat{H}(\widehat{K} \setminus \tau(\alpha))$, with $\lim_{|x| \to \infty} h_\alpha(x) = 0$.

We will show that for every $\alpha \in E$, we have $\|h_\alpha\|_{D_\epsilon} \leq \epsilon$. Indeed let $\Lambda_{\epsilon,\alpha} = D_\epsilon \cap d(\alpha, \sigma(\alpha))$. It is easily seen that $\tau(\alpha)$ is a hole $\Lambda_{\epsilon,\alpha}$. Applying Theorem 15.1 to $f_\epsilon - F(\alpha)$ in $\widehat{H}(D_\epsilon)$, we have

(1) $\|h_\alpha\|_{D_\epsilon} = \|h_\alpha\|_{\Lambda_{\epsilon,\alpha}} \leq \|f_\epsilon - F(\alpha)\|_{\Lambda_{\epsilon,\alpha}}$.

But by definition of $\sigma(\alpha)$, we have:

(2) $\|f_\epsilon - F(\alpha)\|_{\Lambda_{\epsilon,\alpha}} \leq \epsilon$, hence

(3) $\|h_\alpha\|_{D_\epsilon} \leq \epsilon$.

On the other hand, by Theorem 15.1 again, we know that the family $(h_\alpha)_{\alpha \in E}$ tends to 0 when α runs in the countable set E. Let $h := \sum_{\alpha \in E} h_\alpha$. Then $g_\epsilon = f_\epsilon - h$.

By (3) we have

(4) $\|f_\epsilon - g_\epsilon\|_{D_\epsilon} \leq \epsilon$.

Moreover, for every $\alpha \in E$, we see that $\tau(\alpha)$ is not a g_ϵ-hole and of course is not a $(g_\epsilon - F(\alpha))$-hole. By Theorem 15.1 we know that $g_\epsilon - F(\alpha)$ extends to an element of $\widehat{H}(D'')$. Now by (2) and (4) we have

(5) $\displaystyle \lim_{\substack{|x-a| \to \rho(\alpha) \\ |x-a| > \rho(\alpha), x \in D_\epsilon}} |g_\epsilon(x) - f(\alpha)| \leq \epsilon$.

But since g_ϵ belongs to $H(D'')$, and $\tau(\alpha) \subset D''$, we know that

$$\|g_\epsilon - F(\alpha)\|_{\tau(\alpha)} = \lim_{\substack{|x-a| \to \rho(\alpha) \\ |x-a| \neq \rho(\alpha), x \in D''}} |g_\epsilon(x) - F(\alpha)| \leq \epsilon.$$

Then by (5) we obtain

(6) $\|g_\epsilon - F(\alpha)\|_{\tau(\alpha)} \leq \epsilon$.

But by hypothesis we have $|F(x) - F(\alpha)| \leq \epsilon$ whenever $x \in \tau(\alpha) \cap \widehat{D}$, and then by (6) we obtain

(7) $|g_\epsilon - F(\alpha)| \leq \epsilon$ whenever $x \in \tau(\alpha) \cap \widehat{D}$.

This is true for every α and then by (4) and (7) we see that $|g_\epsilon - F(\alpha)| \leq \epsilon$ whenever $x \in \widehat{D}$.

Since ϵ was arbitrarily chosen and since g_ϵ belongs to $\widehat{H}(\widehat{D})$ we see that F also belongs to $\widehat{H}(\widehat{D})$.

Now it only remains to show that the restriction $\widetilde{f}$ of F to D belongs to $H(D)$. We consider the Mittag -Leffler series of F with respect to $\widehat{D}$. The F-holes of $\widehat{D}$ are the $\widehat{T}_m$. According to the notations of Chapter 15, F is of the form $\overline{F}_0 + \sum_{m=1}^{\infty} \overline{F}_{\widehat{T}_m}$. For convenience we put $F_m = \overline{F}_{\widehat{T}_m}$.

First we will prove that for each $m \in \mathbb{N}^*$, the restriction of F_m to $K \setminus T_m$ is an element of $H(K \setminus T_m)$. We fix $m \in \mathbb{N}^*$. Let $(V_i)_{i \in I}$ be the family of the classes of $\widehat{C}(a_m, r_m)$ such that $V_i \cap K \neq \emptyset$ and let $\Sigma = \widehat{C}(a_m, r_m) \setminus (\bigcup_{i \in I} V_i)$. Since the residue class field of $\widehat{K}$ is uncountable, Σ is equal to the union of an uncountable family of classes of $\widehat{C}(a_m, r_m)$, and then the holes of Σ are the V_i, $(i \in I)$. Hence $\widehat{f}$ has a restriction $\overset{\circ}{f}$ to Σ which obviously belongs to $\widehat{H}(\Sigma)$. Moreover, Theorem 15.1 applied to $\overset{\circ}{f}$, with respect to the infraconnected set Σ, gives a term $\overset{\overset{\approx}{\circ}}{f}_{T_m}$ which actually is equal to F_m because T_m is a hole of $\widehat{D}$. Since $\Sigma \cap K = \emptyset$, both u, w have no zero in Σ and therefore satisfy

    (8) $v(\widehat{u}(x)) = v_a(\widehat{u}, -\log r_m)$, whenever $x \in \Sigma$

    (9) $v(\widehat{w}(x)) = v_a(\widehat{w}, -\log r_m)$, whenever $x \in \Sigma$.

By (9), it is seen that $v_a(\widehat{w}, -\log r_m)$ does belong to $|\widehat{K}|$. Thus we can obviously assume

    (10) $v_a(\widehat{w}, -\log r_m) = 0$.

We fix $\epsilon > 0$ satisfying

    (11) $-\log \epsilon > \max(v_a(\widehat{u}, -\log r_m), 0)$.

Let ϕ, ψ belong to $R(D)$ and satisfy $\|u - \phi\|_D \leq \epsilon$, $\|w - \psi\|_D \leq \epsilon$. By (11) and by the properties of the function $v_a(\ .,\mu)$ we see that

$$v_a(u - \phi, -\log r_m) > -\log \epsilon \quad \text{and} \quad v_a(w - \psi, -\log r_m) > -\log \epsilon \quad \text{hence by}$$

(8), (9), (10) we have $v\left(\dfrac{u}{w}(x) - \dfrac{\phi}{\psi}(x)\right) > -\log \epsilon$ whenever $x \in \Sigma$. So we obtain

    (12) $\left\| \overset{\circ}{f} - \dfrac{\phi}{\psi} \right\|_\Sigma \leq \epsilon$.

Next, by considering the Mittag -Leffler series of $\dfrac{\phi}{\psi}$ with respect to D, we put

$$\theta = \overline{\left(\dfrac{\phi}{\psi}\right)}_{T_m}.$$ Then θ is equal to the Mittag-Leffler term of $\dfrac{\phi}{\psi}$ associated to $\widehat{T}_m$, with respect to the infraconnected set $\widehat{D}$, as well as with respect to the

infraconnected set Σ, because T_m and $\widehat{T}_m$ have the same diameter.

By Theorem 15.1 we see that $\left\|F_m - \theta\right\|_\Sigma \leq \left\|\overset{\circ}{f} - \dfrac{\phi}{\psi}\right\|_\Sigma$ hence by (12) we obtain

(13) $\quad \|F_m - \theta\|_\Sigma \leq \epsilon.$

But $F_m - \theta$ belongs to $\widehat{H}(\widehat{K} \setminus \widehat{T}_m)$ hence we have

$$-\log\|F_m - \theta\|_\Sigma = v_a(F_m - \theta, -\log r_m) = -\log\|F_m - \theta\|_{\widehat{K}\setminus\widehat{T}_m},$$

and then, as $K \setminus T_m$ is infraconnected in $\widehat{K} \setminus \widehat{T}_m$, we also have $-\log\|F_m - \theta\|_{K\setminus T_m} = v_a(F_m - \theta, -\log r_m)$. Finally by (13) we have proven that $\|F_m - \theta\|_{K\setminus T_m} \leq \epsilon.$

Since ϵ has been taken arbitrary, we see that F_m belongs to $H(K \setminus T_m)$, and this is true for every $m \in \mathbb{N}^*$.

Now, we just have to show that F_0 also belongs to $H(D)$. Let $(W_i)_{i\in J}$ be the family of the classes of the circle $\widehat{C}(b, \ell)$ such that $W_i \cap K = \emptyset$. We put
$$\Sigma' = \widehat{C}(b, \ell) \setminus \left(\bigcup_{i\in J} W_i\right).$$

Then u, w satisfy

(14) $\quad v(\widetilde{u}(x)) = v_b(\widetilde{u}, -\log \ell)$ whenever $x \in \Sigma'$

(15) $\quad v(\widetilde{w}(x)) = v_b(\widetilde{w}, -\log \ell)$ whenever $x \in \Sigma'.$

As previously, we can assume $v_b(\widetilde{w}, -\log \ell) = 0$. We take again ϕ, ψ. In the same way as when proving (12), here by (14), (15), we check that they satisfy

(16) $\quad \left\|\overset{\circ}{f} - \dfrac{\phi}{\psi}\right\|_{\Sigma'} \leq \epsilon.$

Now, with respect to the infraconnected set D, $\overline{\left(\dfrac{\phi}{\psi}\right)_0}$ is a power series T and then by Theorem 15.1 we have

$$\left\|F_0 - T\right\|_{\Sigma'} \leq \left\|\overset{\circ}{f} - \dfrac{\phi}{\psi}\right\|_{\Sigma'} \leq \epsilon.$$

But at the same time we have
$$-\log\|F_0 - T\|_{\Sigma'} = v_b(F_0 - T, -\log R) = -\log(\|F_0 - T\|_{d(b,\ell)})$$

and this finishes proving that F_0 belongs to $H(d(b,\ell))$. Thus the conclusion of Theorem 39.3 has now been proven when D is closed and bounded.

Finally we just suppose $D \in \mathcal{A}$. If D is not bounded and has no hole, by Corollary 10.8 we have $H(D) = R(D)$ and therefore $H(D)$ is a principal ideal ring. If D is not bounded but has at least one hole $T = d(a, r^-)$, then an inversion σ of center a maps D onto a bounded infraconnected set D^*, such that $H(D)$ is isomorphic to $H(D^*)$. Thus, without loss of generality, we can assume that D

is bounded. So, $\overline{D}$ is closed and bounded. Then by Corollary 10.12 , $H(D)$ is of the form $S(D)^{-1}H(\overline{D})$, with $S(D)$ the set of the polynomials whose zeros lie in $\overline{D} \setminus D$. As we just saw, $H(\overline{D})$ is integrally closed and therefore $H(D)$ is seen to be integrally closed too. This ends the proof of Theorem 39.3.

Example 1:   Let $(a_n)_{n \in \mathbb{Z}}$ be a sequence satisfying

$$(1) \quad \lim_{n \to -\infty} a_n = 0, \quad (2) \quad \lim_{n \to +\infty} |a_n| = 1, \quad (3) \quad \prod_{n=1}^{+\infty} |a_n| > 0.$$

Let $(r_n)_{n \in \mathbb{Z}}$ be a sequence in $\mathbb{R}_+$ such that $0 < r_n \leq |a_n|$, for every $n \in \mathbb{Z}$ and
$$(4) \quad 0 < \lim_{n \to +\infty} r_n < 1.$$

Let $s \geq 1$ and let $D = d(0, s) \setminus \left(\bigcup_{n \in \mathbb{Z}} d(a, r_n^-) \right)$.

The only holes of D are the disks $T_n = d(a_n, r_n^-)\,(n \in \mathbb{Z})$. In particular D admits a Cauchy pierced filter of limit 0, and an increasing pierced filter $\mathcal{F}$ of center 0, of diameter s, and obviously has no other pierced filter. But now, by (3) and (4), and by Proposition 36.5, there exists no T-sequence of the form $(T_n, q_n)_{n \in \mathbb{N}}$, and therefore $\mathcal{F}$ is not a T-filter. Hence D has no T-filter. However D is not open because 0 is not interior to D. Hence, by Corollary 18.11 $H(D)$ is not noetherian. Thus Theorem 39.3 fully applies and shows $H(D)$ to be integrally closed, although it is not noetherian.

Example 2:   In Theorem 36.8 we constructed a closed bounded infraconnected set D with no T-filter, whose interior is empty. Thus $H(D)$ is integrally closed, although no point of D is interior to $\overset{\circ}{\overline{D}}$.

Now we will show algebras $H(D)$ that are not integrally closed.

Lemma 39.4:   *Let D have an increasing T-filter $\mathcal{F}$ of center 0 and diameter r and let $s \in \mathbb{R}$ be such that $r \leq s \leq diam(D)$. Let $f \in A_b(d(0, s^-))$ and let $g \in H_b(D)$ satisfy $\lim_{\mathcal{F}} g(x) = 0$ and $h(x) = 0$ whenever $x \in \mathcal{P}(\mathcal{F})$. The function h defined in D as $h(x) = g(x)f(x)$ whenever $x \in D \cap d(0, s^-)$ and $h(x) = 0$ whenever $x \in D \setminus d(0, s^-)$, belongs to $H(D)$.*

Proof: Let $f(x) = \sum_{n=0}^{\infty} a_n x^n$, let $E = \|f\|_{d(0, s^-)}$, and for every $n \in \mathbb{N}$, let $f_n = \sum_{j=0}^{n} a_j x^j$. We fix $\epsilon \in]0, 1[$. There exists $\rho \in]0, s[$ such that

(1) $\quad |g(x)| \leq \dfrac{\epsilon}{E}$ whenever $x \in D \setminus d(0, \rho)$.

Now let $B = \max(\|g\|_D, 1)$, and let $N \in \mathbb{N}$ be such that $\|f - f_n\|_{d(0,\rho-)} \leq \dfrac{\epsilon}{B} < E$ whenever $n \geq N$. Then, taking $n \geq N$, we have $|g(x)f(x) - g(x)f_n(x)| \leq \epsilon$ whenever $x \in d(0, \rho) \cap D$, and then by (1) we see that $|g(x)f(x) - g(x)f_n(x)| \leq \dfrac{\epsilon}{E}|f(x) - f_n(x)|$. But by the definition of E, we have $|a_j x^j| \leq E$ whenever $j \in \mathbb{N}$, hence $\|f_n\|_{d(0,s-)} \leq E$, and therefore we obtain

(2) $\quad |g(x)f(x) - g(x)f_n(x)| \leq \epsilon$ whenever $x \in d(0, s^-) \setminus d(0, \rho)$.

Next, outside $d(0, s^-)$, we trivially have

(3) $\quad h(x) = g(x)f_n(x) = 0$ whenever $x \in D \setminus d(0, s^-)$.

Thus by (1), (2), (3) we see that $\|h - gf_n\|_D \leq \epsilon$ and therefore h belongs to $H(D)$.

Theorem 39.5: *Let $D \in \mathcal{A}$ be such that $H(D)$ has no divisor of zero. If D has a T-filter $\mathcal{F}$ with a center $a \notin \mathcal{P}(\mathcal{F})$, satisfying at least one of the two following conditions α), β) :*

α) $\quad \mathcal{P}(\mathcal{F}) \neq \emptyset$,

β) $\quad diam(\mathcal{F}) \in |K|$,

then $H(D)$ is not integrally closed.

Proof: Let r be the diameter of $\mathcal{F}$. First we suppose $\mathcal{F}$ increasing, of center a. Since one of the conditions α), β) is satisfied, it is easily seen that there exists $s \in |K|$ such that $r \leq s \leq diam(D)$. Then there exists a mapping ϕ of the form $\lambda x + \nu$, with $\lambda, \nu \in K$, such that $\phi(a) = 0$, and $\phi(d(a, s)) = d(0, 1)$. Let $D' = \phi(D)$. Since the algebras $H(D)$ and $H(D')$ are isomorphic, we can clearly assume $a = 0$ and $s = 1$ without loss of generality. So we have $r \leq 1 \leq diam(D)$. Let $B = d(0, 1^-)$ and let $q \in \mathbb{N}$ be superior or equal to 2, and prime to the residue characteristic p of K in the case when $p \neq 0$. By Theorem 33.7 we know that ψ_q belongs to $A_b(B) \setminus H(B)$. Let $D'' = D \cap B$. We can easily check that ψ_q does not belong to $H(D'')$. Indeed, suppose it does. Since ψ_q is bounded in B, it obviously belongs to $H(\overline{D''})$. But for every hole T of D'' included in B, we have $\overline{(\psi_q)}_T = 0$ because ψ_q obviously belongs to $H(d(0, \rho))$ for every $\rho \in]0, 1[$. Thereby ψ_q belongs to $H(B)$, which is false. This finishes the proof of $\psi_q \notin H(D'')$.

Let $g \in H_b(D)$ satisfy $\lim_{\mathcal{F}} g(x) = 0$ and $g(x) = 0$ whenever $x \in \mathcal{P}(\mathcal{F})$. By Lemma 39.4 the function h defined in D as $h(x) = g(x)\psi_q(x)$ whenever $x \in D \cap B$ and $h(x) = 0$ whenever $x \in D \setminus B$, belongs to $H_b(D)$. Since $H(D)$ has no divisor of zero we can consider its fractions field L. Since $(\psi_q(x))^q = 1 + x$, in $H(D)$ we have $h(x)^q = g(x)^q(1 + x)$ because this relation by definition holds for all

$x \in D \cap B$ and is trivial in $D \setminus B$. Hence in L we have $\left(\dfrac{h(x)}{g(x)}\right)^q = 1 + x$.

But then, if $H(D)$ is integrally closed, we see that $\dfrac{h}{g}$ belongs to $H(D)$, and in particular to $H(D'')$, which is false as we have seen. Hence we have proven that $H(D)$ is not integrally closed when $\mathcal{F}$ is increasing.

Suppose now that $\mathcal{F}$ is decreasing with a center $a \notin \mathcal{P}(\mathcal{F})$. Then a does not belong to D. An inversion of center a maps D onto a set D^* which has an increasing T-filter $\mathcal{F}^*$, and then by Proposition 11.8 $H(D)$ is isomorphic to $H(D^*)$. From above, $H(D^*)$ is not integrally closed, hence $H(D)$ is not either. This ends the proof of Theorem 39.5.

Remark: Theorem 39.5 shows that many T-filters provide us with algebras $H(D)$ that are not integrally closed. So, we might think that an algebra $H(D)$ with no divisor of zero is integrally closed if and only if D has no T-filter. Actually this is far from clear.

Indeed ψ_q is the only one function we have found to be integral over $H(D)$ without lying in $H(D)$. Yet, this no longer holds when one of the hypothesis of Theorem 39.5 is not satisfied. The strangest is that if $\mathcal{F}$ is a decreasing T-filter such that $\mathcal{P}(\mathcal{F})$ is a full disk $d(a, r)$, we can't adapt the proof made when $\mathcal{F}$ is increasing because then, the function $\dfrac{1 + x}{x}$ (which is the composition $(1+x) \circ \dfrac{1}{x}$) does not belong to $H(D)$. This escaped our attention in [32], Theorem 2, and then this puts in doubt Corollary 2. We could characterize the integrally closed algebras $H(D)$ if, for example, we found elements f bounded in D, such that $\sqrt[q]{f}$ does not belong to $H(D)$ but does belong to the fractions field of $H(D)$.

40.   ABSOLUTE VALUES ON $H(D)$

Throughout this chapter, D is infraconnected and belongs to $\mathcal{A}$.

In Chapter 12 we characterized the continuous multiplicative semi-norms of $H(D)$ by means of the circular filters on $H(D)$. Here, assuming $H(D)$ to be a K-algebra, we study which continuous multiplicative semi-norms of $H(D)$ are norms, i.e. are absolute values on $H(D)$. This is a common work with Kamal Boussaf [6]. Of course, this requires $H(D)$ to have no divisors of zero. But then, as a trenscendental extension of the field K, the field of fractions L of $H(D)$ does admit absolute values extending that of K. Hence so does $H(D)$. The problem here is whether such absolute values are continuous with respect to the topology of $H(D)$, i.e. are defined by circular filters on D. So, we will characterize the condition that a circular filter has to satisfy in order that its continuous multiplicative semi-norm is an absolute value, and next, we will characterize the sets D such that at least one of the continuous multiplicative semi-norm is an absolute value.

Theorem 40.1:   *Let $\mathcal{F}$ be a large circular filter on D. Then $_D\varphi_\mathcal{F}$ is not an absolute value if and only if it satisfies one of these conditions:*

a)   *There exists a T-filter $\mathcal{G}$ on D such that $\mathcal{F}$ is secant with $\mathcal{P}(\mathcal{G})$,*

b)   *$\mathcal{F}$ is T-filter.*

Proof :   First, suppose that there exists a T-filter $\mathcal{G}$ satisfying a). By Theorem 37.2. there exists $f \in H(D)$, strictly vanishing along $\mathcal{G}$, equal to 0 in all of $\mathcal{P}(\mathcal{G})$. Hence we have $_D\varphi_\mathcal{F}(f) = 0$ and then $_D\varphi_\mathcal{F}$ is not a norm. Now if $\mathcal{F}$ is a T-filter, then there exists $f \in H(D)$ strictly vanishing along $\mathcal{F}$ and therefore we have $\lim_\mathcal{F} f(x) = 0$, hence $_D\varphi_\mathcal{F}$ is not a norm.

Now we suppose that there exists no T-filter $\mathcal{G}$ satisfying a) and that $\mathcal{F}$ is not a T-filter, and we suppose that $_D\varphi_\mathcal{F}$ is not a norm. Let $f \in H(D) \setminus \{0\}$ satisfying $_D\varphi_\mathcal{F}(f) = 0$. Let $S = diam(\mathcal{F})$. Let $b \in D$ be such that $f(b) \neq 0$.

First, we assume that $\mathcal{F}$ has a center a.

First we suppose that $b \in d(a, S)$. Since $_D\varphi_{a,r}(f) \neq 0$ when r is close to 0, there does exist $s \in]0, S[$ such that $_D\varphi_{b,s}(f) = 0$ and $_D\varphi_{b,r}(f) \neq 0$ whenever $r \in]0, s[$. Hence by Lemma 38.1 f is strictly vanishing along the increasing filter $\mathcal{G}$ of center b and diameter s, and therefore $\mathcal{F}$ is secant with $\mathcal{P}(\mathcal{G})$.

Now we suppose that $|a - b| > S$. Let $t = |a - b|$. If $_D\varphi_{b,t}(f) = 0$, there exists $s \in]0, t]$ such that $_D\varphi_{b,s}(f) = 0$ and $_D\varphi_{b,r}(f) \neq 0$ whenever $r \in]0, s[$, hence by Lemma 38.1 f is strictly vanishing along an increasing T-filter $\mathcal{G}$ of center b and diameter s and therefore $\mathcal{F}$ is secant with $\mathcal{P}(\mathcal{G})$. Now we may assume ${}_D\varphi_{b,t}(f) \neq 0$. But ${}_D\varphi_{b,t}(f) = {}_D\varphi_{a,t}(f)$ and therefore there exists $s \in [S,t[$ such that ${}_D\varphi_{a,s}(f) = 0$ and ${}_D\varphi_{a,r}(f) \neq 0$ whenever $r \in]s,t]$. Hence f is strictly vanishing along a decreasing T-filter $\mathcal{G}$ of center a and diameter s, and $\mathcal{F}$ is secant with $\mathcal{P}(\mathcal{G})$.

Now, assume that $\mathcal{F}$ is a decreasing filter with no center. Let $(D_n)_{n \in \mathbb{N}}$ be a canonical base of $\mathcal{F}$, and for each $n \in \mathbb{N}$, let $D_n = d(a_n, r_n) \cap D$ and let $u_n = {}_D\varphi_{a_n,r_n}(f)$. If there exists $q \in \mathbb{N}$ such that $u_q = 0$, by Proposition 38.5 D admits a T-filter $\mathcal{G}$ such that D_q is included in $\mathcal{P}(\mathcal{G})$, and therefore, $\mathcal{F}$ is obviously secant with $\mathcal{P}(\mathcal{G})$. Hence we can assume that $u_n > 0$ for every $n \in \mathbb{N}$. Now, suppose that there exist $q \in \mathbb{N}$ and $r \in [r_{q+1}, r_q]$ such that ${}_D\varphi_{a_{q+1},r}(f) = 0$. As we just saw, there exists a T-filter $\mathcal{G}$ on D such that circular filter $\mathcal{F}_q$ is secant with $\mathcal{P}(\mathcal{G})$, and therefore so is $\mathcal{F}$.

Thus, without loss of generality, we can assume that ${}_D\varphi_{a_{q+1},r}(f) > 0$ for every $r \in [r_{q+1}, r_q]$, for every $q \in \mathbb{N}$. Hence f is just strictly vanishing along the decreasing filter $\mathcal{F}$ and therefore $\mathcal{F}$ is a T-filter. This finishes proving Theorem 40.1.

Corollary 40.2:   *All the non punctual continuous multiplicative seminorms of $H(D)$ are absolute values if and only if D has no T-filter.*

Definitions and notations: Let $incT(D)$ (resp. $decT(D)$) be the set of the increasisng (resp. decreasing) T-filters on D. We will denote by $\preceq$ the relation defined on $incT(D)$ (resp. $decT(D)$) by $\mathcal{F}_1 \preceq \mathcal{F}_2$ if $\mathcal{C}(\mathcal{F}_2) \subset \mathcal{C}(\mathcal{F}_1)$. This relation is obviously seen to be an order relation on $incT(D)$ (resp. $decT(D)$) .

An increasing (resp. a decreasing) T-filter $\mathcal{F}$ will be said to be *maximal* if it is maximal in $incT(D)$ (resp. in $decT(D)$) with respect to this relation.

We will denote by $\prec$ the strict order associated to $\preceq$ by $\mathcal{F}_1 \prec \mathcal{F}_2$ if $\mathcal{F}_1 \preceq \mathcal{F}_2$ and $\mathcal{F}_1 \neq \mathcal{F}_2$.

We will call *an ascending chain of increasing (resp. decreasing) T-filters* a sequence of increasing (resp. decreasing) T-filters $(\mathcal{F}_n)_{n \in \mathbb{N}}$ such that $\mathcal{F}_n \prec \mathcal{F}_{n+1}$ whenever $n \in \mathbb{N}$.

Let $(\mathcal{F}_n)_{n \in \mathbb{N}}$ be an ascending chain of increasing T-filters. For each $n \in \mathbb{N}$ let $r_n = diam(\mathcal{F}_n)$. Since the sequence $(r_n)_{n \in \mathbb{N}}$ is decreasing, we put $r = \lim_{n \to \infty} r_n$, and then r will be named *the diameter of the chain*.

We put $A := \bigcap_{n \in \mathbb{N}} \mathcal{C}(\mathcal{F}_n)$ and for each $n \in \mathbb{N}$, $D_n := \mathcal{C}(\mathcal{F}_n) \setminus A$. The sequence $(D_n)_{n \in \mathbb{N}}$ is then a base of a filter $\mathcal{F}$ on D of diameter r.

If $r = 0$, since D belongs to $\mathcal{A}$, by Condition B), A is a point a of D, hence $\mathcal{F}$ is the filter of the neighbourhoods of a in D.

If $r > 0$, $\mathcal{F}$ is a decreasing filter on D of diameter r.

In both cases $\mathcal{F}$ will be called *the returning filter of the ascending chain* $(\mathcal{F}_n)_{n\in\mathbb{N}}$.

Now let $(\mathcal{F}_n)_{n\in\mathbb{N}}$ be an ascending chain of decreasing T-filters and let $a \in \mathcal{P}(\mathcal{F}_n)$ for some $n \in \mathbb{N}$. The sequence $(r_n)_{n\in\mathbb{N}}$ is an increasing sequence of limit $r \in]0, +\infty]$, and r will be named *the diameter of the chain*. Since D belongs to $\mathcal{A}$, by Condition A) we notice that $r < +\infty$, and then we will call *the returning filter of the ascending chain* $(\mathcal{F}_n)_{n\in\mathbb{N}}$ the increasing filter $\mathcal{F}$ of center a and diameter r (it is seen that $\mathcal{F}$ is not depending on the point $a \in \mathcal{P}(\mathcal{F}_n)$, whenever $n \in \mathbb{N}$).

Lemma 40.3: *Let $H(D)$ have no divisor of zero. Then $incT(D)$ is totally ordered with respect to the order $\preceq$.*

Proof: Suppose that $\mathcal{F}, \mathcal{G}$ are increasing T-filters on D that are not comparable. We put $A = \mathcal{C}(\mathcal{F})$ and $B = \mathcal{C}(\mathcal{G})$. Then A, B are two disks of D which satisfy neither $A \subset B$, nor $B \subset A$. Hence we have $A \cap B = \emptyset$. As a consequence, $\mathcal{P}(\mathcal{F}) \cup \mathcal{P}(\mathcal{G})$ is equal to D and then, by Corollary 38.12, $H(D)$ has divisors of zero. Hence this contradicts the hypothesis and ends the proof.

Corollary 40.4: *Let $H(D)$ have no divisor of zero, and let $\mathcal{F}$ (resp. $\mathcal{G}$) be an increasing T-filters on D, of diameter r (resp. s). If $r > s$ then $\mathcal{F} \prec \mathcal{G}$. If $r = s$, then $\mathcal{F} = \mathcal{G}$.*

We are now able to characterize the sets D such that $H(D)$ admits continuous absolute values.

Theorem 40.5: *Let $H(D)$ have no divisor of zero. Then $Mult(H(D), \mathcal{U}_D)$ contains no norm if and only if D admits an ascending chain of T-filters $(\mathcal{F}_n)_{n\in\mathbb{N}}$ whose returning filter is either a T-filter or a Cauchy filter.*

Proof : On one hand, we suppose that D admits an ascending chain of T-filters $(\mathcal{F}_n)_{n\in\mathbb{N}}$ whose returning filter $\mathcal{F}$ is either a T-filter or a Cauchy filter and we will prove that $Mult(H(D), \mathcal{U}_D)$ contains no norm. We denote by r the diameter of this ascending chain $(\mathcal{F}_n)_{n\in\mathbb{N}}$, and for each $n \in \mathbb{N}$, we denote by r_n the diameter of $\mathcal{F}_n$. Let $\mathcal{G}$ be a circular filter on D, of diameter $s > 0$. By Theorem 40.1, we only have to show that either $\mathcal{G}$ is a T-filter, or $\mathcal{G}$ is secant with the beach of a T-filter.

First we suppose $\mathcal{F}$ is a Cauchy filter. Then the $\mathcal{F}_n$ are increasing T-filters. Let $q \in \mathbb{N}$ be such that $r_q < s$. Then $\mathcal{G}$ is clearly secant with $\mathcal{P}(\mathcal{F}_q)$.

Now we suppose that $\mathcal{F}$ is a T-filter.Thus, we may assume $\mathcal{G}$ is not secant with $\mathcal{P}(\mathcal{F})$. So, we just have to consider the case when $\mathcal{G}$ is secant with $\mathcal{C}(\mathcal{F})$ and is not equal to $\mathcal{F}$.

First we suppose $\mathcal{F}$ increasing, of center b and diameter r. Hence we have $\mathcal{C}(\mathcal{F}) = d(b, r^-) \cap D$. Then we have $s \in]0, r[$, and then $\mathcal{G}$ admits an element E of diameter $t \in]s, r[$, included in $d(b, r^-) \cap D$. Since $E \cap \mathcal{C}(\mathcal{F}) \neq \emptyset$, given a point $a \in E \cap \mathcal{C}(\mathcal{F})$ we have $E \subset d(a, t)$. Let $q \in \mathbb{N}$ be such that $r_q > \max(t, |a - b|)$. Then E is included in $d(b, r_q)$. But since the $\mathcal{F}_n$ are decreasing, E is included in $\mathcal{P}(\mathcal{F}_q)$. Hence $\mathcal{G}$ is secant with $\mathcal{P}(\mathcal{F}_q)$.

Now, we suppose $\mathcal{F}$ decreasing, of diameter r. Since $\mathcal{G}$ is secant with $\mathcal{C}(\mathcal{F})$, and is not less thin than $\mathcal{F}$, we have $r < s$ and there exists $t \in]r, s[$ and $a \in D$ such that $\mathcal{G}$ is secant with $(K \setminus d(a, t)) \cap D$ while $\mathcal{F}$ is secant with $d(a, t^-) \cap D$. Let $q \in \mathbb{N}$ be such that $r_q < t$ and let a_q be a center of $\mathcal{F}_q$. Then a_q belongs to $d(a, t)$ and we have $\mathcal{P}(\mathcal{F}_q) = (K \setminus d(a, r_q^-)) \cap D$. Hence $\mathcal{G}$ is secant with $\mathcal{P}(\mathcal{F}_q)$ and this finishes showing that $Mult(H(D), \mathcal{U}_D)$ contains no norm.

On the other hand, reciprocally, we suppose that $Mult(H(D), \mathcal{U}_D)$ contains no norm and we will show that D admits an ascending chain of T-filters $(\mathcal{F}_n)_{n \in \mathbb{N}}$ whose returning filter is either a T-filter or a Cauchy filter.

We denote by $\mathcal{R}'$ the set of t diameters of the $\mathcal{F} \in incT(D)$, by $\mathcal{R}''$ the set of the diameters of the $\mathcal{F} \in decT(D)$, and we put $\mathcal{R} = \mathcal{R}' \cup \mathcal{R}''$. Since $H(D)$ has no norm, by Theorem 40.1 $\mathcal{R}$ is not empty. Since D belongs to $\mathcal{A}$, by Condition A) $\mathcal{R}$ is obviously bounded. We put $t = \sup(\mathcal{R})$. Let $a \in D$. We will show that $incT(D) \neq \emptyset$. Indeed, suppose $incT(D) = \emptyset$. First let D be bounded, of diameter S. Any decreasing filter on D has a diameter $r < S$, and therefore the circular filter $\mathcal{G}$ of center a and diameter S is secant with D, but (of course) is not a T-filter on D, and is not secant with the beach of any decreasing T-filter on D. As a consequence, by Theorem 40.1 $_D\varphi_{\mathcal{G}}$ is a norm. Thus we see that D is not bounded. Then, any circular filter $\mathcal{G}$ of center a and diameter $r > t$ is not a T-filter and is not secant with the beach of any T-filter. Finally this shows that $_D\varphi_{\mathcal{G}}$ is a norm again. Thus we see that $incT(D)$ is not empty, and neither is $\mathcal{R}'$.

Now, we put $s = \inf(\mathcal{R}')$. First we suppose $s \in \mathcal{R}'$. Let $\mathcal{T} \in incT(D)$ satisfy $diam(\mathcal{T}) = s$, and let $b \in \mathcal{C}(\mathcal{T})$. Then for every $r \in]0, s[$, the circular filter of center b and diameter r is not a decreasing T-filter, and therefore is secant with the beach of a decreasing T-filter. Hence there exists a decreasing T-filter $\mathcal{F}$ of center b and diameter $\ell \geq s$. But since $H(D)$ has no divisor of zero, $\mathcal{F}$ is not complementary with $\mathcal{T}$, hence we have $s \leq \ell$, i.e. $s \leq \ell < r$. So, we clearly deduce the existence of a sequence of decreasing T-filters $(\mathcal{F}_n)_{n \in \mathbb{N}}$, such that each one admits b as a center and has a diameter r_n satisfying $r_n < r_{n+1} < s$, $\lim_{n \to \infty} r_n = s$.

Therefore the sequence $(\mathcal{F}_n)_{n \in \mathbb{N}}$ is an ascending chain of decreasing T-filters such that $\bigcap_{n \in \mathbb{N}} \mathcal{P}(\mathcal{F}_n) = \mathcal{C}(\mathcal{T})$, hence the returning filter of the ascending chain $(\mathcal{F}_n)_{n \in \mathbb{N}}$ is a T-filter.

Now, we suppose $s \notin \mathcal{R}'$. Let $(\mathcal{T}_n)_{n \in \mathbb{N}}$ be a sequence in $incT(D)$ such that $\lim_{n \to \infty} diam(\mathcal{T}_n) = s$, with $diam(\mathcal{T}_{n+1}) < diam(\mathcal{T}_n)$ for all $n \in \mathbb{N}$. For every $n \in \mathbb{N}$, we put $A_n = \widetilde{\mathcal{C}(\mathcal{T}_n)}$, and $r_n = diam(A_n)$. By Lemma 40.3 the sequence $\mathcal{C}(\mathcal{T}_n)_{n \in \mathbb{N}}$ is strictly decreasing, and so is the sequence $(A_n)_{n \in \mathbb{N}}$. Besides, the sequence $(\mathcal{T}_n)_{n \in \mathbb{N}}$ is an ascending chain of increasing T-filters. Obviously each A_n contains a hole T_n of D.

If $s = 0$, then we have $\delta(b, T_n) \leq r_n$, and therefore b does not belong to $\overset{\circ}{\overline{D}}$, but then, by Condition A) b must belong to $\overline{D}$. Thus, the sequence $(\mathcal{T}_n)_{n \in \mathbb{N}}$ is an ascending chain of increasing T-filters that converges to b, and therefore the returning filter of the ascending chain $(\mathcal{T}_n)_{n \in \mathbb{N}}$ is a Cauchy filter.

Finally, it only remains to consider the case when $s > 0$, with $s \notin \mathcal{R}'$. Let $\mathcal{F}$ be the returning filter of the sequence $(\mathcal{T}_n)_{n \in \mathbb{N}}$. If $\mathcal{F}$ were a T-filter, or were secant with the beach of an increasing T-filter, this increasing T-filter would have a diameter inferior or equal to s. Hence $\mathcal{F}$ either is a decreasing T-filter or is secant with the beach of a decreasing T-filter. Of course, if $\mathcal{F}$ is a T-filter, it is just the returning filter of the chain $(\mathcal{T}_n)_{n \in \mathbb{N}}$. Finally if $\mathcal{F}$ is secant with the beach of a decreasing T-filter $\mathcal{G}$, then we have $diam(\mathcal{G}) \leq s$ because if $diam(\mathcal{G})$ were strictly superior to s, then $\mathcal{G}$ would be complementary to $\mathcal{T}_n$ when n is big enough, and therefore $H(D)$ would have divisors of zero. Hence we have $diam(\mathcal{G}) = s$, and therefore $\mathcal{G}$ is just the returning filter of the sequence $(\mathcal{T}_n)_{n \in \mathbb{N}}$. This finishes proving that D admits an ascending chain of T-filters $(\mathcal{F}_n)_{n \in \mathbb{N}}$ whose returning filter is either a T-filter or a Cauchy filter, and this ends the proof of Theorem 40.5.

41. DISTINGUISHED CIRCULAR FILTERS

Throughout this chapter, D is infraconnected and belongs to $\mathcal{A}$.

The study of the maximal ideals of infinite codimension will require certain particular circular filters named "distinguished circular filters". This technique was defined in [19].

Notations and definitions: If $\mathcal{F}$ is a circular filter on D, of center a and diameter r, we will denote by $\mathcal{Q}(\mathcal{F}, D)$ the set $d(a, r) \cap D$. If $\mathcal{F}$ is a circular filter with no center, we put $\mathcal{Q}(\mathcal{F}, D) = \emptyset$.

Given a monotonous filter $\mathcal{F}$, we will denote by $\check{\mathcal{F}}$ the unique circular filter less thin than $\mathcal{F}$.

A circular filter $\mathcal{F}$ will be said to *surround* a circular filter $\mathcal{G}$ (resp. a monotonous filter $\mathcal{G}$) if $\mathcal{G}$ is secant with $\mathcal{Q}(\mathcal{F}, D)$.

A monotonous filter $\mathcal{F}$ will be said to *surround* a circular filter (resp. a monotonous filter) $\mathcal{G}$ if $\check{\mathcal{F}}$ surrounds $\mathcal{G}$ (resp. $\check{\mathcal{F}}$ surrounds $\check{\mathcal{G}}$).

A monotonous filter or a circular filter $\mathcal{F}$ will be said to *strictly surround* a monotonous or a circular filter $\mathcal{G}$ if $\mathcal{F}$ surrounds $\mathcal{G}$ and if $diam(\mathcal{F}) \neq diam(\mathcal{G})$.

Lemma 41.1 is obvious:

Lemma 41.1:   *Let $\mathcal{F}, \mathcal{G}$ be circular filters on D admitting a commun center a, such that $diam(\mathcal{F}) \leq diam(\mathcal{F})$. Then $\mathcal{G}$ surrounds $\mathcal{F}$.*

Lemma 41.2:   *Let $\mathcal{F}$ be a circular filter on D, of diameter $r > 0$. For every $s \in [r, diam(D)[\cap \mathbb{R}$, there exists a unique circular filter $\mathcal{G}$ of diameter s surrounding $\mathcal{F}$.*

Proof: If $s = r$, we just take $\mathcal{G} = \mathcal{F}$. Suppose $s > r$ and let $d(a, s)$ be a disk such that $d(a, s) \cap D$ belongs to $\mathcal{F}$. Then the circular filter $\mathcal{G}$ of center a and diameter s surrounds $\mathcal{F}$. Now suppose that another circular filter $\mathcal{G}'$ of center b and diameter s also surrounds $\mathcal{F}$. Since $\mathcal{F}$ is secant with both $d(a, s) \cap D$ and $d(b, s) \cap D$ we have $|a - b| \leq s$, and therefore $\mathcal{G} = \mathcal{G}'$.

Definitions: Let $(\mathcal{F}_n)_{n \in \mathbb{N}}$ be a sequence of increasing (resp. decreasing) filters on D. The sequence will be called *a descending chain* if $\mathcal{F}_{n+1} \prec \mathcal{F}_n$ for all $n \in \mathbb{N}$ and if the sequence $(diam(\mathcal{F}_n))_{n \in \mathbb{N}}$ has a limit $r \in]0, +\infty]$.

Let $(\mathcal{F}_n)_{n \in \mathbb{N}}$ be a descending chain of increasing (resp. decreasing) filters. The number $r = \lim_{n \to \infty} diam(\mathcal{F}_n)$ will be called *the diameter of the chain*.

If the $\mathcal{F}_n$ are increasing, the increasing filter of center $a \in \mathcal{C}(\mathcal{F}_0)$ and diameter r will be called *the basement of the chain*.

If the $\mathcal{F}_n$ are decreasing we consider a sequence $(a_n)_{n \in \mathbb{N}}$ such that $a_n \in \mathcal{P}(\mathcal{F}_n) \setminus \mathcal{P}(\mathcal{F}_{n+1})$. Such a sequence is a decreasing distances sequence such that $\lim_{n \to \infty} |a_{n+1} - a_n| = r$. The decreasing filter less thin than the sequence $(a_n)_{n \in \mathbb{N}}$ will be called *the basement of the chain*.

A partition $(A_i)_{i \in I}$ of D will be said to be *T-optimal* if for each $i \in I$, there exists an increasing T-filter $\mathcal{F}_i$ on D which is not surrounded by any increasing T-filter different from $\mathcal{F}_i$, such that $A_i = \mathcal{C}(\mathcal{F}_i)$.

If D admits a T-optimal partition $\mathcal{C}(\mathcal{F}_i)_{i \in I}$, this partition will be said to be *T-specific* if for every decreasing T-filter $\mathcal{G}$ on D there exists $h \in I$ such that $\mathcal{G}$ is secant with $\mathcal{C}(\mathcal{F}_h)$).

Lemma 41.3: *If D admits a T-optimal partition, this T-optimal partition is unique.*

Proof: Suppose D admits two different T-optimal partitions $(\mathcal{F}_i)_{i \in I}$, and $(\mathcal{G}_j)_{j \in J}$. There exists $a \in D$, $h \in I$, $l \in J$ such that $a \in \mathcal{C}(\mathcal{F}_h) \cap \mathcal{C}(\mathcal{G}_l)$ with $\mathcal{F}_h \neq \mathcal{G}_l$. But as a is a center of both $\mathcal{F}_h$, $\mathcal{G}_l$, it is seen that one of these two filters surrounds the other, and this ends the proof.

Theorem 41.4: *The basement of a descending chain of T-filters is a T-filter.*

Proof: Let $(\mathcal{T}_n)_{n \in \mathbb{N}}$ be a descending chain of decreasing T-filters of diameter r, and for each $n \in \mathbb{N}$, let $r_n = diam(\mathcal{T}_n)$, and let a_n be a center of $\mathcal{T}_n$ in $\mathcal{P}(\mathcal{T}_{n+1})$. For each fixed $n \in \mathbb{N}$, let $(T^n_{m,i}, q^{*n}_{m,i})_{1 \le i \le k^{*n}_m}$, $m \in \mathbb{N}$ be a T-sequence that runs the sequence $(\mathcal{T}_n)_{n \in \mathbb{N}}$. Let

$$r^{*n}_m = \delta(a_n, T^n_{m,i}) \ \text{ (whenever } 1 \le i \le k^{*n}_m), \quad \text{let } q^{*n}_m = \sum_{i=1}^{k^{*n}_m} q^{*n}_{m,i} \quad \text{and let}$$

$e^{*n}_m = \gamma_D(a_n, r^{*n}_m, q^{*n}_m)$. Thus, by definition of the T-filters, for each fixed $n \in$

$\mathbb{N}$ for each fixed $t \in \mathbb{N}$, we have $\displaystyle \lim_{m \to \infty} e^{*n}_m \prod_{j=t}^{m-1} \left(\frac{r^{*n}_m}{r^{*n}_m}\right)^{q^{*n}_m} = 0$. Then for each

$n \in \mathbb{N}$, we fix $\sigma(n) \in \mathbb{N}$ such that $r^{*n}_m < r_n$ for all $m \ge \sigma(n)$ and we put

$\displaystyle \lambda_n = \sup_{m \ge \sigma(n)} e^{*n}_m \prod_{j=\sigma(n)}^{m-1} \left(\frac{r^{*n}_m}{r^{*n}_j}\right)^{q^{*n}_m}$. Now, for each $n \in \mathbb{N}$, we can clearly find

$\tau(n) \in \mathbb{N}$ such that

$$\lambda_{n+1} \prod_{j=\sigma(n)}^{\tau(n)} \left(\frac{r^{*n}_m}{r^{*n}_j}\right)^{q^{*n}_m} < \frac{1}{n+1}.$$

Then we will define a weighted sequence and next we will prove it to be a T-sequence.

For each $n \in \mathbb{N}$ we put $s(n) = \sum_{j=0}^{n} \tau(j) - \sigma(j) + 1$. Given $t \in \mathbb{N}$ such that $s(n) \leq t < s(n+1)$ we define a family $(T_{t,i}, q_{t,i})_{i \leq i \leq k_t,\ t \in \mathbb{N}}$ as

$$k_t = k^{*n}_{t-s(n)+\sigma(n)}, \qquad T_{t,i} = T^n_{t-s(n)+\sigma(n),i}, \qquad q_{t,i} = q^{*n}_{t-s(n)+\sigma(n),i} \qquad \text{and}$$

$$q_t = q^{*n}_{t-s(n)+\sigma(n)} = \sum_{i=1}^{k_t} q_{t,i}. \ \text{Besides we put } r_t = r^{*n}_{t-s(n)+\sigma(n)} \ \Sigma_t = C(a_{n+1}, r_t)$$

and $e_t = \gamma_D(a_{n+1}, r_t, q_t)$.   Then we can check that the sequence $(T_{t,i}, q_{t,i})_{1 \leq i \leq k_t, t \in \mathbb{N}}$ is a T-sequence. Indeed when $s(n) \leq t < s(n+1)$ we have

$$e_t \prod_{j=0}^{t} \left(\frac{r_j}{r_t}\right)^{q_j} \ \leq \ e_t \prod_{j=s(n-1)}^{t} \left(\frac{r_t}{r_j}\right)^{q_j} \leq$$

$$\leq e_t \prod_{j=\sigma(n-1)}^{\tau(n-1)} \left(\frac{r^{*n\cdot}_{\sigma(n)+t-s(n)}}{r_j}\right)^{q_j^{*n-1}} \prod_{j=\sigma(n)}^{\sigma(n)+t-s(n)} \left(\frac{r^{*n}_{\sigma(n)+t-s(n)}}{r_j^{*n}}\right)^{q_j^{*n}} \ \leq$$

$$\leq \ e_t \prod_{j=\sigma(n-1)}^{\tau(n-1)} \left(\frac{r^{*n}_{\sigma(n)+t-s(n)}}{r_j}\right)^{q_j^{*n-1}} \leq \frac{1}{n}.$$

Since the sequence $(s(n))_{n \in \mathbb{N}}$ obviously tends to $+\infty$, it is seen that the sequence $(T_{t,i}, q_{t,i})_{1 \leq i \leq k_t, t \in \mathbb{N}}$ is a T-sequence.

In the same way, when considering a descending chain of increasing T-filters, we have a symmetric proof.

Proposition 41.5:   *Let D admit a covering of the form $C(\mathcal{F}_i)_{i \in I}$, with $\mathcal{F}_i$ increasing T-filters of D. Then D admits a T-optimal partition.*

Proof: For each $a \in D$, we denote by $\mathcal{S}_a$ the set of the increasing T-filters on D, of center a, and we put $s(a) = \sup\{diam(\mathcal{F}) \mid \mathcal{F} \in \mathcal{S}_a\}$. We check that the increasing filter of center a and diameter $s(a)$, that we can denote by $\mathcal{G}_a$, is a T-filter. Indeed, if it were not a T-filter, there would be a descending chain of increasing T-filters $(\mathcal{T}_n)_{n \in \mathbb{N}}$ of basement $\mathcal{G}_a$, but then by Theorem 41.4, $\mathcal{G}_a$ is a T-filter. Then it is seen that $C(\mathcal{G}_a) = C(\mathcal{G}_b)$ if and only if $b \in C(\mathcal{G}_a)$, and $C(\mathcal{G}_a) \cap C(\mathcal{G}_b) = \emptyset$ if and only if $b \notin C(\mathcal{G}_a)$. As a consequence, the set $\{C(\mathcal{G}_a) \mid a \in D\}$ forms a partition of E, and by definition each $\mathcal{G}_a$ is not surrounded by any other increasing T-filter. Hence the partition is T-optimal.

Notations and definitions: A circular filter $\mathcal{F}$ on D will be said to be *distinguished* if it satisfies the following two statements:

    i) Either $\mathcal{Q}(\mathcal{F}, D)$ is empty, or it admits a T-specific partition.

    ii) Either $\mathcal{Q}(\mathcal{F}, D) = D$, or D admits a decreasing T-filter thinner than $\mathcal{F}$.

Let $\mathcal{F}$ be a distinguished circular filter on D such that $\mathcal{Q}(\mathcal{F}, D) \neq \emptyset$ and let $\mathcal{C}(\mathcal{F}_i)_{i \in I}$ be the T-specific partition of $\mathcal{Q}(\mathcal{F}, D)$. The family of the increasing T-filters $(\mathcal{F}_i)_{i \in I}$ will be called *the T-family of $\mathcal{F}$*, and the $\mathcal{F}_i$ $(i \in I)$ will be called *the increasing T-filters of $\mathcal{F}$*.

A distinguished circular filter $\mathcal{F}$ on D will be said to be *regular* if

    either $\mathcal{Q}(\mathcal{F}, D) = \emptyset$,

    or $\mathcal{Q}(\mathcal{F}, D) \neq \emptyset$, and all the increasing T-filters of $\mathcal{F}$ have the same diameter.

A distinguished circular filter $\mathcal{F}$ on D that is not regular will be said to be *irregular*.

Let D have a distinguished circular filter $\mathcal{F}$. For every $a \in \mathcal{Q}(\mathcal{F}, D)$, let $\mathcal{R}_a$ be the set of the increasing T-filters of center a, secant with $\mathcal{Q}(\mathcal{F}, D)$. Let $r_a = \sup\{(diam(\mathcal{F}) | \mathcal{F} \in \mathcal{R}_a\}$. By Proposition 41.4 , the increasing filter of center a and diameter r_a, is a T-filter that we will denote by $\mathcal{T}_a$.

Lemma 41.6 is just an obvious remark about the definition of the regular distinguished circular filters.

Lemma 41.6: *Let $\mathcal{F}$ be a regular distinguished circular filter of center a and diameter r such that $\mathcal{Q}(\mathcal{F}, D) \neq \emptyset$, and let $(\mathcal{G}_i)_{i \in J}$ be its T-family. For each $i \in J$, $\mathcal{G}_j$ has a diameter equal to r, and is secant with a class Λ_i of $d(a, r)$. Further, the mapping ψ from J into the set of the classes Λ of $d(a, r)$ such that $\Lambda \cap D \neq \emptyset$ defined as $\psi(i) = \Lambda_i$ is a bijection.*

We will use farther this elementry lemma in topology.

Lemma 41.7: *Let L be a metric space whose distance is denoted by σ and let $(A_i)_{i \in I}$ be a family of closed sets in L. If there exists $r > 0$ such that $\sigma(A_i, A_j) \geq r$ for every $i, j \in I$, such that $i \neq j$, then $\bigcup\limits_{i \in I} A_i$ is closed in L.*

Proof: Let $(x_n)_{n \in \mathbb{N}}$ be a Cauchy sequence in $\bigcup\limits_{i \in I} A_i$ and let $N \in \mathbb{N}$ be such that $\sigma(x_n, x_q) < r$ whenever $n \geq N, q \geq N$. Let $h \in I$ be such that x_N belongs to A_h. Then it is seen that $x_n \in A_h$ for all $n \geq N$, and therefore the sequence $(x_n)_{n \in \mathbb{N}}$ converges in A_h.

Theorem 41.8:   *If K is strongly valued, every distinguished circular filter is regular. If K is weakly valued, there exist closed bounded infraconnected sets with an irregular distinguished circular filter.*

Proof: First we suppose K strongly valued and we suppose that D admits an irregular distinguished circular filter $\mathcal{F}$ of diameter t. Hence $\mathcal{Q}(\mathcal{F}, D)$ admits a T-specific partition $\mathcal{C}(\mathcal{F}_i)_{i \in I}$ with a certain $\mathcal{F}_h$ of diameter $s \in]0, t[$. Let a be a center of $\mathcal{F}_h$. For every $\ell \in]s, t[\cap |K|$, for every $b \in C(a, \ell) \cap D$, b belongs to a certain set of the form $\mathcal{C}(\mathcal{F}_j)$ included in $C(a, \ell)$ because by hypothesis we have $\mathcal{C}(\mathcal{F}_j) \cap \mathcal{C}(\mathcal{F}_h) = \emptyset$. Hence $d(b, \ell^-)$ does contain holes of D. This way, we see that for every $\ell \in]s, t[\cap |K|$, every class of $C(a, \ell)$ contains at least one hole of D. Therefore, by Theorem 36.10, D admits an increasing and a decreasing T-filter of center a and diameter ℓ. Thus, this contradicts the hypothesis "$\mathcal{F}$ is distinguished ".

Now we suppose K weakly valued and we will construct an infraconnected set E with an irregular distinguished circular filter.

First, we will construct a model of sets $\Lambda(u)$ for $u \in]0, 1[$ that will be used in the construction of E. Let $(a_n)_{n \in \mathbb{N}}$ be a sequence in K satisfying for all $n \in \mathbb{N}^*$

$$(1) \qquad \frac{u}{\sqrt{n+1}} < v(a_n) \le \frac{u}{\sqrt{n}}$$

We notice that the sequence $v(a_n)_{n \in \mathbb{N}}$ is strictly decreasing, and obviously positive. Next, for every $n \in \mathbb{N}$, we put $\rho_n = \omega^{\frac{-u\sqrt{n}}{2}}$, $T_n = d(a_n, \rho_n^-)$, and $\Lambda(u) = d(0, 1^-) \setminus \left(\bigcup_{n=0}^{\infty} T_n \right)$. Thus by definition, the only holes of $\Lambda(u)$ which are included in $d(0, 1^-)$ are the T_n and then $\Lambda(u)$ admits a simple pierced filter $\mathcal{T}$ of center 0 and diameter 1. We are going to check that $\mathcal{T}$ is a T-filter. For each $n \in \mathbb{N}$, we put $d_n = |a_n|$. By (1) we check

$$(2) \qquad \sum_{j=1}^{n-1} \log\left(\frac{d_n}{d_j}\right) \ge \sum_{j=1}^{n-1} \frac{u}{\sqrt{j+1}} - \frac{nu}{\sqrt{n}} \ge u(\sqrt{n} - 2\sqrt{2})$$

Hence we have

$$\sum_{j=1}^{n-1} \log\left(\frac{d_n}{d_j}\right) - \log\left(\frac{d_n}{\rho_n}\right) \ge u(\sqrt{n} - 2\sqrt{2}) - \frac{u}{\sqrt{n}} + \frac{u\sqrt{n}}{\sqrt{2}}$$

and then by (2) it is seen that

$$\lim_{n \to \infty} \left(\sum_{j=1}^{n-1} \log\left(\frac{d_n}{d_j}\right) - \log\left(\frac{d_n}{\rho_n}\right) \right) = +\infty$$

Since $\mathcal{T}$ is a simple increasing filter, this is sufficient to prove that it is a T-filter.

Now for $j = 1, \ldots, q$ let $e_j \in T_{n_j}$ with the T_{n_j} not necessarily all different, and let $P(x) = \prod_{j=1}^{q} (x - e_j)$. We want to consider $\left\| \frac{1}{P} \right\|_{\Lambda(u)}$.

First we assume that at least one index ℓ satisfies $n_\ell \geq q$, and then clearly we have

$$\log\Big\|\frac{1}{P}\Big\|_{\Lambda(u)} \geq \Big\|\frac{1}{x - e_\ell}\Big\|_{\Lambda(u)} = -\log(\rho_{n_\ell}) \geq \frac{u\sqrt{q}}{2}$$

Now we assume that $n_j < q$ for all $j = 1, \ldots, q$. Then when $|x| \leq |a_q|$ we have $|P(x)| \leq |a_q|^q$ hence $\log\big\|\frac{1}{P}\big\|_{\Lambda(u)} \geq uq\log(d_q) \geq u\sqrt{q}$. Thus in both cases we have proven $\log\big\|\frac{1}{P}\big\|_{\Lambda(u)} \geq \frac{u\sqrt{q}}{2}$. As a consequence we have

$$(3) \quad \vartheta(\Lambda(u), q) \geq \frac{u\sqrt{q}}{2}$$

Now since $|K|$ is countable, there exists a bijection b from $|K|$ onto $\mathbb{N}$. For every $r \in |K|$ we put $\phi(r) = 4^{b(r)}$, and $B(r) = b(r)^{\phi(r)}$. It is seen that B is injective. The set E we want to construct should satisfy $\log(\gamma_E(0, r, q)) \geq B(r)\sqrt{q}$ for every $r \in]\frac{1}{2}, 1[\cap|K|$ and $q \in \mathbb{N}$. We fix $r \in]0, \frac{1}{2}[\cap|K|$. Since $\mathcal{K}$ is countable we can find a sequence $(\beta_n)_{n\in\mathbb{N}^*}$ in $C(0, r)$ such that the family of disks $(d(\beta_n, r^-))_{n\in\mathbb{N}}$ makes a partition of $C(0, r)$.

For each $n \in \mathbb{N}^*$ we put $\psi_{n,r}(x) = \beta_n(1 + x)$. For every $u \in]0, 1[$ it is seen that the set $\psi_{n,r}(\Lambda(u))$ is an infraconnected subset of $d(\rho_n, r^-)$ which admits an increasing T-filter of center β_n and diameter r.

Now we put $A_{n,r} := \psi_{n,r}(\Lambda(2^{n+2}B(r))), (n \in \mathbb{N})$, $E_r = \bigcup_{n\in\mathbb{N}} A_{n,r}$, and finally $I =]\frac{1}{2}, 1[\cap|K|$, and $E = \bigcap_{r\in I} E_r$. Then E is obviously bounded. Besides, by Lemma 41.7 it is clearly closed because $\delta(A_{r,n}, A_{r',n'}) \geq \frac{1}{2}$ for all $(r, n) \neq (r', n')$.

Finally we check that E is infraconnected. Let $a \in E$, and let (n, r) be the couple such that $a \in A_{r,n}$. If $u \in]0, r[\cap|K|$, there exists $x \in A_{n,r}$ such that $|x - a| = u$. If $u \in [r, 1[\cap|K|$, then any $x \in A_{n,u}$ satisfies $|x - a| = u$. We fix $r \in I$, $q \in \mathbb{N}$ and $\varepsilon > 0$. We will study $\gamma_E(0, r, q)$. By Corollary 34.3 we can find indices $n_1, \ldots, n_k \in \mathbb{N}$ together with strictly positive integers $q_1, \ldots q_k$ satisfying

$$(4) \quad \sum_{j=1}^{k} q_j = q, \quad \text{and}$$

$$(5) \quad \log(\gamma_E(0, r, q)) + \varepsilon \geq \log(\vartheta(A_{r,n_j},j))$$

By Proposition 34.1 we know that for every $s \in \mathbb{N}$ we have $\vartheta(A_{r,n}, s) = \vartheta(\Lambda(2^{n+2}B(r)), s)$. Hence by (3) we have

$$(6) \quad \vartheta(A_{r,n}, s) \geq 2^{n+1}B(r)\sqrt{s}.$$

Hence by (5), (6), we obtain

$$(7) \quad \log(\gamma_E(0, r, q)) + \varepsilon \geq 2^{n_j+1}\sqrt{q_j}B(r) \quad \text{whenever } j = 1, \ldots, k.$$

From (7) we can deduce

$$\sum_{j=1}^{k} 2^{-n_j-1}\left(\frac{\log(\gamma_E(0,r,q))+\varepsilon}{B(r)}\right)^2 \geq \sum_{j=1}^{k} q_j \; = \; q.$$

But since $n_i \neq n_j$ whenever $i \neq j$, we have $\displaystyle\sum_{j=1}^{k} 2^{-n_j-1} < \sum_{n=0}^{\infty} 2^{-n-1} \; = \; 1$. So

we obtain $q \leq \left(\dfrac{\log(\gamma_E(0,r,q))+\varepsilon}{B(r)}\right)^2$. Actually ε was chosen arbitrarily, hence we have

$$(8) \quad q \leq \left(\frac{\log(\gamma_E(0,r,q))}{B(r)}\right)^2.$$

Now, we will show that the circular filter $\mathcal{G}$ of center 0 and diameter 1 is an irregular distinguished circular filter on E. For each couple $(n,r) \in \mathbb{N} \times I$ we denote by $\mathcal{F}_{n,r}$ the increasing T-filter of $A_{n,r}$, that obviously defines a T-filter on E. Clearly the family $(\mathcal{C}(\mathcal{F}_{n,r}))_{(n,r)\in\mathbb{N}\times I}$ makes a partition of E. Hence, to prove that $\mathcal{G}$ is an irregular distinguished circular filter, it is sufficient to show that this partition is T-specific, i.e. none of the $\mathcal{F}_{n,r}$ is surrounded by any T-filter $\mathcal{F}' \neq \mathcal{F}_{n,r}$.

Indeed suppose that certain $\mathcal{F}_{n,\rho}$ is surrounded by a T-filter $\mathcal{F}' \neq \mathcal{F}_{n,\rho}$ and let $r = diam(\mathcal{F}')$. If $\mathcal{F}'$ is decreasing then $\mathcal{P}(\mathcal{F}') = d(0,r)$, hence 0 is a center of $\mathcal{F}'$. If $\mathcal{F}'$ is increasing and different from $\mathcal{F}_{n,\rho}$, then $\mathcal{C}(\mathcal{F}')$ contains $\mathcal{C}(\mathcal{F}_{n,\rho})$ and has a diameter $\rho < r$, hence 0 is a center of $\mathcal{F}'$. Thus we have checked that $\mathcal{F}'$ is a T-filter of center 0 and diameter r.

First we will suppose $\mathcal{F}'$ increasing. Hence E admits an increasing T-sequence of center 0 and diameter r. Let $C(0,r_m)$ be a sequence of circles carrying this T-sequence. So, there exists a sequence of strictly positive integers $(q_m)_{m\in\mathbb{N}}$ such that

$$\lim_{m\to\infty}\left(-\log(\gamma_E(0,r_m,q_m))\right) + \sum_{j=0}^{m-1} q_j \log\left(\frac{r_m}{r_j}\right) = +\infty.$$

Let $a = -\log r_0$. Trivially we have $a \displaystyle\sum_{j=0}^{m-1} q_j \geq \log(\gamma_E(0,r_m,q_m))$ and therefore by (8) we obtain

$$(9) \quad a \sum_{j=0}^{m-1} q_j \geq \sqrt{q_m}\, B(r_m).$$

For every $m \in \mathbb{N}$ we put $h(m) = \sup_{0 \leq j < m} q_j$. By (9) we have

$am\, h(m) \geq \sqrt{q_m}\, B(r_m)$ hence

$$(10) \quad q_m \leq \left(\frac{am\, h(m)}{B(r_m)}\right)^2.$$

Let $m \in \mathbb{N}$ be fixed. Let $w_0,\ldots,w_s$ be the ranks such that $h(w_j) = q_{w_j}$ with $w_0 = h(m)$, $w_1 = h(m-1)$ and $w(s) = h(0)$. By (10) each $h(w_j)$ satisfies

$$h(w_j) \leq \left(\frac{2a\, w_j\, h(w_{j-1})}{B(r_{w_j})}\right)^2, \quad (1 \leq j \leq s).$$

For convenience we put $\lambda(m) = B(r_m)$ ($m \in \mathbb{N}$), and $\theta(n) = 2^n$ ($n \in \mathbb{N}$). By induction on Relations (10) we shortly obtain:

$$(11) \quad q_m \leq \left(\frac{m}{\lambda(m)}\right)^2 \prod_{j=1}^{s} \left(\frac{2a\, w_j}{\lambda(w_j)}\right)^{\theta(j+1)}.$$

Now we notice that $w_j < m$ for every $j = 1,\ldots,s$ and that $\sum_{j=0}^{s} 2^{j+1} \leq 2^m$ because

$s \leq m$. So by (11) we have

$$(12) \quad q_m \leq \frac{(am)^{\theta(m+2)}}{\lambda(m)^2} \quad \text{for every } m \in \mathbb{N}.$$

Let $(\tau_m)_{m \in \mathbb{N}}$ be the sequence defined in $\mathbb{N}$ as $\tau_m = b(r_m)$. Since the mapping b is injective, so is the sequence $(\tau_m)_{m \in \mathbb{N}}$. Therefore, as it takes values in $\mathbb{N}$, it is seen that the inequality $\tau_m \geq m$ holds for infinitely many indices. Hence there exists a subsequence $(\tau_{m_t})_{t \in \mathbb{N}}$ such that $\tau_{m_t} \geq m_t$ whenever $t \in \mathbb{N}$. For convenience we put $\nu(t) = m_t$. By (12) we obtain

$$q_{\nu(t)} \leq \frac{(a\nu(t))^{\theta(\nu(t)+2)}}{\lambda(\nu(t))^2} \leq \frac{(a\nu(t))^{\theta(\nu(t)+2)}}{(\nu(t))^{2\phi(\nu(t))}}.$$

Since the sequence $(\nu(t))_{t \in \mathbb{N}}$ goes to $+\infty$, it is seen that $q_{\nu(t)} = 0$ when t is big enough and then this contradicts the definition of the q_m. So this finishes proving that E does not admit any increasing T-sequence of center 0. Symmetrically, we can show that E does not admit any decreasing T-sequence of center 0. Hence we have proven that the distinguished circular filter is irregular.

42. MAXIMAL IDEALS

OF INFINITE CODIMENSION

Throughout this chapter, D is infraconnected and belongs to $\mathcal{A}$.

Thanks to the distinguished circular filters we will characterize the maximal ideals of infinite codimension in $H(D)$ [19].

Notations: We will denote by $Max_\infty(H(D))$ the set of the maximals ideals of infinite codimension in $H(D)$.

Lemma 42.1:   *Let $\mathcal{F}$ be a distinguished circular filter on D. Then $\mathcal{I}(\mathcal{F})$ is not included in any codimension 1 maximal ideal of $H(D)$.*

Proof: Let $r = diam(\mathcal{F})$. Let $\mathcal{M}$ be a codimension 1 maximal ideal of $H(D)$. By Theorem 16.2 , there exists $a \in D$ such that $\mathcal{M} = \mathcal{I}(a)$. First, suppose $a \in \mathcal{Q}(\mathcal{F}, D)$. Since $\mathcal{F}$ is distinguished, there exists an increasing T-filter $\mathcal{G}$ of center a and diameter $s \in]0, r[$ and therefore there exists $f \in \mathcal{I}_0(\mathcal{G})$, such that $f(a) \neq 0$ and $f(x) = 0$ whenever $x \in D$ such that $|x - a| \geq s$. Hence, f belongs to $\mathcal{I}(\mathcal{F})$ but f does not belong to $\mathcal{I}(a)$. Therefore $\mathcal{I}(\mathcal{F})$ is not included in $\mathcal{I}(a)$.

Now, suppose $a \in D \setminus \mathcal{Q}(\mathcal{F}, D)$. By definition of a distinguished circular filter, the decreasing filter $\mathcal{G}$ thinner than $\mathcal{F}$ is a T-filter. Hence by Theorem 37.2 there exists $f \in \mathcal{I}(\mathcal{G})$ such that $f(a) \neq 0$. But $\mathcal{I}(\mathcal{F}) = \mathcal{I}(\mathcal{G})$, and therefore $\mathcal{I}(\mathcal{F})$ is not included in $\mathcal{I}(a)$. This ends the proof of Lemma 42.1.

Lemma 42.2:   *Let $\mathcal{F}_1$, $\mathcal{F}_2$ be two different distinguished circular filters on D. Then $\mathcal{I}(\mathcal{F}_1)$ and $\mathcal{I}(\mathcal{F}_2)$ are not comparable with respect to the inclusion.*

Proof: First we suppose that $\mathcal{F}_1$ is not surrounded by $\mathcal{F}_2$ and that $\mathcal{F}_2$ is not surrounded by $\mathcal{F}_1$. Hence $\mathcal{F}_1$ is secant with $D \setminus \mathcal{Q}(\mathcal{F}_2, D)$ and $\mathcal{F}_2$ is secant with $D \setminus \mathcal{Q}(\mathcal{F}_1, D)$. Hence $\mathcal{F}_1$ (resp. $\mathcal{F}_2$) admits a decreasing T-filter $\mathcal{G}_1$ (resp. $\mathcal{G}_2$) thinner than it. By thereom 37.2 there exists f_1 (resp. f_2) strictly vanishing along $\mathcal{F}_1$ (resp. $\mathcal{F}_2$) and not vanishing along any other large circular filter secant with $\mathcal{C}(\mathcal{G}_1)$ (resp. $\mathcal{C}(\mathcal{G}_2)$). But $\mathcal{C}(\mathcal{G}_1) = D \setminus \mathcal{Q}(\mathcal{F}_1, D)$, $\mathcal{C}(\mathcal{G}_2) = D \setminus \mathcal{Q}(\mathcal{F}_2, D)$. Hence f_1 (resp. f_2) is not vanishing along $\mathcal{G}_2$ (resp. $\mathcal{G}_1$) and therefore f_1 (resp. f_2) does not belong to $\mathcal{I}(\mathcal{F}_2)$ (resp. $\mathcal{I}(\mathcal{F}_1)$). So $\mathcal{I}(\mathcal{F}_1)$, $\mathcal{I}(\mathcal{F}_2)$ are not comparable.

Now, we suppose that $\mathcal{F}_1$ is surrounded by $\mathcal{F}_2$. Since $\mathcal{F}_1 \neq \mathcal{F}_2$, it is seen that $diam\ (\mathcal{F}_1) < diam\ (\mathcal{F}_2)$ and therefore $\mathcal{Q}(\mathcal{F}_1, D) \neq D$. Hence the decreasing filter $\mathcal{G}_1$ thinner than $\mathcal{F}_1$ is secant with $\mathcal{Q}(\mathcal{F}_2, D)$. By theorem 37.2 there exists $f \in H(D)$ vanishing along $\mathcal{F}_1$ but not vanishing along $\mathcal{F}_2$. Hence f belongs to $\mathcal{I}(\mathcal{F}_1) \setminus \mathcal{I}(\mathcal{F}_2)$. Besides, by definition of the distinguished circular filters, there exists an increasing T-filter $\mathcal{G}$ on $\mathcal{Q}(\mathcal{F}_2, D)$ such that $\mathcal{F}_1$ is secant with $C(\mathcal{G})$. By theorem 37.2 there exists $g \in \mathcal{I}_0(\mathcal{G})$, not vanishing along $\mathcal{F}_1$, and then g belongs to $\mathcal{I}(\mathcal{F}_2) \setminus \mathcal{I}(\mathcal{F}_1)$. Thus $\mathcal{I}(\mathcal{F}_1)$ and $\mathcal{I}(\mathcal{F}_2)$ are not comparable.

Proposition 42.3: *Let $\mathcal{F}$ be a circular filter of diameter r surrounded by a circular filter $\mathcal{G}$ of diameter s. Then $\mathcal{I}(\mathcal{F})$ contains $\mathcal{I}(\mathcal{G})$ if and only if $\mathcal{F}$ is not surrounded by any increasing T-filter of diameter $u \in]r, s]$. $\mathcal{I}(\mathcal{G})$ contains $\mathcal{I}(\mathcal{F})$ if and only if $\mathcal{F}$ is not surrounded by any decreasing T-filter of diameter $u \in [r, s[$.*

Proof: For convenience, we give $\mathcal{F}$ a center α in $\widehat{K}$. Let T be an increasing T-filter of diameter $u \in]r, s]$, surrounding $\mathcal{F}$. By Theorem 37.2 there exists $f \in \mathcal{I}_0(T)$, strictly vanishing along T but not vanishing along $\mathcal{F}$. Since $\mathcal{G}$ is clearly secant with $\mathcal{P}(\mathcal{F})$, we have $f \in \mathcal{I}(\mathcal{G})$, hence $\mathcal{I}(\mathcal{F})$ does not contain $\mathcal{I}(\mathcal{G})$. Reciprocally, we suppose that $\mathcal{I}(\mathcal{F})$ does not contain $\mathcal{I}(\mathcal{G})$. Let $f \in \mathcal{I}(\mathcal{G}) \setminus \mathcal{I}(\mathcal{F})$. So we have $_D\varphi_{\alpha,s}(f) = 0$ and $_D\varphi_{\alpha,r}(f) \neq 0$. By Lemma 38.1 f is strictly vanishing along an increasing filter T, of center α, of diameter $u \in]r, s]$. Then, T is a T-filter that clearly surrounds $\mathcal{F}$.

Now we suppose that T is a decreasing T-filter surrounding $\mathcal{F}$, of diameter $u \in]r, s]$. By Theorem 37.2 there exists $f \in \mathcal{I}_0(T)$ that is not vanishing along $\mathcal{G}$. Since $\mathcal{F}$ is surrounded by T, f is clearly vanishing along $\mathcal{F}$, so $\mathcal{I}(\mathcal{F})$ is not included in $\mathcal{I}(\mathcal{G})$.

Finally we suppose that $\mathcal{I}(\mathcal{G})$ does not contain $\mathcal{I}(\mathcal{F})$. Let $f \in \mathcal{I}(\mathcal{F}) \setminus \mathcal{I}(\mathcal{G})$. So we have $_D\varphi_{\alpha,r}(f) = 0$ and $_D\varphi_{\alpha,s}(f) \neq 0$. By Lemma 38.1 f is strictly vanishing along a decreasing filter T of center α and diameter $u \in]r, s]$. Then T is a decreasing T-filter that surrounds $\mathcal{F}$.

Proposition 42.4: *Let D be bounded. Let $\mathcal{H}$ be a circular filter such that $\mathcal{I}(\mathcal{H})$ is a maximal ideal of infinite codimension. Then $\mathcal{H}$ is surrounded by a distinguished circular filter $\mathcal{G}$ such that $\mathcal{I}(\mathcal{G}) = \mathcal{I}(\mathcal{H})$. Further, if $\mathcal{Q}(\mathcal{H}, D)$ is not empty, it admits a T-specific partition $C(\mathcal{F}_i)_{i \in I}$ and each increasing T-filter $\mathcal{F}_i$ satisfies $\mathcal{I}(\mathcal{F}_i) = \mathcal{I}(\mathcal{F})$.*

Proof: Let $r = diam(\mathcal{H})$, and $E = \mathcal{Q}(\mathcal{H}, D)$.

On the first hand we suppose $\mathcal{Q}(\mathcal{H}, D) \neq \emptyset$ and we will check that E admits a T-specific partition. Indeed for all $a \in E$, there exists $f \in \mathcal{I}(\mathcal{H})$ such that although $_D\varphi_{a,r}(f) = 0$. Hence by Lemma 38.1 there exists an increasing T-filter $\mathcal{G}_a$ of center 0 and diameter $u \in]0,r]$ such that f is strictly vanishing along $\mathcal{G}_a$. Then the set $(\mathcal{C}(\mathcal{G}_a))_{a\in E}$ forms a covering of E, and therefore by Proposition 41.5, E admits a T-optimal partition $\mathcal{C}(\mathcal{F}_i)_{i\in I}$. Next, for every $i \in I$ we put $\rho(i) = diam(\mathcal{F}_i)$. Let $h \in I$. Since the partition $\mathcal{C}(\mathcal{F}_i)_{i\in I}$ is T-optimal, there exists no increasing T-filter of diameter $u \in]\rho(h),r]$, hence by Proposition 42.3 $\mathcal{I}(\mathcal{F}_h)$ contains $\mathcal{I}(\mathcal{H})$ that is a maximal ideal, hence $\mathcal{I}(\mathcal{F}_h) = \mathcal{I}(\mathcal{H})$. Then by Proposition 42.3, $\mathcal{F}_h$ is not surrounded by any decreasing T-filter of diameter $u \in [\rho(h),r[$. Therefore $\mathcal{F}_h$ is not surrounded by any T-filter on $\mathcal{Q}(\mathcal{H},D)$, so the T-optimal partition $\mathcal{C}(\mathcal{F}_i)_{i\in I}$ is T-specific.

On the other hand, we will show that $\mathcal{H}$ is surrounded by a distinguished circular filter $\mathcal{G}$ such that $\mathcal{I}(\mathcal{H}) = \mathcal{I}(\mathcal{G})$. If $\mathcal{Q}(\mathcal{H},D)$ is empty and if $\mathcal{H}$ is a decreasing T-filter, it is just a distinguished circular filter and then we have nothing to prove. So we can suppose that $\mathcal{H}$ is not reduced to a decreasing T-filter. Let $s \in \mathbb{R}_+^*$ be defined as follows.

If $\mathcal{H}$ is not surrounded by any decreasing T-filter, we put just $s = diam(D)$.

If $\mathcal{H}$ is surrounded by a decreasing T-filter on D we denote by $\mathcal{S}$ the set of the decreasing T-filters surrounding $\mathcal{H}$ and we put $s = \inf\{diam(\mathcal{T})|\mathcal{T} \in \mathcal{S}\}$. Suppose that there exists no decreasing T-filter of diameter s that surrounds $\mathcal{H}$. Then by Lemma 41.2 there exists a descending chain of decreasing T-filters $(\mathcal{G}_n)_{n\in\mathbb{N}}$ such that, putting $s_n = diam(\mathcal{G}_n)_{n\in\mathbb{N}}$ we have $s = \lim_{n\to\infty} s_n$. By Theorem 41.3 the basement $\mathcal{B}$ of this chain is a decreasing T-filter of diameter s that surrounds $\mathcal{H}$. Thus there does exist a decreasing T-filter $\mathcal{B}$ surrounding $\mathcal{H}$.

Let $\mathcal{G}$ be the circular filter of diameter s that surrounds $\mathcal{H}$. By hypothesis, on $\mathcal{Q}(\mathcal{G},D)$ there exists no decreasing T-filter surrounding $\mathcal{H}$. Hence by Proposition 42.3 we have $\mathcal{I}(\mathcal{H}) \subset \mathcal{I}(\mathcal{G})$, hence $\mathcal{I}(\mathcal{H}) = \mathcal{I}(\mathcal{G})$ because $\mathcal{I}(\mathcal{H})$ is maximal. But of course, $\mathcal{Q}(\mathcal{G},D)$ is not empty. Finally, we have already proven $\mathcal{Q}(\mathcal{H},D)$ to have a T-specific partition when it is not empty, and now this also holds for $\mathcal{Q}(\mathcal{G},D)$. But as we saw, either $\mathcal{Q}(\mathcal{G},D) = D$, or the decreasing filter thinner than $\mathcal{G}$ is a T-filter. Hence $\mathcal{G}$ is a distinguished circular filter and Proposition 42.4 is proven.

Theorem 42.5:   *For every $\mathcal{M} \in Max_\infty(H(D))$, there exists a unique distinguished circular filter $\mathcal{F}$ on D such that $\mathcal{M} = \mathcal{I}(\mathcal{F})$.*

Proof: First we suppose D closed and bounded. By Lemma 1.16, there exists $\psi \in Mult(H(D),\| \cdot \|_D)$ such that $Ker(\psi) = \mathcal{M}$. Hence by Proposition 42.4 there exists a distinguished circular filter $\mathcal{F}$ on D such that $\mathcal{M} = Ker(_D\varphi_\mathcal{F})$ and then by Lemma 42.2, this distinguished circular filter is unique.

Now we suppose D bounded but not closed. Then we know that $H(D) = S(D)^{-1}H(\overline{D})$, with $S(D)$ the set of the polynomials whose zeros lie in $\overline{D} \setminus D$. Let $\mathcal{J} = \mathcal{M} \cap H(\overline{D})$. We know that $\mathcal{M}$ is equal to $S(D)^{-1}\mathcal{J}$. Let $h \in \mathcal{M}$. It is of the form $\dfrac{f}{Q}$ with $f \in \mathcal{J}$. Let $\mathcal{W}$ be a maximal ideal of $H(\overline{D})$ that contains $\mathcal{J}$ and suppose that $\mathcal{W}$ is of codimension 1. Then there exists $a \in \overline{D}$ such that $f(a) = 0$ whenever $f \in \mathcal{J}$. It is seen that a belongs to D because if $a \in \overline{D} \setminus D$ then $1 \in \mathcal{M}$. Hence $Q(a) \neq 0$ and therefore $h(a) = 0$. But then $\mathcal{M}$ is equal to $\mathcal{I}(a)$ and is a maximal ideal of codimension 1. Thus we have proven that $\mathcal{W}$ is of infinite codimension. Hence there exists a unique distinguished circular filter $\mathcal{F}$ on $\overline{D}$ such that $\mathcal{W} = \mathcal{I}(\mathcal{F})$ and then we have $_D\varphi_{\mathcal{F}}(h) = 0$ for all $h \in \mathcal{J}$. Now let $f \in \mathcal{M}$. Then we have $_D\varphi_{\mathcal{F}}(h) = _{\overline{D}}\varphi'_{\mathcal{F}}(h) = 0$, and therefore $_{\overline{D}}\varphi'_{\mathcal{F}}(f) = 0$. Thus it is seen that $\mathcal{M}$ is included in $\mathcal{I}(\mathcal{F}')$ and therefore is equal to $\mathcal{I}(\mathcal{F}')$. Further, by Lemma 42.2 $\mathcal{F}'$ is obviously unique.

Now we suppose D unbounded. Obviously we may assume that D has at least one hole, and that this hole contains 0. The inversion ℓ defined as $\ell(x) = \dfrac{1}{x}$ maps D onto a bounded infraconnected set $D' \in \mathcal{A}$. Let θ be the mapping from $H(D)$ into $H(D')$ defined as $\theta(f) = f \circ \ell$. So, θ is a K-algebra isomorphism. Let $\mathcal{M}' = \theta(\mathcal{M})$. Then $\mathcal{M}'$ is maximal ideal of infinite codimension in $H(D')$. Hence there exists a distinguished circular filter $\mathcal{F}'$ on D' such that $\mathcal{M}' = \mathcal{I}(\mathcal{F}')$.
First we suppose that 0 does not belong to $\mathcal{Q}(\mathcal{F}', D')$. Let $\mathcal{F} = \ell(\mathcal{F}')$. Then it is easily seen that $\mathcal{F}$ is a distinguished circular filter. Indeed let $(\mathcal{T}'_i)_{i \in I}$ be the T-family of $\mathcal{F}'$ and let $\mathcal{T}'$ be the decreasing T-filter thinner than $\mathcal{F}'$. Since $0 \notin \mathcal{Q}(\mathcal{F}, D)$, by Proposition 35.3 we know that the filter $\mathcal{T} := \ell(\mathcal{T}')$ is a decreasing T-filter on D thinner than $\mathcal{F}$, and the T-filters $\mathcal{T}_i := \ell(\mathcal{T}'_i)$, $(i \in I)$, are increasing filters such that the family $(\mathcal{C}(\mathcal{T}_i))_{i \in I}$ makes a T-specific partition of $\mathcal{Q}(\mathcal{F}, D)$. So $\mathcal{F}$ is a distinguished circular filter. It is seen that $\mathcal{M} = \mathcal{I}(\mathcal{F})$ because given $f \in H(D)$, f satisfies $\lim_{\mathcal{F}} f(x) = 0$ if and only if $\lim_{\mathcal{F}'} f \circ \ell(u) = 0$.

Now we suppose that 0 belongs to $\mathcal{Q}(\mathcal{F}', D')$. Let $\mathcal{T}'_0$ be the increasing T-filter of the T-family of $\mathcal{F}'$ such that $0 \in \mathcal{C}(\mathcal{T}'_0)$. Let $\mathcal{H}'$ be the circular filter less thin than $\mathcal{T}'$, let $\mathcal{H} = \ell(\mathcal{H}')$ and let $\mathcal{T}_0 = \ell(\mathcal{T}')$. Then by Proposition 35.3 $\mathcal{T}_0$ is a decreasing T-filter thinner than $\mathcal{H}$. We will show that $\mathcal{Q}(\mathcal{H}, D)$ admits a T-specific partition. Let $(\mathcal{T}'_i)_{i \in I}$ be the T-family of $\mathcal{F}'$, (with I a set of indices containing 0). For each $i \in I, i \neq 0$, we put $\mathcal{G}_i = \ell(\mathcal{T}'_i)$ and $\mathcal{G}_0 = \mathcal{T}_0$. It is easily seen that $\mathcal{C}(\mathcal{G}_i)_{i \in I}$ is a T-specific partition of $\mathcal{Q}(\mathcal{H}, D)$. Thus $\mathcal{H}$ is a distinguished T-filter on D. We check that $\mathcal{M} = \mathcal{I}(\mathcal{H})$. Indeed we know that $\mathcal{I}(\mathcal{H}) = \mathcal{I}(\mathcal{T}_0)$. But $\mathcal{I}(\mathcal{T}_0) = \{f \circ \ell | f \in \mathcal{I}(\mathcal{T}'_0)\}$ and finally $\mathcal{I}(\mathcal{T}'_0) = \mathcal{I}(\mathcal{F}')$. Thus we have $\mathcal{I}(\mathcal{H}) = \{f \circ \ell | f \in \mathcal{I}(\mathcal{F}')\}$ and therefore $\mathcal{I}(\mathcal{H}) = \theta(\mathcal{I}(\mathcal{F}'))$. This finishes showing that $\mathcal{M} = \mathcal{I}(\mathcal{H})$.

Proposition 42.6: *Let $\mathcal{F}$ be a distinguished circular filter on D. Then $\mathcal{I}(\mathcal{F})$ is a maximal ideal of infinite codimension.*

Proof: Suppose that $\mathcal{I}(\mathcal{F})$ is included in a maximal ideal of codimension 1. By Theorem 16.2 this ideal is of the form $\mathcal{I}(a)$ with $a \in D$. If $a \notin \mathcal{Q}(\mathcal{F}, D)$, then there exists a decreasing T-filter $\mathcal{G}$ thinner than $\mathcal{F}$. Therefore by Theorem 37.2 there exists $f \in \mathcal{I}_0(\mathcal{G})$ such that $f(a) \neq 0$, hence $\mathcal{I}(\mathcal{F})$ is not included in $\mathcal{I}(a)$. If $a \in \mathcal{Q}(\mathcal{F}, D)$ then there exists an increasing T-filter $\mathcal{G}$ on $\mathcal{Q}(\mathcal{F}, D)$ such that $a \in \mathcal{C}(\mathcal{G})$ and therefore by Theorem 37.2 again there exists $f \in \mathcal{I}_0(\mathcal{G})$ such that $f(a) \neq 0$, hence we have the same conclusion. Thus, $\mathcal{I}(\mathcal{F})$ is included in a maximal ideal of infinite codimension $\mathcal{M}$. By Theorem 42.5 there exists a unique distinguished circular filter $\mathcal{H}$ on D such that $\mathcal{M} = \mathcal{I}(\mathcal{H})$. Hence we have $\mathcal{I}(\mathcal{F}) \subset \mathcal{I}(\mathcal{H})$. Therefore by Proposition 42.2 we have $\mathcal{F} = \mathcal{H}$.

Corollary 42.7:   *Let $\mathcal{F}$ be a distinguished circular filter such that $\mathcal{Q}(\mathcal{F}, D) \neq \emptyset$. For every increasing T-filter $\mathcal{F}_i$ of the T-specific partition of $\mathcal{Q}(\mathcal{F}, D)$ we have $\mathcal{I}(\mathcal{F}_i) = \mathcal{I}(\mathcal{F})$.*

Proof: This is a consequence of Proposition 42.5 and Proposition 42.6.

Corollary 42.8:   *The mapping Φ from the set of the distinguished circular filters into $Max_\infty(H(D))$ defined as $\Phi(\mathcal{F}) = \mathcal{I}(\mathcal{F})$ is a bijection.*

Theorem 42.9:   *Let K be strongly valued and let D be closed and bounded. The mapping Θ from $Mult_m(H(D), \| \, . \, \|)$ into $Max(H(D))$ defined as $\Theta(\psi) = Ker\psi$ is a bijection.*

Proof: By Proposition 1.16 we know that Θ is a surjection. Now let $\mathcal{M} \in Max_\infty(H(D))$. Let $\mathcal{F}$ be the distinguished circular filter such that $\mathcal{I}(\mathcal{F}) = \mathcal{M}$, and let $\psi \in Mult_m(H(D), \| \, . \, \|_D)$ be such that $Ker(\psi) = \mathcal{M}$. Let $\mathcal{H}$ be the circular filter on D such that $\psi = {}_D\varphi_{\mathcal{H}}$. Let $s = diam(\mathcal{F})$ and $r = diam(\mathcal{H})$. By Proposition 42.4, $\mathcal{H}$ is surrounded by $\mathcal{F}$ hence we have $r < s$. Let $A \subset D$ be an element of $\mathcal{F}$ of diameter $u \in]r, s[$, and let $b \in A \cap (\mathcal{Q}(\mathcal{H}, D) \setminus \mathcal{Q}(\mathcal{F}, D))$. Let $(\mathcal{F}_i)_{i \in I}$ be the T-family of the $\mathcal{F}$ and let $h \in I$ be such that $b \in \mathcal{C}(\mathcal{F}_h)$. Since $\mathcal{F}_h$ does not surround $\mathcal{F}$ we have $diam(\mathcal{C}(\mathcal{F}_h)) \leq u < r$. But since K is strongly valued this is impossible by Theorem 41.6. Hence we have $\mathcal{F} = \mathcal{H}$, and therefore $\psi = {}_D\varphi_{\mathcal{F}}$. This ends the proof.

Theorem 42.10:   *If K is weakly valued there exists a closed bounded infraconnected set E in K such that $H(E)$ admits multiplicative semi-norms $\psi_1, \psi_2 \in Mult_m(H(E), \| \, . \, \|_E)$, with $\psi_1 \neq \psi_2$ such that $Ker\psi_1 = Ker\psi_2$.*

Proof: By Theorem 41.6 there exists a closed bounded infraconnected set E with an irregular distinguished circular filter $\mathcal{F}$. Let $\mathcal{T}$ belong to the T-family of $\mathcal{F}$, and satisfy $diam(\mathcal{T}) < diam(\mathcal{F})$. Then we have $\mathcal{I}(\mathcal{T}) = \mathcal{I}(\mathcal{F})$. Now given $a \in \mathcal{C}(\mathcal{T})$, it is seen that ${}_D\varphi_{\mathcal{T}}(x - a) = diam(\mathcal{T})$, ${}_D\varphi_{\mathcal{F}}(x - a) = diam(\mathcal{F})$, hence ${}_D\varphi_{\mathcal{T}} \neq {}_D\varphi_{\mathcal{F}}$.

43.   IDEMPOTENT T-SEQUENCES

Let $(T_{m,i}, q_{m,i})_{\substack{1 \leq i \leq h(m) \\ m \in \mathbb{N}}}$ be a T-sequence of diameter r such that for every (m,i), $diam(T_{m,i})$ lies in an interval $[\rho', \rho''] \subset]0, r[$. We will prove that we can replace the $q_{m,i}$ by integers $u_{m,i}$ equal to 0 or 1 so that we obtain an idempotent T-sequence [55].

Lemma 43.1:   *Let $r \in]0, +\infty[$ and let $\rho \in]0, r[$. Let $a_1, \ldots, a_n \in d(0, r^-)$, satisfy $|a_i - a_j| \geq \rho$ whenever $i \neq j$ and let $u_1, \ldots u_n \in]0, 1]$. There exist integers $v_1, \ldots v_n$ equal to 0 or 1 satisfying*

i)   *either $u_i \leq v_i = 1$ or $v_i = 0$*

ii) $0 \leq \displaystyle\sum_{i=1}^{n} v_i - \sum_{i=1}^{n} u_i < 1$

iii) $\displaystyle\sup_{1 \leq i \leq n} \Big(\sum_{\substack{j \neq i \\ 1 \leq j \leq n}} v_j \, \log\Big(\frac{r}{|a_j - a_i|}\Big) \Big) \leq \sup_{1 \leq i \leq n} \Big(\sum_{\substack{j \neq i \\ 1 \leq j \leq n}} u_j \, \log\Big(\frac{r}{|a_j - a_i|}\Big) \Big) +$

$+ 2 \log\Big(\dfrac{r}{\rho}\Big) \Big)$

Proof: First, suppose $n = 2$ and $u_1 \leq u_2$. If $u_1 + u_2 \leq 1$ we put $v_1 = 1$, $v_2 = 0$, and if $u_1 + u_2 > 1$ we put $v_1 = v_2 = 1$. Then in each case, i) and ii) are clearly satisfied . Besides, as $|a_2 - a_1| \geq \rho$, it is seen that

$$\max_{j=1,2} v_j \log\Big(\frac{r}{|a_2 - a_1|}\Big) \leq \log\Big(\frac{r}{\rho}\Big) \leq \max_{j=1,2} u_j \log\Big(\frac{r}{|a_2 - a_1|}\Big) + \log\Big(\frac{r}{\rho}\Big).$$

Now, we will assume the lemma to be established when $n \leq q - 1$ and will prove it when $n = q$. Let $r' = \max_{i \neq j} |a_i - a_j|$ and $\Lambda_1, \ldots, \Lambda_k$ be the classes of $d(a_1, r')$ that contain at least one of the a_h (of course we have $k \geq 2$). In each Λ_h let $(b_{h,i}, s_{h,i})$, with $1 \leq i \leq \ell_h$, be the couples (a_j, u_j) such that $a_j \in \Lambda_h$. Since $k \geq 2$, obviously we have $\ell_h < q$. So, by definition we have

(1) $\displaystyle\sum_{h=1}^{k} \Big(\sum_{i=1}^{\ell_h} s_{h,i} \Big) = \sum_{j=1}^{q} u_j.$

Then it is easily seen that we can find integers $t_h \geq 0$ satisfying

(2) $-1 < t_h - \displaystyle\sum_{i=1}^{\ell_h} s_{h,i} < 1$

$$(3) \quad 0 \le \sum_{h=1}^{k} t_h - \sum_{h=1}^{k} \left(\sum_{i=1}^{\ell_h} s_{h,i} \right) < 1,$$

and then by (1) , Relation (3) actually is equivalent to

$$(4) \quad 0 \le \sum_{h=1}^{k} t_h - \sum_{j=1}^{q} u_j < 1.$$

Since $\ell_h < q$ (for each $h = 1, \ldots, k$) we can apply the Lemma 43.1, supposed true for $n < q$ to each class Λ_h and then we can define a family of integers $(w_{h,i})_{1 \le i \le \ell_h, 1 \le h \le k}$, equal to 0 or 1, satisfying

$$(5) \quad \text{either } 0 \le w_{h,i} - s_{h,i} < 1 \quad \text{or} \quad w_{h,i} = 0$$

$$(6) \quad \begin{cases} \displaystyle\sum_{i=1}^{\ell_h} w_{h,i} = t_h & \text{if} \quad t_h \ge \displaystyle\sum_{i=1}^{\ell_h} s_{h,i} \\[2em] \displaystyle\sum_{i=1}^{\ell_h} w_{h,i} = t_h + 1 & \text{if} \quad t_h < \displaystyle\sum_{i=1}^{\ell_h} s_{h,i} \end{cases}$$

$$(7) \text{ either } l_h = 1, \text{ or } \sup_{1 \le i \le \ell_h} \left(\sum_{\substack{j \ne i \\ j \le \ell_h}} w_{h,j} \, \log\left(\frac{r'}{|b_{h,j} - b_{h,i}|} \right) \right) \le$$

$$\sup_{1 \le i \le \ell_h} \left(\sum_{\substack{j \ne i \\ j \le \ell_h}} s_{h,j} \, \log\left(\frac{r'}{|b_{h,j} - b_{h,i}|} \right) \right) + +2 \log\left(\frac{r'}{\rho} \right)$$

Now we can replace the family $(w_{h,i})_{1 \le i \le \ell_h, 1 \le h \le k}$ by a family $(y_{h,i})_{1 \le i \le \ell_h, 1 \le h \le k}$ satisfying (5) and (7), like the $(w_{h,i})$, and satisfying further

$$(8) \quad 0 \le \sum_{i=1}^{\ell_h} \left(y_{h,i} - s_{h,i} \right) < 1 \text{ whenever } h = 1, \ldots, k.$$

When $t_h \ge \displaystyle\sum_{i=1}^{\ell_h} s_{h,i}$ we put $y_{h,i} = w_{h,i}$ whenever $i = 1, \ldots, \ell_h$.

When $t_h < \displaystyle\sum_{i=1}^{\ell_h} s_{h,i}$, we consider an index j_h such that $w_{h,j_h} = 1$ and we put $y_{h,i} = w_{h,i}$ for all $i \ne j_h$ and $y_{h,j_h} = 0$. Then by (6) it is seen that we have

$$(9) \quad \sum_{i=1}^{\ell_h} y_{h,i} = t_h \text{ whenever } h = 1, \ldots, k$$

and therefore by (3) and (6) we obtain (8) for all $h = 1, \ldots, k$. Besides it is seen that (5) is conserved when replacing $w_{h,i}$ by $y_{h,i}$ because if $y_{h,i} \ne w_{h,i}$ then $y_{h,i} = 0$. Since $y_{h,i} \le w_{h,i}$, Relation (7) is obviously conserved when replacing each $w_{h,i}$ by $y_{h,i}$ for whatever (i,j) such that $1 \le i \le \ell_h, 1 \le h \le k$. Now we can define the family v_j in this way:

For each $j = 1, \ldots, q$, each a_j being also denoted by $b_{h,i}$ (for a certain (h,i)), we put $v_j = y_{h,i}$. Then, by gathering Relations (5), it is seen that the $(v_j)_{1 \leq j \leq n}$ satisfy i) (for $n = q$). By (9) we see that $\sum_{j=1}^{q} v_j = \sum_{h=1}^{k} t_h$ and then by (4), ii) is satisfied.

So, it just remains to check that iii) is also satisfied. According to the notation $(b_{h,i}, y_{h,i})$, Relations iii) becomes

$$\text{iii')} \quad \sup_{1 \leq i \leq \ell_m, \, 1 \leq m \leq k} \left(\sum_{(h,j) \neq (m,i)} y_{h,j} \, \log\left(\frac{r}{|b_{h,j} - b_{m,i}|}\right) \right) \leq$$

$$\leq \sup_{1 \leq i \leq \ell_m, \, 1 \leq m \leq k} \left(\sum_{(h,j) \neq (m,i)} s_{h,j} \, \log\left(\frac{r}{|b_{h,j} - b_{m,i}|}\right) \right) + 2 \log\left(\frac{r}{\rho}\right).$$

But since $|b_{h,j} - b_{m,i}| = r'$ whenever $h \neq m$, we have $\log\left(\frac{r}{|b_{h,j} - b_{m,i}|}\right) = \log\left(\frac{r}{r'}\right)$ each time $h \neq m$. Let $L = \sup_{1 \leq i \leq \ell_m, \, m \leq k} \left(\sum_{(h,j) \neq (m,i)} y_{h,j} \, \log\left(\frac{r}{|b_{h,j} - b_{m,i}|}\right) \right)$

and let $E = \sup_{1 \leq i \leq \ell_m, \, 1 \leq m \leq k} \left(\sum_{(h,j) \neq (m,i)} s_{h,j} \, \log\left(\frac{r}{|b_{h,j} - b_{m,i}|}\right) \right)$. The issue is

just to show that $L \leq E + 2\log\left(\frac{r}{\rho}\right)$. Let $S = \sum_{1 \leq j \leq \ell_k, 1 \leq h \leq k} s_{h,j}$ and $Y =$

$\sum_{1 \leq j \leq \ell_h, 1 \leq h \leq k} y_{h,j}$. Since the $s_{h,j}$ and the $y_{h,j}$ belong to $[0,1]$ it is seen that given any couple (m,i) we have

$$(10) \quad S - 1 \leq \sum_{(h,j) \neq (m,i)} s_{h,j} \leq S \quad \text{and}$$

$$(11) \quad Y - 1 \leq \sum_{(h,j) \neq (m,i)} y_{h,j} \leq Y.$$

Besides by ii) already proven, we have

$$(12) \quad S \leq Y \leq S + 1$$

Let $A = \sup_{1 \leq h \leq k} \left(\sup_{1 \leq i \leq \ell_h} \left(\sum_{j \neq i} y_{h,j} \log\left(\frac{r'}{|b_{m,j} - b_{h,i}|}\right) \right) \right), \quad$ let

$B = \sup_{1 \leq h \leq k} \left(\sup_{1 \leq i \leq \ell_h} \left(\sum_{j \neq i} s_{h,j} \log\left(\frac{r'}{|b_{h,j} - b_{m,i}|}\right) \right) \right) \quad$ and let $\tau \in \mathbb{N}$ be such

$A = \sup_{1 \leq i \leq \ell_\tau} \left(\sum_{j \neq i} y_{\tau,j} \log\left(\frac{r'}{|b_{m,j} - b_{\tau,i}|}\right) \right).$

Now, we can apply Relation iii), (true for $n < q$) to A with $n = \ell_\tau$, and r'

at the place of r. We obtain

$$A \leq \sup_{1 \leq i \leq \ell_r} \left(\sum_{\substack{j \neq i \\ 1 \leq j \leq \ell_r}} s_{r,j} \, \log\left(\frac{r'}{|b_{r,j} - b_{r,i}|}\right)\right) + 2\log\left(\frac{r'}{\rho}\right).$$

Hence with greater reason we have

(13) $A < B + 2\log\left(\dfrac{r'}{\rho}\right)$. Actually, by definition we have

$$L = \sup_{1 \leq i \leq \ell_m, 1 \leq m \leq k} \left(\sum_{(h,j) \neq (m,i)} y_{h,j} \log\left(\frac{r}{r'}\right) + \sum_{(h,j) \neq (m,i)} y_{h,j} \, \log\left(\frac{r'}{|b_{h,j} - b_{m,i}|}\right)\right)$$

But now it is seen that $\displaystyle\sum_{(h,j) \neq (m,i)} y_{h,j} \log\left(\frac{r}{r'}\right) \leq Y \, \log\left(\frac{r}{r'}\right)$ and then by (11) we

have $L \leq Y \, \log\left(\dfrac{r}{r'}\right) + A$. Hence by (13) we obtain

(14) $L \leq Y \, \log\left(\dfrac{r}{r'}\right) + B + 2\log\left(\dfrac{r'}{\rho}\right)$

In the same way, by (10) we have $\displaystyle\sum_{(h,j) \neq (m,i)} S_{h,j} \geq S - 1$ and then

$E \geq (S - 1) \, \log\left(\dfrac{r}{r'}\right) + B$. Hence by (14) we have

$L \leq (Y - S + 1) \, \log\left(\dfrac{r}{r'}\right) + E + 2\log\left(\dfrac{r'}{\rho}\right)$. Finally by (12) we obtain

$L \leq E + 2\log\left(\dfrac{r}{r'}\right) + 2\log\left(\dfrac{r'}{\rho}\right)$ and this ends the proof.

Theorem 43.2: (M.-C. Sarmant)   *Let $(T_{m,i}, q_{m,i})_{\substack{1 \leq i \leq k_m \\ m \in \mathbb{N}}}$ be a T-sequence on D of diameter r satisfying $0 < \rho' \leq diam(T_{m,i}) \leq \rho'' < r$ whenever $i = 1, \ldots, k_m$, $m \in \mathbb{N}$. There exists an idempotent T-sequence of the form $(T_{m,i}, u_{m,i})_{\substack{1 \leq i \leq k_m \\ m \in \mathbb{N}}}$.*

Proof: For each (m,i) $(1 \leq i \leq k_m, m \in \mathbb{N})$ we take $b_{m,i} \in T_{m,i}$ and we put $\rho_{m,i} = diam(T_{m,i})$. It is easily seen that we may assume the T-sequence to be increasing , of center 0. Indeed if the T-sequence is decreasing, of center α, by Proposition 35.3 an inversion of center α transforms it into an increasing T-sequence $(T'_{m,i}, q_{m,i})_{\substack{1 \leq i \leq k_m \\ m \in \mathbb{N}}}$, of radius $\dfrac{1}{r}$ satisfying

$0 < \dfrac{\rho'}{r^2} \leq diam(T'_{m,i}) \leq \dfrac{\rho''}{r^2} < \dfrac{1}{r}$ whenever $i = 1, \ldots, k_m, m \in \mathbb{N}$. Hence if the weighted sequence $(T_{m,i}, u_{m,i})_{\substack{1 \leq i \leq k_m \\ m \in \mathbb{N}}}$ is such that its image by an inversion is a

T-sequence, this weighted sequence $(T_{m,i}, u_{m,i})_{\substack{1 \leq i \leq k_m \\ m \in \mathbb{N}}}$ is a T-sequence. Now if the sequence is decreasing with no center, in $\widehat{K}$, we can consider the weighted sequence $(\tau_{m,i}, q_{m,i})_{\substack{1 \leq i \leq k_m \\ m \in \mathbb{N}}}$, defined as follows: for each (m,i) we put $\tau_{m,i} = \widehat{d}(b_{m,i}, \rho_{m,i}^-)$. Then it is seen that the weighted sequence $(\tau_{m,i}, q_{m,i})_{\substack{1 \leq i \leq k_m \\ m \in \mathbb{N}}}$ is a T-sequence of the set $\widehat{D}$. Hence we find an idempotent T-sequence $(\tau_{m,i}, u_{m,i})$ and therefore the weighted sequence of D $(T_{m,i}, u_{m,i})$ is an idempotent T-sequence of D.

Thus, we assume the T-sequence to be increasing of center 0. Besides without loss of generality we may also assume $|b_{m,i}| > \rho''$ because $\rho'' < R$.

For each $m \in \mathbb{N}$ we put $|b_{m,i}| = d_m \ (1 \leq i \leq k_m)$ $\quad q_m = \displaystyle\sum_{i=1}^{k_m} q_{m,i}$,

$$\theta_m = \log d_m - \left(\frac{1}{q_m}\right) \min_{1 \leq i \leq k_m} \left(q_{m,i} \log \rho'' + \sum_{\substack{j \neq i \\ 1 \leq i \leq k_m}} q_{m,j} \log |b_{m,j} - b_{m,i}|\right) \text{ and}$$

$\phi_m = \displaystyle\sum_{j=1}^{m} q_j(\log d_m - \log d_j) - q_m \theta_m$. Since $d_m > \rho''$ we notice that $\theta_m > 0$ for all $m \in \mathbb{N}$. By definition of a T-sequence, we have $\displaystyle\lim_{m \to +\infty} \phi_m = +\infty$. Hence by Lemma 36.3 we know that the series $\displaystyle\sum_{m=0}^{+\infty} \frac{\log r - \log d_m}{\theta_m}$ diverges. By definition of θ_m, we notice

$$(1) \quad q_m \theta_m \geq q_m \log d_m - \left(q_{m,i} \log \rho'' + \sum_{\substack{j \neq i \\ 1 \leq i \leq k_m}} q_{m,j} \log |b_{m,j} - b_{m,i}|\right), \text{ for every}$$

$i = 1, ..., k_m$. For all $m \in \mathbb{N}$ we put $\mu_m = \dfrac{\log d_m - \log \rho''}{\theta_m}$. Of course we have $\mu_m > 0$. Then by (1), for each $i = 1, \ldots, k_m$, we see that

$$\mu_m \left(\log d_m - \frac{q_{m,i}}{q_m} \log \rho'' - \sum_{\substack{j \neq i \\ 1 \leq j \leq k_m}} \frac{q_{m,j}}{q_m} \log |b_{m,j} - b_{m,i}|\right) \leq \log d_m - \log \rho''.$$

But now we have $\log d_m \geq \log |b_{m,j} - b_{m,i}|$ and therefore

$$\frac{\mu_m q_{m,i}}{q_m}(\log d_m - \log \rho'') \leq \log d_m - \log \rho''.$$

We put $s_{m,i} = \dfrac{\mu_m q_{m,i}}{q_m}$ $\quad (1 \leq i \leq k_m, m \in \mathbb{N})$. Hence each $s_{m,i}$ belongs to $]0,1]$.

By applying Lemma 43.1 to each set $(b_{m,i})_{1\leq i\leq k_m}$ we can find sets of integers $(v_{m,i})_{1\leq i\leq k_m}$, each one equal to 1 or 0 , satisfying

$$(2) \quad 0 \leq \sum_{i=1}^{k_m} v_{m,i} - \sum_{i=1}^{k_m} s_{m,i} < 1$$

$$(3) \quad \sup_{1\leq i\leq k_m} \Big(\sum_{\substack{j\neq i \\ 1\leq j\leq k_m}} v_{m,j} \log\Big(\frac{d_m}{|b_{m,j}-b_{m,i}|}\Big)\Big) \leq$$

$$\leq \sup_{1\leq i\leq k_m} \Big(\sum_{\substack{j\neq i \\ 1\leq j\leq k_m}} s_{m,j} \log\Big(\frac{d_m}{|b_{m,j}-b_{m,i}|}\Big)\Big) + 2\log\Big(\frac{d_m}{\rho'}\Big), \text{ if } k_m \neq 1.$$

For each $m \in \mathbb{N}$, we put $v_m = \sum_{i=1}^{k_m} v_{m,i}$ and $S_m =$

$$\sum_{j=1}^{m} v_j\Big(\log\Big(\frac{d_m}{d_j}\Big)\Big) - \sup_{1\leq i\leq k_m}\Big[v_{m,i}\log\Big(\frac{d_m}{\rho_{m,i}}\Big) + \sum_{\substack{j\neq i \\ 1\leq j\leq k_m}} v_{m,j}\log\Big(\frac{d_m}{|b_{m,j}-b_{m,i}|}\Big)\Big].$$

Since $v_{m,i} \leq 1$, we have

$$(4) \quad v_{m,i}\, \log(\frac{d_m}{\rho_{m,i}}) \leq \log(\frac{r}{\rho'}).$$

Besides, by (3) we see that

$$(5) \quad \sup_{1\leq i\leq k_m}\Big(\sum_{\substack{j\neq i \\ 1\leq j\leq k_m}} v_{m,j}\log\Big(\frac{d_m}{|b_{m,j}-b_{m,i}|}\Big)\Big) \leq$$

$$\leq \sup_{1\leq i\leq k_m}\Big(\sum_{\substack{j\neq i \\ 1\leq j\leq k_m}} s_{m,j}\log\Big(\frac{d_m}{|b_{m,j}-b_{m,i}|}\Big)\Big) + 2\log(\frac{r}{\rho'}).$$

Let $U_m = \sup_{1\leq i\leq k_m} \sum_{\substack{j\neq i \\ 1\leq j\leq k_m}} s_{m,j}\log\Big(\frac{d_m}{|b_{m,j}-b_{m,i}|}\Big)$. By (4) and (5) we have

$$(6) \quad S_m \geq \sum_{j=1}^{m} v_j(\log d_m - \log d_j) - U_m - 3\log(\frac{r}{\rho'}).$$

But since $s_{m,i} = \frac{\mu_m q_{m,i}}{q_m} = (\log d_m - \log \rho'')\frac{q_{m,i}}{\theta_m q_m}$, we see that

$$\sum_{\substack{j\neq i \\ 1\leq j\leq k_m}} s_{m,j}\, \log\Big(\frac{d_m}{|b_{m,i}-b_{m,i}|}\Big) =$$

$$= \frac{\log d_m - \log \rho''}{q_m \theta_m} \sum_{\substack{j \neq i \\ 1 \leq j \leq k_m}} q_{m,j} \log\left(\frac{d_m}{|b_{m,j} - b_{m,i}|}\right) \leq$$

$$\leq \frac{\log r - \log \rho''}{\theta_m q_m} \left[q_{m,i} \log\left(\frac{d_m}{\rho''}\right) + \sum_{\substack{j \neq i \\ 1 \leq j \leq k_m}} q_{m,j} \log\left(\frac{d_m}{|b_{m,j} - b_{m,i}|}\right) \right].$$

Thus we see that

$$q_{m,i} \log\left(\frac{d_m}{\rho''}\right) + \sum_{\substack{j \neq i \\ 1 \leq j \leq k_m}} q_{m,j} \log\left(\frac{d_m}{|b_{m,j} - b_{m,i}|}\right) \leq q_m \theta_m$$

and then finally we obtain

$$\sum_{\substack{j \neq i \\ 1 \leq j \leq k_m}} s_{m,j} \log\left(\frac{d_m}{|b_{m,j} - b_{m,i}|}\right) \leq \log\left(\frac{r}{\rho''}\right).$$

Hence by (6) we have $S_m \geq \sum_{j=1}^{m} v_j(\log d_m - \log d_j) - 4\log\left(\frac{r}{\rho'}\right)$. Thus it is now sufficient to show that the sequence $(S'_m)_{m \in \mathbb{N}}$ defined as

$$S'_m := \sum_{j=1}^{m} v_j(\log(d_m) - \log(d_j)) \text{ tends to } +\infty. \text{ By (2)}, \text{ for each } m \in \mathbb{N} \text{ we have}$$

$$v_n \geq \sum_{i=1}^{k_m} s_{m,i} = \frac{\log d_m - \log \rho''}{\theta_m} \sum_{i=1}^{k_m} \frac{q_{m,i}}{q_m} = \frac{\log d_m - \log \rho''}{\theta_m}.$$

Let $A = \log\left(\frac{d_0}{\rho''}\right)$. Then we have $A > 0$, and $v_m \geq \frac{A}{\theta_m}$ for all $m \in \mathbb{N}$, hence

$S'_m \geq A \sum_{j=1}^{m} \frac{\log d_m - \log d_j}{\theta_j}$. Now, the problem just consists of showing that the

sequence $(X_m)_{m \in \mathbb{N}}$ defined as $X_m := \sum_{j=1}^{m} \frac{\log d_m - \log d_j}{\theta_j}$ tends to $+\infty$.

For each $m \in \mathbb{N}$, let $t(m)$ be the biggest $j \in \mathbb{N}$ such that $\sqrt{\dfrac{d_j}{r}} < \dfrac{d_m}{r}$. Obviously the sequence $(t(m))_{m \in \mathbb{N}}$ goes to $+\infty$. Besides for every $j \leq t(m)$ it is seen that

$$\log d_m - \log d_j > \log r - \log d_m,$$

hence $\log d_m - \log d_j > \dfrac{1}{2}(\log r - \log d_j)$. Hence we have

$$\sum_{j=0}^{t(m)} \frac{\log d_m - \log d_j}{\theta_j} > \frac{1}{2} \sum_{j=0}^{t(m)} \frac{\log r - \log d_j}{\theta_j},$$

and therefore $X_m \geq \dfrac{1}{2} \displaystyle\sum_{j=0}^{t(m)} \frac{\log r - \log d_j}{\theta_j}$.

Since the series $\displaystyle\sum_{j=0}^{m} \frac{\log r - \log d_j}{\theta_j}$ diverges, and since $\displaystyle\lim_{m \to \infty} t(m) = +\infty$, it is seen that $\displaystyle\lim_{m \to \infty} X_m = +\infty$ and this completes the proof.

Remarks: 1) When $\rho' = r$ one can easily construct a T-sequence $(T_m, q_m)_{m \in \mathbb{N}}$ with $T_m = d(b_m, \rho_m^-)$ with $\displaystyle\prod_{j=1}^{\infty} \frac{r}{d_j} < +\infty$ and then there exists no idempotent T-sequence of the form $(T_m, u_m)_{m \in \mathbb{N}}$.

2) When $\rho'' = 0$ it is not known whether there always exists an idempotent T-sequence.

44. *T*-POLAR SEQUENCES

The strong properties of the idempotent T-sequence lead us to study what happens when a center is fixed in each hole of a T-sequence, while their diameter is changing.

Notations: In this chapter and in the following, given a sequence $(b_n)_{n\in\mathbb{N}}$ in K, and $\rho > 0$, we will denote by $\Lambda_\rho((b_n)_{n\in\mathbb{N}})$ the set $K \setminus \left(\bigcup_{n\in\mathbb{N}} d(b_n, \rho^-) \right)$.

Definitions: Let $a \in K$, let $r > 0$, and let $\rho \in]0, r[$. A sequence $(b_n)_{n\in\mathbb{N}}$ will be named *an increasing (resp. a decreasing) polar sequence , of center a, of diameter r and separation ρ* if it satisfies $|b_n - a| \leq |b_{n+1} - a|$
(resp. $|b_n - a| \geq |b_{n+1} - a|$)   for every $n \in \mathbb{N}$,
$$\lim_{n\to+\infty} |b_n - a| = r, \quad \text{and} \quad \inf_{n\neq m} |b_n - b_m| = \rho.$$
Many times a polar increasing (resp. decreasing) sequence will also be denoted in the form $(b_{m,i})_{\substack{1\leq i\leq k(m) \\ m\in\mathbb{N}}}$, with $|b_{m,i} - a| = |b_{m,j} - a|$ whenever $1 \leq i \leq k(m)$, $1 \leq j \leq k(m)$, and $|b_{m,i} - a| < |b_{m+1,j} - a|$ (resp. $|b_{m,i} - a| > |b_{m+1,j} - a|$) whenever $1 \leq i \leq k(m)$, $1 \leq j \leq k(m + 1)$. This form will be called *the graduate form of the polar sequence* and then the sequence $(d_m)_{m\in\mathbb{N}}$ defined as $d_m = |b_{m,j} - a|$ will be called *the monotony of the polar sequence.*

A polar sequence will be said to be *associated to the disk $d(a, r^-)$*   (resp. *to $K \setminus d(a, r)$*) if it is increasing (resp. decreasing), of center a and diameter r.

A polar sequence $(b_n)_{n\in\mathbb{N}}$ will be called *a T-polar sequence* if for some $\sigma \in]0, \rho] \cap]0, r[$, $\Lambda_\sigma((b_n)_{n\in\mathbb{N}})$ admits a T-sequence of the form $(d(b_n, \sigma^-), q_n)_{n\in\mathbb{N}}$, with $q_n \in \mathbb{N}$, $(n \in \mathbb{N})$.

Remarks: Given a polar sequence $(b_n)_{n\in\mathbb{N}}$ in K, of separation $\rho > 0$, for every $\sigma \in]0, \rho]$, $\Lambda_\sigma((b_n)_{n\in\mathbb{N}})$ admits for holes the disks $d(b_n, \sigma^-)$ that satisfy $d(b_n, \sigma^-) \cap d(b_m, \sigma^-) = \emptyset$ whenever $n \neq m$.

Proposition 44.1:   *Let $(b_n)_{n\in\mathbb{N}}$ be a T-polar sequence of diameter r and separation ρ. Then there exists a sequence of integers $(t_n)_{n\in\mathbb{N}}$ such that for every $r \in]0, \rho]$, the sequence $(d(b_n, r^-), t_n)_{n\in\mathbb{N}}$ is an idempotent T-sequence of $\Lambda_r((b_n)_{n\in\mathbb{N}})$.*

Proof: Let $\sigma \in]0, \rho[$, and let $(q_n)_{n \in \mathbb{N}}$ be sequence of natural integers such that the sequence $(d(b_n, \sigma^-), q_n)_{n \in \mathbb{N}}$ is a T-sequence of $\Lambda_\sigma((b_n)_{n \in \mathbb{N}})$. By Theorem 43.2, $\Lambda_\sigma((b_n)_{n \in \mathbb{N}})$ admits an idempotent T-sequence of the form $(d(b_n, \sigma^-), t_n)_{n \in \mathbb{N}}$. Let $(b_{m,i})_{\substack{1 \leq i \leq k(m) \\ m \in \mathbb{N}}}$ be the polar sequence in its graduate

form. For example, suppose the sequence increasing. Then, putting $u_m = \displaystyle\sum_{j=1}^{k(m)} u_{m,j}$

it satisfies

$$\lim_{m \to \infty} \left(\sup_{1 \leq j \leq k(m)} \left[\left(\frac{d_m}{\sigma} \right)^{u_{m,j}} \prod_{\substack{i \neq j \\ 1 \leq i \leq k(m)}} \left(\frac{d_m}{|b_{m,i} - b_{m,j}|} \right)^{u_{m,i}} \right] \prod_{n=1}^{m-1} \left(\frac{d_n}{d_m} \right)^{u_n} \right) = 0.$$

But since each $u_{m,j}$ is equal to 0 or 1, it is obviously seen that when we replace σ by any $r \in]0, \rho[$, there is no change in the limit above. We have a symmetric proof when the sequence is decreasing.

Proposition 44.2: *Let $(b_j)_{j \in \mathbb{N}}$ be a T-polar sequence of diameter r and separation ρ. Let $\sigma \in]\rho, r[$. There exists a subsequence $(b_{n(s)})_{s \in \mathbb{N}}$, which is a T-polar sequence whose separation τ lies in $[\sigma, r[$.*

Proof: By hypothesis the set $\Lambda_\rho((b_j)_{j \in \mathbb{N}})$ admits a T-sequence of the form $(d(b_j, \rho^-), q_j)_{j \in \mathbb{N}}$. For every $j \in \mathbb{N}$, let $V_j = d(b_j, \sigma^-)$. Without loss of generality, we can assume the b_j ordered in such a way that there exists a strictly increasing sequence of integers $n(s)_{s \in \mathbb{N}}$, satisfying $V_{n(s)} = V_j$ if and only if j lies in $\{n(s), ..., n(s+1) - 1\}$. Thus, by construction the sequence $(b_{n(s)})_{s \in \mathbb{N}}$ is a polar sequence of separation $\tau \geq \sigma$. Then we put $w_s = \displaystyle\sum_{j=n(s)}^{n(s+1)-1} q_j$. By

Proposition 35.4 the set $E := \Lambda_\sigma((b_{n(s)})_{s \in \mathbb{N}})$ does admit the weighted sequence $(V_{n(s)}, w_s)_{s \in \mathbb{N}}$ as a T-sequence. But since $\sigma < r$, E admits an idempotent T-sequence of the form $(V_{n(s)}, u_s)_{s \in \mathbb{N}}$ (with u_s, integers equal to 0 or 1). Hence the polar sequence $(b_n(s))_{s \in \mathbb{N}}$ is a T-polar sequence. This ends the proof.

By applying Relation (5) in Chapter 35 to idempotent T-sequences, we obtain this characterization of the T-polar sequences:

Proposition 44.3: *Let $(b_{m,j})_{\substack{1 \leq j \leq k(m) \\ m \in \mathbb{N}}}$ be a polar sequence, in its graduate form, and let $(d_m)_{m \in \mathbb{N}}$ be its monotony. Then the sequence is a T-polar sequence*

if and only if it satisfies

$$\lim_{m\to+\infty}\left(\inf_{1\leq i\leq k(m)}\left(\sum_{\substack{1\leq j\leq k(m)\\ j\neq i}}\log|b_{m,i}-b_{m,j}|\right)-k(m)\log d_m+\right.$$

$$\left.+\sum_{h=0}^{m-1}|\log d_m-\log d_h|_\infty\right)=+\infty.$$

Corollary 44.4: *Let $(b_n)_{n\in\mathbb{N}}$ be an increasing (resp. decreasing) polar sequence of center 0, of diameter r and separation ρ. For every $n\in\mathbb{N}$ let $c_n=\dfrac{1}{b_n}$. The sequence $(c_n)_{n\in\mathbb{N}}$ is a decreasing (resp. an increasing) polar sequence of center 0, of diameter $\dfrac{1}{r}$, and separation $\dfrac{1}{\rho}$. Besides, the sequence $(c_n)_{n\in\mathbb{N}}$ is a T-polar sequence if and only if so is the sequence $(b_n)_{n\in\mathbb{N}}$.*

45. ANALYTIC EXTENSION

THROUGH A T-FILTER

Let f be a power series converging in a disk $d(0,r)$. We will prove that actually, f admits infinitely many extensions to elements F analytic in an unbounded infraconnected set D containing $d(0,r)$, whose holes are located in an annulus $\Gamma(0,r,r')$ and define a well pierced T-sequence [58]. This surprising result will be used in many applications.

We remember that the subsidence of a T-sequence, involved in Lemma 45.1, was defined in Chapter 35.

As defined in Chapter 1, given a real number a, we denote by $Int(a)$ its integral part.

Lemma 45.1:   Let $(T_{m,i}, q_{m,i})_{1\leq i\leq k_m}$ be a T-sequence of diameter r, of piercing $\rho > 0$, and let $A \in\,]0,+\infty[$. There exists a T-sequence $(T_{m,i}, u_{m,i})_{\substack{1\leq i\leq k_m \\ m\in\mathbb{N}}}$ with $u_{m,i} \leq q_{m,i}$ (whenever $i = 1,...,k_m, m \in \mathbb{N}$) whose subsidence ν satisfies $\nu \leq A + 3\log(\dfrac{r}{\rho})$.

Proof: For every $i = 1,\ldots, k_m, m \in \mathbb{N}$, we put $\rho_{m,i} = diam(T_{m,i})$, and take $b_{m,i} \in T_{m,i}$. Let $(d_m)_{m\in\mathbb{N}}$ be the monotony of the T-sequence $(T_{m,i}, q_{m,i})$,

let λ be its subsidence, and for each $m \in \mathbb{N}$, let $q_m = \sum_{j=1}^{k_m} q_{m,j}$,

$$\text{let } e_m = \max_{1\leq i\leq k_m} \left(\frac{d_m}{\rho_{m,i}}\right)^{q_{m,i}} \prod_{\substack{j\neq i \\ 1\leq j\leq k_m}} \left(\frac{d_m}{|b_{m,i} - b_{m,j}|}\right)^{q_{m,j}},$$

and let $\theta_m = \log e_m - \sum_{j=1}^{m} q_j |\log d_m - \log d_j|$. Then we have

(1) $\lim_{m\to+\infty} \theta_m = -\infty$

and

(2) $\lambda = \sup_{m\in\mathbb{N}} \theta_m.$

For every couple $(m,i)_{\substack{1\leq i\leq k_m \\ m\in\mathbb{N}}}$ we put $t_{m,i} = Int(\frac{A}{\lambda}q_{m,i})$ and $u_{m,i} = \frac{A}{\lambda}q_{m,i} - t_{m,i}$.

By Lemma 43.1 there exists a family of integers $(v_{m,i})_{1\leq i\leq k_m}$ all equal to 0 or 1, satisfying

(3)   either $0 \leq v_{m,i} - u_{m,i} < 1$ or $v_{m,i} = 0$.

$$(4)\qquad 0 \leq \sum_{i=1}^{k_m} v_{m,i} - \sum_{i=1}^{k_m} u_{m,i} < 1$$

and

$$(5)\qquad \max_{1\leq i\leq k_m}\Big(\sum_{\substack{j\neq i \\ 1\leq j\leq k_m}} v_{m,j}\,\log\Big(\frac{d_m}{|b_{m,j}-b_{m,i}|}\Big)\Big) \leq$$

$$\leq \max_{1\leq i\leq k_m}\Big(\sum_{\substack{j\neq i \\ 1\leq j\leq k_m}} u_{m,j}\,\log\Big(\frac{d_m}{|b_{m,j}-b_{m,i}|}\Big)\Big) + 2\log\Big(\frac{d_m}{\rho}\Big).$$

Let $s_{m,i} = t_{m,i} + v_{m,i}$ $(1\leq i\leq k_m, m\in\mathbb{N})$,   let $s_m = \sum_{i=1}^{k_m} s_{m,i}$ and let

$$e'_m = \max_{1\leq i\leq k_m}\Big[\Big(\frac{d_m}{\rho_{m,i}}\Big)^{s_{m,i}} \prod_{\substack{j\neq i \\ 1\leq j\leq k_m}} \Big(\frac{d_m}{|b_{m,j}-b_{m,i}|}\Big)^{s_{m,j}}\Big].$$

By (3) we notice that $(t_{m,i} + v_{m,i})\,\log\Big(\frac{d_m}{\rho_{m,i}}\Big) \leq (t_{m,i} + u_{m,i})\,\log\Big(\frac{d_m}{\rho_{m,i}}\Big) +$ $\log\Big(\frac{d_m}{\rho}\Big)$ and then by (5) we have $\log e'_m \leq \max_{1\leq i\leq k_m}\Big[(t_{m,i} + u_{m,i})\,\log\Big(\frac{d_m}{\rho_{m,i}}\Big) +$

$$\sum_{\substack{j\neq i \\ 1\leq j\leq k_m}} (t_{m,j} + u_{m,j})\,\log\Big(\frac{d_m}{|b_{m,j}-b_{m,i}|}\Big) + 3\log\Big(\frac{d_m}{\rho}\Big)\Big].$$

Hence we obtain

$$(6)\qquad \log e'_m \leq \frac{A}{\lambda}\,\log e_m + 3\,\log\Big(\frac{d_m}{\rho}\Big).$$

We will check that the weighted sequence $(T_{m,i}, s_{m,i})_{\substack{1\leq i\leq k_m \\ m\in\mathbb{N}}}$ is a T-sequence

of subsidence $\leq A + 3\,\log\Big(\frac{r}{\rho}\Big)$. Indeed, by (4) we have

$$s_m \geq \sum_{i=1}^{k_m} t_{m,i} + \sum_{i=1}^{k_m} u_{m,i} = \frac{A}{\lambda}q_m, \quad \text{hence} \quad \sum_{j=1}^{m} s_j\Big|\log\Big(\frac{d_m}{d_j}\Big)\Big| \geq \frac{A}{\lambda}\sum_{j=1}^{m} q_j\,\Big|\log\Big(\frac{d_m}{d_j}\Big)\Big|$$

and then by (6) we obtain

$$(7)\qquad \log e'_m - \sum_{j=1}^{m} s_j\Big|\log\Big(\frac{d_m}{d_j}\Big)\Big| \leq \frac{A}{\lambda}\Big(\log e_m - \sum_{j=1}^{m} q_j\,\Big|\log\Big(\frac{d_m}{d_j}\Big)\Big|\Big) + 3\,\log\Big(\frac{r}{\rho}\Big)$$

Now by (1) it is clear that $\displaystyle\lim_{m\to\infty}\left(\log e'_m - \sum_{j=1}^{m} s_j \Big| \log\left(\frac{d_m}{d_j}\right)\Big|\right) = -\infty$ and therefore the weighted sequence we deal with is a T-sequence. Besides by hypothesis we have $\displaystyle\log e_m - \sum_{j=1}^{m} q_j \Big| \log\left(\frac{d_m}{d_j}\right)\Big| \leq \lambda$, hence by (7) the subsidence of the T-sequence $(T_{m,i}, s_{m,j})$ is clearly bounded by $A + 3 \log\left(\frac{r}{\rho}\right)$.

Lemma 45.2. is immediate.

Lemma 45.2:   *Let* $\alpha_1,\ldots,\alpha_q \in K\setminus\{0\}$ *and let* $\displaystyle Q(x) = \prod_{j=1}^{q}\left(1-\frac{x}{\alpha_j}\right) = \sum_{j=0}^{q} a_j x^j.$

Let $\displaystyle m = \min_{1\leq j\leq q}|\alpha_j|$. *Then* $\displaystyle |a_j| \leq \frac{1}{m^j}$ *whenever* $j = 0,\ldots,q.$

Proof: Each a_j is a sum of products of the form $(\alpha_{\sigma(1)},\ldots,\alpha_{\sigma(j)})^{-1}$ with σ any injection from $\{1,\ldots,j\}$ into $\{1,\ldots,q\}$. Each product of this form is obviously bounded by m^{-j}.

Proposition 45.3:   *Let* D *admit an increasing (resp. a decreasing) T-sequence* $(T_{m,i}, q_{m,i})_{\substack{1\leq i\leq k_m \\ m\in\mathbb{N}}}$ *of piercing* $\rho > 0$, *of center* α *and diameter* r.

Let $\displaystyle D' = K\setminus\left(\bigcup_{\substack{1\leq i\leq k_m \\ m\in\mathbb{N}}} T_{m,i}\right)$. *Let* $\mathcal{F}$ *be the T-filter on* D' *associated to this*

T-sequence. For each (m,i) $(1 \leq i \leq k_m$ $m \in \mathbb{N})$, *let* $b_{m,i} \in T_{m,i}$. *Let* $\epsilon \in]0,+\infty[$. *There exists* $\phi_\epsilon \in \mathcal{I}_0(\mathcal{F})$, *meromorphic on each hole* $T_{m,i}$, *satisfying further*

    a) $|\phi_\epsilon(x) - 1| \leq \epsilon$ *whenever* $x \in D \cap d(\alpha, r(1-\epsilon))$ $\left(\textit{resp.}\ x \in D\setminus\right.$
$\displaystyle d\left(\alpha, \left(\frac{r}{1-\epsilon}\right)^{-}\right)\Big),$

    b) $\displaystyle \|\phi_\epsilon\|_{D'} \leq \left(\frac{r}{\rho}\right)^3 + r^3\epsilon,$

    c) $b_{m,i}$ *is a pole of* ϕ_ϵ *of order* $u_{m,i} \leq q_{m,i}$ *and* ϕ_ϵ *has no pole different from* $b_{m,i}$ *in* $T_{m,i}$ *whenever* $i = 1,\ldots k_m$, $m \in \mathbb{N}$.

Proof: We will follow the proof of Proposition 37.2 step to step, assuming further the sequence $(T_{m,i}, q_{m,i})_{\substack{1\leq i\leq k_m \\ m\in\mathbb{N}}}$ to have piercing $\rho > 0$ and making ϕ_ϵ satisfy the strong conditions $a)$ and $b)$. Without loss of generality we may clearly assume $\alpha = 0$. If the Theorem is proven when the T-sequence is increasing, it is immediately generalized to the case when it is decreasing by considering the set $E = \{\frac{1}{x} \mid x \in D \setminus \{0\}\}$. So we will assume the T-sequence to be increasing without loss of generality.

For each $(m,i)_{\substack{1 \leq i \leq k_m \\ m \in \mathbb{N}}}$, we put $\rho_{m,i} = diam(T_{m,i})$ and for each $m \in \mathbb{N}$ we denote by $C(0, d_m)$ the circle of center 0 that contains the holes $(T_{m,i})_{1 \leq i \leq k_m}$. Given any $N \in \mathbb{N}$, we know that the family $(T_{m,i}, q_{m,i})_{\substack{1 \leq i \leq k_m \\ m \geq N}}$ also is a T-sequence. Therefore we may assume $d_m \geq r(1 - \epsilon)$ whenever $m \in \mathbb{N}$ without loss of generality. Let $q \in \mathbb{N}$ satisfy

$(\mathcal{U}_0)$ $(\frac{d_0}{d_1})^q \leq \epsilon.$

By Lemma 45.1 there clearly exists a T-sequence $(T_{m,i}, u_{m,i})_{\substack{1 \leq i \leq k_m \\ m > q}}$ whose sub-sidence λ is inferior or equal to $\log((\frac{r}{\rho})^3 + r^3 \epsilon)$ with $u_{m,i} \leq q_{m,i}$ for all (m,i).

For every $m \in \mathbb{N}$ we put $u_m = \sum_{i=1}^{k_m} u_{m,i}$, and

$$e_m = \sup_{1 \leq i \leq k_m} \left(\frac{d_m}{\rho_{m,i}}\right)^{u_{m,i}} \prod_{\substack{j \neq i \\ 1 \leq j \leq k_m}} \left(\frac{d_m}{|b_{m,j} - b_{m,i}|}\right)^{u_{m,j}}.$$ Since $(T_{m,i}, u_{m,i})_{\substack{1 \leq i \leq k_m \\ m \in \mathbb{N}}}$

is a T-sequence, for every $h \in \mathbb{N}$ we have

$$(1) \quad \lim_{m \to \infty} \left(e_m \prod_{j=h}^{m} \left(\frac{d_j}{d_m}\right)^{u_j}\right) = 0,$$

and for all $m \geq q$ we have

$$(2) \quad \log\left(e_m \prod_{j=q}^{m} \left(\frac{d_j}{d_m}\right)^{u_j}\right) \leq \log\left(\left(\frac{r}{\rho}\right)^3 + r^3 \epsilon\right)$$

By induction we will construct sequences of integers $s(n), \ell(n), w(n)$ satisfying:

$(\mathcal{U}_n)$ $\left(\frac{d_{\ell(n)}}{d_{s(n)}}\right)^{w(n)} < \frac{\epsilon}{n+1}$

$(\mathcal{V}_n)\left(\frac{r}{d_{s(n-1)}}\right)^{w(n-1)} \left(\frac{r}{d_{s(n)}}\right)^{w(n)} e_m \prod_{j=s(n-1)+1}^{m} \left(\frac{d_j}{d_m}\right)^{u_j} < \frac{\epsilon}{n+1}$ whenever $m \geq$

$s(n)$.

Let us suppose we have already defined these 3 sequences up to the rank n. We will construct them a the rank $n + 1$ in this way.

First we choose $\ell(n+1)$ such that $e_m \prod_{j=\ell(n)}^{m} \left(\frac{d_j}{d_m}\right)^{u_j} < \frac{\epsilon}{n+1}$ for all $m \geq \ell(n+1)$.

Then we can choose $w(n+1)$ such that $\left(\dfrac{d_{\ell(n+1)}}{d_{\ell(n+1)+1}}\right)^{w(n+1)} < \dfrac{\epsilon}{n+2}$. By (1) we can choose $s(n+1) > \max(\ell(n+1), s(n)+1+w(n))$ satisfying $(\mathcal{V}_{n+1})$.

Now it just remains to define $s(n), \ell(n), w(n)$ for $n = 1$ and $n = 2$, satisfying $(\mathcal{U}_2)$ and $(\mathcal{V}_2)$ in order to define the sequences for all $n \in \mathbb{N}$. We take $\ell(1) > q$, satisfying $e_m \displaystyle\prod_{j=q+1}^{m} \left(\dfrac{d_j}{d_m}\right)^{u_j} < \dfrac{\epsilon}{2}$ whenever $m \geq \ell(1)$. Then we can choose $w(1)$ such that $\left(\dfrac{d_{\ell(1)}}{d_{\ell(1)+1}}\right)^{w(1)} < \dfrac{\epsilon}{2}$. Next we take $s(1) > \ell(1)$ satisfying

$$(3) \qquad \left(\dfrac{r}{d_{s(1)}}\right)^{w(1)} \prod_{j=q+1}^{m} \left(\dfrac{d_j}{d_m}\right)^{u_j} < \left(\dfrac{\epsilon}{2}\right) \text{ whenever } m \geq s(1).$$

By (1) , we can choose $\ell(2)$ such that $e_m \displaystyle\prod_{j=\ell(2)}^{m} \left(\dfrac{d_j}{d_m}\right)^{u_j} < \dfrac{\epsilon}{3}$ whenever $m > \ell(2)$.

Hence we can choose $w(2)$ satisfying $(\mathcal{U}_2)$ and finally $s(2) > \ell(2)$ satisfying $(\mathcal{V}_2)$. Thus the sequences $s(n), \ell(n), w(n)$ are now defined for all n, satisfying $(\mathcal{U}_n)$ and $(\mathcal{V}_n)$, for all $n \geq 2$.

For every $m \in \mathbb{N}$ we put $Q_m := \displaystyle\prod_{i=1}^{k_m} \left(1 - \dfrac{x}{b_{m,i}}\right)^{u_{m,i}}$ and for each $n \in \mathbb{N}^*$ we put $H_n(x) := \displaystyle\prod_{m=s(n)+1}^{s(n+1)} Q_m$ and $t(n) = deg(H_n)$. We can develop $H_n(x)$ in the form $\displaystyle\sum_{h=0}^{t(n)} a_{n,h}\, x^h$ (with $a_{n,0} = 1$). Since $t(n) = \displaystyle\sum_{m=s(n)+1}^{s(n+1)} u_m$ and since $s(n+1) > s(n)+w(n)$, it is seen that $t(n) > w(n)$. Now we put $G_n(x) := \displaystyle\sum_{h=0}^{w(n)} a_{n,h}\, x^h$, and let $R_n(x) := \dfrac{G_n(x)}{H_n(x)}$. Thus, R_n is defined when $n > 0$. It only remains to define R_0. Let $P(x) = \displaystyle\prod_{m=1}^{q} \left(1 - \dfrac{x}{b_{m,1}}\right)$ and let $H_0(x) = P(x) \displaystyle\prod_{m=q+1}^{s(1)} Q_m(x)$. We can develop $H_0(x) = \displaystyle\sum_{h=0}^{t(0)} a_{0,h}\, x^h$. We put $G_0(x) = \displaystyle\sum_{h=0}^{q} a_{0,h}\, x^h$ and $R_0(x) = \dfrac{G_0(x)}{H_0(x)}$.

By Lemma 45.2 we notice the relation

$$(X_n) \quad |G_n(x)| \leq \max\left(1, \left(\dfrac{|x|}{d_{s(n)}}\right)^{w(n)}\right) \text{ whenever } x \in D, \text{ whenever } n \in \mathbb{N}^*.$$

Next, we check that $(\mathcal{Q}_0)$ $|R_0(x) - 1| \leq \epsilon$ whenever $x \in D \cap d(0, d_0)$. Indeed

we have $|R_0(x) - 1| = \left| \dfrac{\displaystyle\sum_{h=q+1}^{t(0)} a_{0,h}\, x^h}{H_0(x)} \right|$. But $|H(x)| \geq 1$ for all $x \in D \cap d(0, d_0)$

because H_0 has no zero in $d(0, d_1)$. Besides, by Lemma 45.2 we see that

$$\left| \sum_{h=q+1}^{t(0)} a_{0,h}\, x^h \right| \leq \max_{q+1 \leq h \leq t(0)} \frac{|x|^h}{(d_1)^h} \leq \max_{q+1 \leq h \leq t(0)} \left(\frac{d_0}{d_1}\right)^h = \left(\frac{d_0}{d_1}\right)^{q+1}$$

Hence finally by $(\mathcal{U}_0)$ we obtain $|R_0(x) - 1| \leq \epsilon$ for every $x \in D \cap d(0, d_1)$. In the same way we will prove the relations

$(\mathcal{Q}_n)$ $|R_n(x) - 1| \leq \dfrac{\epsilon}{n+1}$ for all $x \in D \cap d(0, d_{\ell(n)})$.

Indeed as $H_n(x)$ has no zero in $d(0, (d_{s(n)+1})^-)$, we have

$(\mathcal{T}_n)$ $|H_n(x)| = 1$ for all $x \in d(0, d_{\ell(n)}) \cap D'$. Besides, by applying Lemma 45.2

to H_n we have $|a_{n,h}| \leq \dfrac{1}{(d_{s(n)+1})^h}$ whenever $h = 0, ..., t(n)$, and therefore when

$x \in d(0, d_{\ell(n)})$ we obtain

$$|H_n(x) - G_n(x)| = \left| \sum_{h=w(n)+1}^{t(n)} a_{n,h}\, x^h \right| \leq$$

$$\leq \max_{w(n)+1 \leq h \leq t(n)} \left(\frac{d_{\ell(n)}}{d_{s(n)+1}}\right)^h = \left(\frac{d_{\ell(n)}}{d_{s(n)+1}}\right)^{w(n)+1} \leq \left(\frac{d_{\ell(n)}}{d_{s(n)}}\right)^{w(n)} \left(\frac{d_{s(n)}}{d_{s(n)+1}}\right)^{w(n)}.$$

But by $(\mathcal{U}_n)$ we have $\left(\dfrac{d_{\ell(n)}}{d_{s(n)}}\right)^{w(n)} \leq \dfrac{\epsilon}{n+1}$ and therefore by $(\mathcal{T}_n)$, we have

proven $(\mathcal{Q}_n)$ $(n \in \mathbb{N})$. Besides, by (X_n), we notice that $|G_n(x)| \leq 1$ whenever $x \in d(0, d_{s(n)})$. So, we have

$(\mathcal{S}_n)$ $|R_n(x)| \leq 1$ for all $x \in D \cap d(0, d_{s(n)})$.

Now we put $f_n(x) = \displaystyle\prod_{j=0}^{n} R_j(x)$. We will prove Relations $(\mathcal{R}_{n,k})$, $k = 1, \ldots, n$

$n \in \mathbb{N}$.

$(\mathcal{R}_{n,k})$ $|f_n(x)| \leq \dfrac{\epsilon}{k+1}$ whenever $x \in D \setminus d(0, d_{s(k)})$, whenever $k = 1, \ldots, n$.

Let us suppose Relations $(\mathcal{R}_{n,k})$ when $n \leq N$ are already proven and let us show them for $n = N + 1$. Since $(\mathcal{R}_{N,k})$ is satisfied, it is seen that it directly implies $(\mathcal{R}_{N+1,k})$ for $k \leq N$. Hence it just remains to prove $(\mathcal{R}_{N+1,N+1})$.

First we suppose $x \in D \setminus d(0, d_{s(N+2)})$. Then we have

$$|H_{N+1}(x)| = \prod_{j=s(N+1)+1}^{s(N+2)} \left(\frac{|x|}{d_j}\right)^{u_j} \geq \prod_{j=s(N+1)+1}^{s(N+2)} \left(\frac{d_{s(N+2)}}{d_j}\right)^{u_j}$$

and therefore by Relation (X_{N+1}) we obtain

$$|R_{N+1}(x)| \leq \left(\frac{r}{d_{s(N+1)}}\right)^{w(N+1)} \prod_{j=s(N+1)+1}^{s(N+2)} \left(\frac{d_j}{d_{s(N+2)}}\right)^{u_j}.$$

Thus by (2) we have $|R_{N+1}(x)| \leq \dfrac{\epsilon}{N+2}$ and therefore by $(\mathcal{R}_{N,N})$ Relation

(α) $|\prod_{N+1}(x)| \leq \dfrac{\epsilon}{N+2}$

holds for all $x \in D$ such that $|x| > d_{s(N+2)}$.

We now suppose $d_{s(N+1)} \leq |x| \leq d_{s(N+2)}$. Actually, for convenience and more generally, we suppose $d_{s(N+1)} \leq |x| < d_{s(N+2)+1}$. For example, let $d_m \leq |x| < d_{m+1}$ with $s(N+1) \leq m < s(N+2)$. It is seen that

$$\frac{1}{|H_{N+1}(x)|} \leq e_m \prod_{j=s(N+1)+1}^{m} \left(\frac{d_j}{|x|}\right)^{u_j}, \quad \text{and}$$

$$|R_{N+1}(x)| \leq e_m \left(\frac{r}{d_{s(N+1)}}\right)^{w(N+1)} \prod_{j=s(N+1)+1}^{m} \left(\frac{d_j}{|x|}\right)^{u_j}.$$

Then we have $|R_n(x)R_{N+1}(x)| \leq$

$$\leq \left(\frac{r}{d_{s(N)}}\right)^{w(N)} \left(\prod_{j=s(N)+1}^{s(N+1)} \left(\frac{d_j}{d_m}\right)^{u_j}\right) e_m \left(\frac{r}{d_{s(N+1)}}\right)^{w(N+1)} \left(\prod_{j=s(N+1)+1}^{m} \left(\frac{d_j}{d_m}\right)^{u_j}\right) =$$

$$= e_m \left(\frac{r}{d_{s(N)}}\right)^{w(N)} \left(\frac{r}{d_{s(N+1)}}\right)^{w(N+1)} \prod_{j=s(N)+1}^{m} \left(\frac{d_j}{d_m}\right)^{u_j}$$

But then by $(\mathcal{V}_{N+1})$ this is inferior or equal to $\dfrac{\epsilon}{N+2}$ by $(\mathcal{V}_{N+1})$ because $m > s(N+1)$. Hence, as $|f_{N-1}(x)| < 1$, we have proven that Relation (α) finally holds for all $x \in D$ such that $|x| \geq d_{s(N+1)}$. This just proves Relation $\mathcal{R}_{N+1,N+1}$.

Thus, it just remains to establish $\mathcal{R}_{1,1}$ in order to start the recurrence. This means $|R_0(x)R_1(x)| \leq \dfrac{\epsilon}{2}$ whenever $x \in D \setminus d(0, d_{s(1)})$ and this is also equivalent to

$$\left|\left(\frac{G_0(x)}{P(x)}\right)\left(\frac{1}{\displaystyle\prod_{m=q+1}^{s(1)} Q_m(x)}\right)\left(\frac{G_1(x)}{\displaystyle\prod_{m=s(1)+1}^{\infty} Q_m(x)}\right)\right| \leq \frac{\epsilon}{2}$$

whenever $x \in D$ such that $|x| > d_{s(1)}$. By construction, the coefficients of G_0 are the coefficients of H_0 of same index when this index runs from 0 to q. But by construction H_0 admits a unique zero in each circle $C(0, d_m)$ for $1 \leq m \leq q$ and no other zero in K. Then it is easily seen that we have

(4) $\quad \left|\dfrac{G_0(x)}{P(x)}\right| \leq \dfrac{1}{\rho}$ whenever $x \in D$,

and in particular

(5) $\quad \left|\dfrac{G_0(x)}{P(x)}\right| = 1$ whenever $x \in D \setminus d(0, d_q)$.

Hence finally we have $|R_0(x)R_1(x)| = \dfrac{|G_1(x)|}{\left|\displaystyle\prod_{m=q+1}^{s(1)} Q_m(x)\right|}$ whenever $x \in D \setminus d(0, d_q)$.

We first consider the case $|x| > d_{s(2)}$. Then we have

$$\frac{1}{\left|\displaystyle\prod_{j=q+1}^{s(2)} Q_j(x)\right|} \leq \prod_{j=q+1}^{s(2)} \left(\frac{d_j}{d_{s(2)}}\right)^{u_j}, \quad \text{hence}$$

(6) $\quad \dfrac{|G_1(x)|}{\left|\displaystyle\prod_{m=q+1}^{s(2)} Q_m(x)\right|} \leq \left(\dfrac{r}{d_{s(1)}}\right)^{w(1)} \displaystyle\prod_{j=q+1}^{s(2)} \left(\dfrac{d_j}{d_{s(2)}}\right)^{u_j}.$

Now it just remains to consider $x \in D$ when $d_{s(1)} \leq |x| \leq d_{s(2)}$. Say $d_m \leq |x| < d_{m+1}$, with $s(1) \leq m \leq s(2)$. We see that

$$\frac{1}{\left|\displaystyle\prod_{m=q+1}^{s(2)} Q_m(x)\right|} \leq \left(\prod_{j=q+1}^{m} \left(\frac{d_j}{d_m}\right)^{u_j}\right) e_m$$

hence

(7) $\quad \dfrac{|G_1(x)|}{\left|\displaystyle\prod_{m=q+1}^{s(2)} Q_m(x)\right|} \leq e_m \left(\dfrac{r}{d_{s(1)}}\right)^{w(1)} \displaystyle\prod_{j=q+1}^{m} \left(\dfrac{d_j}{d_m}\right)^{u_j}.$

Thus by (3), (6), (7) we have $|R_0(x)R_1(x)| \leq \frac{\epsilon}{2}$ for all $x \in D$ such that $|x| > d_{s(1)}$.
This proves $(\mathcal{R}_{1,1})$ and finishes proving all the relations $(\mathcal{R}_{n,k})$. Now by Relations $(\mathcal{Q}_n)$ and $(\mathcal{R}_{n,k})$ it is easily seen that the sequence $(f_n)_{n \in \mathbb{N}}$ converges in $H(D)$ to an element ϕ_ϵ satisfying

(8) $|\phi_\epsilon(x) - R_0(x)| \leq \epsilon$ for all $x \in D \cap d(0, d_0)$ and

$(\mathcal{G}_n)$ $|\phi_\epsilon(x)| = |f_n(x)|$ whenever $x \in D \cap d(0, d_{s(n)})$.

By construction, it is seen that ϕ_ϵ does satisfy Condition $c)$ in Proposition 45.3. Next, by Relations $(\mathcal{Q}_n)$ and (8) , we obtain

(9) $|\phi_\epsilon(x) - 1| \leq \epsilon$ whenever $x \in D \cap d(0, d_0)$

and therefore $|\phi_\epsilon(x) - 1| \leq \epsilon$ whenever $x \in D \cap d(0, r(1 - \epsilon))$, which is just Condition $a)$ in Proposition 45.3.

Now we only have to check Condition $b)$. By Relations $(\mathcal{G}_n)$ and $(\mathcal{R}_{n,k})$ true for $n \geq 1$, we have $|\phi_\epsilon(x)| < 1$ when $|x| > d_{s(1)}$ and therefore it just remains to show $|\phi_\epsilon(x)| \leq (\frac{r}{\rho})^3 + r^3\epsilon$ when $x \in D' \cap d(0, d_{s(1)})$. Actually by Relations $(\mathcal{S}_n)$ and then by (4) it is seen that $|H_n(x)| \leq \frac{r}{\rho}$ whenever $x \in D' \cap d(0, d_q)$ and therefore $|\phi_\epsilon(x)| \leq r^3(\epsilon + \frac{1}{\rho^3})$ whenever $x \in D' \cap d(0, d_q)$.

Now let $x \in D'$ satisfy $d_q \leq |x| \leq d_{s(1)}$. We still have $|R_n(x)| = 1$ for all $n \geq 1$ and then by (5) we obtain $|R_0(x)| = \dfrac{1}{\left| \prod\limits_{m=q+1}^{s(1)} Q_m(x) \right|}$. For example, let $x \in D'$

satisfy $d_m \leq |x| < d_m$ with $q + 1 \leq m \leq s(1)$. Then $|R_0(x)| \leq e_m \prod\limits_{j=q+1}^{m} \left(\dfrac{d_j}{d_m} \right)^{u_j}$,

and therefore by (3) we see that $|R_0(x)| \leq \dfrac{r^3}{\rho^3} + r^3\epsilon$. This finishes showing that ϕ_ϵ satisfies $b)$, and this ends the proof of Proposition 45.3.

Corollary 45.4:   *Let D have an increasing (resp. a decreasing) T-sequence $(T_{m,i}, q_{m,i})_{\substack{1 \leq i \leq k_m \\ m \in \mathbb{N}}}$ of piercing $\rho > 0$, of center α and diameter r. Let $D' = K \setminus (\bigcup\limits_{\substack{1 \leq i \leq k_m \\ m \in \mathbb{N}}} T_{m,i})$. Let $\mathcal{F}$ be the T-filter on D' associated to this T-sequence.*

For each (m, i) $(1 \leq i \leq k_m,\quad m \in \mathbb{N})$, let $b_{m,i} \in T_{m,i}$. Let $\epsilon \in]0, +\infty[$. There exists $\psi_\epsilon \in \mathcal{I}_0(\mathcal{F})$, meromorphic on each hole $T_{m,i}$, satisfying further $\psi_\epsilon(x) = 1$ whenever $x \in \mathcal{P}(\mathcal{F})$ and

a) $|\psi_\epsilon(x)| \leq \epsilon$ whenever $x \in D \cap d(\alpha, r(1-\epsilon))$ $\left(resp.\ x \in D \backslash d\left(\alpha, \left(\frac{r}{1-\epsilon}\right)^-\right)\right)$

b) $\|\psi_\epsilon\|_D \leq \left(\frac{r}{\rho}\right)^3 + r^3 \epsilon$

c) $b_{m,i}$ is a pole of ψ_ϵ of order $u_{m,i} \leq q_{m,i}$.

Proof: Indeed, in Proposition 45.3 , we just consider $\psi_\epsilon = 1 - \phi_\epsilon$.

Proposition 45.5:   Let $r,\ \lambda \in]0, +\infty[$, let D contain $K \backslash d(0, r^-)$ and have an increasing T-sequence $(T_{m,i}, q_{m,i})_{\substack{1 \leq i \leq k_m \\ m \in \mathbb{N}}}$ of piercing $\rho > 0$, of center 0 and diameter r . For each (m,i) $(1 \leq i \leq k_m,\quad m \in \mathbb{N})$, let $b_{m,i} \in T_{m,i}$.

For every $n \in \mathbb{N}$ there exists $g_n \in H(D)$, meromorphic on each hole $T_{m,i}$, admitting $b_{m,i}$ as a pole of order $u_{m,i} \leq q_{m,i}$ and admitting no other pole in $T_{m,i}$, satisfying further $g_n(x) = \dfrac{1}{x^n}$ whenever $x \in K \backslash d(0, r^-)$ and $\|g_n\|_D \leq \dfrac{1}{r^n}(\lambda + (\frac{r}{\rho})^3)$.

Proof: We fix $t \in \mathbb{N}$. Let $\epsilon \in]0, 1[$ satisfy

(1) $\dfrac{\epsilon}{(1-\epsilon)} < 1$

We will construct a finite sequence $(\psi_{n,\epsilon})_{1 \leq n \leq t}$ in $H(D \cup d(0, r(1-\epsilon)))$ satisfying:

(2) $\psi_{n,\epsilon}$ is meromorphic on each hole $T_{m,i}$, admits $b_{m,i}$ as a pole of order $u_{m,i} \leq q_{m,i}$ and no other pole in $T_{m,i}$,

(3) $\psi_{n,\epsilon}(x) = \dfrac{1}{x^n}$ whenever $x \in K \backslash d(0, r^-)$,

(4) $\psi_{n,\epsilon} \in H(D \cup d(0, r(1 - \epsilon)))$,

(5) $\|\psi_{n,\epsilon}\|_D \leq \left(\dfrac{1}{r(1-\epsilon)}\right)^n \left(r^3 \epsilon + \left(\dfrac{r}{\rho}\right)^3\right)$,

(6) $|\psi_{n,\epsilon}(x)| \leq \dfrac{\epsilon}{r^n(1-\epsilon)^n}$ whenever $x \in d(0, r(1-\epsilon))$. By Corollary 45.4 there does exist $\psi_{0,\epsilon} \in H(D)$ satisfying (2), (3), (4), (5), (6) when $n = 0$. So we may assume $\psi_{0,\epsilon}, \ldots, \psi_{n,\eta}$ are already constructed for any $n < t$, and for every $\eta > 0$, satisfying (1) and then, we will define $\psi_{t+1,\epsilon}$.

Let $\eta \in]0, \epsilon[$ satisfy

(7) $\dfrac{\eta}{(1-\eta)^n} < 1$.

We put $\psi_{t+1,\epsilon}(x) = \dfrac{\psi_{t,\eta}(x) - \psi_{t,\eta}(0)}{x - x^{t+1}\psi_{n,\eta}(0)}$. By definition $\psi_{t,\eta}$ satisfies (2), (3) and (4) at the rank t. Clearly $\psi_{t,\eta}(x) - \psi_{t,\eta}(0)$ factorizes in $H(D \cup d(0, r(1 - \eta)))$ in the form $x\theta(x)$. With greater reason, this factorization holds in $H(D \cup d(0, r(1-\epsilon)))$. Besides it is seen that we have

(8) $\|\psi_{t,\eta} - \psi_{t,\eta}(0)\|_D \leq \|\psi_{t,\eta}\|_D$.

By (1), (6), and (7) we have $|x^t \psi_{t,\eta}(0)| < 1$ whenever $x \in d(0,r)$. Hence we have $|x - x^{t+1}\psi_{n,\epsilon}(0)| = |x|$ whenever $x \in d(0,r)$, and therefore

(9) $|\psi_{t+1,\epsilon}(x)| = |\theta(x)|$ for all $x \in d(0, r(1-\epsilon))$.

But, as the norm $\|\cdot\|_{d(0,r(1-\epsilon))} = \varphi_{0,r(1-\epsilon)}$ is multiplicative, we have

$|\theta(x)| \leq \dfrac{1}{r(1-\epsilon)} \|\psi_{t,\eta}\|_{d(0,r(1-\eta))} \leq \dfrac{1}{r(1-\epsilon)}$. Hence by (6), we obtain $|\theta(x)| \leq$

$\dfrac{\eta}{r^{t+1}(1-\epsilon)(1-\eta)^t} \leq \dfrac{\epsilon}{r^{t+1}(1-\epsilon)^{t+1}}$. This is (6) at the rank $t+1$. Besides ,

with greater reason we have $|\theta(x)| \leq \dfrac{\|\psi_{n,\epsilon}\|_D}{r(1-\epsilon)}$ and therefore by (9) we have

(10) $|\psi_{t+1,\epsilon}(x)| \leq \dfrac{\|\psi_{t,\eta}\|_D}{r(1-\epsilon)}$ for all $x \in d(0, r(1-\epsilon))$.

Now when x belongs to $d(0,r) \setminus d(0, r(1-\epsilon))$, by (6) and (1) we see that $|x - x^{n+1}\psi_{t,\eta}(0)| = |x| > r(1-\epsilon)$, hence we have

$|\psi_{t+1,\epsilon}(x)| \leq \dfrac{|\psi_{t,\eta}(x) - \psi_{t,\eta}(0)|}{r(1-\epsilon)} \leq \dfrac{\|\psi_{t,\eta}\|_D}{r(1-\epsilon)}$. So, Relation (10) still holds.

When x belongs to $D \setminus d(0,r)$, since $\psi_{t+1}(x) = \dfrac{1}{x^{t+1}}$, Relation (9) holds trivially and is finally true in all of D. Finally, we have

$\|\psi_{t+1,\epsilon}\|_D \leq \dfrac{\|\psi_{t,\epsilon}\|_D}{r(1-\epsilon)} \leq \left(\dfrac{1}{r(1-\epsilon)}\right)^{t+1}\left(r^3\epsilon + \left(\dfrac{r}{\rho}\right)^3\right)$. This proves Relation (5)

at the rank $t+1$ and finishes the construction of the sequence $\psi_{n,\epsilon}$ satisfying (2), (3), (4), (5), (6) up to the rank $t+1$. Thus, the sequence $(\psi_{n,\epsilon})_{n\in\mathbb{N}}$ is now constructed for every $\epsilon > 0$.

Now for each $n \in \mathbb{N}$, we can choose $\epsilon_n > 0$ such that

$\left(\dfrac{1}{1-\epsilon_n}\right)^n\left(r^3\epsilon_n + \left(\dfrac{r}{\rho}\right)^3\right) < \lambda + \left(\dfrac{r}{\rho}\right)^3$. Then we put $g_n = \psi_{n,\epsilon_n}$. By construction

we have $g_t(x) = \dfrac{1}{x^t}$ whenever $x \in K \setminus d(0, r^-)$. Then by (5), we see that

$\|g_t\|_D \leq \dfrac{1}{r^t}\left(\lambda + \left(\dfrac{r}{\rho}\right)^3\right)$. This ends the proof.

Corollary 45.6. is an obvious consequence.

Corollary 45.6:   *Let r , $\lambda \in]0, +\infty[$, let D contain $d(0,r)$ and have a decreasing T-sequence $(T_{m,i}, q_{m,i})_{\substack{1 \leq i \leq k_m \\ m \in \mathbb{N}}}$ of piercing $\rho > 0$, of center 0 and diameter r. For each (m,i) $(1 \leq i \leq k_m, \quad m \in \mathbb{N})$, let $b_{m,i} \in T_{m,i}$. For each $n \in \mathbb{N}$ there exists $g_n \in H(D)$ meromorphic on each hole $T_{m,i}$, admitting $b_{m,i}$ as a*

pole of order $u_{m,i} \leq q_{m,i}$ and admitting no other pole in $T_{m,i}$, satisfying further
$$g_n(x) = x^n \text{ whenever } x \in d(0,r) \text{ and } \|g_n\| \leq r^n\left(\lambda + \left(\frac{r}{\rho}\right)^3\right).$$

Theorem 45.7 is due to Marie-Claude Sarmant and was first proven by considerations on infinite van der Monde matrices.

Theorem 45.7: (M.-C. Sarmant) *Let $(T_m, q_m)_{m \in \mathbb{N}}$ be an increasing (resp. decreasing) T-sequence on D of piercing $\rho > 0$, of center $\alpha \in D$ and diameter r, and let $\mathcal{F}$ be the increasing (resp. decreasing) T-filter of center α and diameter r. Let $E = \mathcal{P}(\mathcal{F})$ and let $\lambda \in]0, +\infty[$. For every $f \in H(E)$ there exists $F \in H(D)$ meromorphic on each hole T_m, admitting each point (b_m) as a pole of order $u_m \leq q_m$ and no other pole in T_m , satisfying further $F(x) = f(x)$ for all $x \in E$ and $\|F\|_D \leq \|f\|_E\left(\lambda + \left(\frac{r}{\rho}\right)^3\right).$*

Proof: First we suppose $\mathcal{F}$ increasing. There f is of the form $P(x) + \sum_{n=1}^{\infty} a_n x^{-n}$ with $P \in H(D \cup d(0, r^-))$. Hence the issue just consists of showing that the series $g(x) = \sum_{n=1}^{\infty} a_n x^{-n}$ admits an extension G that belongs to $H(D)$, is meromorphic on each hole T_m, admits each b_m as a pole of order $u_m \leq q_m$ and no other pole in T_m. All of these are given by Proposition 45.5. In the same way when $\mathcal{F}$ is decreasing, f is of the form $P + \sum_{n=1}^{\infty} a_n x^n$ with $P \in H(D \cup (K \setminus d(0,r)))$, and then by Corollary 45.6, $\sum_{n=1}^{\infty} a_n x^n$ has an extension G satisfying the statement.

Corollary 45.8:   *Let (b_n) be an increasing (resp. a decreasing) T-polar sequences of center α and diameter r. Let $f \in H(K \setminus d(0, r^-))$ (resp. $f \in H(d(0,r)))$ and for all $\rho > 0$, let $\Lambda_\rho = K \setminus \left(\bigcup_{n \in \mathbb{N}} d(b_n, \rho^-)\right)$. There exists $F \in \bigcap_{\rho > 0} H(\Lambda_\rho)$ such that $f(x) = F(x)$ whenever $x \in K \setminus d(\alpha, r^-)$ (resp. $x \in d(\alpha, r)$), admitting each b_n as a pole of order 1 or 0.*

Throughout this chapter, D is infraconnected and belongs to $\mathcal{A}$.

We will apply Chapter 45 to compare the quotient algebra $\dfrac{H(D)}{\mathcal{I}_0(\mathcal{F})}$ with $H(\mathcal{P}(\mathcal{F}))$ when $\mathcal{F}$ is a T-filter. The conclusions depend on the piercing of $\mathcal{F}$ [58].

Definition: A T-filter will be said to be *well pierced* if it admits a T-sequence whose piercing is not 0.

Lemma 46.1:   *Let $D \in \mathcal{A}$ have a monotonous filter $\mathcal{F}$ such that $\mathcal{P}(\mathcal{F}) \neq \emptyset$. The algebra $\dfrac{H(D)}{\mathcal{I}_0(\mathcal{F})}$ is isomorphic to a subalgebra of $H(\mathcal{P}(\mathcal{F}))$.*

Proof: Let ϕ be the canonical surjection from $H(D)$ onto the algebra $A = \dfrac{H(D)}{\mathcal{I}_0(\mathcal{F})}$. Let ψ be the mapping from $H(D)$ into $\mathcal{P}(\mathcal{F})$ that maps each $f \in H(D)$ to its restriction to $\mathcal{P}(\mathcal{F})$. Clearly we have $Ker(\phi) = Ker(\psi)$ whereas $Im(\psi)$ is a subalgebra of $H(\mathcal{P}(\mathcal{F}))$. Hence A is isomorphic to a subalgebra of $H(\mathcal{P}(\mathcal{F}))$.

Theorem 46.2:   *Let D be closed and bounded and have a well pierced T-filter $\mathcal{F}$ such that $\mathcal{P}(\mathcal{F}) \neq \emptyset$. The algebra $\dfrac{H(D)}{\mathcal{I}_0(\mathcal{F})}$ is isomorphic to $H(\mathcal{P}(\mathcal{F}))$ algebraically and topologically.*

Proof: Let $D' = \mathcal{P}(\mathcal{F})$. By Theorem 45.7, the mapping that maps each $f \in H(D)$ to its restriction $\check{f}$ to D' is a surjection onto $H(D')$. Thus $H(D')$ is algebraically isomorphic to $\dfrac{H(D)}{\mathcal{I}_0(\mathcal{F})}$. We will show that the norm $\|\cdot\|_{D'}$ is equivalent to the quotient norm on $\dfrac{H(D)}{\mathcal{I}_0(\mathcal{F})}$. Let $\|\cdot\|_q$ be this quotient norm.

Let $f \in H(D)$. By definition we have $\|f\|_q = \inf\{\|F\|_D \,|\, F \in H(D), \check{F} = \check{f}\}$, hence obviously $\|f\|_q \geq \|\check{f}\|_{D'}$. Now we will show that there exists a constant $M > 0$ such that $\|f\|_q \leq M\|\check{f}\|_{D'}$. Since $\mathcal{P}(\mathcal{F})$ is not empty, $\mathcal{F}$ has a center, and therefore we may obviously assume 0 to be a center of $\mathcal{F}$.

First we suppose $\mathcal{F}$ increasing of diameter r and piercing $\rho > 0$. Let $\epsilon \in \mathbb{R}_+$. There exists $t \in]0, r[$ such that $|f(x)| \leq_D \varphi_{0,r}(f) + \epsilon$ for every $x \in D \cap \Gamma(0, t, r)$. Since we obviously have $_D\varphi_{0,r}(f) \leq \|f\|_{D'}$, we obtain $\|f\|_{D \cap (K \setminus d(0,t))} \leq \|f\|_{D'} + \epsilon$. In particular, we can choose $t \in]r(1 - \epsilon), r[$. Now by Corollary 45.4 there exists $\psi_\epsilon \in H(D)$ such that $\psi_\epsilon(x) = 1$ whenever $x \in D'$, $|\psi_\epsilon(x)| \leq \epsilon$ whenever

$$x \in D \cap d(0, (r(1 - \epsilon)) \quad \text{and} \quad \|\psi_\epsilon\|_D \leq \left(\frac{r}{\rho}\right)^3 + r^3 \epsilon.$$

Let $F = \psi_\epsilon f$. We see that $|F(x)| \leq \epsilon \|f\|_D$ for all $x \in D \cap d(0, r(1 - \epsilon))$ and that

$$|F(x)| \leq (\|\check{f}\|_{D'} + \epsilon)\left(\left(\frac{r}{\rho}\right)^3 + r^3 \epsilon\right).$$ Since ϵ is arbitrary, it is seen that given $w > 0$

we can find $\epsilon \geq 0$ satisfying both $\quad \epsilon \|f\|_D \leq \|\check{f}\|_{D'} + w$, and

$(\|\check{f}\|_{D'} + \epsilon)\left(\left(\frac{r}{\rho}\right)^3 + r^3 \epsilon\right) \leq \|\check{f}\|_{D'}\left(\left(\frac{r}{\rho}\right)^3 + w\right)$. Thus, by putting $N = \left(\frac{r}{\rho}\right)^3$ we see

that $\|f\|_q \leq N \|\check{f}\|_{D'}$ and this finishes proving the theorem when $\mathcal{F}$ is increasing. We show it symmetrically when $\mathcal{F}$ is decreasing.

When $\mathcal{F}$ is not well pierced, the situation is all different as Theorem 46.3 shows.

Theorem 46.3: *Let D be closed and bounded, with a unique T-filter $\mathcal{F}$ such that $\mathcal{P}(\mathcal{F}) \neq \emptyset$. The set of the holes of D is supposed to be reduced to a sequence $(T_n)_{n \in \mathbb{N}}$ such that $\lim_{n \to \infty} (diam(T_n)) = 0$. Then $H(\mathcal{P}(\mathcal{F}))$ is not isomorphic to*

$$\frac{H(D)}{\mathcal{I}_0(\mathcal{F})}.$$

Proof: For example we assume $\mathcal{F}$ to be decreasing, of center 0 and diameter $r > 0$. Let $D' = \mathcal{P}(\mathcal{F})$ and let $A = \dfrac{H(D)}{\mathcal{I}_0(\mathcal{F})}$. By Lemma 46.1. we may consider A as a subalgebra of $H(D')$. Suppose that the norms $\|\cdot\|_q$ and $\|\cdot\|_D$ are equivalent on A. For each $t \in \mathbb{N}$ we put $f_t(x) = x^t$. We will compare $\|f_t\|_{D'}$ to $\|f_t\|_q$, when t tends to $+\infty$. By definition we have $\|f_t\|_q = \inf\{\|f_t + h_t\|_D \,|\, h \in \mathcal{I}_0(\mathcal{F})\}$. There exists $M \in \mathbb{R}_+$ such that $\|f_t\|_q \leq M \|f_t\|_{D'}$ whenever $t \in \mathbb{N}$. Hence for every $t \in \mathbb{N}$ there exists $h_t \in \mathcal{I}_0(\mathcal{F})$ satisfying

(1) $\quad \|f_t + h_t\|_D \leq (M + 1)\|f_t\|_{D'} = (M + 1)r^t$

For each $t \in \mathbb{N}$ we put $r_t = (\sqrt[t]{M + 1})r$. We now consider $|f_t(x)|$ when $x \in D \setminus d(0, r_t)$. We have $|f_t(x)| > (r_t)^t = (M + 1)r^t \geq \|f_t + h_t\|_D$ hence $|f_t(x)| = |h_t(x)|$ whenever $x \in D \setminus d(0, r_t)$.

Let $I =]-\log(diam(D)), -\log r_t[$. In particular we have $v(h_t, \mu) = v(f_t, \mu) = -t\mu$ for all $\mu \in I$. Let $A = -\log(diam(D))$. Since $\lim_{\mu \to -\log r} v(h_t, \mu) = +\infty$, the function $v(h_t, \cdot)$ is not concave in all of the interval $[-\log r_t, -\log r]$. However, it is obviously concave and equal to $-t\mu$ in $[A, -\log r_t]$, but it is not concave in all $[A, -\log r]$. Let E_t be the set of the $\gamma \in [-\log r_t, -\log r]$ such that the function $v(h_t, \cdot)$ is concave in $[A, \gamma]$ and let $\lambda_t = \inf(E_t)$. The function $v(h_t, \cdot)$ is then continuous and concave in $[A, \lambda_t]$. Therefore it is bounded and piecewise linear in this interval and also in an interval $[A, B]$ with $\lambda_t < B < -\log r$. Since the function $v(h_t, \cdot)$ is concave in $[A, \lambda_t]$ and is equal to $-t\mu$ in $[A, -\log r_t]$, we have

$$(2) \quad v(h_t, \lambda_t) \leq -t\lambda_t = v(f_t, \lambda_t).$$

Since this function is not concave in any interval $[A, \gamma]$ whenever $\gamma \in]\lambda_t, B]$, we have $v'^r(h_t, \lambda_t) > v''(h_t, \lambda_t)$. Hence by Proposition 34.6, $C(0, s_t)$ contains a hole V of D such that

$$(3) \quad \|h_t\|_{C(0,s_t) \cap D} \geq \frac{s_t}{diam(V)} \, {}_D\varphi_{0,s_t}(h_t).$$

Let $m(t)$ be the integer such that $T_{m(t)} = V$. By (2) we have ${}_D\varphi_{0,s_t}(h_t) \geq {}_D\varphi_{0,s_t}(f_t)$ hence by (3) we see that

$$(4) \quad \|h\|_{C(0,s_t) \cap D} \geq \frac{s_t}{\rho_{m(t)}} \, {}_D\varphi_{0,s_t}(f_t) = \frac{(s_t)^{t+1}}{\rho_{m(t)}} \geq \frac{r^{t+1}}{\rho_{m(t)}}.$$

Now it is seen that $\|h_t\|_{C(0,s_t) \cap D} > {}_D\varphi_{0,s_t}(f_t) = \|f_t\|_{C(0,s_t) \cap D}$. Hence we have $\|f_t + h_t\|_{C(0,s_t) \cap D} = \|h_t\|_{C(0,s_t) \cap D}$ and therefore by (4) we obtain

$$(5) \quad \|f_t + h_t\|_{C(0,s_t) \cap D} \geq \frac{r^{t+1}}{\rho_{m(t)}}.$$

But since $\lim_{t \to \infty} r_t = r$ it is seen that $\lim_{t \to \infty} m(t) = +\infty$ hence $\lim_{t \to \infty} \rho_{m(t)} = 0$.

Finally (1) is contradicted by (5) when t is big enough and this finishes the proof of Theorem 46.3. when $\mathcal{F}$ is decreasing. We make a symmetric proof when $\mathcal{F}$ is increasing.

47. MEROMORPHIC PRODUCTS

Meromorphic products were introduced and studied by Marie-Claude Sarmant [57]. Since then, they have proven to be extraordinarily useful to the study of the analytic elements.

In this chapter, we will show that each Motzkin factor of an analytic element f satisfying $\|f - 1\|_D < 1$ actually is a product of meromorphic products. The applications are numerous.

Notations and definitions: $(a_n)_{n\in\mathbb{N}}$, $(b_n)_{n\in\mathbb{N}}$ will denote sequences in K such that $\lim_{n\to\infty} a_n - b_n = 0$. For every $x \in K \setminus \{b_0,\ldots,b_n,\ldots\}$ the product

$$F_m = \prod_{n=0}^{m} \frac{x - a_n}{x - b_n} \quad \text{converges to a limit} \quad \prod_{n\in\mathbb{N}} \frac{x - a_n}{x - b_n}.$$

Such a function defined in $K \setminus \{b_1,\ldots,b_n,\ldots\}$ is called *a meromorphic product associated to the sequence* $(b_n)_{n\in\mathbb{N}}$ [57], [59], [60].

We remember that the sets $\Lambda_\rho((b_n)_{n\in\mathbb{N}})$ were defined in Chapter 44.

Remarks: 1)   Both sequences $(a_n)_{n\in\mathbb{N}}, (b_n)_{n\in\mathbb{N}}$ are not supposed to be injective.

2)   Each a_n is not supposed to be different from certain b_m for some m. Hence a finite product $\prod_{n=0}^{q} \frac{x - a_n}{x - b_n}$ is equal to a meromorphic product by choosing a sequence $(b_n)_{n \geq q}$ and putting $a_n = b_n$ whenever $n \geq q$.

3)   In particular, for $(a_n)_{n\in\mathbb{N}}$, $(b_n)_{n\in\mathbb{N}}$, we can choose polar sequences.

Proposition 47.1:   *Let* $F(x) = \prod_{n\in\mathbb{N}} \frac{x - a_n}{x - b_n}$ *be a meromorphic product. For every* $\rho > 0$, F *belongs to* $H(\Lambda_\rho((b_n)_{n\in\mathbb{N}}))$. *For every* $q \in \mathbb{N}$, *if* $d(b_q,\rho^-)$ *contains finitely many points* b_n, *then* F *is Meromorphic in* $d(b_q,\rho^-)$ *and the set of the poles of* F *in* $d(b_q,\rho^-)$ *is included in set of the the* b_n *that lie in* $d(b_q,\rho^-)$.

Proof: For convenience we put $D = \Lambda_\rho((b_n)_{n\in\mathbb{N}})$. For every $n \in \mathbb{N}$ we have

$$\left\|1 - \frac{x - a_n}{x - b_n}\right\|_D = \left\|\frac{a_n - b_n}{x - b_n}\right\|_D = \frac{|a_n - b_n|}{\rho} \quad \text{because obviously} \quad \left\|\frac{1}{x - b_n}\right\|_D = \frac{1}{\rho}.$$

Hence we have

287

(1)　　$\displaystyle\lim_{n\to\infty}\left\|1-\frac{x-a_n}{x-b_n}\right\|_D=0$

We notice that $\|F_n\|_D$ is bounded and superior or equal to 1 because $\displaystyle\lim_{|x|\to\infty}F_n(x)=1$ whereas F_n is bounded in every set $D\cap d(0,r)$ (whenever $r\geq 0$). Hence by (1), when m is big enough, we see that $\|F_m\|_D$ is a constant C. Then we have $\|F_{m+1}-F_m\|_D\leq C\dfrac{|a_{m+1}-b_{m+1}|}{\rho}$ and therefore the sequence F_m converges to F in $H(D)$. The last statement is obvious.

Proposition 47.2:　　*Let $T=d(a,r^-)$ $(a\in K, r>0)$. Let $(b_n)_{n\in\mathbb{N}^*}$ be a polar sequence of T and for every $\rho\in]0,r[$ let $D_\rho=\Lambda_\rho((b_n)_{n\in\mathbb{N}^*})$. Let $f(x)=\displaystyle\sum_{n=1}^\infty\frac{\theta_n}{x-b_n}$ with $\displaystyle\sup_{n\in\mathbb{N}^*}|\theta_n|<r$ and $\displaystyle\lim_{n\to\infty}\theta_n=0$. Then $1+f(x)$ is equal to a meromorphic product of the form $F(x)=\displaystyle\prod_{n=1}^\infty\frac{x-a_n}{x-b_n}\in H(D_\rho)$. Besides if $|\theta_n|<\rho$ for all $n\in\mathbb{N}^*$ then the sequence $(a_n)_{n\in\mathbb{N}^*}$ may be reordered to satisfy $|a_n-b_n|<\rho$ whenever $n\in\mathbb{N}^*$.*

Proof:　　We put $G(x)=1+f(x)$. Then for every $\rho>0$, G belongs to $H(D_\rho)$, and is meromorphic in each hole $d(b_n,\rho^-)$. We can find $\sigma\in]0,r[$ such that $|\theta_n|<\sigma$ for all n. Then it is seen that in $H(D_\sigma)$ f satisfies $\|f\|_{D_\sigma}<1$, hence
(1)　　$\|G-1\|_{D_\sigma}<1$.
Therefore by Theorem 31.5, in each hole $d(b_n,\sigma^-)$, G admits as many zeros as many poles. Hence we can find a sequence of zeros $(a_n)_{n\in\mathbb{N}^*}$ of G in T, such that
(2)　　$|a_n-b_n|<\sigma$ for every $n\in\mathbb{N}^*$.
Now let $\epsilon\in]0,\sigma[$. By (1) we have $_{D_\sigma}\varphi_{0,l}(G)=1$ for all $l\in]\sigma,r[$. But since G belongs to $H(D_\epsilon)$, it must satisfy $_{D_\sigma}\varphi_{0,l}(G)=_{D_\epsilon}\varphi_{0,l}(G)$ for all $l\geq\sigma$. In particular there exists $s\in]\sigma,r[$ such that $|G(x)-1|<1$ for all $x\in D_\epsilon\cap\Gamma(0,s,r)$. Let $N(\epsilon)$ be such that $|b_n|\geq s$ whenever $n\geq N(\epsilon)$. Then in each hole $d(b_n,\epsilon^-)$ of D_ϵ, with $n\geq N(\epsilon)$, G admits as many zeros as many poles. Besides by (2), it is seen that G admits as many zeros as many poles in $d(0,s)$, and then we have $|a_n|\leq s$ if and only if $n<N(\epsilon)$. Hence we may suppose the a_n are ordered in such a way that $|a_n-b_n|<\epsilon$ for every $n\leq N(\epsilon)$. Since this is true for every $\epsilon\in]0,\sigma[$, this shows that $\displaystyle\lim_{n\to+\infty}a_n-b_n=0$.

Now we can consider the meromorphic product $F(x)=\displaystyle\prod_{n=1}^\infty\frac{x-a_n}{x-b_n}$. Then F is invertible in $H(D_\sigma)$ and meromorphic in each hole $T_n:=d(b_n,\sigma^-)$. We see that in each T_n, both F, G have exactly the same zeros and the same poles (taking multiplicities into account). Hence $\dfrac{G}{F}$ is an element h of $H(D_\sigma)$, and is meromorphic in each T_n but has no pole in T_n. Since $\dfrac{G}{F}$ is bounded and has limit 1 when $|x|$ goes to $+\infty$, we have $\overline{h}_0 = 1$. Since $\dfrac{G}{F}$ has no pole in T_n we see that $\overline{h_{T_n}} = 0$ for every $n \in \mathbb{N}^*$. Hence $G = F$. Finally, when $\rho > \theta_n$ for all $n \in \mathbb{N}^*$, we can just take $\rho = \sigma$, and we see that $|a_n - b_n| < \rho$ for every $n \in \mathbb{N}^*$. This ends the proof of Proposition 47.2.

Theorem 47.3.   *Let* $V = d(a, r^-)$ $(a \in K, r > 0)$, *let* $f \in H(K \setminus V)$ *satisfy* $\lim\limits_{|x| \to \infty} f(x) = 1$ *and* $\|f - 1\|_{K\setminus V} < 1$ *and let* $(b_n)_{n\in\mathbb{N}}$ *be a T-polar sequence of* V. *Let* $\epsilon \in]0, \|f - 1\|_{K\setminus V}[$. *There exists a sequence* $(a_n)_{n\in\mathbb{N}}$ *in T such that* $\lim\limits_{n \to \infty} a_n - b_n = 0$, *satisfying further*

$$|a_n - b_n| < r(\|f - 1\|_{K\setminus V} + \epsilon), \text{ and } \prod_{n=0}^{\infty} \frac{x - a_n}{x - b_n} = f(x) \text{ whenever } x \in K \setminus V.$$

Proof: Let $\rho \in]\dfrac{r}{\sqrt[3]{1 + \frac{\epsilon}{2}}}, 1[$. First, we suppose that the sequence $(b_n)_{n\in\mathbb{N}}$ satisfies

(1) $|b_n - a| > \rho$ whenever $n \in \mathbb{N}$, and

(2) $|b_n - b_m| \geq \rho$ whenever $n \neq m$.

Let $E = K \setminus V$ and $D_\rho = \Lambda_\rho((b_n)_{n\in\mathbb{N}})$. Let $\lambda = \|f - 1\|_E$. For each $n \in \mathbb{N}$ we put $T_n = d(b_n, \rho^-)$. Each T_n is a hole of D_ρ and by hypothesis there exists an idempotent increasing T-sequence of center a and diameter r of the form $(T_n, u_n)_{n\in\mathbb{N}}$ with $u_n = 0$ or 1. Thanks to Relation (1), by Theorem 45.7, there exists $g \in H(D_\rho)$, meromorphic in each T_n, admitting each b_n as a pole of order 1 or 0, and no other pole inside T_n, satisfying further $g(x) = f(x) - 1$ for all $x \in E$, and $\|g\|_{D_\rho} < \lambda((\dfrac{r}{\rho})^3 + \dfrac{\epsilon}{2}) < \lambda(1 + \epsilon) < 1$.

By (2), we notice that in each hole T_n, g only has a simple pole equal to b_n, (and no other pole). Hence, by Theorem 31.6 the Mittag-Leffler series of g in $H_b(D_\rho)$ is of the form $A + \sum\limits_{n=0}^{\infty} \dfrac{\theta_n}{x - b_n}$ with $A \in K$ and $\lim\limits_{n \to \infty} \theta_n = 0$. But since $\lim\limits_{|x| \to +\infty} f(x) = 1$, it is seen that $A = 0$. Then we know that $\|g\|_{D_\rho} = \sup\limits_{n\in\mathbb{N}} \dfrac{|\theta_n|}{\rho}$ and therefore we have

(3) $\displaystyle \sup_{n\in\mathbb{N}} \frac{|\theta_n|}{\rho} < \lambda(1+\epsilon)$, hence $\displaystyle g(x) = \sum_{n=0}^{\infty} \frac{\theta_n}{x - b_n}$.

Now, since $\|g\|_{D_\rho} < 1$, by Proposition 47.2. and by (3) we see that $1 + g(x)$ is equal in $H(D_\rho)$ to a meromorphic product of the form $\displaystyle \prod_{n=0}^{\infty} \frac{x - a_n}{x - b_n}$ with $a_n \in T_n$ $(n \in \mathbb{N})$ satisfying $|a_n - b_n| < \rho\lambda(1+\epsilon) < r\lambda(1+\epsilon)$. Finally we have found a sequence $(a_n)_{n\in\mathbb{N}}$ such that $\displaystyle f(x) = \prod_{n=0}^{\infty} \frac{x - a_n}{x - b_n}$ whenever $x \in D$ and $|a_n - b_n| < r(\|f - 1\|_E + \epsilon)$. Thus we have found the meromorphic product $\displaystyle \prod_{n=0}^{\infty} \frac{x - a_n}{x - b_n}$ satisfying $\displaystyle \prod_{n=0}^{\infty} \frac{x - a_n}{x - b_n} = f(x)$ whenever $x \in E$, provided Relations (1) and (2) are satisfied.

Actually, we can suppose these relations to be satisfied without loss of generality. Indeed, by Proposition 44.2 we can extract from the T-polar sequence $(b_n)_{n\in\mathbb{N}}$ a T-polar subsequence $(b_{n_q})_{q\in\mathbb{N}}$ satisfying $|b_{n_q} - a| > \rho$ whenever $q \in \mathbb{N}$, and $|b_{n_q} - b_{n_s}| \geq \rho$ whenever $s \neq q$. Hence we can find a sequence $(a_{n_q})_{q\in\mathbb{N}}$ in V satisfying $\displaystyle f(x) = \prod_{n=0}^{\infty} \frac{x - a_{n_q}}{x - b_{n_q}}$ whenever $x \in D$ and $|a_{n_q} - b_{n_q}| < r(\|f - 1\|_E + \epsilon)$. Hence we just have to complete the definition of the sequence $(a_n)_{n\in\mathbb{N}}$ by putting $a_n = b_n$ whenever $n \in \mathbb{N}$ such that b_n does not belong to the subsequence $(b_{n_q})_{q\in\mathbb{N}}$, and then, we have obtained the meromorphic product $\displaystyle \prod_{n=0}^{\infty} \frac{x - a_n}{x - b_n}$ satisfying $\displaystyle \prod_{n=0}^{\infty} \frac{x - a_n}{x - b_n} = f(x)$ whenever $x \in E$. By Proposition 44.2 we can extract from the T-polar sequence $(b_n)_{n\in\mathbb{N}}$ a T-polar subsequence $(b_{n_q})_{q\in\mathbb{N}}$ satisfying $|b_{n_q} - a| > \rho$ whenever $q \in \mathbb{N}$, and $|b_{n_q} - b_{n_s}| \geq \rho$ whenever $s \neq q$. Hence without loss of generality we may assume Relations (1), (2) are satisfied. This ends the proof.

Proposition 47.4: *Let D be a closed infraconnected set and let $\phi \in H(D)$ satisfy $\|\phi - 1\|_D < 1$. Let $(T_m)_{m\in\mathbb{N}^*}$ be the sequence of the ϕ-holes of D. Let $(c_m)_{m\in\mathbb{N}^*}$ be a sequence of strictly positive numbers such that $\displaystyle \lim_{n\to\infty} c_m = 0$. Let $a \in D$.*

If D is bounded, of diameter r_0, there exists a strongly copiercing sequence $(a_{n,m}, b_{n,m})_{(n,m)\in\mathbb{N}\times\mathbb{N}}$ associated to D satisfying

$$\lambda\;)\quad \phi(x)=A\prod_{(n,m)\in\mathbb{N}\times\mathbb{N}}\frac{x-a_{n,m}}{x-b_{n,m}}\;\;with\;\;A=\phi(a)\prod_{n=0}^{\infty}\frac{b_{n,0}-a}{a_{n,0}-a}\;\;and$$

$|A-\phi(a)|<\|\phi-\phi(a)\|_D+c_0$.

$\mu)$ *For each* $m\in\mathbb{N}^*$ *, the sequence* $(a_{n,m},b_{n,m})_{n\in\mathbb{N}}$ *is a strongly copiercing sequence associated to* $K\setminus T_m$ *satisfying* $|a_{n,m}-b_{n,m}|\le r_m\Big(\|\phi_{T_m}\|_D+c_m\Big)$ *whereas* $\phi^{T_m}=\displaystyle\prod_{n=0}^{\infty}\frac{x-a_{n,m}}{x-b_{n,m}}.$

$\nu\;)$ *The sequence* $(a_{n,0},b_{n,0})_{n\in\mathbb{N}}$ *is a strongly copiercing sequence associated to* $\widetilde{D}$ *satisfying* $|a_{n,0}-b_{n,0}|\le r_0\Big(\|\phi_0-\phi(a)\|_D+c_0\Big)$ *whereas* $\phi^0=A\displaystyle\prod_{n=0}^{\infty}\frac{x-a_{n,0}}{x-b_{n,0}}.$

If D is not bounded there exists a strongly copiercing sequence $(a_{n,m},b_{n,m})_{(m,n)\in\mathbb{N}\times\mathbb{N}^*}$ *associated to D such that*

$$\lambda'\;)\quad \phi(x)=A\prod_{(n,m)\in\mathbb{N}\times\mathbb{N}^*}\frac{x-a_{n,m}}{x-b_{n,m}}\;\;with\;\;A=\lim_{\left\{\substack{|x|\to\infty\\ x\in D}\right.}\phi(x)$$

$\mu')$ *For each* $m\in\mathbb{N}^*$ *the sequence* $(a_{n,m},b_{n,m})_{n\in\mathbb{N}}$ *is a strongly copiercing sequence associated to* $K\setminus T_m$ *satisfying* $|a_{n,m}-b_{n,m}|\le r_m\Big(\|\phi_m\|_D+c_m\Big)$ *whereas* $\phi^{T_m}=\displaystyle\prod_{n=0}^{\infty}\frac{x-a_{n,m}}{x-b_{n,m}}.$

Proof: Since $\|\phi-1\|_D<1$, by Theorem 32.15 each ϕ^{T_m} is in the form $1+\varphi_m$ with $\varphi_m\in H_0(K\setminus T_m)$ and $\|\varphi_m\|_D<1$, while ϕ^o is in the form $1+\varphi_0$ with $\varphi_0\in H(\widetilde{D})$ and $\|\varphi_0\|_D<1$. With no loss of generality we may obviously assume $a=0$ and $\phi(0)=1$.

Let $m\in\mathbb{N}^*$. By Proposition 47.3 there exists a strongly copiercing sequence $(a_{n,m},b_{n,m})_{n\in\mathbb{N}}$ associated to T_m such that $\dfrac{|a_{n,m}-b_{n,m}|}{r_m}\le\|\varphi_m\|_D+c_m$ and

$$\phi^{T_m}=\prod_{n=0}^{\infty}\frac{x-a_{n,m}}{x-b_{n,m}}.$$

If D is not bounded, the sequence $(a_{n,m},b_{n,m})_{(n,m)\in\mathbb{N}\times\mathbb{N}^*}$ clearly satisfies

$$\lim_{n+m\to\infty}\frac{|a_{n,m}-b_{n,m}|}{r_m}=0$$ because the convergence of the sequence to 0 is uniform with respect to m. Besides, for each couple $(n,m)\in\mathbb{N}\times\mathbb{N}^*$, both $a_{n,m}$, $b_{n,m}$ belong to the hole T_m, hence the sequence is strongly copiercing.

Now we suppose D bounded, of diameter r_0. Let $c_0'\in\,]0,c_0[$ satisfy

(1) $\|\phi^0 - 1\|_D + c_0' < 1.$

Let $t = \dfrac{1}{x}$, let $r = \dfrac{1}{r_0}$, let $T = d(0, r^-)$, let $\psi(t) = \phi^0\left(\dfrac{1}{t}\right)$ and let $E = K \setminus T$. Hence ψ is invertible in $H_b(E)$. Since $\|\phi - 1\|_D < 1$, by Theorem 32.15 it is seen that $\|\phi^0 - 1\|_D < 1$, and therefore $\|\psi - 1\|_E < 1$. But by definition ψ is of the form $B\psi^T$, with $B = \lim_{|t| \to \infty} \psi(t)$. Hence we have

(2) $|B - 1| \le \|\psi - 1\|_E = \|\phi^0 - 1\|_{\widetilde{D}} = \|\phi^0 - 1\|_D.$

Let $\rho \in]r\|\phi^0 - 1\|_D, r[$ satisfy

(3) $\left(\dfrac{r}{\rho}\right)^2 \|\phi^0 - 1\|_D + c_0' < \|\phi^0 - 1\|_D + c_0.$

By Proposition 47.3 there exists a strongly copiercing sequence $(\alpha_n, \beta_n)_{n \in \mathbb{N}}$ associated to E satisfying

(4) $|\alpha_n - \beta_n| < \rho < |\beta_n|$ whenever $n \in \mathbb{N}$

(5) $\dfrac{|\alpha_n - \beta_n|}{r} < \|\psi^T - 1\|_E + c_0'\left(\dfrac{\rho}{r}\right)^2$ whenever $n \in \mathbb{N}$

and such that $\psi(t) = B\psi^T(t) = B \displaystyle\prod_{n=0}^{\infty}\left(\dfrac{t - \alpha_n}{t - \beta_n}\right)$. Now for every $n \in \mathbb{N}$, let

$a_{n,0} = \dfrac{1}{\alpha_n},\ b_{n,0} = \dfrac{1}{\beta_n},\ A = \displaystyle\prod_{n=0}^{\infty}\dfrac{\alpha_n}{\beta_n}$. It is easily seen that we have $\phi^0(x) =$

$BA \displaystyle\prod_{n=0}^{\infty} \dfrac{x - a_{n,0}}{x - b_{n,0}}$. Besides by (4) we have $|\alpha_n - \beta_n| < |\beta_n|$ hence

(6) $|a_{n,0} - b_{n,0}| = \dfrac{|\alpha_n - \beta_n|}{|\alpha_n \beta_n|} = \dfrac{|\alpha_n - \beta_n|}{|\beta_n|^2} < \dfrac{1}{|\beta_n|} = |b_{n,0}|.$

Thus the sequence $(a_{n,m}, b_{n,m})_{(n,m) \in \mathbb{N} \times \mathbb{N}^*}$ is clearly a strongly copiercing sequence associated to D because when $m > 0$ both $a_{n,m}, b_{n,m}$ belong to T_m and when $m = 0$ both $a_{n,m}, b_{n,m}$ belong to $K \setminus \widetilde{D}$ and they satisfy $|a_{n,m} - b_{n,m}| < \delta(b_{n,m}, D)$. Finally , we have constructed the strongly copiercing sequence satisfying $\lambda)$, $\mu)$ and then we just have to check the inequalities given in $\nu)$. By (5) we check that

$\dfrac{|a_{n,0} - b_{n,0}|}{|b_{n,0}|^2} \le r\left(\|\phi^0 - 1\|_D + c_0'\left(\dfrac{\rho}{r}\right)^2\right),$ hence

$\dfrac{|a_{n,0} - b_{n,0}|}{|b_{n,0}|} \le |b_{n,0}| r\left(\|\phi^0 - 1\|_D + c_0'\left(\dfrac{\rho}{r}\right)^2\right).$ Hence by (3), we deduce

(7) $\dfrac{|a_{n,0} - b_{n,0}|}{|b_{n,0}|} \le |b_{n,0}| r\left(\|\phi^0 - 1\|_D + c_0\right)\left(\dfrac{\rho}{r}\right)^2.$

Now by (4) we see that $|b_n| < \dfrac{1}{\rho}$, whence by (7) we obtain

$\dfrac{|a_{n,0} - b_{n,0}|}{|b_{n,0}|} \leq \dfrac{r}{\rho}\Big(\|\phi^0 - 1\|_D + c_0'\big(\dfrac{\rho}{r}\big)^2\Big) \leq \Big(\|\phi^0 - 1\|_D + c_0'\Big)$. As a consequence

we see that the sequence $(A_n)_{n\in\mathbb{N}}$ defined as $A_n = \displaystyle\prod_{j=1}^{n}\Big(\dfrac{b_j}{a_j}\Big)$ satisfies $|A_n - 1| <$

$\|\phi^0 - 1\|_D + c_0$ hence $|A - 1| < \|\phi^0 - 1\|_D + c_0$. This finishes showing ν) and it ends the proof of Proposition 47.4.

Reciprocally, in a meromorphic product, it is useful to recognize the Motzkin factors.

Theorem 47.5:   *Let D be a closed infraconnected set and let*

$\phi = \displaystyle\prod_{n=0}^{\infty} \dfrac{x - a}{x - b} \in H(D)$ *be a meromorphic product satisfying* $\|\phi - 1\|_D < 1$. *Let*

$(T_m)_{m\in\mathbb{N}^*}$ *be the sequence of the ϕ-holes of D. For each $m \in \mathbb{N}^*$, let I_m be*

the set of the $j \in \mathbb{N}$ such that j lies in T_m, and let $I_0 = \mathbb{N} \setminus \big(\displaystyle\bigcup_{m=1}^{\infty} I_m\big)$. Then

$\phi^0 = \displaystyle\prod_{n\in I_0} \dfrac{x - a_n}{x - b_n}$, *and for each $m \in \mathbb{N}^*$, we have* $\phi^{T_m} = \displaystyle\prod_{n\in I_m} \dfrac{x - a_n}{x - b_n}$.

Proof: Indeed, it is shortly checked that for each $m \in \mathbb{N}^*$, we have $\phi^{T_m} =$

$\displaystyle\prod_{n\in I_m} \dfrac{x - a_n}{x - b_n}$ because, putting $h_m = \displaystyle\prod_{n\in I_m} \dfrac{x - a_n}{x - b_n}$, $\dfrac{\phi}{h_m}$ belongs to $H(D\cup T_m)$, and

has no zero inside T_m. As a consequence, by Proposition 32.14, ϕ^0 is obviously

equal to $\displaystyle\prod_{n\in I_0} \dfrac{x - a_n}{x - b_n}$.

Corollary 47.6:   *Let $f \in H(D)$ be a semi-invertible element. There exists $h \in R(D)$ and a strongly copiercing sequence $(a_n, b_n)_{n\in\mathbb{N}}$ associated to D such*

that $f(x) = h(x) \displaystyle\prod_{n=0}^{\infty} \dfrac{x - a_n}{x - b_n}$.

Proof: Indeed, by Chapter 32, f factorizes in the form $h\phi$ with $h \in R(D)$ and $\phi \in H(D)$ satisfying $\|\phi - 1\|_D < 1$. Then Theorem 47.4 fully applies to ϕ and shows f to be the product of a rational function by an infinite product of meromorphic products.

48.　COLLAPSING MEROMORPHIC PRODUCTS

Now we are able to characterize the collapsing meromorphic products.

Theorem 48.1: *Let $V = d(a,r)$ (resp. $V = K \setminus d(a,r)$), let $\rho \in]0,r[$, let $(b_n)_{n\in\mathbb{N}}$ be a sequence V satisfying $\lim\limits_{n\to\infty} |b_n - a| = r$ and let $F(x) = \prod\limits_{n\in\mathbb{N}} \dfrac{x - a_n}{x - b_n}$ be a meromorphic product, collapsing in $H(\Lambda_\rho((b_n)_{n\in\mathbb{N}}))$ along the increasing (resp. the decreasing) filter $\mathcal{F}_\rho$ of center a and diameter r on $\Lambda_\rho((b_n)_{n\in\mathbb{N}})$. Then for every $s \in]0,r[$, F is collapsing in $H(\Lambda_s((b_n)_{n\in\mathbb{N}}))$ along the increasing (resp. the decreasing) filter $\mathcal{F}_s$ of center a and diameter r on $\Lambda_s((b_n)_{n\in\mathbb{N}})$. Besides $\mathcal{F}_s$ is a T-filter and we have* $\lim\limits_{\substack{|x-a|\to r \\ x\in\Lambda_\rho((b_n)_{n\in\mathbb{N}})}} F(x) = 1 = F(t)$ *whenever $t \in K \setminus d(a,r^-)$*

(resp. $\lim\limits_{\substack{|x-a|\to r \\ x\in\Lambda_\rho((b_n)_{n\in\mathbb{N}})}} F(x) = \prod\limits_{n\in\mathbb{N}} \left(\dfrac{a_n}{b_n}\right) = F(t)$ *whenever $t \in d(a,r)$.)*

Proof: For every $s \in]0,r[$, we put $D_s = \Lambda_s((b_n)_{n\in\mathbb{N}})$. Let $\ell = \lim\limits_{\substack{|x-a|\to r \\ x\in D_\rho}} F(x)$.

Hence in $H(D_\rho)$ F satisfies $_{D_\rho}\varphi_{a,r}(F - \ell) = 0$. But then, for every $s \in]0,r[$, in $H(D_s)$ we have $_{D_s}\varphi_{a,r}(F - \ell) = 0$, and therefore F is collapsing along $\mathcal{F}_s$. Since $\mathcal{F}_s$ is the only one pierced filter on D_s, by Corollary 38.2 this shows $F - \ell$ to be strictly vanishing along $\mathcal{F}_s$, and therefore $\mathcal{F}_s$ is a T-filter. Besides by Theorem 38.7, we have $F(x) = \ell$ for all $x \in \mathcal{P}(\mathcal{F}_s)$, hence $F(x) = \ell$ for all $x \in K \setminus d(a,r^-)$ (resp. $x \in d(a,r)$). Finally we see that $\lim\limits_{|x|\to\infty} F(x) = 1$ hence $\ell = 1$ (resp. $F(0) = \prod\limits_{n\in\mathbb{N}} \left(\dfrac{a_n}{b_n}\right)$ hence $\ell = \prod\limits_{n\in\mathbb{N}} \left(\dfrac{a_n}{b_n}\right)$). This finishes the proof of Theorem 48.5.

Definition: Let $V = d(a,r)$ (resp. $V = K \setminus d(a,r)$), let $\rho \in]0,r[$, let $(b_n)_{n\in\mathbb{N}}$ be a sequence satisfying $\lim\limits_{n\to\infty} |b_n - a| = r$ and let $F(x) = \prod\limits_{n\in\mathbb{N}} \dfrac{x - a_n}{x - b_n}$ be a meromorphic product. Then F will be said to be *collapsing* if it is collapsing in $H(\Lambda_s((b_n)_{n\in\mathbb{N}}))$, for any $s \in]0,r[$.

Theorem 48.2: *Let $V = d(a,r^-)$ (resp. $V = K \setminus d(a,r)$). Let $(b_n)_{n\in\mathbb{N}}$ be a T-polar sequence in T. Then there exists collapsing meromorphic products of the* *form* $F(x) = \prod_{n \in \mathbb{N}} \dfrac{x - a_n}{x - b_n}.$

Proof:$\quad$ Let $\rho = \inf_{n \neq m} |b_n - b_m|$. For every $n \in \mathbb{N}$ let $T_n = d(b_n, \rho^-)$ and let $D_\rho = \Lambda_\rho((b_n)_{n \in \mathbb{N}}$. By definition D_ρ admits an increasing (resp. a decreasing) T-filter $\mathcal{F}$ of center a and diameter r. Besides we notice that $b_m \notin T_n$ whenever $m \neq n$. Then we have an idempotent T-sequence (T_n, u_n). Hence by Theorem 45.3 there exists $f \in H(D_\rho)$ strictly vanishing along $\mathcal{F}$, meromorphic in each hole T_n, admitting each b_n as its only pole in T_n, with a multiplicity order equal to 0 or 1. Obviously, we may take f such that $\|f\|_{D_\rho} < 1$, hence by putting $F = 1 + f$ we have $\|F - 1\|_{D_\rho} < 1$. We will show F to be a meromorphic product.

First, suppose $V = d(a, r^-)$. By construction f is of the form $\displaystyle\sum_{n=0}^{\infty} \dfrac{\theta_n}{x - b_n}$. Besides, since $\|f\|_{D_\rho} < 1$, and since $T_n \cap T_m = \emptyset$ whenever $n \neq m$, by Theorem 15.1 we have $\left\|\dfrac{\theta_n}{x - b_n}\right\|_{D_\rho} < 1$, hence $\theta_n < \rho$, whenever $n \in \mathbb{N}$. Hence, by Proposition 47.2, F is a meromorphic product of the form $\displaystyle\prod_{n \in \mathbb{N}} \dfrac{x - a_n}{x - b_n}$ such that a_n belongs to T_n for every $n \in \mathbb{N}$. Since f is vanishing along $\mathcal{F}, F$ is collapsing.

Now suppppose $V = K \setminus d(a, r)$. Without loss of generality we may obviously assume $a = 0$ and $b_n \neq 0$ for all $n \in \mathbb{N}$. Let $r' = \dfrac{1}{r}$ let $V' = d(0, r'^-)$ and for each $n \in \mathbb{N}$ let $\beta_n = \dfrac{1}{b_n}$. Let $\sigma = \inf_{n \neq m} |\beta_n - \beta_m|$. It is seen that $\sigma > 0$. Besides by Corollary 44.4. the sequence $(\beta_n)_{n \in \mathbb{N}}$ is a T-polar sequence in V'. Let $D'_\sigma = \Lambda((\beta_n)_{n \in \mathbb{N}})$. There exists a collapsing meromorphic product $G(u) = \displaystyle\prod_{n \in \mathbb{N}} \dfrac{u - \alpha_n}{u - \beta_n}$ such that $\|G - 1\|_{D'_\sigma} < 1$. For every n we put $a_n = \dfrac{1}{\alpha_n}$.

It is seen that $\lim_{n \to \infty} a_n - b_n = 0$. Let $F(x) = \displaystyle\prod_{n \in \mathbb{N}} \dfrac{x - a_n}{x - b_n}$. We have $G(\dfrac{1}{x}) = \displaystyle\prod_{n \in \mathbb{N}} (\dfrac{b_n}{a_n}) F(x)$ hence we see that $\displaystyle\prod_{n \in \mathbb{N}} (\dfrac{b_n}{a_n}) F(x) = 1$ whenever $|x| \geq r$, hence $F(x) = \displaystyle\prod_{n \in \mathbb{N}} (\dfrac{a_n}{b_n}) = F(0)$ and therefore F is collapsing. This finishes proving Theorem 48.2.

Lemma 48.3:$\quad$ *Let* $V = d(a, r^-)$, *let* $E = K \setminus V$, *let* $(a_n)_{n \in \mathbb{N}}$, $(b_n)_{n \in \mathbb{N}}$ *be sequences in* V *satisfying* $\lim_{n \to \infty} |b_n - a| = \lim_{n \to \infty} |a_n - a| = r$ *and* $\lim_{n \to \infty} b_n - a_n = 0$.

Then the meromorphic product $F(x) = \displaystyle\prod_{n \in \mathbb{N}} \frac{x - a_n}{x - b_n}$ satisfies $F'(x) = 0$ for all $x = E$ if and only if for every $j \in \mathbb{N}^*$ the sequences $(a_n)_{n \in \mathbb{N}}$ and $(b_n)_{n \in \mathbb{N}}$ satisfy $\displaystyle\sum_{n=0}^{\infty} (a_n)^j - (b_n)^j = 0$.

Proof: We can see that $\|F - 1\|_E \leq \displaystyle\sup_{n \in \mathbb{N}} \frac{|b_n - a_n|}{r} < 1$. Hence $\dfrac{F'}{F}$ obviously belongs to $H(E)$. Let $g = \dfrac{F'}{F}$. It is seen that $g(x) = \displaystyle\sum_{n=0}^{\infty} \frac{1}{x - a_n} - \frac{1}{x - b_n}$.

Now, let $s = \max(r, |a|)$, and let $E' = K \setminus d(0, s)$. For each $\alpha, \beta \in V$, and for every $x \in E'$ we have $\dfrac{1}{x - \alpha} - \dfrac{1}{x - \beta} = \displaystyle\sum_{j=0}^{\infty} \frac{\alpha^j - \beta^j}{x^{j+1}} = \sum_{j=1}^{\infty} \frac{\alpha^j - \beta^j}{x^{j+1}}$. Applying this to each term $\dfrac{1}{x - a_n} - \dfrac{1}{x - b_n}$, we obtain $g(x) = \displaystyle\sum_{n=0}^{\infty} \Big(\sum_{j=1}^{\infty} \frac{(a_n)^j - (b_n)^j}{x^{j+1}} \Big)$ for all $x \in E'$. Now, let us fix $x \in E'$. We see that when j tends to $+\infty$, the convergence of $\dfrac{(a_n)^j - (b_n)^j}{x^{j+1}}$ to 0 is uniform with respect to n. Hence we have $g(x) = \displaystyle\sum_{j=1}^{\infty} \Big[\sum_{n=0}^{\infty} \frac{(a_n)^j - (b_n)^j}{x^{j+1}} \Big]$.

But now, this holds for any $x \in E'$. Besides, as F belongs to $H(E)$, its Mittag-Leffler series is the same in $H(E)$ and in $H(E')$, hence this is the Mittag-Leffler series of F in $H(E)$. Now, F is collapsing if and only if g satisfies $g(x) = 0$ for all $x \in E$. Hence we see that $F'(x) = 0$ if and only if the Mittag-Leffler series of g is identically equal to 0, i.e.: $\displaystyle\sum_{n=0}^{\infty} (a_n)^j - (b_n)^j = 0$ for every $j \in \mathbb{N}^*$. This ends the proof.

Theorem 48.4: *K is supposed to have characteristic zero. Let $V = d(a, r^-)$, let $(a_n)_{n \in \mathbb{N}}$, $(b_n)_{n \in \mathbb{N}}$ be sequences in V satisfying $\displaystyle\lim_{n \to \infty} |b_n - a| = \lim_{n \to \infty} |a_n - a| = r$ and $\displaystyle\lim_{n \to \infty} b_n - a_n = 0$. Then the meromorphic product $F(x) = \displaystyle\prod_{n \in \mathbb{N}} \frac{x - a_n}{x - b_n}$ is collapsing if and only if for every $j \in \mathbb{N}^*$ the sequences $(a_n)_{n \in \mathbb{N}}$ and $(b_n)_{n \in \mathbb{N}}$ satisfy $\displaystyle\sum_{n=0}^{\infty} (a_n)^j - (b_n)^j = 0$.*

Proof: Indeed, since K has characteristic zero, by Corollary 13.6, we know that $F'(x)$ is identically zero in $K \setminus V$ if and only if $F(x)$ is a constant in $K \setminus V$, i.e. F is collapsing.

Theorem 48.5: *K is supposed to have characteristic $p \neq 0$. Let $V = d(a, r^-)$, let $E = K \setminus V$, let $(a_n)_{n \in \mathbb{N}}$, $(b_n)_{n \in \mathbb{N}}$ be sequences in V satisfying $\lim\limits_{n \to \infty} |b_n - a| =$*

$= \lim\limits_{n \to \infty} |a_n - a| = r$ and $\lim\limits_{n \to \infty} b_n - a_n = 0$. Let $F(x) = \prod\limits_{n \in \mathbb{N}} \dfrac{x - a_n}{x - b_n}$, let $(e_n)_{n \in \mathbb{N}}$

be a T-polar sequence associated to V. There exists a meromorphic product f of

the form $\prod\limits_{n \in \mathbb{N}} \dfrac{x - c_n}{x - e_n}$ such that $F(x) = (f(x))^p$ whenever $x \in E$ if and only if for

every $j \in \mathbb{N}^*$ the sequences $(a_n)_{n \in \mathbb{N}}$ and $(b_n)_{n \in \mathbb{N}}$ satisfy $\sum\limits_{n=0}^{\infty}(a_n)^j - (b_n)^j = 0$.

Proof: If there exists a meromorphic product f associated to the sequence $(b_n)_{n \in \mathbb{N}}$ such that $(f(x))^p = F(x)$ for all $x \in E$, then obviously we have $F'(x) = 0$ for all $x \in E$, and therefore, by Lemma 48.1, we have

$$(\mathcal{E}_j) \quad \sum_{n=0}^{\infty}(a_n)^j - (b_n)^j = 0 \text{ for every } j \in \mathbb{N}^*.$$

Reciprocally, we suppose Relations $(\mathcal{E}_j)$ are satisfied. By Lemma 48.1 we have $F'(x) = 0$ for all $x \in E$. Hence by Corollary 13.13, there exists $g \in H(E)$ such that $(g(x))^p = F(x)$ for all $x \in E$. Besides, since F is a meromorphic product associated to the sequence $(b_n)_{n \in \mathbb{N}}$, we notice that $\lim\limits_{|x| \to +\infty} F(x) = 1$.

As a consequence, we can choose g such that $\lim\limits_{|x| \to +\infty} g(x) = 1$. Besides, it is seen that $g^p = ((g-1)+1)^p = (g-1)^p + 1$, and therefore we have

(1) $\|F - 1\|_E = (\|g - 1\|_E)^p$, hence $\|g - 1\|_E < 1$.   Let $\epsilon \in]0, 1[$. Then by (1) and by Theorem 47.3, there does exist a meromorphic product f of the form $\prod\limits_{n \in \mathbb{N}} \dfrac{x - c_n}{x - e_n}$ such that $f(x) = g(x)$ whenever $x \in E$, and such that $|e_n - c_n| \leq \sqrt[p]{\|F - 1\|_E} + \epsilon$. This ends the proof.

49. INJECTIVITY, MITTAG-LEFFLER SERIES

AND MOTZKIN PRODUCTS

In this Chapter, D is infraconnected and open.

In Chapter 29, we obtained elementary conditions to provide strictly injective analytic elements of the form $h\phi$, with $h \in H(D)$ a homographic function, and ϕ an analytic element of $H(D)$ close to 1. Here we give two other sufficient conditions on the elements $\phi \in H(D)$ to assure that $h\phi$ is injective in D. The first condition involves the Mittag-Leffler series and the second one involves the Motzkin factors of ϕ. These two conditions are shown to be equivalent to the one obtained in Chapter 29. This was first published in [35].

Given a homographic function f, both $\dfrac{f}{f'}$ and $\dfrac{1}{f'}$ are easily seen to be polynomials of degree at most 2. Proposition 49.1 below will help us apply Condition γ) in theorem 49.4.

Proposition 49.1. *Let $f \in R(D)$ be a homographic function and let $\phi \in R(D)$ be a strongly copiercing homographic function with respect to D. If the pole of ϕ belongs to $K \setminus \widetilde{D}$ then we have*

$$\left\|\frac{f}{f'}\phi'\right\|_D = \left\|\frac{f}{f'}\right\|_D \|\phi'\|_D = \varphi_D\left(\frac{f}{f'}\right)\varphi_D(\phi').$$

If the pole of ϕ belongs to a hole T of D then we have

$$\left\|\frac{f}{f'}\phi'\right\|_D = \varphi_T\left(\frac{f}{f'}\phi'\right) = \left\|\frac{f}{f'}\right\|_T \left(\left\|\frac{1}{\phi'}\right\|_T\right)^{-1} = \left\|\frac{f}{f'}\right\|_T \|\phi'\|_D.$$

Proof: We first suppose that the pole b of ϕ belongs to $K \setminus \widetilde{D}$, hence its zero a also belongs to $K \setminus \widetilde{D}$, and we have $\|\phi'\|_D = \|\phi'\|_{\widetilde{D}} = |\phi'(x)| = \dfrac{|a - b|}{(\delta(b, D))^2}$ whenever $x \in D$. Hence we see that $\left\|\dfrac{f}{f'}\phi'\right\|_D = \left\|\dfrac{f}{f'}\right\|_D \dfrac{|a - b|}{(\delta(b, D))^2} = \left\|\dfrac{f}{f'}\right\|_D\|\phi'\|_D$.

We now suppose that b belongs to a hole $T = d(b, r^-)$ of D, hence so does a. We know that $\dfrac{f}{f'}$ is a polynomial P of degree ≤ 2. Since $\phi' = \dfrac{a - b}{(x - b)^2}$ we see that for every $x \in D$ we have $v(P(x)) \geq v_b(P, v(x - b))$ while $v(\phi'(x)) = -2v(x - b) + v(a - b)$ hence $v(\phi'(x)P(x)) \geq v_b(P, v(x - b)) - 2v(x - b) + v(a - b)$. But as $degP \leq 2$ it is clear that $v_b(P, \mu) \geq v_b(P, -\log r) + 2(\mu + \log r)$. Hence we have $v(\phi'(x)P(x)) \geq v_b(P, -\log r) + v(a - b) + 2(v(x - b) - (-\log r) - v(x - b))$ and then we see that $v(\phi'(x)P(x)) \geq v_b(\phi'P, -\log r)$. But $v_b(\phi'P, -\log r) = -\log \varphi_T(\phi'P)$ hence we have $\varphi_T(\phi'(x)P(x)) \leq \varphi_T(\phi'P)$ whenever $x \in D$ and therefore $\|\phi'P\|_D = \varphi_T(\phi'P) = \varphi_T(\phi') \, \varphi_T(P) = \varphi_T(\phi') \, \|P\|_T$.

Now $\varphi_T(\phi') = \dfrac{|b - a|}{r^2} = \|\phi'\|_D$ hence we have obtained $\|\phi'P\|_D = \|P\|_T \, \|\phi'\|_D$. Since $(\phi')^{-1}$ is a polynomial we also have $\varphi_T((\phi')^{-1}) = \left\|(\phi')^{-1}\right\|_T$ hence $\varphi_T(\phi') = \left(\|\tfrac{1}{\phi'}\|_T\right)^{-1}$ and finally $\varphi_T(\phi') \, \|P\|_T = \left(\|\tfrac{1}{\phi'}\|_T\right)^{-1} \|P\|_T$. This ends the Proof of Proposition 49.1.

Corollary 49.2:   *Let $f \in R(D)$ be a homographic function and let $\phi \in R(D)$ be a strongly copiercing homographic function with respect to D.*

If the pole of ϕ belongs to $K \backslash \widetilde{D}$ then we have $\left\|\dfrac{f}{f'}\phi'\right\|_D = \left\|\dfrac{f}{f'}\right\|_D \|\phi'\|_D$.

If the pole of ϕ belongs to a hole T of D then we have

$$\left\|\frac{f}{f'}\phi'\right\|_D = \left\|\frac{f}{f'}\right\|_T \left(\|\tfrac{1}{\phi'}\|_T\right)^{-1} = \left\|\frac{f}{f'}\right\|_T \|\phi'\|_D.$$

Proposition 49.3:   *Let h be a homographic function and let $\mathcal{F}$ be a circular filter on K, of center α and diameter $r > 0$. Then we have* $\varphi_{\mathcal{F}}\left(\dfrac{h}{h'}\right) \geq r$.

Proof. P still denotes $\dfrac{h}{h'}$. Let $\mu = -\log r$. It is clearly equivalent to show

(a) $v_\alpha(P, \mu) \leq \mu$.

When h is of the form $A(x - \beta)$ we have $P = x - \beta$, hence $v_\alpha(P, \mu) = \min(\mu, v(\beta)) \leq \mu$.

In the same way when h is of the form $\dfrac{A}{x - \gamma}$, we have $P = -(x - \gamma)$ hence

$$v_\alpha(P, \mu) = \inf(\mu, v(\gamma)) \leq \mu.$$

We now suppose h is in the form $A(\dfrac{x - \beta}{x - \gamma})$. Hence we have $P = \dfrac{(x - \beta)(x - \gamma)}{\beta - \gamma}$ and then we obtain

(b) $v_\alpha(P, \mu) = v_\alpha(x - \beta, \mu) + v_\alpha(x - \gamma, \mu) - v(\beta - \gamma)$.

It is seen that $v_\alpha(x - \beta, \mu) \leq \mu$ and $v_\alpha(x - \gamma, \mu) \leq \mu$. Hence, if $v(\beta - \gamma) \geq \mu$, Relation (a) is clearly satisfied. If $v(\beta - \gamma) < \mu$, then we have at least: $v(\beta - \alpha) < \mu$ or $v(\gamma - \alpha) < \mu$.

For example, if $v(\beta - \alpha) < \mu$, we have $v_\alpha(x - \beta, \mu) = v(\beta - \alpha)$ hence by (b) we obtain $v_\alpha(P, \mu) = v(x - \gamma, \mu) \leq \mu$.

In the same way, when $v(\gamma - \alpha) < \mu$ we have $v_\alpha(P, \mu) = v(x - \beta, \mu) \leq \mu$. Relation (a) is then proven in all the cases and this ends the proof of Proposition 49.3.

Theorem 49.4:   *Let $f \in R(D)$ be a homographic function, let $a \in D$ and let $\phi \in H_b(D)$ be semi-invertible and satisfy $|\phi(a)| = 1$. Statements α), β), γ) below are equivalent:*

α)   *For every ϕ-hole T we have $\|\overline{\phi}_T\|_D \left\|\dfrac{f}{f'}\right\|_T < diam(T)$ and if D is bounded, then $\overline{\phi}_0$ satisfies $\|\overline{\phi}_0 - \overline{\phi}_0(a)\|_D \left\|\dfrac{f}{f'}\right\|_D < diam(D)$,*

β)   *ϕ is invertible in $H(D)$ and for every ϕ-hole T we have $\|\phi^T - 1\|_D \left\|\dfrac{f}{f'}\right\|_T < diam(T)$ and if D is bounded ϕ^0 satisfies $\|\phi^0 - \phi^0(a)\|_D \left\|\dfrac{f}{f'}\right\|_D < diam(D)$.*

γ)   *ϕ is in the form $B\displaystyle\prod_{n \in I} \phi_n$ with $B \in C(0,1)$ and with $\left(\phi_n\right)_{(n \in I)}$ a strongly copiercing sequence of homographic functions with respect to D satisfying*

$$\left\|\frac{f}{f'}\phi'_n\right\|_D < 1 \ \text{whenever } n \in I.$$

Besides if Conditions α), β), γ) are satisfied, then ϕ satisfies $\|\phi - \phi(a)\|_D < 1$ and the product $g = f\phi$ is strictly injective in D and satisfies

    (1) $|g(x)| = |f(x)|$

    (2) $|g'(x)| = |f'(x)|$

    (3) $|g(y) - g(x)| = |f(y) - f(x)|$

    (4) $|g(y) - g(x)|^2 = |g'(y)g'(x)| \, |y - x|^2$ *whenever $x, y \in D$.*

Proof: Without loss of generality we may obviously assume $\phi(a) = 1$. If Condition γ) is satisfied, by Theorem 29.9 it is clearly seen that g is strictly injective in D and satisfies (1), (2), (3), (4). Besides, as the sequence $(\phi_n)_{n \in \mathbb{N}}$ is strongly copiercing with respect to D, each ϕ_n satisfies $\|\phi_n - 1\|_D < 1$. Hence we have $\|\phi - B\|_D < 1$, hence $\|\phi - \phi(a)\|_D < 1$. Thus we just have to show that α), β), γ) are equivalent. Let P be the polynomial $\dfrac{f}{f'}$.

First we will check that α) and β) are equivalent. We notice that, by Proposition 49.3, for every hole T of D we have

(5) $\|P\|_T \geq diam(T)$,

and if D is bounded we have

(6) $\|P\|_D \geq diam(D)$.

Assume that $\alpha)$ is satisfied. By (5) and (6) we see that $\|\overline{\phi}_T\|_D < 1$. Besides if D is bounded we see that $\|\overline{\phi}_0 - \overline{\phi}_0(a)\|_D < 1$ while this inequality is trivial if D is not bounded because $\overline{\phi}_0$ is then a constant. Hence by Theorem 32.15 we have $\|\phi^T - 1\|_D = \|\overline{\phi}_T\|_D$ for every ϕ-hole T and $\|\phi^0 - \phi^0(a)\|_D = \|\overline{\phi}_0 - \overline{\phi}_0(a)\|_D$. Hence $\beta)$ clearly comes from $\alpha)$.

In the same way, suppose $\beta)$ is satisfied. By (5) and (6) we have $\|\phi^T - 1\|_D < 1$ and we have $\|\phi^0 - \phi^0(a)\|_D < 1$ if D is bounded , while this last inequality is trivial if D is unbounded because ϕ^0 is a constant. Hence by Theorem 32.15 $\alpha)$ comes from $\beta)$.

Besides, when $\alpha)$, $\beta)$ are satisfied, the set of conditions

$\|\overline{\phi}_T\|_D < 1$, $\|\overline{\phi}_0 - \overline{\phi}_0(a)\|_D < 1$

shows that $\|\phi - \phi(a)\|_D < 1$ hence $\|\phi - 1\|_D < 1$ and then by Proposition 32.14 we have

(7) $\|\phi^T - 1\|_D = \|\overline{\phi}_T\|_D$ for every ϕ-hole T, and

(8) $\|\phi^0 - 1\|_D = \|\overline{\phi}_0 - 1\|_D$

Now we will show that $\beta)$ and $\gamma)$ are equivalent. We suppose that $\beta)$ is satisfied and will show that so is $\gamma)$. Since $\phi(a) = 1$, by (5) and (6) we have $\|\phi - 1\|_D < 1$.

Let $(T_m)_{m \in J}$ be the sequence of the ϕ-holes with $J \subset \mathbb{N}^*$ and for each $m \in J$ let $T_m = d(\beta_m, r_m^-)$. Let $(c_m)_{m \in \mathbb{N}^*}$ be a sequence in $]0, +\infty[$ such that

(9) $\left(\|\phi^{T_m} - 1\|_D + c_m \right) \|P\|_{T_m} < r_m$

whereas $\lim_{m \to \infty} c_m = 0$ if J is infinite. It is seen that we may assume $J = \mathbb{N}^*$ without loss of generality. By Proposition 47.4, for each $m \in \mathbb{N}^*$ there exists a strongly copiercing sequence $(a_{n,m}, b_{n,m})_{n \in \mathbb{N}}$ associated to $K \setminus T_m$, such that $\dfrac{|a_{n,m} - b_{n,m}|}{r_m} \leq \|\phi^{T_m} - 1\|_D + c_m$. Hence by (9) we have $\dfrac{|a_{n,m} - b_{n,m}|}{r_m^2} \|P\|_{T_m} <$ 1, whenever $(n, m) \in \mathbb{N} \times \mathbb{N}^*$. We put $\phi_{n,m}(x) = \dfrac{x - a_{n,m}}{x - b_{n,m}}$, $(n, m) \in \mathbb{N} \times \mathbb{N}^*$. It

is seen that $\varphi_{T_m}(\phi'_{n,m}) = \dfrac{|a_{n,m} - b_{n,m}|}{r_m^2}$ hence $\varphi_{T_m}(\phi'_{n,m} P) < 1$. By Proposition 49.1 we know that $\varphi_{T_m}(\phi'_{n,m} P) = \|\phi'_{n,m} P\|_D$ hence we obtain

(10) $\|\phi'_{n,m} P\|_D < 1$ whenever $(n, m) \in \mathbb{N} \times \mathbb{N}^*$.

Thus if D is not bounded , by Proposition 47.4 we know that ϕ is of the form

$$A \prod_{(n,m)\in \mathbb{N}\times \mathbb{N}^*} \phi_{n,m}$$ with $A \in d(1,1^-)$ and therefore by (10) we see that γ) is clearly satisfied.

We now suppose that D is bounded, of diameter r_0. As $\phi(a) = 1$, by Condition β) we have $\|\phi^0 - 1\|_D \, \|P\|_D < r_0$. Let $c_0 > 0$ be such that

$$(11) \qquad \left(\|\phi^0 - 1\|_D + c_0 \right) \|P\|_D < r_0.$$

By Proposition 47.4 there exists a strongly copiercing sequence associated to $\widetilde{D}, (a_{n,0}, b_{n,0})_{n\in\mathbb{N}}$ satisfying

$$(12) \qquad \frac{|a_{n,0} - b_{n,0}|}{r_0} < \|\phi^0 - 1\|_D + c_0$$

and such that $\phi^0 = B \displaystyle\prod_{n=0}^{\infty} \left(\frac{x - a_{n,0}}{x - b_{n,0}} \right)$ with $B = \displaystyle\prod_{n=0}^{\infty} \left(\frac{b_{n,0}}{a_{n,0}} \right)$.

Since the sequence $(a_{n,0}, b_{n,0})$ is a strongly copiercing sequence associated to D it is seen that we have

$(13) \qquad |B - 1| < 1.$

Now by (11) and (12) we have

$$(14) \qquad \frac{|a_{n,0} - b_{n,0}|}{r_0^2} \, \|P\|_D < 1.$$

We put $\phi_{n,0}(x) = \dfrac{x - a_{n,0}}{x - b_{n,0}} \, (n \in \mathbb{N})$ and see that both $\phi'_{n,o}$, P belong to $H(\widetilde{D})$.

Then, as D is infraconected, we have $\|\phi'_{n,0} P\|_D = \|\phi'_{n,0} P\|_{\widetilde{D}} = \|\phi'_{n,0}\|_{\widetilde{D}} \, \|P\|_{\widetilde{D}} = \|\phi'_{n,0}\|_D \, \|P\|_D$ while $\|\phi'_{n,0}\|_D = \dfrac{|a_{n,0} - b_{n,0}|}{\delta(b_{n,0}, D)^2} \leq \dfrac{|a_{n,0} - b_{n,0}|}{r_0^2}$, hence by (14) we obtain

$(15) \qquad \|\phi'_{n,0} P\|_D = \|\phi'_{n,0}\|_D \, \|P\|_D < 1.$

Obviously the sequence $(a_{n,m}, b_{n,m})_{(n,m)\in\mathbb{N}\times\mathbb{N}}$ is a strongly copiercing sequence associated to D and by (10) , (13) , (15) the family $(\phi_{n,m})_{(n,m)\in\mathbb{N}\times\mathbb{N}}$ satisfies γ). This finishes proving that β) implies γ).

Finally we will show that γ) implies β). For every $n \in \mathbb{N}$, let $\phi_n(x) = \dfrac{x - a_n}{x - b_n}$. Let T be a ϕ-hole of D of diameter r, and let I_T be the set of the $n \in \mathbb{N}$ such that $a_n, b_n \in T$. By Proposition 47.5 we have $\phi^T(x) = \displaystyle\prod_{n\in I_T} \phi_n$ hence it is seen that we have

$$(16) \qquad \|\phi^T - 1\|_D \leq \sup_{n\in I_T} \|\phi_n - 1\|_D = \sup_{n\in I_T} \frac{|a_n - b_n|}{r} \,.$$

But we have

(17) $\qquad \dfrac{|a_n - b_n|}{r} = r\varphi_T(\phi'_n),$

hence we see that $\|\phi^T - 1\|_D \leq r \sup\limits_{n \in I_T} \varphi_T(\phi'_n)$ and therefore we have

(18) $\qquad \|\phi^T - 1\|_D \, \|P\|_T \leq r \left(\sup\limits_{n \in I_T} \varphi_T(\phi'_n) \right) \|P\|_T.$

Now, we have $\varphi_T(\phi'_n) = \varphi_T\left(\left(\frac{1}{\phi'_n} \right)^{-1} \right)$. As $\dfrac{1}{\phi'_n}$ is a polynomial it satisfies

$\varphi_T\left(\dfrac{1}{\phi'_n} \right) = \left\| \dfrac{1}{\phi'_n} \right\|_T$, hence by Proposition 49.1 , we see that $\varphi_T(\phi'_n) \, \|P\|_T = \|\phi'_n P\|_D$. Consequently, by Hypothesis γ) we have

(19) $\quad \varphi_T(\phi'_n) \, \|P\|_T < 1$ whenever $n \in I_T$.

But by (17) the sequence $\left(\varphi_T(\phi'_n) \right)_{n \in I_T}$ either is finite or has limit zero and therefore it reaches its maximum for some $n \in I_T$. Hence by (19) we have $\sup\limits_{n \in I_T} \varphi_T(\phi'_n) \|P\|_T < 1$. Finally, by (18) we obtain $\|\phi^T - 1\|_D \, \|P\|_T < r$, so β) is satisfied when D is not bounded.

We now suppose D is bounded, and then we just have to show that $\|\phi^0 - \phi^0(a)\|_D < r_0$. Actually it is clear that $\|\phi^0 - \phi^0(x)\|_D$ is not depending on x whenever $x \in D$. Hence actually, we have just to show that $\|\phi^0 - \phi^0(a)\|_D < r_0$.

By Proposition 47.5 ϕ^0 is in the form $A \prod\limits_{n \in I_0} \left(\dfrac{x - a_n}{x - b_n} \right)$ where I_0 is the set of the

$n \in \mathbb{N}$ such that $a_n, b_n \in K \setminus \tilde{D}$. For every $n \in I_0$ let $\psi_n = \dfrac{b_n}{a_n}\left(\dfrac{x - a_n}{x - b_n} \right).$

Obviously we have $\lim\limits_{\substack{n \to \infty \\ n \in I_0}} \dfrac{b_n}{a_n} = 1$. Let $\psi(x) = \prod\limits_{n \in I_0} \psi_n(x)$. We see that $\psi(a) = 1$

hence $\phi^0 = \phi^0(a)\psi$ and therefore $\|\phi^0 - \phi^0(a)\|_D = |\phi^0(a)|$ $\|\psi - 1\|_D = \|\psi - 1\|_D$. Hence we just have to show

(20) $\quad \|\psi - 1\|_D \, \|P\|_D < r_0.$

We see that $|\psi_n(x) - 1| = \dfrac{|x(b_n - a_n)|}{|a_n(x - b_n)|} = |x|\dfrac{|b_n - a_n|}{|b_n|^2}$ whenever $x \in D$, $n \in I$.

Hence we obtain

(21) $\quad \|\psi_n - 1\|_D = r_0 \dfrac{|b_n - a_n|}{|b_n|^2}.$

Now since $|\phi'_n(x)| = \dfrac{|a_n - b_n|}{|(x - b_n)^2|} = \dfrac{|a_n - b_n|}{|b_n|^2}$, we have

$|\phi'_n(x)P(x)| = \dfrac{|a_n - b_n|}{|b_n|^2}|P(x)|$ whenever $x \in D$, $n \in I_0$, and

$\|\phi'_n P\|_D = \dfrac{|a_n - b_n|}{|b_n|^2}\|P\|_D$ whenever $n \in I_0$. Hence by (12) we see that

$\|\psi_n - 1\|_D \, \|P\|_D \le r_0 \dfrac{|a_n - b_n|}{|b_n|^2}\,\|P\|_D = r_0\|\phi'_n P\|_D$. But by Hypothesis $\gamma)$ we have $\|\phi'_n P\|_D < 1$. Hence we obtain

(22) $\|\psi_n - 1\|_D \, \|P\|_D < r_0$ whenever $n \in I_0$.

In $H(D)$ it is clear that we have $\displaystyle\lim_{\substack{n \to \infty \\ n \in I_0}} \psi_n = \lim_{\substack{n \to \infty \\ n \in I_0}} \phi_n = 1$, hence by (22) we have

$\displaystyle\sup_{n \in I_0} \|\psi_n - 1\|_D < \dfrac{r_0}{\|P\|_D}$. But we see that $\|\psi - 1\|_D < \dfrac{r_0}{\|P\|_D}$ and therefore (20) is satisfied. This finishes proving that $\gamma)$ implies $\beta)$ and this ends the proof of the Theorem.

Corollary 49.5: *Let D be a bounded (resp. unbounded) open infraconnected set. Let $f \in R(D)$ be a homographic function , let $a \in D$ and let $\varphi \in H_b(D)$ satisfy $\|\overline{\varphi}_T\|_D \, \|\frac{f}{f'}\|_T < diam(T)$ for every φ-hole T and $\|\overline{\varphi}_0\|_D \, \|\frac{f}{f'}\|_D < diam(D)$ (resp. and $|\varphi(a)| < 1$). Then we have $\|\varphi\|_D < 1$ and the function $g := f(1 + \varphi)$ is strictly injective in D and satisfies Conditions (1), (2), (3), (4) of Theorem 49.4.*

Example 1. Let $D = d(0,1) \setminus d(b, \rho^-)$ with $|b| = r \in]0,1[$ and $\rho \in]0, r[$. Let $q \in \mathbb{N}$, let $\varepsilon \in K$ satisfy $|\varepsilon| < \dfrac{\rho^{q+1}}{r}$, let $f(x) = x$ and let $g(x) = x(1 + \dfrac{\varepsilon}{(x-b)^q})$. Then g is strictly injective in D and satisfies Conditions (1), (2), (3), (4). Indeed, we may put $T = d(b, \rho^-)$. Then T is a hole of D and we have $\varphi = \overline{\varphi}_T, \|\overline{\varphi}_T\|_D = \dfrac{|\varepsilon|}{\rho^q} < \dfrac{\rho}{r}$ while $\|\frac{f}{f'}\|_T = r$ hence it is seen that $\|\frac{f}{f'}\|_T \, \|\overline{\varphi}_T\|_D < diam(T)$.

Example 2. Let $E = d(0,1)$, let $(r_n)_{n \in \mathbb{N}}$ be a sequence in $]0,1[$, let $(\beta_n)_{n \in \mathbb{N}}$ be a sequence in $d(0,1)$ such that $|\beta_n - \beta_m| \ge \max(r_n, r_m)$ whenever $n \ne m$, let $T_n = d(\beta_n, r_n^-)(n \in \mathbb{N})$ and let $D = d(0,1) \setminus \left(\bigcup_{n \in \mathbb{N}} T_n\right)$. Let $\alpha \in K \setminus \left(\bigcup_{n \in \mathbb{N}} T_n\right)$, let $\beta \in (K \setminus d(0,1)) \cup T_0$ with $\alpha \ne \beta$, and let $f(x) = \dfrac{x - \alpha}{x - \beta}$. For each $n \in \mathbb{N}$, let

$$\varphi_n(x) = \sum_{j=1}^{\infty} \frac{\epsilon_{n,j}}{(x - \beta_n)^j} \text{ satisfy}$$

(31) $|\epsilon_{n,j}| < \dfrac{|\alpha - \beta| r_n^{j+1}}{|\alpha - \beta_n|\,|\beta - \beta_n|}$

$$(32) \qquad \lim_{n+j \to +\infty} \frac{|\epsilon_{n,j}|}{r_n^j} = 0$$

Hence φ_n is seen to belong to $H(D)$ and satisfy

$$(33) \qquad \|\varphi_n\|_D = \sup_{j \in \mathbb{N}^*} \frac{|\epsilon_{n,j}|}{(r_n)^j}$$

so, the sequence (φ_n) goes to zero. Now let $\varphi_0 = \sum_{j=1}^{\infty} \epsilon_{o,j} x^j$ satisfy

$$(34) \qquad \lim_{j \to \infty} \epsilon_{o,j} = 0 \quad \text{and}$$

$$(35) \qquad |\epsilon_{o,j}| < \frac{|\alpha - \beta|}{\max(|\alpha|, 1) \max(|\beta|, 1)} \qquad \text{whenever } j \in \mathbb{N}^*.$$

Let $\phi = \prod_{n=0}^{\infty}(1 + \varphi_n)$ and let $\psi = 1 + \sum_{n=0}^{\infty} \varphi_n$. We will show that both $f\phi$, $f\psi$ are strictly injective in D and satisfy Conditions (1), (2), (3), (4) of Theorem 49.4 in place of g. Actually we just check that ϕ satisfies Condition β) in Theorem 49.4, while ψ satisfies Condition α). For every $n \geq 1$ we see that α and β do not belong to T_n, hence we have

$$(36) \qquad \left\|\frac{f}{f'}\right\|_{T_n} = \frac{|\alpha - \beta_n| \, |\beta - \beta_n|}{|\alpha - \beta|}.$$

In the same way f satisfies

$$(37) \quad \left\|\frac{f}{f'}\right\|_{\widetilde{D}} = \frac{\max(|\alpha|, 1) \max(|\beta|, 1)}{|\alpha - \beta|}.$$

Now by Proposition 49.3 we have $\left\|\frac{f}{f'}\right\|_{T_n} \geq r_n$. Hence by (36) we have

$\dfrac{|\alpha - \beta| r_n^{j+1}}{|\alpha - \beta_n| \, |\beta - \beta_n|} \leq \dfrac{1}{r_n}$, and then by (31) and (33) we obtain $\|\varphi_n\| < 1$ whenever

$n \in \mathbb{N}^*$. Besides by Proposition 49.3 we have $\left\|\frac{f}{f'}\right\|_{\widetilde{D}} \geq 1$. Hence (as D is

infraconnected) $\left\|\frac{f}{f'}\right\|_D = \left\|\frac{f}{f'}\right\|_{\widetilde{D}} \geq 1$. Now for each $n \in \mathbb{N}^*$, by (31) and (33) we

see that $\|\varphi_n\|_D = \|\varphi_n\|_{T_n} \leq r_n \dfrac{|\alpha - \beta|}{|(\alpha - \beta_n)(\beta - \beta_n)|}$, hence by (36) we have

$$(38) \quad \|\varphi_n\|_D < \frac{r_n}{\left\|\frac{f}{f'}\right\|_{T_n}} \leq 1.$$

In the same way by (35) and (37) we see that

$$(39) \quad \|\varphi_0\|_D < \frac{1}{\left\|\frac{f}{f'}\right\|_D} \leq 1.$$

Hence by (38) and (39) ϕ and ψ satisfy

(40) $\|\phi - 1\|_D = \|\psi - 1\|_D < 1$.

Now for each $n \in \mathbb{N}^*$ we have $\phi^{T_n} = 1 + \varphi_n$ and $\phi^0 = 1 + \varphi_0$ while $\overline{\psi}_{T_n} = \varphi_n$ and $\overline{\psi}_0 = \varphi_0$. Besides, by (38) we see that we have

(41) $\|\phi^{T_n} - 1\|_D \| \dfrac{f}{f'}\|_{T_n} < r_n$ and by (39) we have $\|\phi^0 - 1\|_D \| \dfrac{f}{f'}\|_D < 1$, hence

(42) $\|\phi^0 - \phi^0(0)\|_D \| \dfrac{f}{f'}\|_D < 1$.

Thus ϕ clearly satisfies Condition β in Theorem 49.4. In the same way ψ satisfies

(43) $\overline{\psi}_{T_n} = \varphi_n = \phi^{T_n} - 1$ whenever $n \in \mathbb{N}^*$ and

(44) $\overline{\psi}_0 = \varphi_0 = \phi^0 - 1$. Hence by (40) we have $|\psi(0) - 1| < 1$ and by (41), (42), (43), (44) ψ satisfies Condition α) in place of ϕ.

Theorem 49.4 together with Theorem 27.5 suggests a conjecture designed to characterize the injective analytic elements when D has certain properties.

Definition.   *A set D will be said to be hyper-infraconnected if for every $a \in \widetilde{D}$, for every $r \in [\delta(a, D) , \delta(a, K \setminus \widetilde{D})] \cap |K|$ all the classes of $C(a, r)$ are included in D but finitely many ones.*

Example: A disk $d(a, r)$ is hyper-infraconnected. But if $r \in |K|$ then $d(a, r^-)$ is not. Every affine infraconnected set is hyper-infraconnected.

Here is the first conjecture (given in [35]).

Conjecture 1.   *Let D be an open hyper-infraconnected bounded (resp. unbounded) set and let $g \in H(D)$. Then g is strictly injective if and only if it is in the form $f \prod_{i \in I} \phi_i$ with f a homographic function in $R(D)$ and $(\phi_i)_{i \in I}$ a strongly copiercing sequence of homographic functions with respect to D, satisfying $\| \dfrac{f}{f'} \phi_i'\|_D < 1$ whenever $i \in I$.*

Thus when D is hyper-infraconnected , this Conjecture suggests that an element $g \in H(D)$ is injective if and only if it has a factorization $f\phi$ satisfying Condition γ). Since α), β), γ) are actually equivalent , this conjecture seems to be somewhat strenghtened by this equivalence.

Besides, as it was noticed in [35] , when D is hyper-infraconnected, the condition $"\| \dfrac{f}{f'} \phi_i'\|_D < 1"$ is equivalent to the apparently weaker condition $"|f(x)\phi_i'(x)| < |f'(x)|$ for all $x \in D"$ because in a strongly infraconnected set, in particular in a hyper-infraconnected set, given an element $g \in H(D)$ such that $\|g\|_D \in |K|$, g does reach its maximum at a point of D.

Now we notice that the injective analytic elements we know in an infraconnected set (hyper-infraconnected or not) do satisfy the finite increasing Relation (4) again. For example so does $(1+x)^q$ (with q prime to p), in $d(0,1^-)$.

Reciprocally when a not constant analytic element satisfies Relation (4) in an infraconnected set D, it is clearly injective in D. Indeed let $f \in H(D)$ satisfy Relation (4). If f' admits any zero a in D, by Relation (4) we have $f(x) = f(a)$ for every x in D. Hence, if f is not a constant, f' has no zero and therefore f is injective by Relation (4). Finally if Conjecture 1 is true, we see that an analytic element f which is not constant in the union of a chained family of hyper-infraconnected open sets, is injective if and only if it satisfies Relation (4).

However, this equivalence has no chance to hold any more when D is not infraconnected. Indeed, let D be not infraconnected and let a, b belong to D in such a way that certain annulus $\Gamma(a, r', r")$ satisfy $\Gamma(a, r', r") \cap D = \emptyset$ with $0 < r' < r" < |a - b|$. Let $f(x) = \alpha(x - a)$ with $0 < |\alpha| < 1$ for $|x - a| \le r'$ and let $f(x) = x - a$ for $|x - a| \ge r"$. We know that $f \in H(D)$. Besides f is seen to be injective in D . However we have $|f(b) - f(a)| = |b - a|$ while $|f'(a)f'(b)| = |\alpha| < 1$ hence Relation (4) is not satisfied.

Besides we can easily construct certain infraconnected open sets D with injective elements $f \in H(D)$ which do not satisfy Relation (4). Indeed, for example we assume the residue characteristic p different from 2. Let $\rho \in]0, 1[$. For every $r \in [\rho, 1] \cap |K|$ we may choose $a_r \in d(1, 1^-), b_r \in d(-1, 1^-)$ satisfying $|a_r - 1| = |b_r + 1| = r$ and $|a_r + b_r| = r$. Now, we put

$$E = \bigcup_{\rho \le r \le 1} (d(a_r, r^-) \bigcup d(b_r, r^-)). \text{ Let } f(x) = x^2. \text{ We check that } f \text{ is injective in}$$

E and satisfies $|f(a_r) - f(b_r)| = r$ while $|a_r - b_r| = 1$. As $|f'(x)| = 1$ whenever $x \in E$, f does not satisfy Relation (4).

An analytic element f of $H(D)$ is said to be *a bianalytic element in D* if f is injective in D and such that the reciprocal function f^{-1} defined in $f(D)$ belongs to $H(f(D))$. Here the counter-example $f(x) = x^2$ we have just constructed is not bianalytic in E. Finally, the Theorem above suggests Conjecture 2.

Conjecture 2.   *Let $a \in D$, let $f \in R(D)$ be a homographic function and let $\phi \in H_b(D)$ satisfy $\phi(a) \ne 0$ and Conditions $\alpha), \beta), \gamma)$ in Theorem 49.4. Then $f\phi$ is a bianalytic element of $H(D)$.*

50. ANALYTIC FUNCTIONS

AND ANALYTIC ELEMENTS

We can construct analytic elements vanishing along a T-filter by using an analytic function not bounded in a disk $d(0, r^-)$. This process also provides us with analytic elements in certain infraconnected set D, equal to 0 in a disk $d(0, r)$, and to 1 in an annulus $\Delta(0, 1, s)$, (with $r < 1 < s$). Most of these results come from [56], in particular Theorems 50.2 and 50.5.

Notations: Throughout this chapter ρ, r are numbers such that $0 < \rho < r$. We will denote by $A_u(d(0, r^-)$ the subset of $A(d(0, r^-))$ that consists of the functions $f \in A(d(0, r^-))$ that are unbounded in $d(0, r^-)$.

Let $f \in A(d(0, r^-))$. Let $\mathcal{Z}$ be the set of the zeros of f in $d(0, r^-)$. We will denote by $\mathcal{D}(f, \rho)$ the set of the $x \in d(0, r^-)$ such that $|x - \alpha| \geq \rho$ for every $\alpha \in \mathcal{Z}$.

Theorem 50.1:   *Let $f \in A_u(d(0, r^-))$ satisfy $f(0) \neq 0$. Let $(r_n)_{n \in \mathbb{N}^*}$ be a strictly increasing sequence of limit r such that $\bigcup_{m=1}^{\infty} C(0, r_m)$ contains all the zeros of f. Let $D = K \setminus \left(\bigcup_{m=1}^{\infty} C(0, r_m) \right)$. Let $g \in A(d(0, r^-)$ satisfy*

$$\lim_{|x| \to r, x \in D \cap d(0, r^-)} \frac{g(x)}{f(x)} = 0.$$

Let h be the function defined in D as $h(x) = \dfrac{g(x)}{f(x)}$ when $x \in D \cap d(0, r^-)$ and $h(x) = 0$ when $x \notin d(0, r^-)$. Then h belongs to $H(D)$.

Proof: Let $D' = D \cap d(0, r^-)$, let $f(x) = \sum_{n=0}^{\infty} a_n x^n$, $g(x) = \sum_{n=0}^{\infty} b_n x^n$ and for all $m \in \mathbb{N}$, let $P_m(x) = \sum_{n=0}^{m} a_n x^n$ and $Q_m(x) = \sum_{n=0}^{m} b_n x^n$. Let $\alpha \in D \cap d(0, r^-)$. Since f has no zero in $C(0, |\alpha|)$, by Theorem 23.3 there exists a unique integer $q = N^+(f, v(\alpha)) = N^-(f, v(\alpha))$ such that $|f(\alpha)| = |a_q \alpha^q| > |a_n||\alpha|^n$ whenever $n \neq q$, and therefore we have:

$$(1) \quad |f(\alpha)| = \sup_{n \in \mathbb{N}} |a_n||\alpha|^n$$

(2) $|P_m(\alpha)| = |a_q|\,|\alpha|^q$ for every $m \geq q$.

We fix $\epsilon > 0$, we put $E_\epsilon = D' \setminus d(0, s(\epsilon))$, and take $s(\epsilon) \in]0, r[$ such that $|h(x)| \leq \epsilon$ whenever $x \in E_\epsilon$. Now, for every $m \in \mathbb{N}$, we have

$$\left|\frac{Q_m}{P_m} - \frac{g}{f}\right| \leq \max\left(\left|\frac{Q_m(x) - g_m(x)}{P_m(x)}\right|, \left|\frac{g(x)}{f(x)}\right|\left|\frac{f(x) - P_m(x)}{P_m(x)}\right|\right).$$

Then by (1) and (2) it is shortly checked that there exists $t \in \mathbb{N}$ such that $\left\|\dfrac{Q_m}{P_m} - \dfrac{g}{f}\right\|_{E_\epsilon} \leq \epsilon$ for every $m \geq t$. In the same way, by (1) and (2) there clearly exists $s \in \mathbb{N}$ such that $\left\|\dfrac{Q_m}{P_m} - \dfrac{g}{f}\right\|_{D \cap d(0, s(\epsilon))} \leq \epsilon$, and this finishes proving that h belongs to $H(D')$. But then, since h is vanishing along the increasing filter of center 0 and diameter r, by Theorem 21.8 h belongs to $H(D)$.

Theorem 50.2:   *Let $f \in A_u(d(0, r^-))$ satisfy $f(0) \neq 0$ and* $\displaystyle \lim_{\substack{|x| \to r \\ x \in \mathcal{D}(f, \rho)}} \frac{g(x)}{f(x)} = 0$

for every $\rho \in]0, r[$. Then $\dfrac{g}{f}$ belongs to $H(\mathcal{D}(f, \rho))$ for every $\rho > 0$, and the function $h(x) = 1 + \dfrac{g}{f}$ is equal to a meromorphic product collapsing along the increasing filter of center 0 and diameter r on $D(f, \rho)$.

Proof: Let $D = \mathcal{D}(f, \rho)$, and for each $s \in]0, r[$, let $D_s = \mathcal{D}(f, \rho) \cap d(0, s)$. By hypothesis we have $\displaystyle \lim_{\substack{|x| \to r^- \\ x \in D_s}} h(x) = 1$. For each $s \in]0, r[$, clearly h belongs to $H(D_s)$ and is meromorphic in each hole of D_s. Let $\ell \in]0, r[$ be such that $\|h - 1\|_{\Gamma(0, \ell, r) \cap D} < 1$. We now consider a hole $T = d(\alpha, \rho^-)$ of D_s included in $\Gamma(0, \ell, r)$. Then h is clearly meromorphic in T. By construction of D there exists a set $\Gamma(\alpha, \rho, \rho')$ included in $\Gamma(0, \ell, r) \cap D$ and therefore we have $\|h(x) - 1\|_{\Gamma(\alpha, \rho, \rho')} < 1$. Hence, by Lemma 31.5 h admits as many zeros as many poles in $d(\alpha, \rho')$ (taking multiplicities into account).

In the same way, by considering the set $\Gamma(0, \ell, r) \cap D$, h is meromorphic in each hole $d(a, \ell^-)$ included in $d(0, \ell)$, and therfore h admits as many zeros as many poles in each one of these holes and also in $d(0, \ell)$ (taking multiplicities into account). Thus the sequence of the zeros (α_n) of h and the sequences of the poles (β_n) of h satisfies this property: there exists $N(\rho) \in \mathbb{N}$ such that both α_n, β_n belong to $d(0, \ell)$ whenever $n \leq N(\rho)$ and if $n > N(\rho)$ there exists a hole T of $\mathcal{D}(f, \rho)$ such that both α_n, β_n belong to T. Hence we have $|\alpha_n - \beta_n| < \rho$ whenever $n > N(\rho)$. Since ρ was chosen abitrary, it is seen that $\displaystyle \lim_{n \to \infty} \alpha_n - \beta_n = 0$.

Now we can consider the meromorphic product $\psi(x) = \displaystyle\prod_{n=0}^{\infty} \frac{x - \alpha_n}{x - \beta_n}$. Obviously, for each $\rho > 0$, ψ belongs to $H(\mathcal{D}(f, \rho))$. Let $(C(0, r_m))_{m \in \mathbb{N}}$ be the sequence of the circles that contain at least one zero of f, and let $E = K \setminus \left(\bigcup_{n \in \mathbb{N}} C(0, r_m) \right)$. Obviously G belongs to $H(D)$. By Theorem 50.1 so does $\dfrac{g}{f}$, and then so does h. By Theorem 21.8, h has continuation to an element $\widetilde{h}$ of $H(E)$ such that $\widetilde{h}(x) = 1$, whenever $x \in K \setminus d(0, r^-)$. Besides it is seen that both $\widetilde{h}$, ψ have no zero in E, and are invertible elements of $H(E)$, because they are not vanishing along the only one pierced filter of E, that is the increasing filter of center 0 and diameter r. Let $G = \dfrac{\widetilde{h}}{\psi}$. Then G belongs to $H(E)$ and is meromorphic in each hole of E because so are both h, ψ. Further, G has no pole in any hole of E, because by construction, h, ψ have the same zeros and the same poles (taking multiplicities into account). Hence G belongs to $H(K)$. But as $\displaystyle\lim_{|x| \to +\infty} h(x) = \lim_{|x| \to +\infty} \psi(x) = 1$, we have $G = 1$, and this ends the proof of Theorem 50.2.

Proposition 50.3: *Let r_n be a sequence in $]0, r[$ such that $r_n < r_{n+1}$, $\displaystyle\lim_{n \to \infty} r_n = r$. Let $a_0 \in K$ and for all $n \in \mathbb{N}^*$ let $a_{n+1} \in K$ be such that $|a_{n+1}| = \dfrac{|a_n|}{r_n}$.*

Let $f(x) = \displaystyle\sum_{n=0}^{\infty} a_n x^n$. Then f belongs to $A(d(0, r^-))$. When x belongs to $d(0, r_0^-)$ we have $|f(x)| = |a_0| > |a_n|\,|x|^n$ for every $n > 0$. When x belongs to $\Gamma(0, r_q, r_{q+1})$ $(q \in \mathbb{N})$ we have $|f(x)| = |a_q|\,|x|^q > |a_n|\,|x|^n$ for every $n \neq q$. Further, f is bounded if and only if $\displaystyle\prod_{n=0}^{\infty} \frac{r_n}{r} \neq 0$. In each circle $C(0, r_q)$, f admits a unique zero of order 1 and has no other zero in $d(0, r^-)$. Moreover, if f is not bounded, then it satisfies $\displaystyle\lim_{\substack{|x| \to r \\ x \in \mathcal{D}(f, \rho)}} |f(x)| = +\infty$ for every $\rho \in]0, r[$.

Proof: It is seen that f belongs to $A(d(0, r^-))$ because $\displaystyle\lim_{n \to \infty} \left| \frac{a_{n+1}}{a_n} \right| = \lim_{n \to \infty} \frac{1}{r_n} = \frac{1}{r}$. When $x \in d(0, r_0^-)$ we have $|a_0| = |a_1| r_0 = |a_2| r_0 r_1 = \ldots = |a_n| r_0 \ldots r_{n-1} > |a_n x^n|$ for all $x \in d(0, r_0^-)$. Hence $|f(x)| = |a_0| > |a_n|\,|x|^n$ for all $n \in \mathbb{N}^*$, for all $x \in d(0, r_0^-)$. Now let $q \in \mathbb{N}$ and let $x \in \Gamma(0, r_q, r_{q+1})$. For each $j = 0, \ldots, q-1$

we have $|a_q| = \dfrac{|a_{q-1}|}{r_{q-1}} = \dfrac{|a_{q-2}|}{r_{q-2}r_{q-1}} = \ldots = \dfrac{|a_{q-j}|}{r_{q-j}\ldots r_{q-1}}$. Hence it is seen that

$|a_q| > \dfrac{|a_n|}{|x|^{q-n}}$ and then $|a_q||x|^q > |a_n||x|^n$ for each $j = 0,\ldots,q-1$. In the

same way, we check that $|a_q| = |a_{q+j}|r_q\ldots r_{q+j-1} > |a_{q+j}|\,|x|^j$ and therefore $|a_q|\,|x|^q > |a_n|\,|x|^n$ for each $n > q$. Since $r_n < r$, obviously for every $n \in$ $\mathbb{N}$, f belongs to $H(d(0,r_n))$. Moreover, f satisfies $N^+(f,-\log r_q) = q+1$, $N^-(f,-\log r_q) = q$. Hence f admits a unique zero in $C(0,r_q)$. Further, it is seen that f is bounded if and only if $\sup\limits_{n\in\mathbb{N}} |a_n|r^n$ is bounded. But we have

$|a_n|r^n = |a_{n-1}|\dfrac{r^n}{r_{n-1}} = \ldots = |a_0|\prod\limits_{j=0}^{n}\dfrac{r}{r_j}$. Thus the sequence $|a_n|r^n$ is bounded if

and only if so is $\prod\limits_{j=0}^{\infty}\dfrac{r}{r_j}$. But this product is bounded if and only if $\prod\limits_{j=0}^{\infty}\dfrac{r_j}{r} \neq 0$.

Finally we suppose f unbounded and will check that $\lim\limits_{\substack{|x|\to r \\ x\in\mathcal{D}(f,\rho)}} |f(x)| = +\infty$ for

every $\rho \in]0,r[$. Indeed, for every $s < r$ we have $\|f\|_{d(0,s)} = d(0,s)\varphi_{0,s}(f) =$ $\sup\limits_{n\in\mathbb{N}} |a_n|s^n$, hence $\lim\limits_{\mu\to-\log r} v(f,\mu) = -\infty$. But as f has at most one zero in each circle $C(0,s)$, it is seen that we have $v(f(x)) \leq v(f,v(x)) + \log s - \log\rho$ for all $x \in \mathcal{D}(f,\rho)$, hence $\lim\limits_{\substack{|x|\to r \\ x\in\mathcal{D}(f,\rho)}} v(f(x)) = -\infty$. This ends the proof of Proposition

50.3.

Theorem 50.4:   *Let r_n be a sequence in $]0,r[$ such that $r_n < r_{n+1}$, $\lim\limits_{n\to\infty} r_n = r$. Let $a_0 \in K$ and for all $n \in \mathbb{N}^*$ let $a_{n+1} \in K$ be such that $|a_{n+1}| = \dfrac{|a_n|}{r_n}$. Let*

$$f(x) = \sum_{n=0}^{\infty} a_n x^n. \ \text{Then } \frac{1}{f} \text{ belongs to } H(\mathcal{D}(f,\rho)).$$

Proof: Indeed, by Propostion 50.3 f admits a unique zero α_n in each circle $C(0,r_n)$, $(n \in \mathbb{N})$, and it satisfies $\lim\limits_{\substack{|x|\to r \\ x\in\mathcal{D}(f,\rho)}} |f(x)| = +\infty$ for every $\rho \in]0,r[$.

Hence by Theorem 50.2, $1 + \dfrac{1}{f}$ is a meromorphic product, and therefore belongs

to $H(\mathcal{D}(f,\rho))$. Hence so does $\dfrac{1}{f}$.

Definition: Given a disk V of the form $d(a,r^-)$ in K, we will call *peripheral* of V any class of $C(a, r)$.

Theorem 50.5:   *Let $r, s \in \mathbb{R}_+$ satisfy $0 < r < s$ and let $(r_n)_{n \in \mathbb{N}}, (s_n)_{n \in \mathbb{N}}$ be sequences in $]r, s[\cap |K|$ satisfying $r < r_{n+1} < r_n < s_n < s_{n+1} < s$, $r_0 = s_0$,*

$$\lim_{n \to \infty} r_n = r, \quad \lim_{n \to \infty} s_n = s, \quad \prod_{n=0}^{\infty} \left(\frac{r_n}{r} \right) = \prod_{n=0}^{\infty} \left(\frac{s}{s_n} \right) = 0. \quad \textit{There exists an open}$$

closed infraconnected set D that contains $K \setminus \left(\bigcup_{n=0}^{\infty} C(0, r_n) \cup C(0, s_n) \right)$ such that for every hole T, only finitely many peripherals of T are not included in D, together with an element $f \in H_b(D)$ satisfying

$$f(x) = 0 \text{ whenever } x \in d(0, r)$$
$$f(x) = 1 \text{ whenever } x \in K \setminus d(0, s^-).$$

Proof: Let $a_0 = b_0 = 1$ and let $(a_n)_{n \in \mathbb{N}}$, $(b_n)_{n \in \mathbb{N}}$ be sequences in K satisfying $|a_{n+1}| = r_n |a_n|$ and $|b_{n+1}| = \dfrac{|b_n|}{s_n}$. Let $g(x) = \sum_{n=0}^{\infty} \dfrac{a_n}{x^n}$, let $h(x) =$

$\sum_{n=0}^{\infty} b_n x^n$, and for every $n \in \mathbb{N}$, let $g_n = \sum_{j=0}^{n} \dfrac{a_j}{x_j}$, $h_n = \sum_{j=0}^{n} b_j x^j$.

First we check that $g \in A(K \setminus d(0, r))$ and $h \in A(d(0, s^-))$, because we have $\lim_{n \to \infty} \left| \dfrac{a_{n+1}}{a_n} \right| = r$, $\lim_{n \to \infty} \left| \dfrac{b_{n+1}}{b_n} \right| = \dfrac{1}{s}$. Next, it is easily checked that the sequence $\dfrac{|a_n|}{r^n}$ (resp. $|b_n| s^n$) is strictly decreasing (resp. increasing). By Proposition 50.3 we have,

(1) $|h(x)| = |b_0| = 1 > |b_n| |x|^n$ whenever $x \in d(0, r_0^-)$

(2) $|g(x)| = |a_0| = 1 > \dfrac{|a_n|}{|x|^n}$ whenever $x \in K \setminus d(0, s_0)$

(3) $|h(x)| = |b_q| |x|^q > |b_n| |x|^n$ for all $n \neq q$, whenever $x \in \Gamma(0, s_q, s_{q+1})$ $(q \in \mathbb{N})$. Besides, for each $q \in \mathbb{N}$, by Proposition 20.7 there exist finitely many classes $L_{q,j}$ $(1 \leq j \leq \tau_q)$ of $C(0, s_q)$ such that

(4) $|h(x)| = |b_q| s_q$ whenever $x \in C(0, s_q) \setminus \left(\bigcup_{j=1}^{\tau_q} L_{q,j} \right).$

By (3) and (4) it is seen that we have

(5) $v(h(x)) = v(h, v(x))$ whenever $x \in D' \cap d(0, s^-)$.

Besides, since $b_0 = 1$, we have $v(h, \mu) < 0$ for all $\mu < -\log s_0$. In particular, we have

(6) $|h(x)| \geq 1$ whenever $x \in D'$.

Now, let $m \in \mathbb{N}$ be fixed, and consider h_m. By (3), we have

(7) $|h_m(x)| = |h(x)| = |b_q x^q|$ whenever $x \in \Gamma(0, s_q, s_{q+1})$, for each $q < m$,
and of course
(8) $|h_m(x)| = |b_m x^m|$ for every $x \in K \setminus d(0, r_m)$.
Besides, in each circle $C(0, s_q)$ $(0 \leq q \leq m)$, the equality $|h_m(x)| = |b_q x^q|$ is true in all the classes but finitely many ones. Hence, since the sequence $(h_n)_{m \in \mathbb{N}}$ converges uniformly to h in each disk $d(0, l)$ $(0 < l < s)$, by (5), (6) it is seen that for each $q \in \mathbb{N}$ there exist finitely many classes $W_{q,j}$, $(1 \leq j \leq u_q)$ of $C(0, s_q)$ such that the equalities $v(h_m(x)) = v(h_m, v(x))$ and $v(h(x)) = v(h, v(x))$ hold

for all $x \in C(0, s_q) \setminus \left(\bigcup_{j=1}^{u_q} W_{q,j} \right)$, whenever $m \in \mathbb{N}$.

Let $D' = K \setminus \left(\bigcup_{q \in \mathbb{N}} \left(\bigcup_{j=1}^{u_q} W_{q,j} \right) \right)$. By (5), (6), (7), (8) we see

(9) $v(h_m(x)) = v(h_m, v(x)) \leq 0$ whenever $m \in \mathbb{N}$, whenever $x \in D'$, and
(10) $v(h(x)) = v(h, v(x)) \leq 0$ whenever $x \in D'$.
In particular, we have
(11) $v(h_m(x)) = v(h_m, v(x)) < 0$ for all $m \in \mathbb{N}$, whenever $x \in (K \setminus d(0, s_0)) \cap D'$,
(12) $v(h(x)) = v(h, v(x)) < 0$ whenever $x \in (d(0, s^-) \setminus d(0, s_0)) \cap D'$.

At the same way, symmetrically, for every $q \in \mathbb{N}$, there exist finitely many classes $V_{q,j}(1 \leq j \leq t_q)$ of $C(0, r_q)$ such that $v(g_m(x)) = v(g_m, v(x))$ and

$v(g(x)) = v(g, v(x))$ are true for all $x \in C(0, r_q) \setminus \left(\bigcup_{j=1}^{t_q} W_{q,j} \right)$, whenever $m \in \mathbb{N}$.

Then, putting $D'' = K \setminus \left(\bigcup_{q \in \mathbb{N}} \left(\bigcup_{j=1}^{t_q} V_{q,j} \right) \right)$, we have

(13) $v(g_m(x)) = v(g_m, v(x)) \leq 0$ whenever $m \in \mathbb{N}$, whenever $x \in D''$,
(14) $v(g(x)) = v(g, v(x)) \leq 0$ whenever $x \in D''$.
In particular, we have
(15) $v(g_m(x)) = v(g_m, v(x)) < 0$ whenever $m \in \mathbb{N}$, whenever $x \in d(0, r_0^-) \cap D''$,
(16) $v(g(x)) = v(g, v(x)) < 0$ whenever $x \in (d(0, r_0^-) \setminus d(0, r)) \cap D''$.
Now we take $E = D' \cap D''$ and we consider the function $\ell(x) = g(x) + h(x)$ in $D \cap \Gamma(0, r, s)$. By (12) and (16) we have
(17) $|\ell(x)| > 1$ whenever $x \in (E \cap \Gamma(0, r, s)) \setminus \Delta(0, r_0, s_0)$.
Since the set $F := \Delta(0, r_0, s_0) \cap E$ has no pierced filter, the restriction of ℓ to F is seen to be a quasi-invertible element of $H(F)$. Hence there are finitely many circles $C(0, \rho_k)$ $(1 \leq k \leq e)$ and in each one, finitely many classes $(T_{k,j})_{1 \leq j \leq w_k}$ such that

(18) $v(\ell(x)) = v(\ell, v(x))$ whenever $x \in F \setminus \left(\bigcup_{k=1}^{e} \bigcup_{j=1}^{w_k} T_{k,j} \right)$.

Now we put $D = E \setminus \left(\bigcup_{k=1}^{e} \left(\bigcup_{j=1}^{w_k} T_{k,j} \right) \right)$. The only holes of D are seen to be the

$(V_{q,i})$ $(1 \leq i \leq u_q, q \in \mathbb{N})$,   the $(W_{q,j})$ $(1 \leq j \leq u_q, \ q \in \mathbb{N})$,   and the $(T_{k,j})$ $(1 \leq j \leq w_k, \ k = 1, \ldots, e)$. Further, we notice that by construction of D', D'', E, F, for each hole of such a set, there only are finitely many peripherals which are not included in this set. So this property is clearly conserved on D. Now, by (17), (18) we have

(19) $v(\ell(x)) = v(\ell, v(x)) \leq 0$ whenever $x \in D \cap \Gamma(0, r, s)$.

Now we define f in D as $f(x) = \dfrac{h(x)}{g(x) + h(x)}$ when $x \in D \cap \Gamma(0, r, s)$, $f(x) = 1$ when $x \in K \setminus d(0, s^-)$ and $f(x) = 0$ when $x \in d(0, r)$. For every $n \in \mathbb{N}$ we put $\ell_n = g_n + h_n$, $f_n = \dfrac{h_n}{\ell_n}$ and we will prove that the sequence $(f_n)_{n \in \mathbb{N}}$ converges to

f in $H(D)$. For each $n \in \mathbb{N}$ we put $D_n = D \cap \Gamma(0, r_n, s_n)$. Since $\prod_{n=0}^{\infty} \left(\dfrac{s}{s_n} \right) = 0$,

by Proposition 50.3, h is not bounded in $d(0, s^-)$. Hence we have

(20) $\lim_{n \to \infty} \|h\|_{D_n} = +\infty$.

But actually by (1) we have $\|h\|_{D_n} = |b_n|(s_n)^n$. Hence by (20) we have

(21) $\lim_{n \to \infty} |b_n|(s_n)^n = +\infty$.

Symmetrically we have

(22) $\lim_{n \to \infty} \dfrac{|a_n|}{(r_n)^n} = +\infty$.

We fix $\epsilon > 0$. By (21), (22) we can take $N \in \mathbb{N}$ such that $\dfrac{|a_N|}{(r_N)^N} > \dfrac{1}{\epsilon}$ and

$|b_N|(s_N)^N > \dfrac{1}{\epsilon}$. With greater reason, for all $n \geq N$ we have $|b_n|(s_n)^n > \dfrac{1}{\epsilon}$,

and therefore

(23) $|h(x)| \geq \dfrac{1}{\epsilon}$ for all $x \in D \setminus d(0, r_N)$, whenever $n \geq N$.

Symmetrically, we obtain

(24) $|g(x)| \geq \dfrac{1}{\epsilon}$ for all $x \in D \cap d(0, (s_N)^-)$, whenever $n \geq N$.

But by (1), (2) we have $|h(x)| = |h_m(x)| = 1$ for every $m \in \mathbb{N}$, for all $x \in d(0, s_0^-)$ and as $|g(x)| = |g_m(x)| = 1$ for every $m \in \mathbb{N}$, for all $x \in K \setminus d(0, r_0)$, and therefore by (23) we have

(25) $|\ell_n(x)| = |h_n(x)| \geq \dfrac{1}{\epsilon}$ for every $n > N$, for all $x \in d(0, s_0^-)$,

and symmetrically

(26) $\quad |\ell_n(x)| = |g_n(x)| \geq \dfrac{1}{\epsilon}$ for every $n > N$, for all $x \in D \cap d(0,(s_N)^-)$.

By (23), (24), (25), (26), we have

(27) $\quad |f_n(x) - f(x)| \leq \epsilon$ for every $x \in D \setminus \Delta(0, r_0, s_0)$.

On the other hand, since $|\ell(x)| \geq 1$ for all $x \in D$, it is easily seen that the sequence $(f_n)_{n \in \mathbb{N}}$ converges uniformly to f in $H(D \cap \Delta(0, r_0, s_0))$. In particular there exists $t \geq N$ such that

(28) $\quad |f_n(x) - f(x)| \leq \epsilon$ for every $n \geq t$, whenever $x \in D \cap \Delta(0, r_0, s_0)$.

Thus, (27) and (28) finish showing the sequence $(f_n)_{n \in \mathbb{N}}$ to converge to f in $H(D)$, and this ends the proof of Proposition 50.5.

Remark: Let $f \in A_u(d(0, r^-))$ satisfy $f(0) \neq 0$. Let $(r_n)_{n \in \mathbb{N}^*}$ be a strictly increasing sequence of limit r such that $\displaystyle\bigcup_{m=1}^{\infty} C(0, r_m)$ contains all the zeros of f.

Let $D = K \setminus \left(\displaystyle\bigcup_{m=1}^{\infty} C(0, r_m) \right)$. Then by Theorem 23.3 we have $v(fx)) = v(f, v(x))$ for all $x \in D$, hence $\displaystyle\lim_{|x| \to r, x \in D \cap d(0, r^-)} \dfrac{1}{f(x)} = 0$, and then by Theorem 50.1 $\dfrac{1}{f}$ belongs to $H(D)$. Let $h = \dfrac{1}{f}$. Of course h has no zero in D, hence $h(D)$ is included in the set $E = K^*$. Let $\phi(u) = \dfrac{1}{u} \in R(E)$. Then we see that $\phi(h(x)) = f(x)$ whenever $x \in D$, and therefore $\phi \circ h$ does not belong to $H(D)$, although $h(D) \subset E$. This proves that the composition of analytic elements, sometimes, is not an analytic element.

51. INFINITE VAN DER MONDE MATRICES

We study the kernel of infinite van der Monde matrices and show close connections with the zeroes of power series. We study when such a matrix is invertible [4]. These results will be used in the next chapter to obtain interpolation processes for analytic functions.

Notations: Given K-vector spaces E, F, $\mathcal{L}(E, F)$ will denote the space of the K-linear mappings from E into F.

$\mathcal{E}$ will denote the K-vector space of the sequences in K, and $\mathcal{E}_0$ will denote the subspace of the bounded sequences. The identically zero sequence will be denoted by (0).

$\mathcal{E}_1$ will denote the set of the sequences $(a_n)_{n \in \mathbb{N}}$ such that $\limsup\limits_{n \to \infty} \sqrt[n]{|a_n|} \leq 1$. So $\mathcal{E}_1$ is seen to be a subspace of $\mathcal{E}$ isomorphic to the space $A(d(0, 1^-))$, and obviously contains $\mathcal{E}_0$.

Let $\mathbf{M}_\infty$ be the set of the infinite matrices $(\lambda_{i,j})$ with coefficients in K.

$\delta_{i,j}$ will denote the Kronecker symbol. I_∞ will denote the infinite identical matrix defined as $\lambda_{i,j} = \delta_{i,j}$.

In this chapter, $(a_n)_{n \in \mathbb{N}}$ will denote an injective sequence in $d(0, 1^-)$ such that $a_n \neq 0$ for every $n > 0$. We will denote by $\mathcal{M}(a_n)$ the infinite matrix $M = (\lambda_{i,j})$ defined as $\lambda_{i,j} = a_i^j$, $(i, j) \in \mathbb{N} \times \mathbb{N}$.

A matrix $M = (\lambda_{i,j}) \in \mathbf{M}_\infty$ will be said to be *bounded* if there exists $A \in \mathbb{R}_+$ such that $|\lambda_{i,j}| \leq A$ whenever $(i, j) \in \mathbb{N} \times \mathbb{N}$.

M will be said to be *line-vanishing* if for each $i \in \mathbb{N}$, we have $\lim\limits_{j \to \infty} \lambda_{i,j} = 0$.

A line-vanishing matrix M is seen to define a K-linear mapping ψ_M from $\mathcal{E}_0$ into $\mathcal{E}$. So the matrix $M = \mathcal{M}(a_n)$ clearly defines a K-linear mapping ϕ_M from $\mathcal{E}_1$ into $\mathcal{E}$, because given a sequence $(b_n) \in \mathcal{E}_1$, the series $\sum\limits_{n=0}^{\infty} b_n a_j^n$ is obviously convergent.

Lemmas 51.1 and 51.2 are immediate:

Lemma 51.1: *Let $M \in \mathbf{M}_\infty$ be line-vanishing. Then the following three statements are equivalent:*

ψ_M is continuous

ψ_M *is an endomorphism of* $\mathcal{E}_0$
M *is bounded* .

In particular, Lemma 51.1 applies to matrices of the form $\mathcal{M}(a_n)$.

Lemma 51.2: *Let* $M = \mathcal{M}(a_n)$ *and let* $(b_n)_{n\in\mathbb{N}} \in \mathcal{E}_1$. *Then* $(b_n)_{n\in\mathbb{N}}$ *belongs to* $Ker\phi_M$ *if and only if the analytic function* $f(t) = \sum_{n=0}^{\infty} b_n t^n$ *admits each point* a_j *for zero.*

Theorem 51.3: *Let* $M = \mathcal{M}(a_n)$. *Then* $Ker\phi_M \neq \{(0)\}$ *if and only if* $\lim_{n\to\infty} |a_n| = 1$. *Besides* $Ker\psi_M \neq \{(0)\}$ *if and only if* $\prod_{n=0}^{\infty} |a_n| > 0$.

Proof: Let $\mathbf{b} = ((b_n)_{n\in\mathbb{N}}) \in \mathcal{E}_1 \backslash \{(0)\}$ and let $f(t) = \sum_{n=0}^{\infty} b_n t^n \in A(d(0, 1^-))$.

First we suppose $Ker\phi_M \neq \{(0)\}$ and therefore we can assume $\mathbf{b} \in Ker\phi_M$. Then, by Lemma 51.2, f satisfies $f(a_j) = 0$ for every $j \in \mathbb{N}$. But for every $r \in]0, 1[$, we know that f belongs to $H(d(0, r))$ and has finitely many zeros in $d(0, r)$. Hence we have $\lim_{n\to\infty} |a_n| = 1$.

Reciprocally, let the sequence (a_n) satisfy $\lim_{n\to\infty} |a_n| = 1$. By Proposition 25.5 we know that there exists a non identically zero analytic function $f(t) = \sum_{n=0}^{\infty} b_n t^n \in A(d(0, 1^-))$ which admits each a_j as a zero. Hence we have $\sum_{n=0}^{\infty} b_n a_j^n = 0$, and of course the sequence $(b_n)_{n\in\mathbb{N}}$ belongs to $\mathcal{E}_1$, hence to $Ker\phi_M$.

Now we suppose that $Ker\psi_M \neq (0)$ and we assume that the sequence (b_n) belongs to $Ker\psi_M$. In particular $Ker\phi_M \neq (0)$ and therefore $\lim_{n\to\infty} |a_n| = 1$. Without loss of generality we may clearly assume $|a_n| \leq |a_{n+1}|$ for all $n \in \mathbb{N}$. Besides, by definition we have $|a_1| > 0$. By Theorem 13.9 we know that $\inf_{n\in\mathbb{N}} v(b_n) = \lim_{\mu\to 0^+} v(f, \mu) = \lim_{|x|\to 1, x\in D} v(f(x)) = -\log \|f\|_{d(0,1^-)}.$

Now for each $\mu > 0$, let $q(\mu)$ be the unique integer such that $v(a_n) \geq \mu$ for every $n \leq q(\mu)$ and $v(a_n) < \mu$ for every $n > q(\mu)$. By Theorem 23.13, we have

(1) $\quad v(f, \mu) - v(f, v(a_1)) \leq \sum_{j=2}^{q(\mu)} \mu - v(a_j) + 2(\mu - v(a_1))$. Since $v(f, \mu)$ is bounded

when μ approaches 0, by (1) it is seen that $\sum_{j=1}^{\infty} v(a_j)$ must be bounded and therefore we have $\displaystyle\prod_{n=1}^{\infty} |a_n| > 0$.

Reciprocally we suppose $\displaystyle\prod_{n=1}^{\infty} |a_n| > 0$. We can easily check that $\displaystyle\lim_{n\to\infty} |a_n| = 1$, and then we can assume $|a_n| \leq |a_{n+1}|$ for all $n \in \mathbb{N}$ without loss of generality.

For each $j \in \mathbb{N}$ we put $P_j(x) = \displaystyle\prod_{m=1}^{j} (1 - \frac{x}{a_m})$. By Theorem 25.5 there exists $f \in A(d(0, 1^-))$ (f not identically zero) satisfying

    (2) $f(a_m) = 0$ for all $m \in \mathbb{N}$, and

    (3) $v(f, \mu) \geq v(P_{q(\mu)}, \mu) - 1$ for all $\mu > 0$.

Now we notice that if $\mu_1 > \mu_2 > 0$ then we have $v(P_{q(\mu_1)}, \mu_1) = v(P_{q(\mu_2)}, \mu_1)$ and then we see that $\displaystyle\lim_{\mu\to 0+} v(P_{q(\mu)}, \mu) = \sum_{j=1}^{\infty} v(a_j)$. But by hypothesis, we have $\displaystyle\sum_{j=1}^{\infty} v(a_j) < +\infty$ and therefore by (3), $v(f, \mu)$ is bounded in $]0, +\infty[$.

Let $f(t) = \displaystyle\sum_{n=0}^{\infty} b_n t^n$. By Theorem 13.9 the sequence (b_n) is bounded and by (2) it clearly belongs to $Ker.\psi_M$. This finishes the proof of Theorem 51.3.

Theorem 51.4: *Let* $\mathbf{b} = (b_n)_{n\in\mathbb{N}} \in \mathcal{E}_0$. *There exists an injective sequence* $(\alpha_n)_{n\in\mathbb{N}}$ *in* $d(0, 1^-)$ *such that* $\mathbf{b} \in Ker\psi_{\mathcal{M}(\alpha_n)}$ *if and only if* $\mathbf{b}$ *satisfies* $|b_j| < \displaystyle\sup_{n\in\mathbb{N}} |b_n|$ *for all* $j \in \mathbb{N}$.

Proof: By Theorem 23.10 the function $f(t) = \displaystyle\sum_{n=0}^{\infty} b_n t^n \in A(d(0, 1^-))$ admits infinitely many zeros in $d(0, 1^-)$ if and only if $|b_j| < \displaystyle\sup_{n\in\mathbb{N}} |b_n|$ for every $j \in \mathbb{N}$. Then the conclusion comes from Lemma 51.2.

Definitions and notations: Let $(a_n)_{n\in\mathbb{N}}$ be a polar sequence associated to $d(0, 1^-)$, of separation ρ. We will denote by $\mathcal{O}_r(a_n)$ the set $\Lambda_r((a_n)_{n\in\mathbb{N}})\cap d(0, 1^-)$ and by $\mathcal{O}(a_n)$ the set $\mathcal{O}_\rho(a_n)$.

Let $\mathbf{a} = (a_n)_{n\in\mathbb{N}}$ and $\mathbf{b} = (b_n)_{n\in\mathbb{N}}$ be two sequences in K. We will denote by $\mathbf{a} * \mathbf{b}$ the convolution product $(c_n)_{n\in\mathbb{N}}$ defined as $c_n = \displaystyle\sum_{j=0}^{n} a_j b_{n-j}$.

Lemma 51.5: *Let $f \in A(d(0,1^-))$ have a polar sequence of zeros $(b_n)_{n\in\mathbb{N}}$ associated to $d(0,1^-)$, and satisfy $\lim\limits_{\substack{|x|\to 1^- \\ x\in\mathcal{O}(b_n)}} |f(x)| = +\infty$. Then $\dfrac{1}{f}$ belongs to $H(\mathcal{O}(b_n))$.*

Proof: Let $r = \inf\limits_{m\neq n} \{|b_n - b_m|\}$. By hypothesis, we have $\lim\limits_{\substack{|x|\to 1^- \\ x\in\mathcal{D}(f,r)}} |f(x)| = +\infty$.

Now, let $s \in]0,r[$, and consider $f(x)$ when x lies in certain annulus $\Delta(b_n, s, r)$. For convenience, we put $a = b_n$, and take $l \in]|a|, 1[$. By hypothesis b_n is the only zero of f in $d(b_n, r)$, hence we have $|f(x)| \leq \dfrac{r_d(0,l)\varphi_{a,r}(f)}{s}$. As a consequence, it is seen that for every $\ell \in [r, 1[$, we have $\inf\limits_{x\in\mathcal{D}(f,s)\cap C(0,\ell)} |f(x)| \geq \dfrac{r_d(0,l)\varphi_{a,r}(f)}{s}$, and therefore $\lim\limits_{\substack{|x|\to 1^- \\ x\in\mathcal{D}(f,s)}} |f(x)| = +\infty$. Since s has been taken arbitrary, we can apply Theorem 49.2, and then $\dfrac{1}{f}$ belongs to $H(\mathcal{O}(b_n))$.

Theorem 51.6: *Let $(\alpha_n)_{n\in\mathbb{N}}$ be a polar sequence associated to $d(0,1^-)$ such that there exists $g \in A(d(0,1^-))$ satisfying*
 (i) α_n is a zero of order 1 of g for all $n \in \mathbb{N}$.
 (ii) $g(x) \neq 0$ whenever $x \in d(0,1^-)\backslash\{\alpha_n | n \in \mathbb{N}\}$.
 (iii) $\lim\limits_{\substack{|x|\to 1^- \\ x\in\mathcal{O}(\alpha_n)}} |g(x)| = +\infty$.
Let $M = \mathcal{M}(\alpha_n)$. Then ψ_M is injective but its image does not contain $\mathcal{E}_0$. Also there exists $P = (\lambda_{i,j}) \in \mathcal{M}_\infty$ (not unique) satisfying
(1) P is line-vanishing.
(2) $\lim\limits_{n\to\infty} \lambda_{n,j}\alpha_h^n = 0$ for all $(j, h) \in \mathbb{N} \times \mathbb{N}$.
(3) $\sum\limits_{n=0}^{\infty} \lambda_{n,j}\alpha_h^n = \delta_{j,h}$ for all $(j, h) \in \mathbb{N} \times \mathbb{N}$.
(4) $MP = PM = I_\infty$.
(5) $P(\mathbf{b}) \in \mathcal{E}_1$ for all $\mathbf{b} \in \mathcal{E}_0$.
(6) $MP(\mathbf{b}) = \mathbf{b}$ for all $\mathbf{b} \in \mathcal{E}_0$.
(7) ψ_P is injective.
Let (ν_n) be a sequence in K such that $|\nu_0| \geq |\nu_n|$ for every $n > 0$. For every $j \in \mathbb{N}$, let $(\mu_{n,j})_{n\in\mathbb{N}}$ denote the sequence $\left(\dfrac{1}{\sum_{m=0}^{\infty} \nu_m \alpha_j^m}\right)((\lambda_{n,j}) * (\nu_n))$. Then the matrix $Q = (\mu_{i,j})$ also satisfies properties $(1) - (7)$ and is not equal to P for infinitely many sequences (ν_n).

Proof: We may obviously assume $|\alpha_n| \leq |\alpha_{n+1}|$ and therefore $\alpha_n \neq 0$ whenever $n > 0$. Since g is not bounded in $d(0, 1^-)$, by Theorem 13.9 we have $\lim_{\mu \to 0+} v(g, \mu) = -\infty$, and then by Corollary 23.14 we know that the sequence of zeros (α_n) satisfies $\prod_{n=1}^{\infty} |\alpha_n| = 0$, hence ψ_M is injective. Now we look for P. Since g admits each α_j as a simple zero, it factorizes in $A(d(0, 1^-))$ in the form $\psi_j(x)(1 - \frac{x}{\alpha_j})$ and we have $\psi_j(\alpha_j) \neq 0$. We put $g_j(x) = \frac{\psi_j(x)}{\psi_j(\alpha_j)}$. Then g_j belongs to $A(d(0, 1^-))$ and may be written as $\sum_{n=0}^{\infty} \lambda_{n,j} x^n$. We denote by P the matrix

$$
\begin{pmatrix}
\lambda_{00} & \lambda_{01} & \cdots & \lambda_{0n} & \cdots \\
\lambda_{10} & \lambda_{11} & \cdots & \lambda_{1n} & \cdots \\
\vdots & \vdots & \ddots & \vdots & \ddots \\
\lambda_{j0} & \lambda_{j1} & \cdots & \lambda_{jn} & \cdots \\
\vdots & \vdots & \ddots & \vdots & \ddots
\end{pmatrix}
$$

and we will show this satisfies Properties $(1) - (7)$.

For convenience, we put $D = \mathcal{O}(\alpha_n)$. Since $\lim_{\substack{|x| \to 1^- \\ x \in D}} |g(x)| = +\infty$, by Lemma 51.5, we know that $\frac{1}{g}$ belongs to $H(D)$. For each $n \in \mathbb{N}$, we put $u_n = \frac{x^n}{g}$. Then in $H(D)$, u_n has a Mittag-Leffler series of the form $\sum_{j=0}^{\infty} \frac{\beta_{j,n}}{1 - \frac{x}{\alpha_j}}$. Now we put $\theta_j = \psi_j(\alpha_j)$ and we have $g(x) = \theta_j g_j(x)(1 - \frac{x}{\alpha_j})$. We will compute the $\beta_{j,n}$. Let $v_{j,n} = (1 - \frac{x}{\alpha_j})u_n$. Then we have $v_{j,n}(\alpha_i) = \frac{\alpha_j^n}{g_j(\alpha_j)\theta_j}$. But since $g_j(\alpha_j) = 1$ whenever $j \in \mathbb{N}$, we see that $\beta_{j,n} = \frac{\alpha_j^n}{\theta_j}$, hence $x^n g(x) = \sum_{i=0}^{n} \frac{\alpha_j^n}{\theta_j(\frac{1-x}{\alpha_j})}$. We notice that $\left\| \frac{\alpha_j^n}{1 - \frac{x}{\alpha_j}} \right\|_D = \frac{|\alpha_j|^{n+1}}{\rho}$ and then we have $\lim_{j \to \infty} |\theta_j| = +\infty$, because the sequence of the terms $\frac{x^n}{g(x)}$ must tend to 0. Now we have $x^n = \frac{\sum_{j=0}^{n} \alpha_j^n g(x)}{\theta_j(1 - \frac{x}{\alpha_j})}$, while $g_j(x) =

$\dfrac{g(x)}{\theta_j(1 - \frac{x}{\alpha_j})}$. Since $g_j(x) = \displaystyle\sum_{n=0}^{\infty} \lambda_{n,j} x^n$, we obtain (8) $x^n = \displaystyle\sum_{j=0}^{\infty} \alpha_j^n (\sum_{h=0}^{\infty} \lambda_{h,j} x^h)$.

In particular, (8) holds in every disk $d(0,r)$ with $r \in\,]0,1[$. But then we know

that $\|g_j\|_{d(0,r)} = \displaystyle\sup_{h\in\mathbb{N}} |\lambda_{j,h}| r^h \leq \dfrac{\|\psi_j\|_{d(0,r)}}{|\theta_j|}$. Now, we have $\|\phi_j\|_{d(0,r)} \leq \|g\|_{d(0,r)}$

as soon as $|\alpha_i| > r$ because then $\left\|\dfrac{1}{(1 - \frac{x}{\alpha_j})}\right\|_{d(0,r)} = 1$ and therefore the sequence

$(\|\phi_j\|_{d(0,r)})_{j\in\mathbb{N}}$ is bounded. Then the family $(|\lambda_{h,j}| r^h)_{j,h\in\mathbb{N}}$ tends to zero when j tends to $+\infty$, uniformly with respect to h. In particular, P is line-vanishing. For each $h \in \mathbb{N}$, we put $s_h = \sup_{j\in\mathbb{N}} |\lambda_{h,j}|$. We will show

(9) $\displaystyle\limsup_{h\to+\infty} (s_h)^{1/h} \leq 1$.

Indeed this is equivalent to showing that for every $r \in\,]0,1[$, we have

(10) $\displaystyle\lim_{h\to\infty} s_h r^h = 0$.

Let $r \in\,]0,1[$ and let $\epsilon > 0$. Since the family $(|\lambda_{h,j}| r^h)_{j,h\in\mathbb{N}}$ tends to zero uniformly with respect to h when j tends to $+\infty$, there clearly exists an integer N such that $|\lambda_{h,j}| r^h < \epsilon$ whenever $j > N$, whenever $h \in \mathbb{N}$, hence for every $h \in \mathbb{N}$, we have $s_h r^h \leq \displaystyle\max_{1\leq j\leq N} |\lambda_{h,j}| r^h$. But for each fixed $i \in \mathbb{N}$, we know that

$\displaystyle\lim_{h\to\infty} |\lambda_{h,j}| r^h = 0$, hence $\displaystyle\lim_{h\to\infty} (\max_{1\leq j\leq N} |\lambda_{h,j}| r^h) = 0$. This finishes showing (10). Therefore (9) is proven and so is (2). Now, we can apply the limits inversion theorem and, then, by (8), we have

(11) $x^n = \displaystyle\sum_{h=0}^{\infty} (\sum_{j=0}^{\infty} \alpha_j^n \lambda_{h,j}) x^h$, whenever $x \in d(0,r)$.

Actually this is true for all $r \in\,]0,1[$ and therefore (11) holds for all $x \in d(0,1^-)$.

Hence we have $\displaystyle\sum_{j=0}^{\infty} \alpha_j^n \lambda_{h,j} = 0$ whenever $n \neq h$ and $\displaystyle\sum_{j=0}^{\infty} \alpha_j^n \lambda_{n,j} = 1$. So (3) is

satisfied.

Thus we have proven that $PM = I_\infty$. Now we check that $MP = I_\infty$. For

every $h \neq j$, we have $g_j(\alpha_h) = g(\alpha_h) = 0$, hence $\displaystyle\sum_{h=0}^{\infty} \alpha_h^n \lambda_{h,j} = 0$. Besides, it is

seen that $g_j(\alpha_j) = 1$, hence $\displaystyle\sum_{n=0}^{\infty} \alpha_j^n \lambda_{n,j} = 1$. So we conclude that $MP = I_\infty$ and

this finishing proving (4).

Now, we will check that $P(\mathbf{b}) \in \mathcal{E}_1$ for all $\mathbf{b} \in \mathcal{E}_0$. Let $\mathbf{b} := (b_n)_{n\in\mathbb{N}} \in$

$\mathcal{E}_0$, let $\mathbf{a} := (a_n)_{n \in \mathbb{N}} = P(\mathbf{b})$ and let $f(t) = \displaystyle\sum_{n=0}^{\infty} a_n t^n$. For each $j \in \mathbb{N}$ we

put $f_j(t) = \displaystyle\sum_{m=0}^{j} b_m g_m(t)$. Then f_j belongs to $A(d(0,1^-))$ for all $j \in \mathbb{N}$. Let

$r \in]0,1[$. Like the family $|\lambda_{n,j}| r^n$, the family $|\lambda_{n,j} b_j| r^n$ tends to zero uniformly with respect to n when j tends to $+\infty$. That way, in $H(d(0,r))$ we have $\displaystyle\lim_{j \to \infty} \|f - f_j\|_{d(0,r)} = 0$ and therefore f belongs to $H(d(0,r))$. This is true for all $r \in]0,1[$ and therefore f belongs to $A(d(0,1^-))$. Hence $P(\mathbf{b}) \in \mathcal{E}_1$. This shows (5).

Let us show (6). Let $\mathbf{b} := (b_0, \ldots, b_n, \ldots)$ be a bounded sequence. Let $\mathbf{a} = P\mathbf{b}$, and let $\mathbf{a} = (a_0, \ldots, a_n, \ldots)$. We will show

$$(12) \quad \limsup_{n \to \infty} |a_n|^{1/n} \leq 1.$$

Without loss of generality, we may assume $|b_j| \leq 1$, whenever $j \in \mathbb{N}$. Then we have $|a_n| \leq \sup_{j \in \mathbb{N}} |\lambda_{n,j}| = s_n$, therefore $\displaystyle\limsup_{n \to \infty} |a_n|^{1/n} \leq \limsup_{n \to \infty} s_n^{1/n} \leq 1$. Now,

by (12), it is seen that for all $j \in \mathbb{N}$, the series $\displaystyle\sum_{n=0}^{\infty} a_n \alpha_j^n$ is convergent and

therefore we may consider $M\mathbf{a} = M(P\mathbf{b})$. By definition, for each $i \in \mathbb{N}$, we have $a_i = \displaystyle\sum_{j=0}^{\infty} \lambda_{i,j} b_j$. Let $M\mathbf{a} = (x_h)_{h \in \mathbb{N}}$. For each $h \in \mathbb{N}$ we have

$$x_h = \sum_{m=0}^{\infty} \alpha_h^m a_m = \sum_{m=0}^{\infty} \alpha_h^m \Big(\sum_{j=0}^{\infty} \lambda_{m,j} b_j \Big).$$

Let $r = |\alpha_h|$. As we have seen, the family $|\lambda_{m,j} b_j| r^m$ tends to 0 when m tends to $+\infty$, uniformly with respect to j. Hence by the Limits Inversion Theorem, we have

$$\sum_{m=0}^{\infty} \alpha_h^m \Big(\sum_{j=0}^{\infty} \lambda_{m,j} b_j \Big) = \sum_{j=0}^{\infty} b_j \Big(\sum_{m=0}^{\infty} \lambda_{m,j} \alpha_h^m \Big).$$

Hence by (3), we see that $x_j = b_j$ and this finishes proving (6). Then by (6) ψ_P is clearly injective.

Finally we will prove the last statement of the theorem. Let $\phi(x) = \displaystyle\sum_{n=0}^{\infty} \nu_n x^n$.

The function ϕ belongs to $A(d(0,1^-))$ and is invertible in $A(d(0,1^-))$ thanks to the inequality $|\nu_0| > |\nu_n|$ whenever $n > 0$. Hence the function $G(x) = g(x)\phi(x)$

is easily seen to satisfy i), ii), iii), iv) like g. Then G factorizes in $A(d(0,1^-))$ and may be written as $\phi_j(x)(\dfrac{1-x}{\alpha_j})$ with $\phi_j(x) = \psi_j(x)\phi(x)$. Hence we put

$$G_j(x) = \frac{\phi_j(x)}{\phi_j(\alpha_j)} = \frac{g_j(x)\phi(x)}{\phi(\alpha_j)}.$$ Now it is clearly seen that the power series of G_j

is $\displaystyle\sum_{n=0}^{\infty} \mu_{n,j}x^n$. By definition, the matrix Q satisfies the same properties as P. But when ϕ is not a constant function, for each fixed $j \in \mathbb{N}$, we do not have $\mu_{n,j} = \lambda_{n,j}$ for all $n \in \mathbb{N}$. Hence Q is different from P. As a consequence we see that ψ_M is not surjective, it would be an automorphism of $\mathcal{E}_0$ and therefore ψ_P would also be an automorphism of $\mathcal{E}_0$ and it would be unique. This ends the proof of Theorem 51.6.

Defintion: Given a polar sequence $(\alpha_n)_{n\in\mathbb{N}}$ associated to $d(0,1^-)$, we will call *inverse of* $\mathcal{M}(\alpha_n)$ any matrix $P \in \mathcal{M}_\infty$ satisfying (1), (2), (3), (4), (5), (6), (7).

Remarks. 1.The proof of Theorem 51.6 takes inspiration from that of Lemma 3 in [58]. However, in this lemma, the matrix considered, roughly, was P. Here the matrix we consider is a van der Monde matrix M and we look for P.
2. Given M, the matrix P depends on g and therefore is not unique, satisfying $(1) - (7)$. Indeed $\mathcal{M}_\infty$ is not a ring because the multiplication of matrices is not always defined and even when it is defined, is not always associative. As a consequence, if P, P' satisfy $MP = MP' = PM = P'M = I_\infty$, we cannot conclude $P' = P$. Actually we can consider $\phi_M \circ \psi_P \in \mathcal{L}(\mathcal{E}_0, \mathcal{E})$ and then this is the identity in $\mathcal{E}_0$. Next we can consider $\psi_{P'} \circ \psi_M \in \mathcal{L}(\mathcal{E}_0, \mathcal{E}_1)$ and this is the identity in $\mathcal{E}_0$. But we cannot consider $\psi_{P'} \circ (\phi_M \circ \psi_P)$ because $\psi_{P'}$ is not defined in $\mathcal{E}_1$. In the same way, we cannot consider $(\psi_{P'} \circ \psi_M) \circ \psi_P$ because $\psi_{P'} \circ \psi_M$ is only defined in $\mathcal{E}_0$. We consider the matrix P and look for "inverses" M such that $MP = PM = I_\infty$. Suppose that there exists a bounded matrix $M' \neq M$ such that $PM' = M'P = I_\infty$. Now we can consider $\phi_{M'} \circ (\psi_P \circ \psi_M) \in \mathcal{L}(\mathcal{E}_0, \mathcal{E})$. Since $\psi_P \circ \psi_M$ is the identity in $\mathcal{E}_0$, then $\phi_{M'} \circ (\psi_P \circ \psi_M)$ is equal to $\psi_{M'}$. Next we can consider $(\phi_{M'} \circ \psi_P) \circ \psi_M \in \mathcal{L}(\mathcal{E}_0, \mathcal{E})$. Since $\phi_{M'} \circ \psi_P$ is the identity on $\mathcal{E}_0$, we have $(\phi_{M'} \circ \psi_P) \circ \psi_M = \psi_M$ and therefore $\psi_M = \psi_{M'}$ hence $M = M'$.
3. Let $P, Q \in \mathcal{M}_\infty$ satisfy $(1) - (7)$. Let $\mathcal{E}' = \psi_P(\mathcal{E}_0)$, let $\mathcal{E}'' = \psi_Q(\mathcal{E}_0)$. Then the restriction of ϕ_M to $\mathcal{E}'$ (resp. $\mathcal{E}''$) is just the reciprocal of ψ_P (resp. ψ_Q).

Conjecture: Under the hypothesis of Theorem 51.3, every matrix satisfying properties $(1) - (7)$ is of the form $\mu_{n,j} = \left(\dfrac{1}{\sum_{m=0}^{\infty} \nu_m \alpha_j^m}\right)((\lambda_{n,j}) * (\nu_n))$.

52. p-ADIC ANALYTIC INTERPOLATION

We will use the results on infinite van der Monde matrices to construct an analytic function f in the disk $d(0,1^-)$ satisfying a sequence of equalities of the form $f(\alpha_n) = b_n$, with $(\alpha_n)_{n \in \mathbb{N}}$ a polar sequence associated to $d(0,1^-)$, and $(b_n)_{n \in \mathbb{N}}$ a bounded sequence in K. This is a common work with J. Araujo [4].

Notation: For each $q \in \mathbb{N}^*$, $\mathcal{G}(q)$ will denote the group of the q-th roots of 1.

Lemma 52.1: *Let $(a_n)_{n \in \mathbb{N}}$ be a sequence in $d(0,1^-)$ such that $\lim\limits_{n \to \infty} |a_n| = 1$. For each $s \in \mathbb{N}$, there exists a prime integer $q > p$ and $\zeta \in \mathcal{G}(q)$ such that $\left|\zeta^h a_s - a_j\right| = \max(|a_s|, |a_j|)$ for every $j \in \mathbb{N}$, for every $h = 1, \ldots, q-1$.*

Proof:.   Let $r = |a_s|$. Since $\lim\limits_{n \to \infty} |a_n| = 1$, the circle $C(0,r)$ contains finitely many terms of the sequence $(a_n)_{n \in \mathbb{N}}$. Without loss of generality we may assume $|a_n| < r$ whenever $n < l$, $|a_n| > r$ whenever $n > t$ and $|a_n| = r$, whenever $n = l, \ldots, t$ (with obviously $l \leq s \leq t$). Whatever $q \in \mathbb{N}$, $\zeta \in \mathcal{G}(q)$ are, it is seen that we have $\left|\zeta^h a_s - a_j\right| = |a_s|$ for all $j < l$ and $\left|\zeta^h a_s - a_j\right| = |a_j|$ for all $j > t$. In the residue class field $\mathcal{K}$ of K, for every $j = l, \ldots, t$, we denote by γ_j the class of $\dfrac{a_j}{a_s}$. There does exist a prime integer $q > p$ such that the polynomial $x^q - 1$ admits none of the γ_j $(l \leq j \leq t)$ as a zero. Hence, for every q-root ν of 1 in $\mathcal{K}$, we have $\nu^h \neq \gamma_j$ whenever $j = l, \ldots, t$, whenever $h = 1, \ldots, q-1$. Now let ζ be a q-th root of 1 in K. Then by classical properties of the polynomials, we have $\left|\dfrac{\zeta^h - a_j}{a_s}\right| = 1$, hence $\left|\zeta^h a_s - a_j\right| = |a_s| = r$ whenever $h = 1, \ldots, q-1$, whenever $j = l, \ldots, t$. This completes the proof of Lemma 52.1.

Lemma 52.2: *Let $(a_n)_{n \in \mathbb{N}}$ be a polar sequence associated to $d(0,1^-)$, of separation ρ. There exists a sequence $(b_n)_{n \in \mathbb{N}}$ in $d(0,1^-)$ satisfying:*

(1) $\lim\limits_{n \to \infty} |b_n| = 1$.

(2) $|b_n - b_m| \geq \rho$ *whenever $n \neq m$.*

(3) $(a_n)_{n \in \mathbb{N}}$ *is a subsequence of $(b_n)_{n \in \mathbb{N}}$,*

(4)   *there exists a sequence $(q_n)_{n \in \mathbb{N}}$ of prime integers different from p satisfying $\lim\limits_{n \to \infty} q_n = +\infty$, such that for every $m \in \mathbb{N}$, $\zeta \in \mathcal{G}(q_n)$, ζb_n is another term of the sequence $(b_n)_{n \in \mathbb{N}}$,*

(5)   *There exists $f \in A(d(0,1^-)))$ admitting each b_n as a simple zero and having no other zero in $d(0,1^-)$, satisfying* $\displaystyle \lim_{\substack{|x|\to 1^- \\ x \in \mathcal{O}(b_n)}} |f(x)| = +\infty.$

Proof: First we will construct a sequence (b'_n) satisfying (1), (2), (3), (4). Let (q_j) be a strictly increasing sequence of prime integers strictly bigger than p and, for each $j \in \mathbb{N}$, let $\displaystyle s_j = \sum_{i=0}^{j} q_i$, let $\zeta_j \in \mathcal{G}(q_j)\backslash\{1\}$ and let $b'_{\zeta_j + h} = \zeta_j^h a_j$ $(0 \le h \le q_j - 1)$. We will show that a good choice of the sequence (q_j) enables us to obtain

(6) $|b'_n - b'_m| = \max(|b'_n|, |b'_m|)$ for every couple (n,m) satisfying $n \ne m$ and $(n,m) \ne (s_i, s_j)$ whenever $(i,j) \in \mathbb{N} \times \mathbb{N}$.

In other words $|b'_n - b'_m| = \max(|b'_n|, |b'_m|)$ must be true at all time except when $n = m$ and when (b'_n, b'_m) is equal to some couple (a_{s_i}, a_{s_j}). For each $t \in \mathbb{N}$, let $F_t = \{s_0, s_1, \ldots, s_t\}$ and let E_t be $\{0,1,\ldots,s_t - 1\}\backslash F_t$. Assume that $q_0, q_1, \ldots, q_{t-1}$ have been chosen to satisfy the following properties (α_t) and (β_t)

(α_t) $|b'_n - a_{s_j}| = \max(|b'_n|, |a_{s_j}|)$ for all $j \in \mathbb{N}$, for all $n \in E_t$.

(β_t) $|b'_n - b'_m| = \max(|b'_n|, |b'_m|)$ for all $(n,m) \in E_t \times E_t$ such that $n \ne m$. We will choose q_t such that both (α_{t+1}), (β_{t+1}) are satisfied. Indeed, by Lemma 52.1 we can take a prime integer u such that, given $\zeta_t \in \mathcal{G}(u)$, we have $|\zeta_t^h a_t - a_j| = \max(|a_t|, |a_j|)$ for all $j \in \mathbb{N}$, for all $h = 1, \ldots, u-1$, $|\zeta_t^h a_t - b'_n| = \max(|a_t|, |b'_n|)$ for all $n < s_t$, for all $h = 1, \ldots, u-1$. Thus we can take $q_t = u$ and we see that both (α_{t+1}), (β_{t+1}) are satisfied. Hence we can construct the sequence $(q_t)_{t \in \mathbb{N}}$ by induction and, therefore, the sequence $(b'_n)_{n \in \mathbb{N}}$ satisfying (6) is now constructed. Then it is easily checked that the sequence $(b'_n)_{n \in \mathbb{N}}$ so obtained satisfies (1), (2), (3), (4). Now let $\{r_0, \ldots, r_n, \ldots\} = \{|a_j| : j \in \mathbb{N}\}$ and let $D = \mathcal{O}(b_n)$. The infinite product $\displaystyle g(x) = \prod_{j=0}^{\infty}\left(1 - \left(\frac{x}{a_j}\right)^{q_j}\right)$ converges in $A(d(0,1^-))$ and has no zero in $d(0,r) \cap D$ because, by construction of the sequence (b'_n), each zero of g is one of the points b'_m for some $m \in \mathbb{N}$. Hence it is seen that we have $|g(x)| \ge 1$ for every $x \in d(0,1^-) \backslash \left(\displaystyle\bigcup_{n=0}^{\infty} C(0, r_n)\right)$. For each $n \in \mathbb{N}$, we put $\Sigma_n = D \cap C(0, r_n)$, $\displaystyle \tau_n = \inf_{x \in \Sigma_n} |g(x)|$, and we take $\sigma_n \in]r_n, r_{n+1}[\cap |K|$, $c_n \in C(0, \sigma_n)$, and $u_n > \min(p,n)$ a prime integer such that $\tau_n \left(\dfrac{r_{n+1}}{\sigma_n}\right)^{u_n} > n+1$.

Since $\displaystyle \lim_{n \to \infty} u_n = +\infty$, it is seen that the infinite product $\displaystyle h(x) = \prod_{n=0}^{\infty}\left(1 - \left(\frac{x}{c_n}\right)^{u_n}\right)$

converges in $A(d(0, 1^-))$. Let $D' = d(0, 1^-) \setminus ((c_n), \rho)$ and let $D'' = D' \cap D$. Let

$$h(x) = \sum_{n=0}^{\infty} \lambda_n x^n$$ and, for each $r \in (0, 1)$, let $W(r) = \sup_{n \in \mathbb{N}} |\lambda_n| r^n$. Each zero of

h is simple and is of the form ζc_n with $\zeta \in \mathcal{G}(u_n)$. Hence by Theorem 23.13 h satisfies

(7) $\quad |h(x)| \geq W(|x|) \dfrac{\rho}{|x|}$ for all $x \in D'$.

Now, let $x \in D'' \cap \Delta' r_n, r_{n+1}$. By Theorem 23.13 we have $W(|x|) \geq$ $W(r_n)\left(\dfrac{r_n}{\sigma_{n-1}}\right)^{u_n}$, and therefore by (7), we have

(8) $\quad |h(x)| \geq \left(\dfrac{r_n}{\sigma_{n-1}}\right)^{u_n} \dfrac{\rho}{|x|}$

Finally, by (8) we obtain $|g(x)h(x)| \geq |h(x)|\tau_n > n$ for all $x \in D''$ such that $r_n \leq |x| < r_{n+1}$, and this shows

(9) $\quad \lim_{\substack{|x| \to 1 \\ x \in D''}} |g(x)h(x)| = +\infty.$

Now let $(b''_n)_{n \in \mathbb{N}}$ be the sequence of the zeros of g. Clearly $(b''_n)_{n \in \mathbb{N}}$ satisfies (1) and (4) and also satisfies $|b''_n - b'_m| = \max(|b''_n|, |b'_m|)$ whenever $n, m \in \mathbb{N}$ and $|b''_n - b''_m| = \max(|b''_n|, |b''_m|)$ whenever $n \neq m$. Now we put $b_{2n} = b'_n$ and $b_{2n+1} = b''_n$. The sequence $(b_n)_{n \in \mathbb{N}}$ clearly satisfies (1), (2), (3), (4) and also satisfies (5) because the zeros of h are the b''_n while those of g are the b'_n. Thus the zeros of f are just the b_n, and then, by (9), we have $\lim_{\substack{|x| \to 1 \\ x \in \mathcal{O}(b_n)}} |f(x)| = +\infty.$

This ends the proof.

Theorem 52.3: (J. Araujo, A. Escassut)   *Let $(\alpha_n)_{n \in \mathbb{N}}$ be a polar sequence associated to $d(0, 1^-)$. There exists a polar sequence $(\gamma_n)_{n \in \mathbb{N}}$ associated to $d(0, 1^-)$ such that $(\alpha_n)_{n \in \mathbb{N}}$ is a subsequence of (γ_n) satisfying:*
for every inverse matrix P of $\mathcal{M}(\gamma_n)$ obtained in Theorem 51.5 and for every bounded sequence $\mathbf{b} = (b_n)_{n \in \mathbb{N}}$ of K, the sequence $\mathbf{a} = P(\mathbf{b}) := (a_n)_{n \in \mathbb{N}}$ de-
fines an analytic function $f(x) = \sum_{n=0}^{\infty} a_n x^n \in A(d(0, 1^-))$ such that $f(\gamma_j) = b_j$
whenever $j \in \mathbb{N}$.

Proof: By Lemma 52.2, there exists a polar sequence $(\gamma_n)_{n \in \mathbb{N}}$ associated to $d(0, 1^-)$ such that $(\alpha_n)_{n \in \mathbb{N}}$ is a subsequence of $(\gamma_n)_{n \in \mathbb{N}}$ together with an analytic function $g \in A(d(0, 1^-))$ admitting each γ_m as a simple zero and having no other zero in $d(0, 1^-)$, satisfying $\lim_{\substack{|x| \to 1^- \\ x \in \mathcal{O}(\gamma_n)}} |g(x)| = +\infty$ with $\rho = \inf_{n \neq m} |\gamma_n - \gamma_m|$.

Then, by Theorem 51.6, the matrix $M = \mathcal{M}(\gamma_n)$ admits line-vanishing inverses M' satisfying $M(M'(\mathbf{b})) = \mathbf{b}$ for all bounded sequence $\mathbf{b} = (b_n)_{n\in\mathbb{N}}$. Let $\mathbf{a} := (a_n)_{n\in\mathbb{N}} = M'(\mathbf{b})$. Thus we have $M(\mathbf{a}) = \mathbf{b}$ and therefore $\sum_{n=0}^{\infty} a_n \gamma_j^n = b_j$ whenever $j \in \mathbb{N}$. This ends the proof of Theorem 52.3.

53. ANALYTIC ELEMENTS

WITH A ZERO DERIVATIVE

In Chapter 19, we saw that if an infraconnected open set D satisfies $\overset{\circ}{\overline{D}} \subset D$, then every element of $H(D)$ whose derivative is identically zero is a constant. Here we will construct an infraconnected open set D with a non constant element $f \in H(D)$ whose derivative is identically zero [24].

Proposition 53.1 : *Let D be an infraconnected closed open bounded set , let $a \in D$, let $f \in H(D)$, let ε and $r \in \mathbb{R}_+^*$ be such that $d(a,r) \subset D$ and $|f(x) - f(a)| \leq \varepsilon$ for all $x \in d(a,r)$. Let $\rho \in]0,r]$. There exists an infraconnected closed open bounded set $D' \subset D$ such that:*
(i) $d(a,\rho) \subset D'$,
(ii) $D \setminus d(a,r^-) \subset D'$,
(iii)   *for every hole T of D' included in $\Gamma(a,\rho,r)$ the disk $d(a,\rho)$ is included in a peripheral of T and only finitely many peripherals of T are holes of D',*
(iv)   *there exists $g \in H(D')$ satisfying*
   $\alpha)$ $g(x) = f(a)$ *for all* $x \in d(a,\rho)$,
   $\beta)$ $g(x) = f(x)$ *for all* $x \in D' \setminus d(a,r^-)$,
   $\gamma)$ $\|g - f\|_D \leq \epsilon$.

Proof: Let S be the diameter of D. By Theorem 50.4, thanks to a suitable change of variable, there exists an infraconnected closed open bounded set A containing $d(a,\rho)$ and $d(a,S) \setminus d(a,r^-)$, such that every hole T of A is a class of a circle $C(a,\sigma)$ and such that only finitely many classes of $C(a,\sigma)$ are holes of A, and there exists $u \in H(A)$ such that $u(x) = 0$ for all $x \in A \setminus d(a,r^-)$, $u(x) = 1$ for all $x \in d(a,\rho)$, and $\|u\|_A = 1$. But then, $d(a,\rho)$ is clearly included in a peripheral of T and there are finitely many peripherals of T that are holes of A. Let $D' = A \cap D$. Thus, the holes of D' included in $\Gamma(a,\rho,r)$ satisfy the assertion of Proposition 53.1. Besides D' is obviously infraconnected. Now let $g = (1 - u) f + u f(a)$. Clearly g belongs to $H(D')$ and satisfies
$g(x) = f(x)$ for every $x \in D' \setminus d(a,r^-)$,
$g(x) = f(a)$ for every $x \in d(a,\rho^-)$, and
$|g(x) - f(x)| = |u(x)(f(a) - f(x))| \leq |f(a) - f(x)| \leq \varepsilon$ for every $x \in d(a,r^-)$,
hence $\|g - f\|_{D'} \leq \varepsilon$. So, Proposition 53.1 is proven.

Notations: Let L be a complete algebraically closed subfield of K. Given $a \in L$, and $r > 0$, we put $d_L(a,r) = d(a,r) \cap L$, $d_L(a,r^-) = d(a,r^-) \cap L$, $C_L(a,r) = C(a,r) \cap L$. Besides, given a set A in L, we denote by $R_L(A)$ the L-algebra $R(A) \cap L(x)$ and by $H_L(A)$ the closure of $R_L(A)$ in $H(A)$ (i.e., the space of the analytic elements on A, when the ground field is L instead of K).

Lemma 53.2: *Let L be a complete algebraically closed subfield of K. Let A be an infraconnected open set in L of the form $\bigcup\limits_{n=0}^{\infty} d_L(a_n, r_n)$ and let $B = \bigcup\limits_{n=0}^{\infty} d(a_n, r_n)$. Every element $f \in H_L(A)$ has a unique extension $\widehat{f} \in H(B)$ such that $\widehat{f}(x) = f(x)$ whenever $x \in A$.*

Proof: By Theorem 10.5 there exist $g \in H_{Lb}(A)$ and $h \in R_L(A)$ such that $f = g + h$. It is easily seen that the Mittag-Leffler series of g on the infraconnected set A also converges on the infraconnected set B in K. The uniqueness comes from the fact that an element $\ell \in H(B)$ whose restriction to A is identically zero has a Mittag-Leffler series that is identically zero.

Lemma 53.3: *K contains a separable complete algebraically closed subfield L such that the restriction of the absolute value to L is not trivial.*

Proof: Since K has characteristic 0, it contains a subfield F that is algebraically isomorphic to $\mathbb{Q}$. So we can consider that $\mathbb{Q} \subset K$ and therefore $\mathbb{Q}$ is provided with the absolute of K.

If the residue characteristic is $p \neq 0$, by Lemma 8.2 we know that the absolute value of $\mathbb{Q}$ is $|\cdot|_p$, and therefore K contains a subfield L isomorphic to $\mathbb{C}_p$ that is complete, algebraically closed and provided with a dense countable subset.

If the residue characteristic is 0, then by Lemma 8.2 $\mathbb{Q}$ is provided with the trivial absolute value. Hence it is discrete in K and therefore complete. Then by Theorem 6.4 the absolute value on its algebraic closure E in K is also trivial. Let $x \in K$ be such that $0 < |x| < 1$, let J be the algebraic closure of $E(x)$ in K and let L be the topological closure of J in K. By theorem 6.10, we know that L is also algebraically closed. Hence it is complete and algebraically closed. Since E is clearly countable, so are $E(x)$ and J, hence L is separable. (Actually, we know that L is the completion of the field of Puiseux series $\bigcup\limits_{n=1}^{\infty} E((x^{1/n}))$).

Theorem 53.4: *There exists a bounded infraconnected open set D such that $H(D)$ contains a bounded not constant element f with an identically zero derivative. Moreover f also belongs to $H(\overline{D})$.*

Proof: By Lemma 53.3 there exists a complete separable algebraically closed subfield L such that the absolute value induced on L by the absolute value of K is not trivial. In addition, since L is algebraically closed, we know that its absolute value is dense in $\mathbb{R}_+$.

We suppose first that $K = L$.

Let $D_0 = d(0,1)$, let $f_0(x) = x$ $(x \in D_0)$ and let $(a_n)_{n \in \mathbb{N}}$ be a sequence in D_0 such that $\{(a_n)_{n \in \mathbb{N}}\}$ is dense in D_0. Let us suppose that we have already constructed up to rank q:

a decreasisng sequence of closed open bounded infraconnected sets $(D_n)_{n \in \mathbb{N}}$,

a sequence $(f_n)_{n \in \mathbb{N}}$ of analytic elements $f_n \in H(D_n)$,

and a sequence of disks $(\Lambda_n)_{n \in \mathbb{N}}$ (with Λ_n of the form $d(\alpha_n, r_n)$), satisfying the following conditions for $i, j, n \leq q$:

     (1) $\Lambda_i \cap \Lambda_j = \emptyset$ for all $i \neq j$,

$$(2) \ \bigcup_{j=1}^{n} \Lambda_j \subset D_n ,$$

     (3) for every hole T of D_n, only finitely many peripherals of T are holes of D_n and one of the peripherals that are not included in D_n contains sets Λ_j with $j \leq n$,

     (4) the lower bound λ_n of the set of diameters of the holes of D_n is not zero and every hole of D_n which is not a hole of D_{n-1} has a diameter strictly smaller than λ_{n-1},

$$(5) \text{ either } a_n \text{ belongs to } \bigcup_{j=1}^{n} \Lambda_j \text{ or } a_n \text{ belongs to a hole of } D_n,$$

     (6) $f_n(x)$ is equal to a constant α_n for all $x \in \Lambda_n$,

     (7) $\alpha_n \in \Lambda_n$

     (8) $f_n(x) = f_j(x)$ whenever $x \in \Lambda_j$, for every $n \geq j$,

     (9) $f_n(x) = f_0(x) = x$ whenever $x \in \Lambda_j$, for every $n < j$,

$$(10) \ \|f_n - f_{n-1}\|_{D_n} \leq \frac{1}{n}.$$

We will define D_{q+1}, Λ_{q+1}, f_{q+1} such that the sequence $(D_n, \Lambda_n, f_n)_{1 \leq n \leq q+1}$ again satisfies (1) to (10) up to rank $q+1$. Let n_{q+1} be the smallest of the integers m such that a_m belongs neither to $\bigcup_{j=1}^{q} \Lambda_j$ nor to any hole of D_q. For convenience we set $\alpha = a_{n_{q+1}}$. Let $r \in]0, \lambda_q[$ satisfy

$$(11) \ d(\alpha, r) \subset D_q \text{ and } d(\alpha, r) \cap \left(\bigcup_{n-1}^{q} \Lambda_n \right) = 0$$

and $|f_q(x) - f_q(\alpha)| \leq \dfrac{1}{q+1}$ whenever $x \in d(\alpha, r)$.

Since the absolute value of L is dense, there exists $\rho \in]0, r[$. Further, there exists a closed open bounded infraconnected set $D_{q+1} \subset D_q$ containing the disk $d(\alpha, \rho)$ such that $D_{q+1} \setminus d(\alpha, r^-) = D_q \setminus d(\alpha, r^-)$ and such that for every hole T of D_{q+1} included in $\Gamma(\alpha, \rho, r)$, only finitely many peripherals of T are holes of D_{q+1}, but one of the peripherals of T that is not included in D_{q+1} contains Λ_{q+1}. Next, there exists $f_{q+1} \in H(D)$ satisfying

(12) $\|f_q - f_{q+1}\|_{D_{q+1}} \leq \dfrac{1}{q+1}$,

(13) $f_{q+1}(x) = f_q(x)$ whenever $x \in D_{q+1} \setminus d(\alpha, \rho^-)$,

(14) $f_{q+1}(x) = f_q(\alpha)$ whenever $x \in d(\alpha, \rho)$.

We put $\Lambda_{q+1} = d(\alpha, r)$ and will check that the (D_n, Λ_n, f_n) satisfy the ten basic relations (1),...,(10) for $n \leq q+1$.

Relation (1) is clearly satisfied thanks to (12). Relation (2) is obvious because $\Lambda_{q+1} \subset D_{q+1}$. Relation (5) is satisfied by definition of α. Relation (4) is satisfied for $n = q+1$ because all the holes of D_{q+1} that are not holes of D_q, by construction have a diameter bigger than ρ. Hence we have $\lambda_{q+1} > 0$ and every hole of D_{q+1} which is not a hole of D_q is included in $d(\alpha, r)$, hence it has a diameter strictly smaller than λ_q.

We now check (3) for $n = q+1$. If T is a hole of D_{q+1} included in $K \setminus d(\alpha, r^-)$ then T is a hole of D_q hence (3) is true by the induction hypothesis. If T is a hole of D_{q+1} included in $d(\alpha, r^-)$ then T satisfies (3) by construction of D_{q+1} because Λ_{q+1} is in a peripheral of T and only finitely many peripherals of T are holes of D_{q+1}. Thus (3) is satisfied for $n = q+1$, and then it only remains to prove the relations (6) to (10) for $n \leq q+1$.

Relation (6) comes from (14), (8). Relation (9) comes from (13) and from the induction hypothesis. Relation (10) is due to (12). Finally we will check (7) at the rank $q+1$. By (9) we have $f_q(x) = f_0(x) = x$ when $x \in \Lambda_{q+1}$, hence by (14) α_{q+1} is the point α chosen in Λ_{q+1} and this finishes the proof of (7).

The induction is initiated by the definition of $D_0 = d(0,1)$ and $f_0(x) = x$. Thus the sequence (D_n, Λ_n, f_n) satisfying the ten basic relations (1),...,(10) is defined for all $n \in \mathbb{N}$. By (2), we notice

(15) $\displaystyle \bigcup_{j=1}^{n} \Lambda_j \subset \bigcap_{j=1}^{\infty} D_n$.

Now we consider a hole T of a set D_q. Let σ be the diameter of T and let χ be a peripheral of T that is not a hole of D_q. We will prove that χ necessarily contains a set Λ_n. Indeed suppose that χ contains no set Λ_n. Since χ is not a hole of D_q there exists a disk A included in $D_q \cap \chi$. Obviously, A contains a point a_t of the sequence $(a_n)_{n \in \mathbb{N}}$. Then a_t does not belong to any Λ_n because if it did,

either Λ_n would be contained in χ,

or Λ_n would contain χ and also T, which is absurd because by (15) we have $\Lambda_n \subset D_q$.

But then, since a_t does not belong to any Λ_n, by (5) a_t belongs to a hole T' of a set D_s with $s > q$. Let σ' be the diameter of T'. By (4) we have $\sigma' < \lambda_q < \sigma$. By (3) one of the peripherals χ' of T' contains a set Λ_n and since $\sigma' < \sigma$, we see that $\chi' \not\subseteq \chi$ and $\Lambda_n \subset \chi$. That finishes showing that χ contains a set Λ_n.

Now, let $D = \bigcup_{n=1}^{\infty} \Lambda_n$ and let us show that D is infraconnected. Indeed suppose that D is not infraconnected and let $a, b \in D', r', r'' \in \mathbb{R}_+^*$ with $0 < r' < r'' < |b - a|$ be such that $\Gamma(a, r', r'') \cap D = \emptyset$. There exists a point a_t of the sequence $(a_n)_{n \in \mathbb{N}}$ in $\Gamma(a, r', r'')$. By (5), since $a_t \notin D, a_t$ belongs to a hole $T = d(w, \rho^-)$ of a set D_q. Hence, from the above, every peripheral of T, contains a set Λ_n except possibly a finitely many ones.

On the other hand, all the peripherals of T are included in $\Gamma(a, r', r'')$ except possibly one. Therefore T has peripherals χ included in $\Gamma(a, r', r'')$ which contain sets Λ_n, and then this just contradicts the hypothesis that $\Gamma(a, r', r'') \cap D = \emptyset$.

Thus D is infraconnected. By construction D is open included in $\bigcap_{n=0}^{\infty} D_n$ (actually we could prove that $\overline{D} = \bigcap_{n=0}^{\infty} D_n$). All the f_n belong to $H(D)$ and by (10) the sequence $(f_n)_{n \in \mathbb{N}}$ converges to a limit $f \in H(D)$ that satisfies $f(x) = \alpha_n$ whenever $x \in \Lambda_n$, with $\alpha_n \in \Lambda_n$. Thus f is not constant in D. Also by (10) we see that the sequence $\|f_n\|_D$ does converge. Hence f is bounded in D and therefore it belongs to $H(\overline{D})$ because its Mittag-Leffler series on D has no pole in $\overline{D} \setminus D$. This finishes the proof of the theorem for $K = L$.

We now suppose that $L \neq K$. By the foregoing we can construct a sequence of disks $(\Lambda_n)_{n \in \mathbb{N}}$ in L, of the form $\Lambda_n = d_L(a_n, r_n)$, such that the set $A = \bigcup_{n=1}^{\infty} \Lambda_n$ is infraconnected and open in L and such that $H_L(A)$ has an element f not constant in D but satisfying $f' = 0$. Let $\widehat{A} = \bigcup_{n=1}^{\infty} d(a_n, r_n)$. Then $\widehat{A}$ is infraconnected and open in K. By Lemma 53.2, f has a unique extension $\widehat{f} \in H(\widehat{A})$ such that $\widehat{f}(x) = f(x)$ whenever $x \in A$. Obviously $\widehat{f}$ is not constant in $\widehat{A}$. On the other hand $\widehat{f}'(x)$ is equal to 0 in all of A. Indeed, for each $n \in \mathbb{N}$, the restriction of

$\widehat{f'}$ to $d(a_n, r_n)$ belongs to $H(d(a_n, r_n))$ and $\widehat{f'}$ is equal to 0 in all of $d_L(a_n, r_n)$. Hence it is identically zero. Finally, it is easily seen that $\|\widehat{f}\|_{\widehat{A}} = \|f\|_A$ because for every $n \in \mathbb{N}$, we have $\|\widehat{f}\|_{d(a_n, r_n)} = \|f\|_{d_L(a_n, r_n)}$. This ends the proof of Theorem 53.4.

As a conclusion the infraconnected sets play roughly the same part in the theory of ultrametric analytic elements as the connected sets do in the theory of holomorphic functions as long as we consider algebras of analytic elements.

Now consider a bounded open connected set D in $\mathbb{C}$. A function holomorphic in D and continuous in $\overline{D}$ is the limit of a sequence of rational functions with no pole in D, convergent for the uniform convergence on D. But no function holomorphic in D and continuous in $\overline{D}$ has an identically zero derivative different from the constant functions. From that point of view, the infraconnected open sets do not correspond to the connected open sets in complex analysis.

54. GENERALITIES ON

THE DIFFERENTIAL EQUATION $y' = fy$ IN $H(D)$

In this and in the following chapters K has characteristic 0, D denotes an open infraconnected set, and f is an element of $H(D)$.

Notations : We will denote by $\mathcal{E}(f)$ the differential equation $y' = fy$ with $y \in H(D)$ and by $\mathcal{S}(f)$ the K-vector space of the solutions $h \in H(D)$.

By classical results, we know that $\mathcal{S}(f)$ may be reduced to $\{0\}$. (For example, if D is the disk $|x| \leq 1$, it is easily seen that the equation $y' = y$ has no solution in $H(D)$). In the next chapters we will study carefully whether this dimension may be bigger than 1. Here we will only give general and elementary results about this problem, and next we will study what happens when a solution has a zero, or a pole.

Theorem 54.1:   *Let $D \in \mathcal{A}$. If $\mathcal{E}(f)$ admits at least one solution g invertible in $H(D)$, then $\mathcal{S}(f)$ has dimension 1.*

Proof :   Let $g \in \mathcal{S}(f)$ be invertible in $H(D)$, and let h be another solution. We check that $\dfrac{h}{g}$ is a constant in $H(D)$. Indeed, since D belongs to $\mathcal{A}$, $\dfrac{h}{g}$ belongs to $H(D)$. Then $\left(\dfrac{h}{g}\right)' = \dfrac{h'g - hg'}{g^2} = \dfrac{fhg - hfg}{g^2} = 0$. But then, since K has characteristic zero, by Corollary 19.6 $\dfrac{h}{g}$ is a constant in D.

Theorem 54.2:   *We assume that $\mathcal{E}(f)$ admitss at least one solution g non identically equal to zero. Then g has no isolated zero in D. Besides, if D belongs to $\mathcal{A}$, then*
  a)   *either g is invertible in $H(D)$,*
  b)   *or g is strictly vanishing along a T-filter on D.*

Proof: We assume that g has an isolated zero a in D. Since D is open we know that g factorizes in the form $(x - a)^q h(x)$ with $h \in H(D)$ and $h(a) \neq 0$, hence $g' = (x - a)^{q-1} (qh + (x - a)h')$, hence $qh = (x - a)(f - h')$. Since $q \neq 0$,

334

this contradicts the hypothesis $h(a) \neq 0$. Hence g has no isolated zero in D. Now we suppose that g is not invertible. Since it has no isolated zero, it is not quasi-invertible , and since D is open and belongs to $\mathcal{A}$, by Theorem 22.11 g is not quasi-minorated. Hence by Theorem 38.2 g is strictly vanishing along a T-filter on D. This ends the proof.

Corollary 54.3:   *Let D belong to $\mathcal{A}$ and have no T-filter, and let $f \in H(D)$ be such that $\mathcal{S}(f) \neq \{0\}$. Then $\mathcal{S}(f)$ has dimension 1.*

Remark: In particular Corollary 54.3 applies to sets with finitely many holes.

Now we will study what happens when f is meromorphic and has a pole.

Theorem 54.4:   *Let $a \in \overline{D} \setminus D$, and let $h \in \mathcal{S}(f)$, be meromorphic (resp. holomorphic) and have a pole (resp. a zero) of order q at a. Then f is meromorphic at a, and admits a as a pole of order 1. Further the residue of f at a is equal to $-q$ (resp. q).*

Proof: Indeed, let $h(x) = \dfrac{g(x)}{(x-a)^q}$, (resp. let $h(x) = g(x)(x-a)^q$), with

$g \in H(D \cup \{a\})$ and $g(a) \neq 0$. Then for all $x \in D$ we have $f(x) = \dfrac{-q}{x-a} + \dfrac{g'(x)}{g(x)}$

(resp. $f(x) = \dfrac{q}{x-a} + \dfrac{g'(x)}{g(x)}$). Since g is holomorphic at a, there exists a disk

$d(a,r)$ such that $g(x)$ belongs to $H(d(a,r))$ and has no zero in $d(a,r)$. Therefore

it is seen that $\dfrac{g'(x)}{g(x)}$ also belongs to $H(d(a,r))$. Consequently, f admits a as

a pole of order 1. Besides, we see that $f + \dfrac{q}{x-a}$ (resp. $f - \dfrac{q}{x-a}$) obviously

belongs to $H(D)$ and is holomorphic at a, like $\dfrac{g'(x)}{g(x)}$. Hence the residue of f at

a is $-q$, (resp. q).

Corollary 54.5:   *If $\mathcal{S}(f) \neq \{0\}$ then a solution h has finitely many zeros in $\overline{D} \setminus D$.*

Lemma 54.6:   *Let D belong to $\mathcal{A}$, let T be a hole of D. Let $f \in H_b(D)$ and let $h \in \mathcal{S}(f)$ admitting a Motzkin factor in the hole T. Then the Mittag-Leffler term of f associated to T is equal to $\dfrac{(h^T)'}{h^T}$.*

Proof: Let $g = h^T$, and let $\ell = \dfrac{h}{g}$. By definition, ℓ belongs to $H(D \cup T)$ and

has no zero in T. Besides $\dfrac{g'}{g}$ obviously belongs to $H(D)$, hence so does $\dfrac{\ell'}{\ell}$. Hence

ℓ' clearly belongs to $H(D)$. But then, by Theorem 19.4, we have $\overline{(\ell')}_T = (\overline{\ell_T})'$,

and therefore $(\ell')_T = 0$. Hence ℓ' belongs to $H(D \cup T)$, and finally so does $\dfrac{\ell'}{\ell}$.

Next, by Lemma 32.1, $\dfrac{g'}{g}$ belongs to $H_0(K \setminus T)$. Hence by Corollary 15.3 we see

that $\overline{f}_T = \dfrac{g'}{g}$. This ends the proof.

Theorem 54.7:   *Let D belong to $\mathcal{A}$, let T be a hole of D and $h \in \mathcal{S}(f)$ be
meromorphic in T. Then so is f. Given an extension $\widetilde{h}$ of h meromorphic in T,*

f admits an extension $\widetilde{f} = \dfrac{\widetilde{h}'}{\widetilde{h}}$ whose poles in T are simple and are the poles and

the zeros of $\widetilde{h}$ in T, according to Theorem 54.4.

Proof: Since $\overline{f}$ is equal to $\dfrac{(h^T)'}{h^T}$, it belongs to $R_0(K \setminus T)$, and therefore f is
clearly meromorphic in T. The conclusion comes from theorem 54.4.

Theorem 54.8:   *Let D belong to $\mathcal{A}$, let $h \in \mathcal{S}(f)$, and let T be a hole of D.
If f has continuation to an element $\widetilde{f}$ of $H(D \cup T)$ then h has continuation to
an element $\widetilde{h} \in H(D \cup T)$, satisfying $\widetilde{h} \in \mathcal{S}(\widetilde{f})$, in $H(D \cup T)$.*

Proof: Indeed, suppose that this is not true. Without loss of generality, we
may assume $T = d(0, r^-)$. Then we have $_D\varphi_{0,r}(h) \neq 0$ because if $_D\varphi_{0,r}(h) = 0$,
by Theorem 21.8 h has continuation to an element of $H(D \cup T)$ such that $h(x) = 0$
whenever $x \in T$. Then by Theorem 10.5 and Proposition 20.2 there exists annuli
$\Gamma(a_j, s', s'')_{(1 \leq j \leq q,)}$ with (1) $|a_j| = |a_j - a_k| = r$ whenever $j \neq k$, $1 \leq j \leq q$, $1 \leq$

$k \leq q$, such that $f, h, \dfrac{1}{h}$ are bounded in the set $D' := (D \bigcap (\bigcap_{j=1}^{q} \Gamma(a_j, s', s'')))$.

Hence $f, h, \dfrac{1}{h}$ belong to $H(\overline{D'})$. Clearly by (1) T is a hole of $\overline{D'}$, and then in

$H(\overline{D'})$, h admits a Motzkin factor g in the hole T. Hence in $H(\overline{D'})$ f admits a

Mittag-Leffler term $\overline{f}_T$ and then by Lemma 54.6 we have $\overline{f}_T = \dfrac{g'}{g}$. Since f has

continuation to an element $\widetilde{f}$ of $H(D \cup T)$ then we have $\overline{f}_T = 0$, and therefore
$g' = 0$, hence $g = 1$. Finally, the Mittag-Leffler term of h associated to T, (with

respect to D') is equal to 0. Hence h belongs to $H(D' \cup T)$, and therefore by Theorem 15.10, it belongs to $H(D \cup T)$.

55. THE DIFFERENTIAL EQUATION $y' = fy$

IN ALGEBRAS $H(D)$

In this chapter D is infraconnected, open and belongs to $\mathcal{A}$, f denotes an element of $H(D)$, and $\mathcal{E}(f)$, $\mathcal{S}(f)$ are defined like in Chapter 54.

Definition: Let $g \in H(D)$. We will call *the support* of g the set Σ of the $x \in D$ such that $g(x) \neq 0$, and Σ will be said to be *reinforced* if for every $a, b \in \Sigma$, the function in μ defined as $v_a(f, \mu)$ is bounded in the interval $[v(a - b), +\infty[$.

Proposition 55.1: *Assume that $H(D)$ has no divisor of zero. Then every $f \in H(D) \setminus \{0\}$ has a reinforced support.*

Proof: Let $f \in H(D)$, let Σ be the support of f, and $a, b \in \Sigma$. We will show that $v_a(f, \mu)$ is bounded when $\mu \in [v(a - b), +\infty[$. Indeed assume that it is not. By hypothesis we have $f(a) \neq 0$, hence there exists $\gamma \in \mathbb{R}$ such that $v_a(f, \mu) = v(f(a))$ whenever $\mu \geq \gamma$. Suppose that the function in μ, $v_a(f, \mu)$ is not bounded in $[v(a - b), +\infty[$. Since $v_a(f, \cdot)$ is a continuous function, there exists $\lambda \geq v(a - b)$ such that $v_a(f, \mu) < +\infty$ whenever $\mu > \lambda$ and $v_a(f, \lambda) = +\infty$. Hence by Lemma 38.1 D has an increasing T-filter $\mathcal{F}$ of center a and diameter $r = \omega^{-\lambda}$.

Assume first $v_a(f, v(a - b)) < +\infty$. Then there exists $\nu \in]v(a - b), \lambda]$ such that $v_a(f, \mu) < +\infty$ whenever $\mu \in]v(a - b), \nu[$, and $v_a(f, \nu) = +\infty$. By Lemma 38.1 this means that D has a decreasing T-filter $\mathcal{G}$ of center a and diameter $\omega^{-\nu} > r$. Then $\mathcal{G}$ is complementary to $\mathcal{F}$, which contradicts the hypothesis "$H(D)$ has no divisor of zero". Then we have proven that $v_a(f, v(a - b)) = +\infty$, and $v_b(f, v(a - b)) = +\infty$. Reasoning as above, we can show the existence of an increasing T-filter $\mathcal{G}$ of center b and diameter $s < \omega^{-v(a-b)}$, hence $\mathcal{G}$ is complementary to $\mathcal{F}$, which contradicts again the hypothesis "$H(D)$ has no divisor of zero". Thus, $v_a(f, \mu)$ is finally bounded on $[v(a - b), +\infty[$ and this ends the proof of Proposition 55.1.

Proposition 55.2: *Let $f \in H(D)$ and assume that the support Σ of f is reinforced. Then for every couple $(a, b) \in \Sigma \times \Sigma$, there exists an open bounded* *infraconnected set $\Omega_a^b \subset \Sigma$ with $a, b \in \Omega_a^b$ and $\Omega_a^b \in \mathcal{A}$, together with a number $t > 0$ such that $|f(x)| \geq t$ whenever $x \in \Omega_a^b$.*

Proof: Let $r = |a-b|$. By hypothesis there exists $B \in \mathbb{R}_+$ such that $v_a(f, \mu) \leq B$ and $v_b(f, \mu) \leq M$ for all $\mu \geq v(a - b)$. The equality $v(f(x)) = v_a(f, v(x - a))$ (resp. $v(f(x)) = v_b(f, v(x - b)))$ is true in all $D \cap d(a, r)$ (resp. $D \cap d(b, r)$), except may be in finitely many circles of center a (resp. b) and radii $\rho \leq r$.

Let $C(a, \rho_i)_{1 \leq i \leq m}$ (resp. $C(b, \sigma_j)_{1 \leq j \leq n}$) be the circles of center a (resp. b) that contains points $x \in D$ such that $v(f(x)) \neq v_a(f, v(x - a))$ (resp. $v(f(x)) \neq v_b(f, v(x - b)))$ and let $\Lambda_a^b = (d(a, r^-) \cap D) \setminus \left(\bigcup_{i=1}^{m} C(a, \rho_i) \right)$ (resp.

$$\Lambda_b^a = (d(b, r^-) \cap D) \setminus \left(\bigcup_{j=1}^{n} C(b, \sigma_j) \right).$$

Then Λ_a^b (resp. Λ_b^a) is clearly infraconnected open and bounded. Moreover by hypothesis we have $v(f(x)) = v_a(f, v(x - a)) \leq B$ in all Λ_a^b and $v(f(x)) = v_b(f, v(x - b)) \leq B$ on all Λ_b^a. We put $\Omega_a^b = \Lambda_a^b \cup \Lambda_b^a$. Then we have $v(f(x)) \leq B$ whenever $x \in \Omega_a^b$ hence we can take $t = \omega^{-B}$ in order to obtain the relation $|f(x)| \geq t$ in Ω_a^b.

Next, Ω_a^b is clearly open. Besides by Lemma 2.8 Ω_a^b is infraconnected because both Λ_a^b, Δ_b^a are infraconnected sets such that $\widetilde{\Lambda_a^b} = \widetilde{\Lambda_b^a} = d(a, r)$. Finally it just remains to check that Ω_a^b belongs to $\mathcal{A}$. Assume that $\Omega_a^b \notin \mathcal{A}$. Since Ω_a^b is bounded, it doesn't satisfy condition B), and therefore there exists $\alpha \in \left(\overline{\Omega_a^b} \right) \setminus \Omega_a^b$ together with a sequence of holes $(T_n)_{n \in \mathbb{N}}$ of Ω_a^b such that $\lim_{n \to \infty} \delta(\alpha, T_n) = 0$. We put $r_n = \delta(\alpha, T_n)$ $(n \in \mathbb{N})$. It is seen that Ω_a^b is closed in D because so are both Λ_a^b, Λ_b^a. Hence α belongs to $\overline{D} \setminus D$. Let $\rho = \min_{1 \leq i \leq m} \rho_i$, let $\sigma = \min_{1 \leq j \leq n} \sigma_j$ and let $s = \min(\rho, \sigma)$. As soon as $r_n < s$, it is seen that T_n is a hole of D because then $d(\alpha, r_n) \cap D$ is included in Ω_a^b. Hence, α does not belong to $\left(\overset{\circ}{\overline{D}} \right)$ and then this contradicts the hypothesis $D \in \mathcal{A}$. Thus, Ω_a^b does belong to $\mathcal{A}$ and Proposition 55.2 is now proven.

Proposition 55.3: *Let $f \in H(D)$. We assume that $\mathcal{E}(f)$ has a solution g whose support is reinforced. For every $h \in \mathcal{S}(f)$ there exists $\lambda \in K$ such that $h(x) = \lambda g(x)$, whenever $x \in \Sigma$.*

Proof: Let $h \in \mathcal{S}(f)$. Since D is open, Σ is clearly open in K, hence for every $a \in \Sigma$ there exists a disk $\Lambda(a)$ included in Σ. Let $(\mathcal{E}_a(f))$ be the equation $y' =$

$f(x)y$ for $x \in \Lambda(a)$. Then $(\mathcal{E}_a(f))$ has non zero solutions (like the restriction of g to $\Lambda(a)$), hence by Corollary 54.3, the space of its solutions has dimension one. As a consequence, there exists $\lambda(a) \in K$ such that $h(x) = \lambda(a)g(x)$ whenever $x \in \Lambda(a)$. Now, it only remains to show that $\lambda(a)$ is constant when a runs in Σ. Let us fix a and b in Σ. By Proposition 55.2 there exists $t > 0$ together with an open bounded infraconnected set $\Omega_a^b \subset \Sigma$, with $\Omega_a^b \in \mathcal{A}$ and with $a, b \in \Omega_a^b$ such that $|g(x)| \geq t$ whenever $x \in \Omega_a^b$. The restriction $\tilde{g}$ of g to Ω_a^b is then invertible in $H(\Omega_a^b)$. Hence the restriction $\widetilde{h/g}$ of h/g to Ω_a^b is a locally constant element of $H(\Omega_a^b)$. Since Ω_a^b is closed open infraconnected it satisfies Condition B), hence by Theorem 19.5 we know that h/g is a constant in $H(\Omega_a^b)$, hence $(h/g)(b) = (h/g)(a)$ and then Proposition 55.3 is proven.

Theorem 55.4: *If $H(D)$ has no divisor of zero, then $\mathcal{E}(f)$ has dimension 0 or 1.*

Proof: Assume that $\mathcal{E}(f)$ has a not identically zero solution g. By Proposition 55.1, the support Σ of g is reinforced. Let h be another not identically zero solution. Since $H(D)$ has no divisor of zero, the support Σ' of h does have common points with Σ. By Proposition 55.3 there exists $\lambda \in K$ such that $h(x) = \lambda g(x)$ whenever $x \in \Sigma$. Since $\Sigma \cap \Sigma' \neq \emptyset$, λ can't be zero. Hence $h(x) \neq 0$ whenever $x \in \Sigma$, therefore $\Sigma' \supset \Sigma$. By the same reasonning we just have $\Sigma' \subset \Sigma$, hence $\Sigma' = \Sigma$. The relation $h(x) = \lambda g(x)$ is then true on Σ, and it is trivially true on $D \setminus \Sigma$ where $h(x) = g(x) = 0$. Theorem 55.4 is then proven .

56.   THE EQUATION $y' = fy$

IN ZERO RESIDUE CHARACTERISTIC

In this chapter, the field K is supposed to have residue characteristic equal to zero. As in Chapter 55, D is infraconnected, open and belongs to $\mathcal{A}$, f denotes an element of $H(D)$, and $\mathcal{E}(f)$, $\mathcal{S}(f)$ are defined like in Chapter 54.

We will show that when the residue characteristic is zero, the dimension of $\mathcal{S}(f)$ is never greater than 1 [26]. Before proving this result we have to establish propositions mainly dedicated to the behaviour of the valuation function $v(f, \mu)$ in the particular case when the residue characteristic is zero.

For convenience, we will call *affine function* in an interval I, a function of the form $h(x) = Ax + B$.

Lemma 56.1:   *Let r and $s \in \mathbb{R}_+$ with $0 < r < s$ and let D be $\Gamma(0, r, s)$. Let μ belong to $] - \log s, - \log r[$ and let f be a Laurent series $\displaystyle\sum_{-\infty}^{+\infty} a_n x^n \in H(D)$ such that $v(f, \mu) = v(a_q) + q\mu$ with $q \neq 0$. Then $v(f, \mu) = v(f', \mu) + \mu$.*

Proof: $\displaystyle f'(x) = \sum_{-\infty}^{+\infty} n a_n x^{n-1}$. Hence $v(f', \mu) = \inf_{n \in \mathbb{Z}} v(n a_n) + (n - 1)\mu$. Since the residue characteristic of K is zero, we have $v(n a_n) = v(a_n)$ for every $n \neq 0$, hence $\displaystyle\inf_{n \in \mathbb{Z}} v(n a_n) + (n-1)\mu = v(q a_q) + (q-1)\mu = v(a_q) + (q-1)\mu = v(f, \mu) - \mu$.

Lemma 56.2:   *Let r', r'' be numbers such that $0 < r' < r''$ and let $h(x)$ be a rational function in $K(x)$ such that $v(h, \mu)$ is not constant in any interval included in $[r', r'']$. Then $v(h', \mu) = v(h, \mu) - \mu$ whenever $\mu \in [- \log r'', - \log r']$.*

Proof: Since the function $v(h, \mu)$ is continuous in μ, it is enough to prove the relation in $] - \log r'', - \log r'[$. Let $\sigma \in] - \log r'', - \log r'[$ and let $s = \omega^{-\sigma}$. We will prove the relation at σ by considering $t \in]s, r''[$ such that h has no pole in $\Gamma(0, s, t)$. Then $h(x)$ is equal to a Laurent series $\displaystyle\sum_{-\infty}^{+\infty} a_n x^n$ and we can apply Lemma 56.1 which shows that the relation is true for every $\mu \in] - \log t, \sigma[$. By continuity the relation then is true at σ.

Proposition 56.3: D *is supposed to be bounded, of diameter S, such that 0 belongs to $\widetilde{D}$. Let $r = \delta(0, D)$ and let $r', r'' \in \mathbb{R}^+$ be such that $0 < r' < r'' \leq S$ and $r \leq r'$. Let $f \in H(D)$. We assume that the function in μ $v(f, \mu)$ is bounded in the interval $I = [\log r'', -\log r']$ and is not constant in any interval $J \subset I$. Then we have $v(f, \mu) = v(f', \mu) + \mu$ whenever $\mu \in I$.*

Proof: Let M be the upper bound of $v(f, \mu)$ in I and let $t = \omega^{-M} \min(1, \frac{1}{S})$. By Theorem 19.8 there exists $h \in R(D)$ satisfying

(1) $\|f - h\|_D < t$, together with

(2) $\|f' - h'\|_D < t$.

Relation (1) also implies $v(f - h, \mu) > M \geq v(f, \mu)$ hence

(3) $v(f, \mu) = v(h, \mu)$ whenever $\mu \in I$.

Then the function $v(h, \mu)$ in μ is not constant in any interval included in I. Hence by Lemma 56.2, we have

(4) $v(h', \mu) = v(h, \mu) - \mu$ whenever $\mu \in I$.

On the other hand , by (2) we have $v(f' - h', \mu) > M + \log R > M - \mu$, hence $v(f' - h', \mu) > v(h', \mu)$ and therefore $v(f', \mu) = v(h', \mu)$ whenever $\mu \in I$. Then relations (3) and (4) show that $v(f', \mu) = v(f, \mu) - \mu$ whenever $\mu \in I$.

Proposition 56.4:   *Let D be bounded with a T-filter $\mathcal{F}$ and let $f \in H(D)$. We assume the equation $y' = fy$ to admit a solution g strictly vanishing along $\mathcal{F}$. Then f is not vanishing along $\mathcal{F}$.*

Proof: We will first assume that $\mathcal{F}$ is increasing, of center a and diameter S. We can obviously assume $a = 0$. Since g is strictly vanishing along $\mathcal{F}$, there exists $\lambda > -\log S$ such that $\lim\limits_{\mu \to -\log S} v(g, \mu) = +\infty$ with $v(g, \mu) < +\infty$ for $\mu \in] - \log S, \lambda]$, and then there exists a sequence of couples $(\lambda'_n, \lambda''_n)$ with $-\log S > \lambda'_n > \lambda''_n$, $\lim\limits_{n \to +\infty} \lambda''_n = \lim\limits_{n \to +\infty} \lambda'_n = -\log S$ and such that $(\frac{d}{d\mu})v(f, \mu)$ exists and is strictly negative whenever $\mu \in [\lambda'_n, \lambda''_n]$. By Proposition 56.3 we know that $v(g', \mu) = v(g, \mu) - \mu$ whenever $\mu \in [\lambda'_n, \lambda''_n]$, therefore $v(f, \mu) = -\mu$ whenever $\mu \in [\lambda'_n, \lambda''_n]$. Thus $v(f, \mu)$ does not tend to $+\infty$ when μ approaches $-\log S$, which proves that f is not vanishing along $\mathcal{F}$.

In the case when $\mathcal{F}$ is decreasing we can use the same proof by choosing a center of $\mathcal{F}$ (we can take it in $\widehat{K}$ if required).

Proposition 56.5:   *Let $0 \in D$. We assume the function $v(f, \mu)$ in μ to be affine in an interval $I = [\lambda', \lambda'']$ and $v(h, \mu) < +\infty$ whenever $\mu \in I$. Then for every $h \in \mathcal{S}(f) \setminus \{0\}$ the function $v(h, \mu)$ is also affine in I.*

Proof: Suppose that $v(h, \mu)$ is not affine in I. There exists a point $\sigma \in]\lambda', \lambda''[$ such that $v'^r(h,, \sigma) \neq v''(\varphi, \sigma)$. By Theorem 20.8 σ belongs to $v(K)$. Hence with no loss of generality we can suppose $\sigma = 0$ by performing a suitable change of variable.

We will first construct an interval $J = [\mu', \mu'']$ with $\mu' < 0 < \mu''$, such that the function $v(h, \mu)$ is affine in both $[\mu', 0]$ and $[0, \mu'']$ and such that $v(h, \mu)$ is bounded in J. Since $v(h, \mu) < +\infty$ whenever $x \in I$, there exist $\mu_1, \mu_2 \in I$ with $\mu_1 < 0 < \mu_2$ such that $v(h, \mu)$ is bounded by a number L in the interval $J = [\mu_1, \mu_2]$.

We can obviously choose μ_1, μ_2 close enough to 0 in order to have the function $v(h, \mu)$ affine in each of the intervals $[\mu_1, 0]$ and $[0, \mu_2]$ because it is bounded in J, hence piecewise linear in J. Since $v''g(h, 0) \neq v'^r(h, 0)$, the function $v(h, \mu)$ is not constant in at least one of the two intervals $[\mu_1, 0]$ and $[0, \mu_2]$.

For example suppose first it is not constant in $[\mu_1, 0]$. Since it is affine, it is not constant in any one of the intervals included in $[\mu_1, 0]$ and then we can apply Proposition 56.3 which proves that $v(h', \mu) = v(h, \mu) - \mu$ whenever $\mu \in [\mu_1, 0]$. Therefore $v(h', \mu)$ is bounded in $[\mu_1, 0]$ by a number L_1'. In addition, since $v(h', 0) = v(h, 0)$, there exists $\mu'' \in [\mu_2, 0]$ such that $v(h', \mu)$ is bounded in $[0, \mu'']$ by a number L_2'. We put $\mu' = \mu_1$ and $L' = \max(L_1', L_2')$. The function $v(h', \mu)$ in μ is then upper bounded by L' in $[\mu', \mu'']$ while $v(h, \mu)$ is upper bounded by L.

In the same way, if we suppose $v(h, \mu)$ not constant in $[0, \mu_2]$ we have a symmetric construction and therefore we finally have an upper bound L' for $v(h', \mu)$ in the interval $[\mu', \mu'']$ in all cases.

Now we set $B = \max(L, L')$. By definition h satisfies

(1) $v(h, \mu) \leq B$ whenever $\mu \in J$,

(2) $v(h', \mu) \leq B$ whenever $\mu \in J$.

Now there exists $\psi \in R(D)$ satisfying $\|h - \psi\|_D < \omega^{-B}$ and $\|h' - \psi'\|_D < \omega^{-B}$ and therefore we have

(3) $v(h - \psi, \mu) > B$ whenever $\mu \in J$,

together with

(4) $v(h' - \psi', \mu) > B$ whenever $\mu \in J$.

Then (1) and (3) imply

(5) $v(h, \mu) = v(\psi, \mu)$ whenever $\mu \in J$

while (2) and (4) imply $v(h', \mu) = v(\psi', \mu)$ whenever $\mu \in J$, hence

$v(\dfrac{\psi'}{\psi}, \mu) = v(\dfrac{h'}{h}, \mu) = v(f, \mu)$ whenever $\mu \in J$, which proves that $v(\dfrac{\psi'}{\psi}, \mu)$ is affine

in J, in the form $q\mu + b$ with $q \in \mathbb{Z}$. Now ψ factorizes in the form $(\dfrac{P}{Q})\theta$ where P

and Q are monic polynomials that have all their zeros in $C(0, 1)$ and θ belongs to $R(D)$ and has no zero in $C(0,1)$. Since $v'^r(h,0) \neq v'^l(h,0)$, by (5) we have $v'^r(\psi,0) \neq v'^l(\psi,0)$, so that P and Q do not have the same number of zeros in $C(0,1)$. Let $P(x) = \sum_{j=0}^{m} \alpha_j x^j$ and let $Q(x) = \sum_{j=0}^{n} \beta_j x^j$. Then $m \neq n$ and

(6) $\quad \alpha_m = \beta_n = 1.$

On the other hand, since P and Q have all of their zeros in $C(0,1)$ we see that $|\alpha_j| \leq 1$ whenever $j = 0, \ldots, m$, $|\beta_j| \leq 1$ whenever $j = 0, \ldots, n$ and

(7) $\quad v(P,O) = v(Q,0) = 0.$

Now $P'Q - PQ'$ is a polynomial of the form $\sum_{j=0}^{m+n-1} \lambda_j x^j$ with $|\lambda_j| \leq 1$ whenever $j = 0, \ldots, m+n-1$. By (6) we have $\lambda_{m+n-1} = m - n$, hence $|\lambda_{m+n-1}| = 1$. Then $v(P'Q - PQ', 0) = 0$, hence by (7) we see that

(8) $\quad v\Big(\dfrac{P'Q - PQ'}{PQ}, 0\Big) = 0.$

Now we will show that

(9) $\quad v\Big(\dfrac{\theta'}{\theta}, 0\Big) > 0.$

Since θ has neither any zero nor any pole in $C(0,1)$, there exist r', r'' such that $r' < 1 < r''$ and such that θ has neither any zero nor any pole in $\Gamma(0, r', r'')$. Therefore θ is equal to a Laurent series $\sum_{-\infty}^{+\infty} a_n x^n$ convergent in $\Gamma(0, r', r'')$. Moreover, there exists $t \in \mathbb{Z}$ such that $|a_t|\,|x|^t > |a_n|\,|x|^n$ whenever $x \in \Gamma(0, r', r'')$. Let us factorize θ in the form $x^t \gamma$. Then in $\Gamma(0, r', r'')$ γ is equal to a Laurent series $\sum_{-\infty}^{+\infty} b_n x^n$ with $b_0 = a_t$ and then we see that

$$v(\gamma', 0) = \inf_{n \in \mathbb{Z}} v(n b_n) = \inf_{n \neq 0} v(b_n) > v(b_0) = v(\gamma, 0)$$

from which we obtain $v(\dfrac{\gamma'}{\gamma}, 0) > 0$. As $\dfrac{\theta'}{\theta} = \dfrac{\gamma'}{\gamma} + \dfrac{t}{x}$ we see that

$v(\dfrac{\theta'}{\theta}, 0) = v(\dfrac{\gamma'}{\gamma}, 0) + v(\dfrac{t}{x}, 0) = v(\dfrac{\gamma'}{\gamma}, 0) > 0$ which finally shows (9).

Now, let us consider $\dfrac{\psi'}{\psi} = \dfrac{\theta'}{\theta} + \Big(\dfrac{P'Q - PQ'}{PQ}\Big)$. By (8) and (9) we have

$$v\Big(\dfrac{\theta'}{\theta}, 0\Big) > v\Big(\dfrac{P'Q - PQ'}{PQ}, 0\Big)$$

and therefore there exists an interval $\mathcal{U} = [-\rho, \rho]$ such that

$$v\left(\frac{\theta'}{\theta}, \mu\right) > v\left(\frac{P'Q - PQ'}{PQ}, \mu\right) \text{ whenever } \mu \in \mathcal{U}. \text{ Then we have}$$

$$(10) \quad v\left(\frac{\psi'}{\psi}, \mu\right) = v\left(\frac{P'Q - PQ'}{PQ}, \mu\right) \text{ whenever } \mu \in \mathcal{U}.$$

We put $g(x) = \dfrac{P'Q - PQ'}{PQ}$. Since $P(x)Q(x)$ has exactly $m + n$ zeros in $C(0, 1)$

and $P'Q - PQ'$ has at most $m + n - 1$ zeros in $C(0, 1)$ we see that

$$(11) \quad v'^r(g, 0) > v'^l(g, 0).$$

Finally by (10) we have $v'^r(\frac{\psi'}{\psi}, 0) = v'^r(g, 0)$ and $v'^l(\frac{\psi'}{\psi}, 0) = v'^l(g, 0)$ hence by

(1) we obtain $v'^r(\frac{\psi'}{\psi}, 0) > v'^l(\frac{\psi'}{\psi}, 0)$ which contradicts the fact $v(\frac{\psi'}{\psi}, \mu)$ is an

affine function in J. This finishes proving Proposition 56.5.

Theorem 56.6: *If $\mathcal{S}(f)$ is not reduced to $\{0\}$, it has dimension one and every not identically zero solution is invertible in $H(D)$.*

Proof: Assume that $\mathcal{E}(f)$ admits a not identically zero solution g and assume that g is not invertible in $H(D)$. By Theorem 54.2, g is strictly vanishing along a T-filter $\mathcal{F}$ on D and, by Proposition 56.4, f is not strictly vanishing along $\mathcal{F}$.

For example suppose that $\mathcal{F}$ is increasing, of center 0 and diameter S and assume first f is vanishing along $\mathcal{F}$. Since f is not strictly vanishing along $\mathcal{F}$, there exists $r \in]0, S[$ such that $v(f, \mu) = +\infty$ whenever $\mu \in [-\log S, -\log r]$ hence we have
$(1) \quad v(g', \mu) = +\infty$ whenever $\mu \in [-\log S, -\log r]$.
But since g is strictly vanishing along $\mathcal{F}$, there exists an interval J included in $[-\log S, -\log r]$ such that $v(g, \mu)$ is affine and not constant in J and then by Proposition 56.3. we know that $v(g', \mu) = v(g, \mu) - \mu$, which contradicts (1). Thus, f is not vanishing along $\mathcal{F}$. Now we know that the function $v(f, \mu)$ is affine in an interval $[-\log S, \lambda]$. By Proposition 56.5 the function $v(g, \mu)$ is also affine in $[-\log S, \lambda]$ and this contradicts the hypothesis that g is strictly vanishing along $\mathcal{F}$.

In case $\mathcal{F}$ is decreasing we can perform a similar proof (by taking a center in $\widehat{K}$ if required). Thus g is invertible in $H(D)$ and then by Theorem 54.1 we know that $\mathcal{S}(f)$ has dimension 1.

57.   THE EQUATION $y' = fy$ IN C_p

WITH f NOT QUASI-INVERTIBLE

In this Chapter and in the following ones, we suppose that K has characteristic zero, and residue characteristic $p \neq 0$.

D is supposed to be infraconnected, open, closed and bounded.

We will construct an open closed bounded infraconnected set D together with elements $f \in H(D)$ such that $\mathcal{S}(f)$ have dimension greater than 1, and even an infinite dimension.

Lemma 57.1:   *Assume D to have a sequence of increasing T-filters $(\mathcal{F}_n)_{n \in \mathbb{N}}$ such that $\mathcal{C}(\mathcal{F}_n) \cap \mathcal{C}(\mathcal{F}_m) = \emptyset$ whenever $n \neq m$. Let $f \in H(D)$ be such that $f(x) = 0$ whenever $x \in \bigcap_{n \in \mathbb{N}} \mathcal{P}(\mathcal{F}_n)$ and assume there exists a sequence $(f_n)_{n \in \mathbb{N}}$ in $H(D)$ such that $f_n \in \mathcal{I}_0(\mathcal{F}_n)$ and $f(x) = f_n(x)$ whenever $x \in \mathcal{C}(\mathcal{F}_n)$ for every $n \in \mathbb{N}$. Then the series $\sum_{n \in \mathbb{N}} f_n$ converges to f in $H(D)$.*

Proof: Suppose first the series $\sum_{n \in \mathbb{N}} f_n$ converges to a limit h in $H(D)$. Then h is clearly equal to f because by definition when $x \in \bigcap_{n \in \mathbb{N}} \mathcal{P}(\mathcal{F}_n)$, we have $f_n(x) = 0$ for all n, hence $h(x) = f(x) = 0$. When $x \in \mathcal{C}(\mathcal{F}_q)$ we have $x \in \mathcal{P}(\mathcal{F}_n)$ for all $n \neq q$ hence $h(x) = f_q(x) = f(x)$. Now we only have to prove that the series $\sum_{n \in \mathbb{N}} f_n$ converges and that means to prove that the sequence $\|f_n\|_D$ goes to zero. Suppose the sequence $\|f_n\|_D$ admits a subsequence $\|(f_{n_q})_{q \in \mathbb{N}}\|_D$ such that $\|f_{n_q}\|_D \geq \lambda$ whenever $q \in \mathbb{N}$, with $\lambda > 0$. Consider now the sequence of T-filters $(\mathcal{F}_{n_q})_{q \in \mathbb{N}}$ and for each $q \in \mathbb{N}$ let $a_q \in \mathcal{C}(\mathcal{F}_{n_q})$ be such that $|f(a_q)| \geq \lambda$.

Suppose first we can extract a subsequence $(a_{q_m})_{m \in \mathbb{N}}$ satisfying

(1) $|a_{q_m} - a_{q_j}| = r$ whenever $j \neq m$.

We set $b_m = a_{q_m}$, and $\mathcal{G}_m = \mathcal{F}_{n_{q_m}}$. With no loss of generality we can obviously assume $b_0 = 0$. We know that $v(f, -\log r) = v(f(b_m))$ is true for all m except maybe finitely many ones. Hence we see that $v(f, -\log r) \leq -\log \lambda$. Let $A =$

$v(f, -\log r)$. We know that $v(f(x)) = A$ is true in $C(0, r) \cap D$ except (maybe) in finitely many disks $d(b_m, r^-)$. Consider $m \in \mathbb{N}$ such that $v(f(x)) = A$ for $x \in d(b_m, r^-) \cap D$. Then trivially we have $\lim\limits_{\mathcal{G}_m} v(f(x)) = A < +\infty$ hence that contradicts the hypothesis $f \in \mathcal{I}(\mathcal{F}_n)$ for all n. Hence the sequence $(a_{q_m})_{m \in \mathbb{N}}$ satisfying (1) does not exist.

Thus we can extract a subsequence $(a_{q_m})_{m \in \mathbb{N}}$ such that the sequence $|a_{q_{n-1}} - a_{q_n}|$ is strictly monotonous, of limit r. We put again $\mathcal{G}_m = \mathcal{F}_{n_{q_m}}$ $(m \in \mathbb{N})$ and $b_m = a_{q_m}$.

Suppose first $r = 0$. The sequence $(b_m)_{m \in \mathbb{N}}$ converges to a limit α in D and we can assume $\alpha = 0$. Since f is continuous and satisfies $|f(\alpha_{q_m})| \geq \lambda$, there exists $\rho > 0$ such that $|f(x)| \geq \dfrac{\lambda}{2}|$ for $|x| \leq \rho$, hence $|f(x)| \geq \dfrac{\lambda}{2}$ for $x \in C(\mathcal{G}_m)$ when m is big enough and then this contradicts $f \in \mathcal{I}(\mathcal{F}_n)$ for all $n \in \mathbb{N}$. Hence $r \neq 0$. Suppose now the sequence $|b_{m+1} - b_m|$ is strictly increasing. We can obviously assume $b_0 = 0$ hence the sequence $(|b_m|)_{m \in \mathbb{N}}$ is strictly increasing, of limit r. By hypothesis, we have $_D\varphi_{0,r}(f) \geq \lambda$. Hence, there exists $s \in]0, r[$ such that $|f(x)| \geq \lambda$ for every $s \in D \cap \Gamma(0, s, r)$. Let $N \in \mathbb{N}$ be such that $|b_m| \geq \lambda$ whenever $m \geq N$. Since the $\mathcal{G}_m$ are two by two complementary, $C(\mathcal{G}_m)$ is included in $d(b_m, r^-)$, and therefore this contradicts the hypothesis $f \in \mathcal{I}(\mathcal{G}_m)$ whenever $m \geq N$.

We obtain a similar contradiction when this sequence is strictly decreasing of limit $r \neq 0$ and this finishes proving Lemma 57.1.

Proposition 57.2: *Assume that D has a sequence of increasing T-filters $(\mathcal{F}_n)_{m \in \mathbb{N}}$ (resp. finitely many increasing T-filters $(\mathcal{F}_n)_{m \in I}$) such that $C(\mathcal{F}_n) \cap C(\mathcal{F}_m) = \emptyset$ whenever $n \neq m$. Assume that for each $n \in \mathbb{N}$ (resp. $n \in I$) there exists $f_n \in \mathcal{I}(\mathcal{F}_n)$ such that $\mathcal{E}(f_n)$ has a solution $g_n \neq 0$ in $\mathcal{I}_0(\mathcal{F}_n)$ and such that the series $\sum\limits_{n \in \mathbb{N}} f_n$ converges in $H(D)$. Let $f = \sum\limits_{m \in \mathbb{N}} f_n$ (resp. $f = \sum\limits_{m \in I} f_n$) .*

The family $(g_n)_{n \in \mathbb{N}}$ (resp. $(g_n)_{n \in I}$) is a linearly free family of solutions of $\mathcal{E}(f)$. Moreover if the $(\mathcal{F}_n)_{n \in \mathbb{N}}$ (resp. $(\mathcal{F}_n)_{n \in I}$) are the only T-filters on D, then for every solution g of $\mathcal{E}(f)$, there exists a unique sequence $(\lambda_n)_{n \in \mathbb{N}}$ in K such that the series $\sum\limits_{n \in \mathbb{N}} \lambda_n g_n$ converges to g in $H(D)$ (resp. $\{g_1, ..., g_n\}$ is a base of the space $\mathcal{S}(f)$.)

Proof: The proof is obviously easier, and follows the same way, when the set of T-filters $(\mathcal{F}_n)_{n \in I}$ is finite. So we will just consider the case when we have a sequence of T-filters $(\mathcal{F}_n)_{n \in \mathbb{N}}$. The family $(g_n)_{n \in \mathbb{N}}$ is clearly linearly free because the supports of the g_n are two by two disjointed. We will show that if D has no T-filter but the $(\mathcal{F}_n)_{n\in\mathbb{N}}$, every solution g of $\mathcal{E}(f)$ is the sum of a series $\sum_{n\in\mathbb{N}} \lambda_n g_n$. Let $g \in \mathcal{S}(f)$ and first consider $g(x)$ when $x \in \mathcal{C}(\mathcal{F}_q)$ for certain $q \in \mathbb{N}$. Let E_q be the differential equation $\mathcal{E}(\overline{\mathcal{F}}_q)$ defined in $\mathcal{C}(\mathcal{F}_q)$ by the restriction $\overline{f}_q$ of f_q to $\mathcal{C}(\mathcal{F}_q)$. We know that $f(x) = f_q(x)$ whenever $x \in \mathcal{C}(\mathcal{F}_q)$ and then every $h \in \mathcal{S}(f)$ has a restriction to $\mathcal{C}(\mathcal{F}_q)$ that is a solution of E_q. Since D has no T-filter except the $(\mathcal{F}_n)$, and since the $\mathcal{C}(\mathcal{F}_n)$ are two by two disjointed, $\mathcal{C}(\mathcal{F}_q)$ has no T-filter except $\mathcal{F}_q$. Hence $H(\mathcal{C}(\mathcal{F}_q))$ has no divisor of zero and therefore by theorem 55.4, E_q has a space of solutions of dimension 1. Hence there exists $\lambda_q \in K$ such that $g(x) = \lambda_q g_q(x)$ whenever $x \in \mathcal{C}(\mathcal{F}_q)$. Now by hypothesis g_q belongs to $\mathcal{I}_0(\mathcal{F}_q)$, hence g belongs to $\mathcal{I}(\mathcal{F}_q)$. We will show that $g(x) = 0$ whenever $x \in \bigcap_{n\in\mathbb{N}} \mathcal{P}(\mathcal{F}_n)$. Indeed let $\Lambda = \mathcal{C}(\mathcal{F}_q) \cup \left(\bigcap_{n\in\mathbb{N}} \mathcal{P}(\mathcal{F}_n)\right)$. Clearly $\mathcal{F}_q$ is secant to Λ and then Λ has no T-filter complementary to $\mathcal{F}_q$. Hence by Theorem 38.13, we know that $g(x) = 0$ whenever $x \in \mathcal{P}(\mathcal{F}_q) \cap \Lambda$, and therefore $g(x) = 0$ whenever $x \in \bigcap_{n\in\mathbb{N}} \mathcal{P}(\mathcal{F}_n)$. Finally we have proven that either x belongs to a set $\mathcal{C}(\mathcal{F}_n)$ and then $g(x) = \lambda_n g_n(x)$, or x does not belong to $\bigcup_{n\in\mathbb{N}} \mathcal{C}(\mathcal{F}_n)$, hence $x \in \bigcap_{n\in\mathbb{N}} \mathcal{P}(\mathcal{F}_n)$, and then $g(x) = 0$. By Lemma 57.1 g is the sum of the convergent series $\sum_{n\in\mathbb{N}} \lambda_n g_n$. The sequence (λ_n) is clearly unique. Indeed suppose $g \in \sum_{n\in\mathbb{N}} \mu_n g_n$. Then when $x = \mathcal{C}(\mathcal{F}_n)$ we have $g(x) = \mu_n g_n(x)$ because $g_m(x) = 0$ for all $m \neq n$ hence $\mu_n = \lambda_n$.

Proposition 57.3:   *Let $h \in A_u(d(0, 1^-))$, let $(C(0, r_m))_{m\in\mathbb{N}}$, with $r_m < r_{m+1}$ and $\lim_{m\to+\infty} r_m = r$ be the sequence of the circles of center 0 that contain at least one zero of h. Let D satisfy*

$\quad$ (1)$\quad$ $d(0, 1) \subset \tilde{D}$,

$\quad$ (2)$\quad$ $D \subset K \setminus \left(\bigcup_{m\in\mathbb{N}} C(0, r_m)\right)$

Moreover, we assume that h satisfies $\quad \lim\limits_{\substack{|x|\to 1 \\ x\in D\cap d(0,1^-)}} \dfrac{h'(x)}{h(x)} = 0$. Let ϕ and f be the

functions defined in D by $\phi(x) = \dfrac{1}{h(x)}, f(x) = \dfrac{h'(x)}{h(x)}$ whenever $x \in D \cap d(0, 1^-)$

and $\phi(x) = f(x) = 0$ whenever $x \in D \setminus d(0,1^-)$. Then both ϕ, f belong to $H(D)$ and ϕ belongs to $\mathcal{S}(f)$.

Proof: Since $\lim\limits_{\substack{|x| \to 1 \\ x \in D}} |h(x)| = +\infty$ by Theorem 49.2 we know that ϕ belongs to

$H(D)$. In the same way, since $\lim\limits_{\substack{|x| \to 1 \\ x \in D}} \left(\dfrac{h'(x)}{h(x)} \right) = 0$, by Theorem 49.2 we know that

f belongs to $H(D)$. Then ϕ obviously satisfies $\mathcal{E}(f)$.

Proposition 57.4: *Let $(a_n)_{\in \mathbb{N}}$ be the sequence defined as follows :*

$a_0 \neq 0$

$a_n = 1$ *when n is neither 0 nor in the form p^m for any $m \in \mathbb{N}$.*

$a_n = \frac{1}{n}$ *when n is in the form p^m for some $m \in \mathbb{N}$.*

Let $h(x) = \sum\limits_{n=0}^{\infty} a_n x^n$. Then h is convergent in $d(0,1^-)$. Let Λ be the set of the

$x \in d(0,1^-)$ such that $v(h(x)) = v(h, v(x))$ and assume that $D \cap d(0,1^-) = \Lambda$.
Then

(1) $\quad h(x) \neq 0$ *for all $x \in D$.*

(2) $\quad \lim\limits_{\substack{|x| \to 1 \\ x \in D}} |h(x)| = +\infty$

(3) $\quad \lim\limits_{\substack{|x| \to 1 \\ x \in D}} \dfrac{h'(x)}{h(x)} = 0$

Moreover the increasing filter $\mathcal{F}$ of center 0 and diameter 1 is a T-filter and is the only one T-filter on Λ. In addition, the functions f and ϕ defined in Λ as

$$f(x) = -\frac{h'(x)}{h(x)} \quad \text{and} \quad \phi = \frac{1}{h} \quad \text{belong to } H(\Lambda) \text{ and they both are strictly vanishing}$$

along $\mathcal{F}$. Further ϕ is a solution of $\mathcal{E}(f)$ and it generates $\mathcal{S}(f)$.

Proof: First, it is easily seen that h converges for $|x| < 1$ because we check that $\lim\limits_{n \to \infty} \sqrt[n]{|a_n|} = 1$. Next, the relation $v(h(x)) = v(h, v(x))$ is true in all

$d(0,1)$ except in a set included in a union of circles in the form $\bigcup\limits_{n=1}^{\infty} C(0, r_n)$ with

$\lim\limits_{n \to \infty} r_n = 1$. The set Λ is obviously infraconnected open and closed. As $h(0) = a_0 \neq 0$, (1) is satisfied by the definition of Λ because $v(h(x)) = v(h, v(x)) < +\infty$, for all $x \in \Lambda$. Since the sequence $|a_n|$ is not bounded, we know that $\lim\limits_{\mu \to 0^+} v(h, \mu) = +\infty$, hence Relation (2) is clearly satisfied. Now h' is bounded in

$d(0,1)$. Indeed we have $h'(x) = \sum_{n=1}^{\infty} n a_n x^{n-1}$ with $|n a_n| = |n| \leq 1$ when n is not of the form p^m and $|p^m a_{p^m}| = p^{-m} p^m = 1$. Hence we have $|n a_n| \leq 1$ whenever $n \in \mathbb{N}$. Finally we have $|h'(x)| \leq 1$ in $d(0,1)$ and then (3) is clearly satisfied. On the other hand, by definition f, ϕ, ϕ' are strictly vanishing along the increasing filter $\mathcal{F}$ of center 0 and diameter 1. Hence $\mathcal{F}$ is a T-filter. By the definition of Λ it is clearly seen that Λ has no T-filter different from $\mathcal{F}$. By proposition 57.3, both ϕ, f belong to $H(\Lambda)$ and by construction they are strictly vanishing along $\mathcal{F}$. Then ϕ obviously is a solution of $\mathcal{E}(f)$. Besides, since $\mathcal{F}$ is the only T-filter on Λ, by Theorem 38.10 $H(\Lambda)$ has no divisor of zero, hence $\mathcal{S}(f)$ has dimension ≤ 1. Since $\phi \neq 0$, we see that ϕ generates $\mathcal{S}(f)$, and this finishes proving Proposition 57.4.

Theorem 57.5: *Let $n \in \mathbb{N}$. There exists an infraconnected closed open bounded set D and elements $f \in H(D)$ strictly vanishing along n increasing T-filters two by two complementary such that $dim\, \mathcal{S}(f) = n$.*

Proof: Consider again the set Λ obtained in Proposition 57.4. Let S be > 1, and let $a_1, \ldots, a_n$ be points in $d(0, S)$ such that $|a_i - a_j| \geq 1$ whenever $i \neq j$. For all $i = 1, \ldots, n$ let $A_i = a_i + \Lambda = \{a_i + x \mid x \in \Lambda\}$, let $D_i = (d(0, S) \setminus d(a_i, 1^-)) \bigcup A_i$ and let $D = \bigcap_{i=1}^{n} D_i$. By Proposition 57.4, the set D has increasing T-filters $\mathcal{F}_i$ of center a_i and diameter 1 (obtained from $\mathcal{F}$ by translation of a_i). Hence for every $i = 1, \ldots, n$ there exists $f_i \in \mathcal{I}_0(\mathcal{F}_i)$ $(f \neq 0)$ such that $\mathcal{E}(\mathcal{F}_i)$ admits solutions $g_i \in \mathcal{I}_0(\mathcal{F}_i)$, $(g_i \neq 0)$ strictly vanishing along $\mathcal{F}_i$. Since $\mathcal{F}_i$ has diameter 1 and since $|a_i - a_j| \geq 1$ for every $i \neq j$ the $\mathcal{C}(\mathcal{F}_i)$ satisfy $\mathcal{C}(\mathcal{F}_i) \cap \mathcal{C}(\mathcal{F}_j) = \emptyset$ for $i \neq j$ and the only T-filters on D are the $\mathcal{F}_i$ $(1 \leq i \leq n)$. Now let $f = \sum_{i=1}^{n} f_i$. It is easily seen that $f(x) = f_i(x)$ whenever $x \in \mathcal{C}(\mathcal{F}_i)$ and

$$f(x) = 0 \text{ whenever } x \in \bigcap_{i=1}^{n} \mathcal{P}(\mathcal{F}_i).$$

For each $i = 1, \ldots n$, g_i is a solution of $\mathcal{E}(f_i)$ and therefore it is a solution of $\mathcal{E}(f)$.

Finally, by Proposition 57.2, the $(g_i)_{1 \leq i \leq n}$ are linearly independent. As $\mathcal{F}_1, \ldots, \mathcal{F}_n$ are the only T-filters on D, the family $\{g_1, \ldots, g_n\}$ is then a base of the space of solutions by the last statement of Proposition 57.2.

Theorem 57.6: *There exists an infraconnected open closed bounded set D with a sequence of increasing T-filters $(\mathcal{F}_n)_{n \in \mathbb{N}}$ two by two complementary, and elements $f \in H(D)$ strictly vanishing along each one of the $\mathcal{F}_n$ such that $\mathcal{S}(f)$ is isomorphic to the space of the sequence $(\lambda_n)_{n \in \mathbb{N}}$ such that $\lim_{n \to \infty} \lambda_n = 0$.*

Proof: The proof will roughly follow the same process as in Theorem 57.5 with a sequence of filter $\mathcal{F}_n$ instead of a finite set of $(\mathcal{F}_j)$. Let S be > 1 and consider a sequence $(a_n)_{n\in\mathbb{N}}$ in $d(0,S)$ such that $|a_n - a_m| = 1$ for $n \neq m$. For every $n \in \mathbb{N}$ we take a power series h_n be defined like h in Proposition 57.4, specifying here $h_n(0) = p^{-n}$, and we define A_n as the set of the $x \in d(0,1^-)$ such that $v(h_n(x)) = v(h_n, v(x))$. Then we have $v(h_n, \mu) \leq -n$ for all $\mu > 0$, hence $v(h_n(x)) \leq -n$ for all $x \in A_n$. Therefore we have

$$(1) \quad \left| \frac{h'_n(x)}{h_n(x)} \right| \leq p^{-n} \text{ whenever } x \in A_n.$$

Now we set $\Lambda_n = a_n + A_n$, $B = d(0,S) \setminus \left(\bigcup_{n=1}^{\infty} d(a_n, 1^-) \right)$ and $D = B \bigcup \left(\bigcup_{n=1}^{\infty} \Lambda_n \right)$.

Obviously B has no T-filter, and each Λ_n has an increasing T-filter $\mathcal{F}_n$ of center a_n and diameter 1. Each $\mathcal{F}_n$ induces a T-filter on D and we will still denote it by $\mathcal{F}_n$. Then D has no T-filter other than the $\mathcal{F}_n$. Now we put $f_n(x) = \dfrac{h'_n(x - a_n)}{h_n(x - a_n)}$ for $x \in \Lambda_n$ and $f_n(x) = 0$ for $x \in D \setminus \Lambda_n$. By (1) we have $\|f_n\|_D = p^{-n}$ and then the series $\displaystyle\sum_{n=1}^{\infty} f_n$ converges to a limit $f \in H(D)$ such that $f(x) = f_n(x)$ whenever $x \in \Lambda$ and $f_n(x) = 0$ whenever $x \in D \setminus \left(\bigcup_{n=1}^{\infty} \Lambda_n \right)$. By Proposition 57.3, for each $n \in \mathbb{N}^*$, there exists a solution $g_n \in \mathcal{S}(f_n) \bigcap \mathcal{I}_0(\mathcal{F}_n)) \setminus \{0\}$ and we can obviously choose the g_n in order to satisfy

$$(2) \quad \|g_n\|_D = 1.$$

By Proposition 57.2 the set $\{g_n \mid n \in \mathbb{N}\}$ is linearly free and every solution g of $\mathcal{E}(f)$ may be written of a unique matter in the form $\displaystyle\sum_{n=1}^{\infty} \lambda_n g_n$ (with $\lambda_n \in K$).

Then $\mathcal{S}(f)$ is isomorphic to a subspace of $K^{\mathbb{N}}$. On the other hand by (2) a series $\displaystyle\sum_{n=1}^{\infty} \lambda_n g_n$ converges in $H(D)$ if an only if $\lim_{n\to\infty} \lambda_n = 0$. If $(\lambda_n)_{n\in\mathbb{N}}$ is such a sequence , the series $\displaystyle\sum_{n=1}^{\infty} \lambda_n g_n$ converges to a limit g. Thus $\mathcal{S}(f)$ is isomorphic to the space of the sequence $(\lambda_n)_{n\in\mathbb{N}}$ such that $\lim_{n\to\infty} \lambda_n = 0$ and this ends the proof of Theorem 57.6.

58. THE EQUATION $y' = fy$ IN C_p

WITH f QUASI-INVERTIBLE

In this chapter as in chapter 57, we suppose that K has characteristic zero, and residue characteristic $p \neq 0$.

D is supposed to be infraconnected, open, closed and bounded.

In chapter 57 we saw that there does exist an infraconnected open closed bounded set with a T-filter $\mathcal{F}$ and an element f vanishing along $\mathcal{F}$ such that the solutions of $\mathcal{S}(f)$ are also vanishing along $\mathcal{F}$. Thanks to such T-filters, for every $n \in \mathbb{N}$ we could construct infraconnected closed open bounded sets D with $f \in H(D)$ such that $\mathcal{S}(f)$ has dimension n, and we even constructed sets D with $f \in H(D)$ such that $\mathcal{S}(f)$ is isomorphic to the space of the sequence of limit zero. This suggested that a situation where the solutions of $\mathcal{E}(f)$ were not invertible in $H(D)$ should be associated to a non quasi-invertible element f, and so should be spaces $\mathcal{S}(f)$ of dimension greater than one.

Here we will prove that this connection does not hold by constructing an infraconnected closed open bounded set D with a T-filter $\mathcal{F}$ and a quasi-invertible element $f \in H(D)$ such that $\mathcal{E}(f)$ has solutions strictly vanishing along $\mathcal{F}$ [28].

Next, for every fixed integer t, an extension of that construction will provide us with a set D and a quasi-invertible $f \in H(D)$ such that $dim(\mathcal{S}(f)) = t$.

Proposition 58.1: *Let $(b_m)_{m \in N}$ be a sequence in $d(0, 1^-)$ such that $|b_m| < |b_{m+1}|$, and let $(t_m)_{m \in \mathbb{N}}$ be a sequence of integers of the form p^{q_m} where $(q_m)_{m \in \mathbb{N}}$ is a sequence of integers, satisfying*

$(1) \quad \lim_{m \to \infty} q_m = +\infty,$

$(2) \quad |t_1| > |t_m| \quad whenever \quad m \geq 2,$

$(3) \quad \lim_{m \to \infty} \left| \dfrac{b_m}{b_{m+1}} \right|^{t_{m+1}} = 0.$

Let S be ≥ 1, and let $D = d(0, S) \setminus \left(\bigcup_{m=1}^{\infty} d(b_m, |b_m|^-) \right)$. For each $m \in \mathbb{N}^*$ let

$$h_m = \prod_{j=1}^{m} \frac{1}{\left(1 - \dfrac{x}{b_j}\right)^{t_j}} \in R(D). \text{ Then the sequence } (h_m) \text{ converges in } H(D) \text{ to a}$$

limit h that is strictly vanishing along an increasing T-filter $\mathcal{F}$ of center 0 and

352

diameter 1. Besides h belongs to $\mathcal{I}_0(\mathcal{F})$. The series $\displaystyle\sum_{m=1}^{\infty} \frac{t_m}{(b_m - x)}$ converges in $H(D)$ to a limit f quasi-invertible in $H(D)$ and h is a solution of $\mathcal{E}(f)$.

Proof : Since $\displaystyle\lim_{m\to\infty} |\frac{b_m}{b_{m+1}}| = 0$ we have $\displaystyle\lim_{m\to\infty} (t_{m+1} \log |\frac{b_{m+1}}{b_m}|) = +\infty$. Thus we can easily define a sequence of integers ℓ_m such that $\displaystyle\lim_{m\to\infty} (q_m - \ell_m) = +\infty$ and $\displaystyle\lim_{m\to\infty} (p^{\ell_{m+1}} \log |\frac{b_{m+1}}{b_m}|) = +\infty$. We put $s_m = p^{\ell_m}$, $w_m = |\frac{t_m}{s_m}|$, $e_m = |\frac{b_{m-1}}{b_m}|^{\ell_m}$. Then we have $\displaystyle\lim_{m\to\infty} w_m = \lim_{m\to\infty} e_m = 0$. As the holes of D are of the form $d(b_m, |b_m|^-)$ it is easily seen that

$$(4) \quad \left\| \frac{1}{1 - \dfrac{x}{b_j}} \right\|_D \leq 1.$$

We consider $|h_{m+1}(x) - h_m(x)|$ when $|x| \geq |b_m|$. We have

$$(5) \quad |h_m(x)| \leq \prod_{j=1}^{m-1} \frac{1}{|1 - \dfrac{x}{bj}|^{t_j}} \leq e_m.$$

In the same way, we have $|h_{m+1}(x)| \leq e_m$ hence

$$(6) \quad |h_{m+1}(x) - h_m(x)| \leq e_m.$$

Now we consider $h_{m+1}(x) - h_m(x)$ when $|x| < |b_m|$ and we put

$$u(x) = \frac{1}{\left(1 - \dfrac{x}{b_{m+1}}\right)^{t_{m+1}}} - 1 = -\frac{\displaystyle\sum_{j=1}^{t_{m+1}} \binom{t_{m+1}}{j} \left(-\dfrac{x}{b_{m+1}}\right)^j}{\left(1 - \dfrac{x}{b_{m+1}}\right)^{t_{m+1}}}.$$

Then it is clear that $|u(x)| \leq \displaystyle\max_{1 \leq j \leq t_m} \left| \binom{t_{m+1}}{j} \right| \cdot \left| \frac{b_m}{b_{m+1}} \right|^j$. Thus for $1 \leq j \leq s_{m+1}$, as $|j| \geq |s_{m+1}|$, by Lemma 38.3 we obtain $\left| \binom{t_{m+1}}{j} \right| \leq \left| \frac{b_m}{b_{m+1}} \right|$.

Now for $j > s_{m+1}$ we see that $\left| \frac{b_m}{b_{m+1}} \right|^j \leq \left| \frac{b_m}{b_{m+1}} \right|^{s_{m+1}} = e_m$ and then every term $\binom{t_{m+1}}{j} \left(-\dfrac{x}{b_{m+1}}\right)^j$ is upper bounded by $\max (w_{m+1}, e_m)$ and therefore $|u(x)| \geq \max(w_{m+1}, e_m)$ whenever $x \in D \cap d(0, |b_m|)$.

Finally by (6) we see that $\|h_{m+1} - h_m\|_D \leq \max(w_{m+1}, e_m)$ and then the sequence h_m converges in $H(D)$ to the convergent infinite product

$$h(x) = \prod_{j=1}^{\infty} \frac{1}{\left(1 - \dfrac{x}{b_j}\right)^{t_j}}.$$

By (3) and by the definition of D it is easily seen that the increasing filter $\mathcal{F}$ of center 0 and diameter 1 is a T-filter and it is the only T-filter on D.

On the other hand, by (5) we have $|h(x)| \leq e_m$ whenever $x \in D \setminus d(0, |b_m|^-)$ and therefore h is clearly vanishing along $\mathcal{F}$. Since $\mathcal{F}$ is the only T-filter on D, by Theorem 38.7 h is strictly vanishing along $\mathcal{F}$ and it satisfies $h(x) = 0$ whenever $x \in \mathcal{P}(\mathcal{F})$, hence $h \in \mathcal{I}_0(\mathcal{F})$.

Now we consider the series $\displaystyle\sum_{j=1}^{\infty} \frac{t_j}{(b_j - x)}$. Since $\displaystyle\lim_{m \to \infty} |t_m| = 0$, by (4) we see that this series converges to a limit $f \in H(D)$. Moreover, it is easily seen that $\displaystyle\lim_{\substack{|x| \to 1, |x| \neq 1 \\ x \in D}} \left| \frac{t_j}{b_j - x} \right| = |t_j|$ for every $j \in \mathbb{N}^*$ hence, by (2), we have $\displaystyle\lim_{\substack{|x| \to 1, |x| \neq 0 \\ x \in D}} |f(x)| = |t_1|$. Hence f is not vanishing along $\mathcal{F}$. Since $\mathcal{F}$ is the only T-filter, by Theorem 38.2 f is quasi-minorated, and since D is closed and open, by Theorem 14.7 f is quasi-invertible. We now check that h is a solution of $\mathcal{E}(f)$. Of course we have $\delta(D, k \setminus D) > 0$, and therefore, by Theorem 19.1 h' belongs to $H(D)$. Further, by Theorem 19.4, we see that the sequence h'_m converges to h' in $H(D)$. On the other hand, it is easily seen that

$$h'_m = \left(\sum_{j=1}^{m} \frac{t_j}{\left(1 - \dfrac{x}{b_j}\right)^{t_j}}\right) h_m = h_m \sum_{j=1}^{m} \frac{t_j}{b_j - x}$$

hence

$$\lim_{m \to \infty} h'_m = h\left(\sum_{j=1}^{\infty} \frac{t_j}{b_j - x}\right) = hf.$$

Therefore h is a solution of $\mathcal{E}(f)$, and this ends the proof of Proposition 58.1.

Corollary 58.2:   *There exist an infraconnected open closed bounded set D with a T-filter $\mathcal{F}$ and quasi-invertible elements $f \in H(D)$ such that $\mathcal{E}(f)$ has solutions strictly vanishing along $\mathcal{F}$ and such that $\mathcal{S}(f)$ has dimension 1.*

Thanks to Proposition 58.1 we will construct open sets D with an element $f \in H(D)$ such that $\mathcal{S}(f)$ has dimension $n > 1$.

Notations: As previously, given $a \in \mathbb{R}$, we denote by $Int(a)$ the integral part of a, i.e., the unique $t \in \mathbb{N}$ such that $t \le a < t+1$.

Lemma 58.3: *Let q, n be integers such that $0 < n < q$. Then*
$$\left| \frac{q!}{n!} \right| \le p^{1 - \frac{(q-n)}{p}} .$$

Proof: $\dfrac{q!}{n!}$ has $q - n$ consecutive factors. It is easily seen that among these $q - n$ factors, the number of them that are multiple of p, is at least $Int(\dfrac{q-n)}{p})$ and therefore $v(\dfrac{q!}{n!}) \ge Int(\dfrac{q-n)}{p}) > \dfrac{(q-n)}{p} - 1.$

Lemma 58.4: *Let $r \in [p^{\frac{-1}{p}}, 1[$, let $\epsilon \in]0, \dfrac{1}{p}[$ and let $g(x) = \sum_{-\infty}^{+\infty} a_n x^n$ be a Laurent series convergent for $|x| = r$, such that $\sup_{n \in \mathbb{Z}} |a_n| r^n = |a_q| r^q$ with $q < 0$. Then g does not satisfy the inequality*

(1) $\left| \dfrac{g'(x)}{g(x)} - 1 \right| < \epsilon$ *whenever* $x \in C(0, r)$.

Proof: We suppose that g satisifies (1) and we put $M = |a_q| r^q$. By (1) it is easily seen that

(2) $|n a_n - a_{n-1}| r^{n-1} \le \epsilon M$ for every $n \in \mathbb{Z}$.

If $q = -1$, Relation (2) gives $\dfrac{|-a_{-1}|}{r} \le \epsilon \dfrac{|a_{-1}|}{r}$ hence $\ell = 0$. We will suppose $q < -1$ and we will prove that for every $n = q+1, q+2, \ldots, -2, -1$, we have

(3) $|a_n| = \dfrac{|(-h-1)!| \, |a_q|}{|(-q-1)!|}$

Indeed, suppose it has been proven up to the rank t with $q \le t < -1$. We will prove it at the rank $t+1$. By (2) we have

(4) $|(t+1) a_{t+1} - a_t| r^t \le \epsilon |a_q| r^q$

hence $|(t+1) a_{t+1} - a_t| \le \frac{\epsilon |a_q|}{r^{t-q}}$ and then by (4) we have

(5) $|(t+1) a_{t+1} - a_t| \le \dfrac{\epsilon |a_t| |(-q-1)!|}{r^{t-q} |(-t-1)!|}.$

By Lemma 58.3 we know that $\left| \dfrac{(-q)!}{(-t)!} \right| \le p^{1 - \frac{(t-q)}{p}}.$

Since $r \ge p^{\frac{-1}{p}}$, we see that $r^{t-q} \ge p^{\frac{-(t-q)}{p}}$. Hence $\left| \dfrac{(-q)!}{(-t)!} \right| \le p r^{t-q}$ and therefore $\epsilon \left| \dfrac{(-q)!}{(-t)!} \right| \le r^{t-q}$. Then by Relation (5) we obtain

(6) $|(t+1)a_{t+1} - a_t| < |a_t|$

hence $|(t+1)a_{t+1}| = |a_t|$, and therefore $|a_{t+1}| = \left|\dfrac{a_t}{t+1}\right|$. Then, by (3) which we suppose to be true at the rank t, Relation (3) at the rank $t+1$ is obvious. So Relation (3) is now proven for every $n = q+1, \ldots, -1$.

Now for $n = 0$, Relation (2) gives $|a_{-1}|r^{-1} \leq \epsilon |a_q| r^q$, hence by (4) we have $\dfrac{|a_q|}{|(-q-1)!|} \leq \epsilon r^{q+1} |a_q|$ and therefore

(7) $\epsilon \dfrac{|(-q-1)!|}{r^{q+1}} \geq 1.$

But we know that $r^{q+1}|(-q-1)!| \leq p^{\frac{-(q+1)}{p}} < \dfrac{1}{\epsilon}$ and then (7) is impossible. This ends the proof of Lemma 58.4.

Theorem 58.5:   *For every $t \in \mathbb{N}$ there exists an infraconnected closed open bounded set D and quasi-invertible elements $f \in H(D)$ such that $dim(\mathcal{S}(f)) = t$.*

Proof: Let $w_1, \ldots, w_t$ be points in $d(0,1)$ such that $w_1 = 0$, $|w_i - w_j| = 1$ whenever $i \neq j$. Let $r \in]0,1[$ and let $(b_m)_{m \in \mathbb{N}}$ be a sequence in $d(0, r^-)$ such that $|b_m| < |b_{m+1}|$ and $\lim\limits_{m \to \infty} |b_m| = r$. Let $(q_m)_{m \in \mathbb{N}}$ be a sequence of integers such that $q_1 < q_m$ for all $m > 1$, $\lim\limits_{m \to \infty} q_m = +\infty$ and $\lim\limits_{m \to \infty} \prod\limits_{j=1}^{m-1} \left|\dfrac{b_j}{b_m}\right|^{(p^{q_j})} = 0.$

Let $T_m = d(b_m, |b_m|^-)$, let $u_m = p^{q_m}$ and let $A = d(0, r^-) \setminus \left(\bigcup\limits_{m=1}^{\infty} T_m \right)$. The weighted sequence (T_m, q_m) is easily seen to be a T-sequence. Let $\mathcal{T}$ be the increasing T-filter of center 0 and diameter r on A. First, we will construct an infraconnected closed open set Ω included in $d(0,1)$, of diameter 1, satisfying the following conditions :

(1) $\Omega \cap d(0, r^-) = A.$

(2) Ω has an increasing T-filter $\mathcal{F}$ of center 0 and diameter 1.

(3) Ω has a decreasing T-filter $\mathcal{G}$ of center 0 and diameter $s \in]r, 1[.$

(4)   The only T-filters of Ω are $\mathcal{T}, \mathcal{F}, \mathcal{G}.$

(5)   There exists χ and $\psi \in H(\Omega)$ such that

$$\chi(x) = 1, \quad \psi(x) = 0 \qquad \text{for} \quad x \in \Omega \cap d(0, s)$$

and

$$\chi(x) = 0, \quad \psi(x) = 1 \qquad \text{for} \quad x \in \Omega \setminus d(0, 1^-).$$

Indeed we can take sequences $(r'_n)_{n \in \mathbb{N}}$, $(r''_n)_{n \in \mathbb{N}}$ in $|K|$ satisfying

$$s < r'_{n+1} < r'_n < r''_n < r''_{n+1} < 1,$$

$$\lim_{n\to\infty} r'_n = s, \ \lim_{n\to\infty} r''_n = 1, \ \prod_{n=0}^{\infty} r''_n = 0, \ \prod_{n=0}^{\infty} \frac{s}{r'_n} = 0.$$

Let $\Lambda = d(0,1) \setminus \left(\bigcup_{n=0}^{\infty} C(0,r'_n) \cup C(0,r''_n) \right)$. By Proposition 50.5 there exists $\chi \in H(\Lambda)$ satisfying $\chi(x) = 1$ whenever $x \in d(0,s)$ and $\chi(x) = 0$ whenever $x \in C(0,1)$. Let $\psi = 1 - \chi$ and let Ω be the set $A \cup (\Lambda \setminus d(0,r^-))$. Clearly, Ω admits three T-filters:

- the filter $\mathcal{F}$ on A,

- the increasing filter $\mathcal{F}$ of center 0 and diameter 1 (χ is strictly vanishing along $\mathcal{F}$),

- the decreasing filter $\mathcal{G}$ of center 0, of diameter 1, (ψ is strictly vanishing along, $\mathcal{G}$).

It is easily seen that these three T-filters are the only T-filters on Ω. So, Ω, χ, ψ are defined. Now we put $F(x) = \left(\dfrac{\sum_{m=1}^{\infty} u_m}{(1 - \frac{x}{b_m})} \right)$ and $f_a(x) = \chi(xF(x) + \psi(x)$.

Then we have $f_1(x) = F(x)$ whenever $x \in \Omega \cup d(0,s)$, and $f_1(x) = 1$ whenever $x \in \Omega \setminus d(0,1^-)$. We can deduce that f_1 is a quasi-invertible element in $H(\Omega)$. On one hand, by Proposition 58.1, f is not vanishing along $\mathcal{T}$ and along $\mathcal{G}$, hence neither is f_1. On the other hand, as $f_1(x) = 1$ when $|x| = 1$, f_1 is not vanishing along $\mathcal{F}$. Hence f_1 is not vanishing along any one of the three T-filters on Ω, therefore it is quasi-invertible in $H(\Omega)$. By Proposition 58.1, $\mathcal{E}(f_1)$ has a solution

$$g_1 = \prod_{m=1}^{\infty} \frac{1}{(1 - \frac{x}{b_m})^{u_m}}.$$

Now, for each $j = 2, \ldots t$, we put $\Omega_j = w_j + \Omega = \{x + w_j | x \in \Omega\}$ and we take $f_j \in H(\Omega_j)$ defined as $f_j(x + w_j) = f_1(x)$. In Ω_j the equation $\mathcal{E}(f_j)$ has a solution g_j defined as $g_j(x + w_j) = g_1(x)$. Let $D = \bigcap_{j=1}^{t} \Omega_j$ and let $f(x) =$

$\prod_{j=1}^{t} f_j(x) \in H(D)$. Obviously, we have $f(x) = f_j(x)$ when $|x - w_j| < 1$ and $f(x) = 1$ when $|x - w_l| = 1$ for every $l = 1, \ldots, t$, $l \neq j$. Each one of the f_j is quasi-invertible in $H(D)$ hence so is f. Besides each g_j ($1 \leq j \leq t$) is a solution of $\mathcal{E}(f)$. Indeed, when $|x - w_j| < 1$ we have $g'_j(x) = f_j(x)g_j(x) = f(x)g_j(x)$, and when $|x - w_j| = 1$, we have $g_j(x) = 0$. Next, the g_j clearly have supports two by two disjointed, hence they are linearly independant, and then this shows $\mathcal{S}(f)$ have dimension $n \geq t$.

We will end the proof by showing that $\{g_1, \ldots, g_t\}$ generates $\mathcal{S}(f)$. For each $j = 1, \ldots, t$, we put $D_j = d(w_j, 1) \cap D$ and $B_j = d(w_j, s^-)$. Let $D' = D \backslash (\bigcup_{j=1}^{t} D_j)$.

By definition of f we see that $f(x) = 1$ for all $x \in D'$. Besides, $d(\alpha, 1^-)$ is included in D' for every $\alpha \in D'$. But it is well known that the equation $y' = y$ has no solution y in $H(d(\alpha, 1^-))$, except the zero solution. Let $h \in \mathcal{S}(f)$. For every $x \in D'$, the restriction of h to $d(\alpha, 1^-)$ is a solution of the equation $y' = y$ that belongs to $H(d(\alpha, 1^-))$, hence we have $h(x) = 0$ for all $x \in D'$. Since D' is equal to $d(0, 1) \backslash (\bigcup_{j=1}^{t} d(w_j, 1^-))$ we have

$$(7) \quad v(h, 0) = +\infty.$$

Now we consider $h(x)$ when $x \in B_1$. Since $D_1 = \Omega \cap d(0, 1^-)$, the three T-filters $\mathcal{T}, \mathcal{F}, \mathcal{G}$ on Ω are secant with D_1 and they are the only T-filters on D_1. Then $\mathcal{F}$ is the only T-filter on B_1 because both $\mathcal{F}, \mathcal{G}$ are not secant with $d(0, s)$. Hence the algebra $H(B_1)$ has no divisor of zero. Consider the restriction $\widetilde{f}_1$ (resp. $\widehat{f}_1$) of f to D_1(resp. B_1). In $H(B_1)$, by Theorem 55.6 the space $\mathcal{S}(\widehat{f}_1)$ has dimension one hence there exists $\lambda_1 \in K$ such that $h(x) = \lambda_1 g_1(x)$ whenever $x \in B_1$. Since $g_1 \in \mathcal{I}(\mathcal{F})$, this implies $h(x) = 0$ whenever $x \in \Gamma(0, r, s)$ hence $v(h, -\log s) = +\infty$. We will deduce that $v(h, \mu) = +\infty$ whenever $\mu \in [0, -\log s]$. Indeed, suppose this is not true. Then h is strictly vanishing along an increasing T-filter of center 0 and diamater $\sigma > s$, hence actually h is strictly vanishing along $\mathcal{F}_i$. Since $\lim\limits_{\substack{|x| \to 1, |x| \neq 1 \\ x \in D}} (1 - \chi(x)) = \lim\limits_{\substack{|x| \to 1, |x| \neq 1 \\ x \in D}} \psi(x) = 1$, there exists $u \in]s, 1[$ such that

$$(8) \quad \left| \frac{h'(x)}{h(x)} - 1 \right| \geq \frac{1}{p^2} \text{ for } x \in D \cap \Gamma(0, u, 1).$$

On the other hand, it is easily seen that $h(x)$ is equal to a Laurent series in each annulus $\Gamma(0, r''_n, r''_{n+1})$ and for every $u < 1$ there exists an interval $[r', r''] \subset]u, 1[$ such that the function $v(h, \mu)$ is strictly decreasing in $[-\log r'', -\log r']$ and such that $h(x)$ is equal to a Laurent series $\sum\limits_{-\infty}^{+\infty} a_n x^n$. Let $\rho \in]r', r''[$ in $\Gamma(0, r', r'')$.

Since $v(h, u)$ is strictly decreasing in $[-\log r'', -\log r']$ there exists $q < 0$ such that $|a_q| \rho^q = \sup\limits_{n \in \mathbb{Z}} |a_n| \rho^n$. Then h satisfies the hypothesis of Lemma 58.4 and therefore relation (7) is impossible. Hence we have $v(h, \mu) = +\infty$ for every μ in $[0, -\log s]$. It follows that $h(x) = 0$ for every $x \in \Gamma(0, s, 1)$ because if there existed a point $\alpha \in \Gamma(0, s, 1)$ with $h(\alpha) \neq 0$, α should be a center of an increasing

T-filter, h would be vanishing along this T-filter, but the unique T-filter of center α is $\mathcal{F}$ and we have just seen that h is not vanishing along $\mathcal{F}$.

Thus we have now proven that $h(x) = 0$ for all $x \in B_1$ such that $r \leq |x| < 1$. Since $g_1(x) = 0$ whenever $x \in \Gamma(0, r, 1)$, the relation $h(x) = \lambda_1 g_1(x)$ is then true in all of B_1. In the same way, for each $j = 2, \ldots, t$, we can show that there exists $\lambda_j \in K$ such that $h(x) = \lambda_j g_j(x)$ for every $x \in B_j$ and then the equality

$$h(x) = \sum_{j=1}^{t} \lambda_j g_j(x) \text{ holds in } \bigcup_{j=1}^{t} B_j, \text{ and of course in } D', \text{ hence finally it holds in}$$

all D. This finishes proving that $\{g_1, \ldots, g_t\}$ is a base of $\mathcal{S}(f)$, and this ends the proof of Theorem 58.5.

Remark: On the contrary to the case when f is not quasi-invertible, here we are not able to construct an infraconnected open closed set D and an $f \in H(D)$ such that $\mathcal{S}(f)$ has infinite dimension. This suggests this conjecture :

Conjecture :   *If f is quasi-invertible, $\mathcal{S}(f)$ has finite dimension.*

59.   RESIDUES AND EQUATION $y'=fy$

As in Chapters 57 and 58 we suppose that K has characteristic zero.
D will denote an open closed infraconnected set.

We remember that the residue of an analytic element on a hole was introduced in chapter 17. Here we use residues to find integrals and solutions of the equation $y' = fy$. This is a common work with Marie-Claude Sarmant. Many results were given in [29] and [30].

Definition: Let $f \in H(D)$. Then f will be said to be *integrable* in $H(D)$ if there exists $F \in H(D)$ such that $F' = f$.

Theorem 59.1:   *Let f be an integrable element in $H(D)$. For every hole T of D, we have $res(f,T) = 0$.*

Proof: Let $F \in H(D)$ be such that $F' = f$. Let T be a hole of D. First we assume that F belongs to $H_0(K \setminus T)$. Then F is of the form $\displaystyle\sum_{n=1}^{\infty} \frac{\lambda_n}{(x-a)^n}$ while $f = F' = -\displaystyle\sum_{n=1}^{\infty} \frac{n\lambda_n}{(x-a)^{n+1}}$, hence $res(f,T) = 0$ and then $res(f,V) = 0$ for every hole V other than T because f belongs to $H_0(K \setminus T)$.

Now, we come back to the general case. By Theorem 19.4 we know that $\overline{f}_{T_n} = (\overline{F}_{T_n})'$ for each $n \in \mathbb{N}^*$ and $\overline{f}_0 = (\overline{F}_0)'$. Then by the results already shown when $f \in H_0(K \setminus T)$ we see that for every hole T of D we have $res(\overline{F'_{T_n}}, T) = 0$ whenever $n \geq 1$, (and obviously $res(\overline{F'_0}, T) = 0$) hence finally $res(f,T) = 0$.

Lemma 59.2:   *Let $f \in H(D)$ be such that $res(f,T) = 0$ for all the holes T of D, but finitely many of them $T_1, \ldots T_q$, let α_j and $\beta_j \in T_j$ and let $\lambda_j = res(f,T_j)$ $(1 \leq j \leq q)$.*

If $f - \displaystyle\sum_{j=0}^{q} \frac{\lambda_j}{x - \alpha_j}$ is integrable in $H(D)$ then so is $f - \displaystyle\sum_{j=1}^{q} \frac{\lambda_j}{x - \beta_j}$.

Proof: Clearly, we may assume $q = 1$ with no loss of generality and then we put $\lambda = \lambda_1$, $\alpha = \alpha_1$, $\beta = \beta_1$, and $T = T_1$. Then we have

$$f - \frac{\lambda}{x - \beta} = f - \frac{\lambda}{x - \alpha} + \lambda\Big(\frac{1}{x - \alpha} - \frac{1}{x - \beta}\Big)$$

360

Hence we just have to check that $\dfrac{1}{x-\alpha} - \dfrac{1}{x-\beta}$ has an integral in $H(D)$.

Let r be the diameter of T. Since $|\alpha - \beta| < r$, $Log\left(1 + \dfrac{\beta - \alpha}{x - \beta}\right)$ belongs to $H_0(K \setminus T)$, because $\left|\dfrac{\beta - \alpha}{x - \beta}\right| \leq \dfrac{|\beta - \alpha|}{r}$ whenever $x \in K \setminus T$. Thus $Log\left(\dfrac{x - \alpha}{x - \beta}\right)$ belongs to $H(D)$ and it is an integral of $\dfrac{1}{x - \alpha} - \dfrac{1}{x - \beta}$. This ends the proof of Lemma 59.2.

Notation: We will denote by r_1 the radius of convergence of the exponential. Hence, if K has residue characteristic zero we have $r_1 = 1$, and if K has residue characteristic p, we have $\log r_1 = -\dfrac{1}{p-1}$.

Proposition 59.3: *Let $(a_n, b_n)_{n \in \mathbb{N}^*}$ be a strongly copiercing sequence associated to D and let g be the meromorphic product $\displaystyle\prod_{n=1}^{\infty} \left(\dfrac{x - a_n}{x - b_n}\right) \in H(D)$. We assume that g' belongs to $H(D)$. Let $f = \dfrac{g'}{g}$. Then for every $r \in {]}0, r_1{[}$ there exists $t(r) \in \mathbb{N}$ and $F \in H(D)$ such that $\|F\|_D < r$, $\dfrac{dF}{dx} \in H(D)$, and*

$$f = \sum_{n=1}^{t(r)} \left(\frac{1}{x - a_n} - \frac{1}{x - b_n}\right) + \frac{dF}{dx}.$$

Proof : Let I_0 be the set of the $n \in \mathbb{N}^*$ such that $b_n \in \widetilde{D}$ and let $(T_m)_{m \in \mathbb{N}^*}$ be the sequence of the holes that contain at least one term of the sequence $(b_n)_{n \in \mathbb{N}^*}$. For every $m \in \mathbb{N}^*$, let $I_m = \{n \in \mathbb{N} | b_n \in T_m\}$. By Proposition 47.4 g factorizes in the form $\displaystyle\prod_{m=0}^{\infty} g_m$ with $g_m(x) = \displaystyle\prod_{n \in I_m} \left(\dfrac{x - a_n}{x - b_n}\right)$ and $\displaystyle\lim_{m \to \infty} \|g_m - 1\|_D = 0$. Since the sequence $(a_n, b_n)_{n \in \mathbb{N}^*}$ is strongly copiercing, each g_m obviously satisfies $\|g_m - 1\|_D < 1$ and then, by Theorem 11.2, we know that $Log \circ (g_m)$ does belong to $H(D)$, for every $m \in \mathbb{N}^*$. Further, when m is big enough we have $\|g_m - 1\|_D < r_1$, hence $\|Log \circ (g_m)\|_D < 1$ and therefore the series $\displaystyle\sum_{m=0}^{\infty} Log \circ (g_m)$ converges in $H(D)$. More precisely, $Log \circ (g_0)$ belongs to $H(\widetilde{D})$ and for each $m \in \mathbb{N}^*$ $Log \circ (g_m)$ belongs to $H(K \setminus T_m)$. Hence the series

$\displaystyle\sum_{m=0}^{\infty} Log \circ (g_m)$ is just the Mittag-Leffler series of $Log \circ (g)$ on the infraconnected set D. Since g' belongs to $H(D)$, obviously so does $\dfrac{d}{dx}(Log \circ (g)) = \dfrac{g'}{g}$ and then, by Theorem 19.4, the Mittag-Leffler series of $\dfrac{d}{dx}(Log \circ (g))$ is just $\displaystyle\sum_{m=0}^{\infty}\dfrac{g'_m}{g_m}$.

Since the sequence $(a_n, b_n)_{n\in \mathbb{N}^*}$ is strongly copiercing, there exists $t(r) \in \mathbb{N}$ such that $\left\|\dfrac{x - a_n}{x - b_n} - 1\right\|_D < \min(r, r_1)$ for every $n > t(r)$ and then by Theorem 28.3 we know that $\left\|Log\left(\dfrac{x - a_n}{x - b_n}\right)\right\|_D < r$. Let $F = \displaystyle\sum_{n=t(r)+1}^{\infty} Log\left(\dfrac{x - a_n}{x - b_n}\right)$. We have $\|F\|_D < r$ and

$$f = \sum_{n=1}^{t(r)}\left(\frac{1}{x - a_n} - \frac{1}{x - b_n}\right) + \frac{dF}{dx}.$$ This ends the proof of Proposition 59.3.

Lemma 59.4: *Let T be a hole of D and let $a, b \in T$. Then for every hole V of D we have* $res\left(\dfrac{1}{x - a_n} - \dfrac{1}{x - b_n}, V\right) = 0$.

Proof: If $V = T$, we have $res\left(\dfrac{1}{x - a}, T\right) = 1 = res\left(\dfrac{1}{x - b}, T\right)$ hence $res\left(\dfrac{1}{x - a} - \dfrac{1}{x - b}, V\right) = 0$. If $V \neq T$, then $res\left(\dfrac{1}{x - a}, V\right) = res\left(\dfrac{1}{x - b}, V\right) = 0$, hence $res\left(\dfrac{1}{x - a} - \dfrac{1}{x - b}, V\right) = 0$.

Lemma 59.5: *Let (a_n, b_n) be a strongly copiercing sequence associated to D such that the series* $\displaystyle\sum_{n=1}^{\infty}\dfrac{1}{x - a_n} - \dfrac{1}{x - b_n}$ *converges in $H(D)$ and let f be its sum. For every hole T of D we have $res(f, T) = 0$.*

Proof: By Theorem 17.6 it is seen that the mapping ϕ defined on $H(D)$ by $\phi(f) = res(f, T)$ is continuous. Next, by Lemma 59.4 we have $res\left(\dfrac{1}{x - a_n} - \dfrac{1}{x - b_n}, T\right) = 0$ for every $n \in \mathbb{N}^*$, hence $res(f, T) = 0$.

Theorem 59.6: *Let $f \in H(D)$ and let $r \in]0, r_1[$. The equation $\mathcal{E}(f)$ has solutions g invertible in $H(D)$ if and only if f is of the form*

$$\sum_{i=1}^{q}\frac{u_i}{x - c_i} + \sum_{n=1}^{t(r)}\left(\frac{1}{x - a_n} - \frac{1}{x - b_n}\right) + \frac{dF}{dx} \quad where$$

    i)   *each c_i belongs to a hole T_i, with $T_j \neq T_i$ whenever $j \neq i$, $u_i \in \mathbb{Z}$, $u_i =$ res(f, T_i)*

    ii)   *the finite sequence $(a_n, b_n)_{1 \leq n \leq t(r)}$ is a strongly copiercing sequence associated to D, and depends on r*

    iii) $F \in H(D)$ *and* $\|F\|_D \leq r$

    iv)   *for every hole T of D other than $T_1, \ldots, T_q$,* res$(f, T) = 0$.

Moreover if i), ii) ,iii), iv) are satisfied, $\mathcal{S}(f)$ is the space of the functions:

$$\lambda \Big(\prod_{i=1}^{q} (x - c_i)^{u_i} \Big) \Big(\prod_{n=1}^{t(r)} \Big(\frac{x - a_n}{x - b_n} \Big) \Big) exp(F(x)) \ \textit{with} \ \lambda \in K.$$

Proof: We first suppose that i), ii), iii), iv) are satisfied. Then $exp \circ F$ belongs to $H(D)$ and it is easily seen that the functions

$$\lambda \prod_{i=1}^{q} (x - c_i)^{u_i} \prod_{n=1}^{t(r)} \Big(\frac{x - a_n}{x - b_n} \Big) \ exp(F(x)) \ (\text{with } \lambda \in K)$$

are solutions of $\mathcal{E}(f)$. They are invertible elements of $H(D)$ and then by Theorem 54.1 the space of solutions of $\mathcal{E}(f)$ has dimension 1, hence all the solutions have this form.

    Conversely, we assume $\mathcal{E}(f)$ to have a solution g invertible in $H(D)$. Then by Corollary 47.6 g factorizes as $F(x)h(x)$ with $F \in R(D)$, F of the form $\prod_{i=1}^{q} (x - c_i)^{u_i}$, each c_i belongs to a hole T_i of D, $T_j \neq T_i$ whenever $j \neq i$, u_i belongs to $\mathbb{Z}^*$, and with h a meromorphic product $\prod_{n=1}^{\infty} \Big(\frac{x - a_n}{x - b_n} \Big)$ where $(a_n, b_n)_{n \in \mathbb{N}^*}$ is a strongly copiercing sequence associated to D. Then we have $f = \dfrac{G'}{G} + \dfrac{h'}{h}$. By Proposition 59.3, there exist $t(r) \in \mathbb{N}$ and $F \in H(D)$ such that $\|F\|_D < r$, $\dfrac{dF}{dx} \in H(D)$ and such that $\dfrac{h'}{h} = \displaystyle\sum_{n=1}^{t(r)} \Big(\frac{1}{x - a_n} - \frac{1}{x - b_n} \Big) + \dfrac{dF}{dx}$. We put

$$\phi = \sum_{n=1}^{t(r)} \Big(\frac{1}{x - a_n} - \frac{1}{x - b_n} \Big) + \frac{dF}{dx}.$$

By Lemma 59.5 we see that res$(\phi, T) = 0$ for every hole T of D, res$\Big(\dfrac{G'}{G}, T_i \Big) = u_i$ for every $i = 1, \ldots, q$ and res$\Big(\dfrac{G'}{G}, T \Big) = 0$ for every hole T different from $T_1, \ldots, T_q$. We see that Conditions ii) and iii) are already satisfied and then it only remains to show that res$\Big(\dfrac{dG}{dx}, T \Big) = 0$ for every hole T, in order to conclude that both i) and iv) are also satisfied.

Let T be a hole, and let ρ be its diameter. Let $I(T)$ be the set of the $n > t(r)$ such that both a_n, b_n belongs to T. The series $\displaystyle\sum_{n \in I(T)} \frac{1}{x - a_n} - \frac{1}{x - b_n}$ converges to a limit ψ and then by Lemma 59.5, we see that

(1) $\quad res(\psi, T) = 0$.

On the other hand it is easily seen that $\psi = \dfrac{d}{dx}\Big(\displaystyle\sum_{n \in I(T)} Log\big(\frac{x - a_n}{x - b_n}\big)\Big)$ because

$$\Big\|\frac{d}{dx}\big(Log\big(\frac{x - a_n}{x - b_n}\big)\big)\Big\|_D = \Big\|\frac{1}{x - an} - \frac{1}{x - b_n}\Big\|_D = \frac{|a_n - b_n|}{\rho} \text{ whenever } n \in I(T).$$

Moreover when $n > t(r)$ and $n \notin I(T)$, $Log\big(\dfrac{x - a_n}{x - b_n}\big)$ does belong to $H(D \cup T)$,

hence $\displaystyle\sum_{n > t(r)} Log\big(\frac{x - a_n}{x - b_n}\big)$ belongs to $H(D \cup T)$ and so does $\dfrac{dF}{dx} - \psi$. Hence

$res\big(\dfrac{dF}{dx} - \psi, T\big) = 0$. Finally by (1) we have $res\big(\dfrac{dF}{dx}, T\big) = 0$. This is true for every hole T of D and therefore Conditions i) and iv) are now proven, and Theorem 59.6 is proven too.

Remark: In the general case when D is not well pierced we can't assert that the series $\displaystyle\sum_{n=1}^{\infty} \frac{1}{x - a_n} - \frac{1}{x - b_n}$ converges in $H(D)$ because the convergence of the series is not uniform in D in the general case. This is the reason why we have to consider separately ψ and $\dfrac{dF}{dx} - \psi$.

When D is well pierced the situation is simpler as is shown Lemma 59.7.

Lemma 59.7: *Let D be a well pierced closed bounded infraconnected set and let $(a_n, b_n)_{n \in \mathbb{N}^*}$ be a copiercing sequence associated to D. Let h be a meromorphic product $\displaystyle\prod_{n=1}^{\infty} \big(\frac{x - a_n}{x - b_n}\big)$. Then h' belongs to $H(D)$ and the series*

$$\sum_{n=1}^{\infty} \frac{1}{x - a_n} - \frac{1}{x - b_n} \text{ converges to } \frac{h'}{h} \text{ in } H(D).$$

Proof: Let $\lambda > 0$ be such that every hole of D has diameter $\rho \geq \lambda$. By Theorem 19.1, we know that for every $f \in H(D)$, f' belongs to $H(D)$ and we have $\|f'\|_D \leq \dfrac{1}{\lambda}\|f\|_D$. For every $q \in \mathbb{N}^*$ we put $h_q = \displaystyle\prod_{n=1}^{q} \frac{x - a_n}{x - b_n}$. Then

$$\frac{h'_q}{h_q} = \sum_{n=1}^{q} \frac{1}{x - a_n} - \frac{1}{x - b_n}.$$ Now we have $\|h_q - 1\|_D < 1$ and $\|h - 1\|_D < 1$ hence $|h(x)| = |h_q(x)| = 1$ for all $x \in D$, and then

$$\left| \frac{h'_q(x)}{h_q(x)} - \frac{h'(x)}{h(x)} \right| = |h'_q(x)h(x) - h'(x)h_q(x)| \leq$$

$$\leq \max(|h'_q(x) - h'(x)|\,|h(x)|, |h'(x)|\,|h(x) - h_q(x)|).$$

But $|h(x)| = 1$, $\|h\|_D \leq \dfrac{1}{\lambda}$ and $\|h_q - h'\|_D \leq \dfrac{1}{\lambda}\|h_q - h\|_D$. So it is seen that $\left| \dfrac{h'_q(x)}{h_q(x)} - \dfrac{h'(x)}{h(x)} \right| \leq \dfrac{1}{\lambda}\|h_q - h\|_D$ whenever $x \in D$ and then we have $\left\| \dfrac{h'_q}{h_q} - \dfrac{h'}{h} \right\|_D \leq \dfrac{1}{\lambda}\|h_q - h\|_D$. Then the sequence $\left(\dfrac{h'_q}{h_q} \right)_{q \in \mathbb{N}}$ converges to $\dfrac{h'}{h}$ and that ends the proof of Lemma 59.7.

Definition: Let $f \in H(D)$ be such that $res(f, T) = 0$ for all the holes T of D, but finitely many of them $T_1, \ldots, T_q$. Then f will be said to be *quasi-integrable*

in $H(D)$ if, given $\alpha_j \in T_j$ $(1 \leq j \leq q)$, $f - \displaystyle\sum_{j=1}^{q} \frac{res(f, T_j)}{x - a_j}$ is integrable in $H(D)$.

(According to Lemma 59.2, this does not depend on the choice of the α_j in T_j respectively.)

Theorem 59.8: *Let D be well pierced and let $f \in H(D)$. The following statements* a), b) *are equivalent:*

a) *f expands in the form* $\displaystyle\sum_{i=1}^{q} \frac{u_i}{x - c_i} + \sum_{n=1}^{\infty} \frac{1}{x - a_n} - \frac{1}{x - b_n}$ *where $u_i \in \mathbb{Z}^*$, c_i belongs to a hole T_i of D with $T_j \neq T_i$ whenever $j \neq i$, and where the sequence $(a_n, b_n)_{n \in \mathbb{N}^*}$ is a strongly copiercing sequence associated to D,*

b) *The equation $\mathcal{E}(f)$ admits solutions invertible in $H(D)$.*

Moreover if these conditions are satisfied, then we have
$res(f, T_i) = u_i$, whenever $i = 1, \ldots, q$,
$res(f, T) = 0$ for every hole T different from $T_1, \ldots, T_q$,

and the solutions of $\mathcal{E}(f)$ are the functions $\quad g(x) = \lambda \left(\displaystyle\prod_{i=1}^{q} (x - c_i)^{u_i} \right) \left(\prod_{n=1}^{\infty} \frac{x - a_n}{x - b_n} \right)$

with $\lambda \in K$.

Proof: We first assume a) to be satisfied. Since the sequence $(a_n, b_n)_{n \in \mathbb{N}^*}$ is a strongly copiercing sequence associated to D, the meromorphic product $h(x) = \prod_{n=1}^{\infty} \dfrac{x - a_n}{x - b_n}$ obviously converges in $H(D)$. We put $P(x) = \prod_{i=1}^{q} (x - c_i)^{u_i}$ and $g(x) = P(x)h(x)$. Obviously, both P, h are invertible in $H(D)$. Hence so is g. Besides, we have $\dfrac{g'(x)}{g(x)} = f(x)$ whenver $x \in D$, hence h is a solution of $\mathcal{E}(f)$. Since g is invertible, the space of the solutions has dimension 1. Hence if a) is satisfied, the solutions of $\mathcal{E}(f)$ are the functions $\lambda \Big(\prod_{i=1}^{q} (x - c_i)^{u_i} \Big) \Big(\prod_{n=1}^{\infty} \dfrac{x - a_n}{x - b_n} \Big)$ with $\lambda \in K$. Then we know that $\Big(\sum_{i=1}^{q} \dfrac{u_i}{x - c_i}, T_j \Big) = u_j$ and $res\Big(\sum_{i=1}^{q} \dfrac{u_i}{x - c_i}, T \Big) = 0$ for every hole T different from $T_1, \ldots T_q$. Hence, by Lemma 59.6, we have

$$res\Big(\sum_{n=1}^{\infty} \frac{1}{x - a_n} - \frac{1}{x - b_n}, T \Big) = 0 \text{ for every hole } T \text{ of } D. \text{ So we have proven a)}$$

implies b) and the additional properties.

     Conversely, we assume $\mathcal{E}(f)$ to have an invertible solution $g \in H(D)$, and will show that a) is satisfied. By Corollary 47.6 g factorizes in the form $P(x)h(x)$ where h is a meromorphic product $\prod_{n=1}^{\infty} \dfrac{x - a_n}{x - b_n}$ with $(a_n, b_n)_{n \in \mathbb{N}^*}$ a strongly copiercing sequence associated to D, and P is a rational function in the form $\prod_{i=1}^{q} (x - c_i)^{u_i}$, with $u_i \in \mathbb{Z}^*$, each c_i belongs to a hole T_i of D such that $T_j \neq T_i$ whenever $j \neq i$. By Lemma 59.5 $\dfrac{h'}{h}$ is equal to $\sum_{n=1}^{\infty} \dfrac{1}{x - a_n} - \dfrac{1}{x - b_n}$ hence

$$f = \sum_{i=1}^{q} \frac{u_j}{x - c_i} + \sum_{n=1}^{\infty} \frac{1}{x - a_n} - \frac{1}{x - b_n}, \text{ and this finishes proving Theorem 59.8.}$$

Corollary 59.9:   *Let $f \in H(d(0,1))$. Then $\mathcal{S}(f)$ is not reduced to $\{0\}$ if and only if there exists a strongly copiercing sequence $(a_n, b_n)_{n \in \mathbb{N}^*}$ associated to $d(0,1)$ such that $f(x) = \sum_{n=1}^{\infty} \dfrac{1}{x - a_n} - \dfrac{1}{x - b_n}.$*

     Henceforth we suppose K has residue characteristic p different from 0. Hence we have $r_1 = p^{\frac{-1}{p-1}}$.

Theorem 59.10: *Let $\lambda \in K$. There exists a strongly copiercing sequence $(a_n, b_n)_{n \in \mathbb{N}^*}$ associated to $d(0,1)$ such that $\lambda = \sum_{n=1}^{\infty} \dfrac{1}{x - a_n} - \dfrac{1}{x - b_n}$ for all $x \in d(0,1)$ if and only if $|\lambda| < r_1$.*

Proof: First, suppose that there exists a strongly copiercing sequence $(a_n, b_n)_{n \in \mathbb{N}^*}$ associated to $d(0,1)$ such that $\lambda = \sum_{n=1}^{\infty} \dfrac{1}{x - a_n} - \dfrac{1}{x - b_n}$ for all $x \in d(0,1)$. The meromorphic product $g(x) = \prod_{n=1}^{\infty} \dfrac{x - a_n}{x - b_n}$ obviously belongs to $H(d(0,1))$, and satisfies $\dfrac{g'}{g} = \lambda$. Hence $g(x)$ is a power series converging in $d(0,1)$. On the other hand, when $|x| < \dfrac{r_1}{|\lambda|}$ it is seen that $g(x)$ is of the form $\alpha exp(\lambda x)$, with $\alpha \in K$. But the radius of convergence of this series is just $\dfrac{r_1}{|\lambda|}$. Since g must converge in $d(0,1)$, we have $\dfrac{r_1}{|\lambda|} \geq 1$, hence finally $|\lambda| \leq r_1$.

Now, reciprocally, we assume $|\lambda| \leq r_1$ and put $g(x) = exp(\lambda x)$. Then both g, g' belong to $H(d(0,1))$. Of course, we have $\dfrac{g'(x)}{g(x)} = \lambda$ whenever $x \in d(0,1)$. But by Proposition 47.4 there exists a strongly copiercing sequence $(a_n, b_n)_{n \in \mathbb{N}^*}$ associated to $d(0,1)$ such that $g(x) = \prod_{n=1}^{\infty} \dfrac{x - a_n}{x - b_n}$. As a consequence, it is seen that we have $\lambda = \dfrac{g'(x)}{g(x)} = \sum_{n=1}^{\infty} \dfrac{1}{x - a_n} - \dfrac{1}{x - b_n}$ for every $x \in d(0,1)$.

Theorem 59.11: *Let D be well pierced and let $f \in H(D)$. Then f is integrable (resp. quasi-integrable) in $H(D)$ if and only if f is of the form*

$$\frac{1}{p^h} \sum_{n=1}^{\infty} \frac{1}{x - a_n} - \frac{1}{x - b_n} \quad \left(resp. \sum_{i=1}^{q} \frac{\lambda_i}{x - \alpha_i} + \frac{1}{p^h} \sum_{n=1}^{\infty} \frac{1}{x - a_n} - \frac{1}{x - b_n}\right)$$

with $h \in \mathbb{N}$, $(a_n, b_n)_{n \in \mathbb{N}^*}$ a copiercing sequence associated to D (resp. and where each α_i belongs to a hole T_i of D with $T_i \neq T_j$ whenever $i \neq j$, and $\lambda_i = res(f, T_i)$).

Proof: We first suppose f to be of the form $\sum_{n=1}^{\infty} \dfrac{1}{x - a_n} - \dfrac{1}{x - b_n}$ with

$(a_n, b_n)_{n \in \mathbb{N}^*}$ a strongly copiercing sequence associated to D. We will show that f is integrable in $H(D)$. Let g be the meromorphic product $\displaystyle\prod_{n=1}^{\infty} \frac{x - a_n}{x - b_n}$.

Then we have $f = \dfrac{g'}{g}$. By Proposition 59.3, for every $r > 0$, f is of the form

$$\sum_{n=1}^{t(r)} \frac{1}{x - a_n} - \frac{1}{x - b_n} + \frac{dF}{dx} \text{ with } F \in H(D) \text{ and } \|F\|_D < r. \text{ We know that}$$

$$\sum_{n=1}^{t(r)} Log \frac{x - a_n}{x - b_n} \in H(D) \text{ hence } f \text{ clearly appears to be the derivative of an ele-}$$

ment of $H(D)$, and therefore is integrable. More generally, if f has the form

$$\sum_{i=1}^{q} \frac{\lambda_j}{x - \alpha_i} + \sum_{n=1}^{\infty} \frac{1}{x - a_n} - \frac{1}{x - b_n} \text{ with each } \alpha_i \text{ in a hole } T_i \text{ such that } T_i \neq T_j$$

whenever $i \neq j$, then f is clearly quasi-integrable in $H(D)$.

Now we suppose f to be quasi-invertible. Let $f = \displaystyle\sum_{i=1}^{q} \frac{\lambda_j}{x - \alpha_i} + \frac{dF}{dx}$ with

$F \in H(D)$. Let $s \in \mathbb{N}$ be such that $\|p^s F\|_D < r_1$. The equation $\mathcal{E}(p^s \dfrac{dF}{dx})$

clearly has solutions $g \in H(D)$, of the form $\lambda exp(p^s F)$, and then $p^s \dfrac{dF}{dx}$ has

the form announced in Theorem 59.6. But by hypothesis $\dfrac{dF}{dx}$ only has zero

residues, hence $p^s \dfrac{dF}{dx}$ is in the form $\displaystyle\sum_{n=1}^{\infty} \frac{1}{x - a_n} - \frac{1}{x - b_n}$ with $(a_n, b_n)_{n \in \mathbb{N}^*}$ a

strongly copiercing sequence associated to D. Thus a quasi-integrable f has the form announced in Theorem 59.11. Finally, if f is integrable in $H(D)$, then by Lemma 59.5 f only has zero residues, hence $\lambda_i = 0$ whenever $i = 1, \ldots, q$. Hence,

f is just of the form $\displaystyle\sum_{n=1}^{\infty} \frac{1}{x - a_n} - \frac{1}{x - b_n}$. This finishes proving Theorem 59.11.

Corollary 59.12: *Let D be well pierced. For every $f \in H(D)$ there exists an integer $s \in \mathbb{N}$ and a strongly copiercing sequence $(a_n, b_n)_{n \in \mathbb{N}^*}$ associated to D such that $f(x) = \dfrac{1}{p^s} Log\Big(\displaystyle\prod_{n=1}^{\infty} \frac{x - a_n}{x - b_n}\Big).$*

60.   EQUATION $g' = fg$ WITH $g^n \in H(D)$

In this chapter we put $B = d(0, 1^-)$ and f belongs to $H(B)$.

Here we will consider solutions g of the equation $\mathcal{E}(f)$ that are not necessarily elements of $H(B)$, but are elements of $A(B)$. The results given in this Chapter are mainly due to Marie-Claude Sarmant and were first published in [30].

Notations: As in Chapters 8, 28 and 59 if the residue characteristic of K is $p \neq 0$, we put $r_1 = p^{\frac{-1}{p-1}}$.

The space of the solutions $g \in A(B)$ of $\mathcal{E}(f)$ will be denoted by $\mathcal{T}(f)$.

Lemma 60.1: $\mathcal{T}(f)$ *has dimension 0 or 1. If $\mathcal{S}(f) \neq \{0\}$, then $\mathcal{S}(f) = \mathcal{T}(f)$.*

Proof: By Chapter 55, it is obviously seen that $\mathcal{T}(f)$ can't have dimension greater than 1. Indeed, let $g \in \mathcal{T}(f) \setminus \{0\}$, and let $h \in \mathcal{T}(f)$. For every $r \in$ $]0, 1[$, both g, h belong to $H(d(0, r))$, and therefore there exists $\lambda \in K$ such that $h(x) = \lambda g(x)$ whenever $x \in d(0, r)$. Hence of course, λ does not depend on r. Therefore we have $h = \lambda g$ in $A(d(0, 1^-))$.

Now, suppose $\mathcal{S}(f) \neq \{0\}$, let $g \in \mathcal{S}(f) \setminus \{0\}$, and let $h \in \mathcal{T}(f)$. Since $\mathcal{T}(f)$ has dimension at most one, and contains $\mathcal{S}(f)$ that has dimension one, the spaces are obviously equal.

Notations: We will denote by $\sqrt{H(B)}$ the subring of $H(B)$ consisting of the $g \in A(B)$ such that g^n belongs to $H(B)$ for some $n \in \mathbb{N}^*$. Besides, given $g \in H(B)$, the set of the integers $n \in \mathbb{Z}$ such that $g^n \in H(B)$ is seen to be an ideal of $\mathbb{Z}$.

Now let $f \in H(B)$ be such that $dim(\mathcal{T}(f)) = 1$. For every $g \in \mathcal{T}(f)$, the ideal $n \in \mathbb{Z}$ such that $g^n \in H(B)$ actually does not depend on the solution $g \neq 0$. We will denote by $J(f)$ this ideal.

Let $q \in \mathbb{N}^*$ be an integer prime to the residue characteristic p of K when this residue characteristic is different from 0. In Chapter 33 we have defined $\psi_q(x)$ in B, and here we will keep this definition.

α) **We first suppose $p = 0$.**

Let $\phi \in A(B)$. We know that ϕ has integrals in $A(B)$ and we will denote by ψ the one that satisfies $\psi(0) = 0$. Then, denoting by θ the one that satisfies

$\theta(0) = 0$, we have $\|\theta\|_B = \|\phi\|_B$. The differential equation $y' = \phi y$ admits solutions different from zero in $A(B)$ if and only if $\|\phi\|_B \leq 1$, and then when $\|\phi\|_B \leq 1$, these solutions are seen to be the functions $\lambda exp(\theta(x))$.

Thanks to Theorem 59.8 we can easily characterize the $f \in H(B)$ such that $T(f) \cap \sqrt{H(B)}$ is not reduced to $\{0\}$.

Theorem 60.2: *The following statements are equivalent.*

i) *There exists $g \in T(f)$ and $N \in \mathbb{N}^*$ such that $g^n \in H(B)$*

ii) $N f$ *is of the form* $\displaystyle\sum_{i=1}^{q} \frac{u_i}{x - c_i} + \sum_{n=1}^{\infty} \frac{1}{x - a_n} - \frac{1}{x - b_n}$

where each c_i belongs to a hole T_i of B with $T_j \neq T_i$ whenever $j \neq i$, each u_i belongs to $\mathbb{Z}^*$ and $u_i = res(f, T_i)$ and where the sequence (a_n, b_n) is a strongly copiercing sequence associated to B.

Proof: We first suppose n to belong to $J(f)$. Then by Theorem 59.8 $N f$ is in the form announced in ii).

Now we assume N to be in the form announced in ii). By Theorem 59.8, $\mathcal{E}(N f)$ admits invertible solutions g in $H(D)$, hence $|g(x)|$ is constant in B. Let $h = \dfrac{g}{g(0)}$. Since $h(0) = 1$ it is easily seen that $|h(x) - 1| < 1$ whenever $x \in B$, hence the function w defined in B as $w(x) = \psi_N(h(x))$ belongs to $A(B)$. Hence f clearly belongs to $T(f) \cap \sqrt{H(B)}$ and then by Lemma 60.1 we have $S(N f) = T(N f)$, which finishes the proof of Theorem 60.2.

Corollary 60.3: $J(f)$ *is equal to the set of the integers N satisfying ii) in Theorem 60.1.*

β) **We now suppose** $p \neq 0$.

Theorem 60.4 now is a direct application of Theorem 59.8.

Theorem 60.4: *We assume $g \in T(f) \cap \sqrt{H(B)} \setminus \{0\}$. Let $N \in \mathbb{N}^*$. The following conditions a) and b) are equivalent*

a) $g^N \in H(B)$,

b) *f is of the form* $\dfrac{1}{N}\Big(\displaystyle\sum_{i=1}^{q} \frac{u_i}{x - c_i} + \sum_{n=1}^{\infty} \frac{1}{x - a_n} - \frac{1}{x - b_n}\Big)$

where the sequence (a_n, b_n) is a strongly copiercing sequence associated to B and where each c_i belongs to a hole T_i of B with $T_j \neq T_i$ whenever $j \neq i, u_i \in \mathbb{Z}^*, \dfrac{u_i}{N} \in \mathbb{Q}^* \cap \mathbb{Z}_p$, $u_i = Nres(f, T_i)$ and $res(f, T) = 0$ for every hole T of B different from $T_1, ..., T_q$.

Proof: We first suppose a) is satisfied. Let $\ell = g^N$. Then g is a solution of $\ell(Nf)$ in $H(B)$. Hence Nf is in the form a) in Theorem 59.8:

$$\sum_{i=1}^{q} \frac{u_i}{x - c_i} + \sum_{n=1}^{\infty} \frac{1}{x - a_n} - \frac{1}{x - b_n},$$

hence f is clearly in the form b) in Theorem 60.4.

Now we suppose b) is satisfied. Let $g \in \mathcal{T}(f)$. Then $g^N \in \mathcal{T}(Nf)$. By Theorem 59.8, we have $\mathcal{S}(Nf) \neq \{0\}$. Hence by Lemma 60.1, g^N belongs to $\mathcal{S}(f)$ and that ends the proof of Theorem 60.4.

Definition. f will be said to have *finitely many residues, all of them in a set* $A \subset K$, if for every hole T of $B, res(f,T) = 0$ except for finitely many holes T such that $res(f,T) \in A$.

In terms of quasi-integrability we obtain the following Theorem 60.4 which applies to the Frobenius Structure.

Theorem 60.5: *Let f be such that $\mathcal{T}(f) \neq \{0\}$. There exists $N \in \mathbb{N}$ such that $g^N \in H(B)$ if and only if f is quasi-integrable and has finitely many residues all of them in* $\mathbb{Q}$.

Proof: If $\mathcal{T}(f)$ has a non zero solution g such that $g^N \in H(B)$, by Theorem 59.8 and Theorem 59.9 we know that Nf is quasi-integrable and by Theorem 60.4, f has finitely many residues all of them in $\mathbb{Q} \cap \mathbb{Z}_p$.

Now we suppose f quasi-integrable with finitely many residues, all of them in $\mathbb{Q}$. Since $\mathcal{E}(f)$ has non zero solutions $g \in A(B)$, we know that each residue λ_i of f satisfies $|\lambda_i| \leq 1$ hence $\lambda_i \in \mathbb{Q} \cap \mathbb{Z}_p$. Then by Theorem 59.9, f is of the form b) in Theorem 60.4. Hence by Theorem 60.4 there exists an integer N such that $g^N \in H(B)$ and that ends the proof of Theorem 60.5.

We will now look for a sufficient condition for $\mathcal{E}(f)$ to admit a non zero solution $g \in A(B)$ such that a certain power g^N belongs to $H(B)$.

Theorem 60.6: *Let f belong to $A(B)$ and be of the form* $\displaystyle\sum_{i=1}^{s} \frac{\lambda_i}{x - \alpha_i} + \frac{dF}{dx}$

with $\lambda_i \in \mathbb{Q}^* \cap \mathbb{Z}_p$, $|\alpha_i| \geq 1(q \leq i \leq s)$ and $F \in A(B)$ satisfying $\|F\|_B \leq r_1$. Then $\mathcal{T}(f) \neq \{0\}$. Moreover if $F \in H(B)$, then for every $N \in \mathbb{N}$ such that $\|NF\|_B < r_1$ and $N\lambda_i \in \mathbb{Z}(1 \leq i \leq s)$, we have $g^N \in H(B)$.

Proof: By considering $\phi = F - F(0)$ which satisfies the same hypothesis as F we can assume $F(0) = 0$ with no loss of generality. Then by Lemma 23.18, the function $\ell(r)$ defined as $\ell(r) = \|F\|_{d(0,r)}$ is strictly increasing in the interval $]0,1[$. Therefore we have $|F(x)| < r_1$ whenever $x \in B$. Hence $F(d(0,1^-))$ is included in $d(0, r_1^-)$. Then, by Theorem 23.19 $\exp(F(x))$ belongs to $A(B)$, and therefore to $T(f)$. Now let q be the smallest common multiple of the denominators of the $\lambda_i (1 \le i \le s)$ and let $v_i = q\lambda_i \in \mathbb{Z}$ $(1 \le i \le s)$. We see that $\prod_{i=1}^{s}(x - \alpha_i)^{v_i}$

belongs to $R(B)$ and is a solution of the equation $y' = \left(\sum_{i=1}^{s} \dfrac{v_i}{x - \alpha_i}\right)y$. Therefore

the function $h(x) = \prod_{i=1}^{s}\left(\dfrac{\alpha_i - x}{\alpha_i}\right)^{v_i} \exp(qF(x))$ belongs to $A(B)$ and is a solution

of $\mathcal{E}(qf)$. Since $|\alpha_i| \ge 1$ whenever $i = 1, ..., s$ we have $|g(x) - 1| < 1$ for all $x \in B$. On the other hand, since the λ_i belong to $\mathbb{Z}_p$, q is prime to p, hence the function ℓ defined in B as $\ell(x) = \psi_q(h(x))$ belongs to $A(B)$ and therefore to $T(f)$.

Finally if $F \in H(B)$ and if $\|N\ell\|_B < r_1$ then we know that $\exp(N(F)) \in H(B)$. Obviously $\left(\dfrac{\alpha_1 - x}{\alpha_1}\right)$ belongs to $R(B)$ $(1 \le i \le s)$, so ℓ^n belongs to $H(B)$ and this ends the proof of Theorem 60.6.

The following conjecture is now suggested by Theorem 60.6 and by the case when $h \in K$ with $|f| = r_1$.

Conjecture: $T(f) \cap \sqrt{H(B)}$ *is not reduced to* $\{0\}$ *if and only if f is of the form*

$$\sum_{j=1}^{s} \frac{u_j}{x - c_j} + \frac{dF}{dx} \quad \text{with } u_j \in \mathbb{Q} \cap \mathbb{Z}_p \ (1 \le i \le s), \ |c_j| \ge 1, \ F \in H(B), \text{ and}$$

$\|F\|_B < r_1.$

The very simple example $f = \dfrac{1}{p}\left(\dfrac{1}{x - a} - \dfrac{1}{x - b}\right)$ helps us understand the prob-

lem. If $|a - b| \le \dfrac{r_1}{p}$, the function $Log\dfrac{x - a}{x - b}$ belongs to $H(B)$, and satisfies

$|Log(\dfrac{x - a}{x - b})| < |a - b|$ for all $x \in B$, hence $\dfrac{1}{p}Log\dfrac{x - a}{x - b}$ belongs to $A(B)$. Then

the function $g(x) = exp\left(\dfrac{1}{p}Log\dfrac{x - a}{x - b}\right)$ belongs to $A(B)$. So g belongs to $T(f)$,

and satisfies $g^p \in H(B)$.

Now if $\dfrac{r_1}{p} < |a - b| \le r_1$, $\mathcal{E}(pf)$ still admits $\ell(x) = \dfrac{b}{a}\left(\dfrac{x - a}{x - b}\right)$ for solution. But

$T(f)$ is reduced to $\{0\}$. Indeed let $g \in T(f)\backslash\{0\}$. Then $g^p(x) = \ell(x) = \dfrac{b}{a}\left(\dfrac{x - a}{x - b}\right) =$

$\dfrac{b}{a}(1 + \dfrac{b-a}{x-a})$ and then by Corollary 26.2 we know that $\ell(B) = d(1, |b-a|^-)$. Moreover, without loss of generality we can assume $g(0) = 1$ because $\ell(0) = 1$. Then $g(B)$ is a disk $d(1, r^-)$ and it is easily seen that $r > r_1$. Indeed we know that $|(1+u)^p - 1| = \dfrac{1}{p}|u|$ whenever $|u| < r_1$, hence if $r \le r_1$ then $|\ell(x) - 1| \le \dfrac{r_1}{p}$ whenever $x \in B$ and therefore $|b - a| \le \dfrac{r_1}{p}$, which contradicts the hypothesis. Hence we have $r > r_1$. But then, we know that the function in u defined as $(1 + u)^p$ is not injective in $d(0, r^-)$ if $r > r_1$. Hence the function $(g(x))^p$ is not injective in B, while ℓ is. That finishes proving that $\mathcal{E}(f)$ has no solution $g \in A(B)$ but zero.

Lemma 60.7:   *Let r, r' be such that $0 < r < r'$. Let $T = d(0, r^-)$. Let $T' = d(0, r'^-)$ and let $f \in H_0(K \setminus T)$ be such that $res(f, T) = 0$. The sequence $((\dfrac{r}{r'})^n \dfrac{1}{|n-1|})(n \ge 2)$ has limit zero and there exists $F \in H_0(K \setminus T')$ such that the restriction of f to $K \setminus T'$ is equal to F' in $H_0(K \setminus T')$ and satisfies*

$$\|F\|_{K \setminus T'} \le \|f\|_{K \setminus T} \sup_{n \ge 2} \dfrac{1}{|n-1|}(\dfrac{r'}{r})^n.$$

Proof: Since $|n-1| \ge \dfrac{1}{n-1}$ the sequence $(\dfrac{r}{r'})^n$ obviously tends to 0. Let $A = \sup_{n \ge 2}(\dfrac{r}{r'})^n \dfrac{1}{|n-1|}$. Since $f \in H_0(K \setminus T)$ and since $res(f, T)(f) = 0$, $f(x)$ is a Laurent series of the form $\displaystyle\sum_{n=2}^{\infty} \dfrac{\lambda_n}{x^n}$ with $\displaystyle\lim_{n \to \infty} \dfrac{|\lambda_n|}{r^n} = 0$. Let $\theta_{n-1} = \dfrac{\lambda_n}{n-1}$. Then $\dfrac{|\theta_{n-1}|}{r^n} \le r \, A(\dfrac{|\lambda_n|}{r^n})$, hence $\displaystyle\lim_{n \to \infty} \dfrac{|\mu_n|}{r'^{n-1}} = 0$ and then the series $F(x) = \displaystyle\sum_{n=1}^{\infty} \dfrac{\mu_u}{x^n}$ belongs to $H_0(K \setminus T')$ and satisfies $F'(x) = f(x)$ whenever $x \in K \setminus T$.

Theorem 60.8:   *Let D be a set of the form $d(0, S) \setminus (\bigcup_{i \in I} d(\alpha_i, r))$ with $|\alpha_i - \alpha_j| = |\alpha_i| = 1$ whenever $i \ne j$ and $r < 1 < S$. Let $f \in H(D)$ have finitely many non zero residues. There exists a set $D' = d(0, S') \setminus (\bigcup_{i \in I} d(\alpha_i, r'^-))$ with $r < r' < 1 < S' < S$ such that f is quasi-integrable in $H(D')$.*

Proof: With no loss of generality, we may obviously assume f to have all its residues equal to zero. Let $A = \sup_{n \ge 2} \dfrac{1}{|n-1|}(\dfrac{r}{r'})^{n-1}$. For each $i \in I$ we put

$T_i = d(\alpha_i, r^-)$, $T_i' = d(\alpha_i, r')$ and we see that there exists $F_i \in H_0(K \setminus T_i)$ such that $F_i'(x) = \overline{f}_{T_i}(x)$ whenever $x \in K \setminus T_i'$ and $\|F_i\|_{D'} \le Ar\|\overline{f}_{T_i}\|_D$. So, the series $\sum_{i \in I} F_i$ converges in $H(D')$.

In the same way, as $\lim_{n \to \infty} \left(\dfrac{S'}{S}\right)^n \dfrac{1}{|n-1|} = 0$, the power series $\overline{f}_0(x) = \sum_{n=0}^{\infty} \alpha_n x^n$ admits an integral in the form $F_0(x) = \sum_{n=0}^{\infty} \dfrac{n}{n+1} x^{n+1}$ that converges in $d(0, S') = \widetilde{D}$. Let $F = F_0 + \sum_{i \in I} F_i$. By Theorem 19.4 we know that $F' = f$ and that ends the proof of Theorem 60.8.

Remark. It is easily seen that Theorem 60.8 has no reciprocal. We can easily construct an element f quasi-integrable in $H(B)$ which has no extension to any set of this kind.

61.   THE p-ADIC FOURIER TRANSFORM

Notations :   Here we assume $K = \mathbb{C}_p$. As in Chapter 8, for every integer $s \in \mathbb{N}$, A_s will denote the multiplicative group of the p^s -th roots of 1, and $A = \bigcup_{s \in \mathbb{N}} A_s$ will denote the multiplicative group of all the p^s-th roots of 1 for any $s \in \mathbb{N}$.

As in Chapter 8, we define $r_s \in]0,1[$ by $-\log r_s = \dfrac{1}{p^{s-1}(p-1)}$ $(s \in \mathbb{N}^*)$.

According to Chapter 8, for every $s \in \mathbb{N}^*$ we have

a) $A_s \setminus A_{s-1} \subset C(1, r_s)$

b)   given any primitive p^s root α of 1 $A_s = \bigcup_{j=0}^{p-1} \alpha^j A_{s-1}$

c)   given any $\alpha \in A_s$, the mapping θ defined in A_s by $\theta(x) = x - \alpha$ is an isometric bijection from A_s onto A_s.

Given n such that $1 < n < s$, $\phi_{s,n}$ denotes the canonical surjection from A_s onto $\dfrac{A_s}{A_n}$.

Lemma 61.1 is easily deduced from Properties a), b), c).

Lemma 61.1:   *Let $n, s \in \mathbb{N}$ satisfy $0 < n < s$. The quotient group $\dfrac{A_s}{A_n}$ is provided with a distance δ' defined by $\delta'(\phi_{s,n}(x), \phi_{s,n}(y)) = |x - y|$ whenever $x, y \in A_s$.*

Notations: In $A_s \setminus A_{s-1}$ there exists a subset $E_{s,n}$ isometric to $\dfrac{A_s}{A_n}$, satisfying $diam(E_{s,n}) = r_{s-1}$, such that $E_{s,n}$ has p^{s-n} elements.

As in Chapter 28, given two topological groups $(A, +)$ and $(B, *)$ we denote by $\mathcal{H}om(A, B)$ the group of the continuous homomorphisms from A into B.

Lemma 61.2:   *Let $\gamma \in A_s$. The group homomorphism ϕ_γ from $(\mathbb{Z}, +)$ into $(\mathbb{C}_p^*, \cdot)$ defined as $\phi_\gamma(n) = \gamma^n$ is continuous with respect to the p-adic absolute value on $\mathbb{Z}$ and on $\mathbb{C}_p^*$. Then ϕ_γ has continuation to a continuous group homomorphism from $(\mathbb{Z}_p, +)$ into $(\mathbb{C}_p^*, \cdot)$*

Proof: Indeed we have $\phi_\gamma(n) = 1$ when $|n| < p^{-s}$, hence the group homomorphism ϕ_γ is continuous at 0 and therefore is continuous. As a consequence, by continuity ϕ_γ has continuation to a continuous group homomorphism from $(\mathbb{Z}_p, \cdot)$ into $(\mathbb{C}_p^*, \cdot)$.

Notation :   Given $\gamma \in A$, we will denote by ψ_γ^* the unique continuous group homomorphism from $(\mathbb{Z}_p, +)$ into $(\mathbb{C}_p^*, \cdot)$ such that $\phi_\gamma^*(n) = \gamma^n$ whenever $n \in \mathbb{Z}$.

Lemma 61.3 :   *The mapping Ψ from A into $\mathcal{H}om((\mathbb{Z}_p, +), (\mathbb{C}_p^*, \cdot))$ defined as $\Psi(\gamma) = \phi_\gamma^*$ is a group isomorphism.*

Proof :   First we check that Ψ is injective. Indeed, if $\gamma \in Ker(\Psi)$ then we have $\gamma^n = 1$ whenever $n \in \mathbb{Z}$ hence $\gamma = 1$. Second, we check that Ψ is surjective. Let $\theta \in \mathcal{H}om((\mathbb{Z}_p, +), (\mathbb{C}_p^*, \cdot))$ and let $\gamma = \theta(1)$. It is seen that $\theta(n) = \gamma^n$ whenever $n \in \mathbb{Z}$, therefore by continuity we have $\theta = \phi_\gamma^*$.

Notations: In order to simplify the notations, henceforth we put $\gamma^n = \phi_\gamma^*(n)$ for all $n \in \mathbb{Z}_p$.

Let $\mathcal{J}$ denote the filter of the complementaries of the finite subsets of A. Let $L^1(A)$ denote the $\mathbb{C}_p$-Banach vector space of the functions f from A into $\mathbb{C}_p$ such that $\lim_{\mathcal{J}} f(x) = 0$, provided with the norm $\| \cdot \|$ of uniform convergence on A.

Given $f, g \in L^1(A)$, the series $\displaystyle\sum_{\lambda \in A} (f(\lambda)\, g(\gamma \lambda^{-1}))$ is seen to converge. We denote by $f * g$ the mapping from A into $\mathbb{C}_p^*$ defined by $f * g(\gamma) = \displaystyle\sum_{\lambda \in A} f(\lambda)\, g(\gamma \lambda^{-1})$.

Thus, $*$ is a convolution on $L^1(A)$.

Lemma 61.4 :   *$L^1(A)$ is provided with a structure of $\mathbb{C}_p$-Banach algebra whose multiplication is $*$.*

Proof :   First we check that $f * g$ belongs to $L^1(A)$ whenever $f, g \in L^1(A)$. Indeed let $\epsilon \in]0, +\infty[$. Every finite subset of A is clearly included in a finite subgroup of A. Hence there exists a finite subgroup B of A such that $|f(\lambda)|\, \|G\| \leq \epsilon$ and $|g(\lambda)|\, \|f\| \leq \epsilon$ whenever $\lambda \in A \setminus B$. Hence when $\gamma \in A \setminus B$ we see that either $\lambda \in A \setminus B$ or $\gamma \lambda^{-1} \in A \setminus B$ and then we have $|f(\lambda)\, g(\gamma \lambda^{-1})| \leq \epsilon$ whenever $\lambda \in A$. This shows that $\lim_{\mathcal{J}} f * g(\gamma) = 0$. Thus $*$ is an internal law in $L^1(A)$.

Finally, formal calculations show that $L^1(A)$ is a K-algebra with this law as a multiplication.

Notation :   Let $\mathcal{C}(\mathbb{Z}_p, \mathbb{C}_p)$ be the Banach algebra of the continuous functions from $\mathbb{Z}_p$ into $\mathbb{C}_p$. For every $f \in L^1(A)$, let $\mathcal{F}(f)$ be the mapping from $\mathbb{Z}_p$ into $\mathbb{C}_p^*$ defined as

$$\mathcal{F}(f)(n) = \sum_{\gamma \in \Gamma} f(\gamma)\, \gamma^n.$$

Lemma 61.5 :   *For all $f \in L^1(A)$, $\mathcal{F}(f)$ belongs to $C(\mathbb{Z}_p, \mathbb{C}_p)$.*

Proof :   Let $u \in \mathbb{Z}_p$. Let $\epsilon \in]0, +\infty[$ and let A_s be a subgroup of A such that $|f(\gamma)| \leq \epsilon$ whenever $\gamma \in A \setminus A_s$. It is seen that $\gamma^{(pq)} = 1$ whenever $\gamma \in A_s$ and $q \geq s$. As a consequence we have $\gamma^n = \gamma^u$ for all $\gamma \in A_s$ when $|n - u| \leq \dfrac{1}{p^s}$ and therefore $|f(n) - f(u)| \leq \sup\limits_{\gamma \in A \setminus A_s} |f(\gamma)| \leq \epsilon$. Thus we have checked that $\mathcal{F}(f) \in C(\mathbb{Z}_p, \mathbb{C}_p)$.

Definition :   For all $f \in C(\mathbb{Z}_p, \mathbb{C}_p)$, $\mathcal{F}(f)$ will be named *the Fourier Transform of f with respect to $\mathbb{Z}_p$.*

The problem on whether $\mathcal{F}$ is injective was asked by Bernard de Mathan and simultaneously got two different solutions in 1973 ([3], [17], [39]). Actually, Yvette Amice showed this problem to be equivalent to a problem of T-filter.

Notations :   Let $\rho \in]0, p^{-\frac{1}{(p-1)}}[$ and let $D_\rho = \mathbb{C}_p \setminus (\bigcup\limits_{\gamma \in A} d(\gamma, \rho^{-1}))$. Let $\mathcal{G}$ be the increasing filter of center 1 and radius 1 on D_ρ. By Properties a), b), the set $A_s \setminus A_{s-1}$ consists of $(p-1)p^{s-1}$ points γ satisfying $v(\gamma - 1) = \dfrac{1}{p^{s-1}(p-1)}$ and $v(\gamma - \lambda) \leq \frac{1}{(p-1)}$ whenever $\gamma, \lambda \in A$. Thus it is seen that $\mathcal{G}$ is a pierced filter.

Theorem 61.6:   **(Y. Amice)**   *If there exists an idempotent T-sequence associated to $\mathcal{G}$, then $\mathcal{F}$ is not injective.*

Proof : We assume that there exists an idempotent T-sequence associated to $\mathcal{G}$. We notice that $\mathcal{P}(\mathcal{G}) = \mathbb{C}_p \setminus d(1, 1^-)$. Since the $d(\gamma, r^-)$ are the only holes of D_r, by Theorem 37.2 there exists $g \in H(D_r)$, strictly vanishing along $\mathcal{G}$, equal to 0 in all of $\mathcal{P}(\mathcal{G})$, meromorphic on each hole $d(\gamma, r^-)$, admitting each $\gamma \in A$ as a pole of order 1 or 0 and having no other pole in $d(\gamma, r^-)$. Hence g is of the form $\sum\limits_{\gamma \in A} \dfrac{a_\gamma}{1 - \gamma x}$ with $\lim\limits_{\mathcal{J}} a_\gamma = 0$, and certain $a_\lambda \neq 0$. So we have

$$(1) \quad \sum_{\gamma \in A} \frac{a_\gamma}{1 - \gamma x} = \sum_{\gamma \in A} a_\gamma \sum_{n=0}^{\infty} (\gamma x)^n.$$

Further, when $x \in d(0, 1^-)$, the series $\sum\limits_{\gamma \in A} a_\gamma \sum\limits_{n=0}^{\infty} (\gamma x)^n$ is clearly equal to

$\sum_{n=0}^{\infty}\left(\sum_{\gamma\in A}a_\gamma\,\gamma^n\right)x^n$. This is a power series that by (1) is identically equal to zero, whenever $x\in d(0,r^-)$. Hence we have

(2) $\quad \sum_{\gamma\in A}a_\gamma\,\gamma^n=0$ for all $n\in\mathbb{N}$.

Now , let $f\in C(\mathbb{Z}_p,\mathbb{C}_p)$ be defined as $f(n)=\sum_{\gamma\in A}a_\gamma\,\gamma^n$. Since certain a_γ are different from zero, f is not identically zero. But by (2) we see that $\mathcal{F}(f)(n)=0$ for all $n\in\mathbb{N}$. Actually $\mathbb{N}$ is dense in $\mathbb{Z}_p$, and therefore we see that f is identically zero in $\mathbb{Z}_p$. This ends the proof of Theorem 61.6 .

Theorem 61.7: *$\mathcal{G}$ admits an idempotent T-sequence.*

Proof: For each $m\in\mathbb{N}^*$, we put $S_m=C(1,r_m)\cap D$, we denote by u_m the integral part of $\log m$, and we put $q_m=p^{m-1-u_m}$. We know that $A_m\setminus A_{m-1}\subset S_m$ and then by Lemma 61.1 E_{m,u_m} has q_m elements. Let $E_{m,u_m}=(\alpha_{m,j})_{1\leq j\leq q_m}$ and for every $j=1,...,q_m$ let $T_{m,j}=d(\alpha_{m,j},\rho^-)$. We will prove the weighted sequence $(T_{m,j},1)_{1\leq j\leq q_m}$, $m\in\mathbb{N}$ to be an idempotent T-sequence. For each $m\in\mathbb{N}^*$, let Q_m be the q_m-degree monic polynomial whose zeros are the $(\alpha_{m,j})_{1\leq j\leq n}$, each one being a simple zero, and let $\lambda_m=\|\frac{1}{Q_m}\|_{S_m}(r_m)^{q_m}$. We will prove the sequence (λ_m) to be bounded. By Lemma 61.3 there exists $\alpha\in E_{m,u_m}$ such that $\|\frac{1}{Q_m}\|_{S_m}=\varphi_{\alpha,\rho}\left(\frac{1}{Q}\right)$. Besides we have $(r_m)^{q_m}=\varphi_{\alpha,r_m}(Q_m)$ hence $\|\frac{1}{Q_m}\|_{S_m}(r_m)^{q_m}=\frac{\varphi_{\alpha,r_m}(Q_m)}{\varphi_{\alpha,\rho}(Q_m)}$ and therefore $\lambda_m\leq\frac{\varphi_{\alpha,r_m}(Q_m)}{\varphi_{\alpha,\rho}(Q_m)}$. In terms of valuations we obtain

(1) $\quad -\log\lambda_m\geq v_\alpha(Q_m,-\log\,r_m)-v_\alpha(Q_m,-\log\,\rho)$.

We will compute $v_\alpha(Q_m-\log\,r_m)-v_\alpha(Q_m)-\log\,\rho)$. We may put it in the form

(2) $\quad \sum_{h=v_m+1}^{m}v_\alpha(Q_m,-\log r_h)-v_\alpha(Q_m,-\log r_{h-1})+v_\alpha(Q_m,-\log r_{u_m})$

$-v_\alpha(Q_m,-\log\rho)$

Since E_{m,u_m} is isometric to $\dfrac{A_m}{A_{u_m}}$, it is seen that Q_m admits exactly p^{h-u_m} zeros (taking multiplicities into account) inside $d(\alpha,r_h)$ and therefore we have

$$v_\alpha(Q_m,-\log\,r_h)-v_\alpha(Q_m,-\log\,r_{h-1})=(\log\,r_{h-1}-\log\,r_h)p^{h-1-u_m}.$$

But actually

$$\log\, r_{h-1} - \log\, r_h = \frac{1}{p-1}\left(\frac{1}{p^h} - \frac{1}{p^{h-1}}\right) = -\frac{1}{p^h}.$$

So we have

(3) $v_\alpha(Q_m, -\log\, r_h) - v_\alpha(Q_m, -\log\, r_{h-1}) = -p^{-u_m-1}.$

Besides since α is the only zero of Q_m in $d(\alpha, r_{u_m}^-)$ we have

(4) $v_\alpha(Q_m, -\log\, r_{u_m}) - v_\alpha(Q_m, -\log\, \rho) = \log\, \rho - \log\, r_{u_m} =$

$= \log\, \rho + \dfrac{1}{(p-1)p^{u_m-1}}.$

Hence by (2) , (3), (4) we obtain

(5) $v_\alpha(Q_m, -\log\, r_m) - v_\alpha(Q_m, -\log\, \rho)$

$= -(m - u_m - 1)p^{-u_m-1} + \log\, \rho + \dfrac{1}{(p-1)p^{u_m-1}}.$

Actually by definition we have $p^{u_m-1} > m$ and then by (5) we obtain

(6) $v_\alpha(Q_m, -\log\, r_m) - v_\alpha(Q_m, -\log\, \rho) > -\log\, \rho - 1.$

Thus by (1) and (6) , the sequence (λ_m) is clearly bounded. Then, by Lemma 35.1, it is seen that the weighted sequence $(T_{m,j}, 1)_{1 \le j \le q_m, m \in \mathbb{N}}$ is a T-sequence if and only if

(7) $\displaystyle\lim_{m \to \infty}\left(\prod_{j=1}^{m}\left(\frac{r_j}{r_m}\right)^{q_j}\right) = 0$

For convenience we consider

$$B_m = \sum_{j=1}^{m} q_j(\log\, r_m - \log\, r_j) = \sum_{j=1}^{m} p^{j-u_j}\left(\frac{1}{(p-1)(p^{j-1})} - \frac{1}{(p-1)p^{m-1}}\right)$$

$$= \frac{1}{p-1}\left(\sum_{j=1}^{m} p^{-u_j+1} - \frac{p^{j-u_j}}{p^{m-1}}\right).$$

By definition we have $p^j < j$ hence

(8) $\displaystyle\sum_{j=1}^{m} p^{-u_j+1} > p\sum_{j=1}^{m}\frac{1}{j}.$

Besides , it is seen that

(9) $\displaystyle\sum_{j=1}^{m}\frac{p^{j-u_j}}{p^{m-1}} < \sum_{j=1}^{m} p^{j-m+1} < \frac{1}{p-1}.$

Thus by (8) and (9) we see that $\lim_{m \to \infty} B_m = +\infty$ and therefore (7) is true. This ends the proof of Theorem 61.7 .

Corollary 61.8: $\mathcal{F}$ *is not injective* .

REFERENCES

[1]   **AMICE, Y.**   *Les nombres p-adiques*, P.U.F. (1975).

[2]   **AMICE, Y.**   *Dual d'un espace $H(D)$ et transformation de Fourier*, Groupe d'étude d'Analyse Ultramétrique (IHP), 1973-74, n5, Paris.

[3]   **AMICE, Y. et ESCASSUT, A.**   *Sur la non injectivité de la transformation de Fourier relative a $\mathbb{Z}_p$* , C.R.A.S. Paris, A, 278, p583-585, (1974).

[4]   **ARAUJO, J. and ESCASSUT, A.**   *p-adic analytic interpolation* Annales Mathématiques Blaise Pascal 2, n1 p29. (1995).

[5]   **BOUSSAF, K.**   *Motzkin factorization in algebras of analytic elements*, Annales Mathématiques Blaise Pascal 2, n1, p73 (1995).

[6]   **BOUSSAF, K. and ESCASSUT, A.**   *Absolute values on algebras of analytic elements*, to appear in Annales Mathématiques Blaise Pascal 2, n2 (1995).

[7]   **CHRISTOL, G.**   *Modules Differentiels et Equations Differentielles p-adiques*, Queen's Papers in Pure and Applied Mathematics, n66., Queen's University, Kingston, Ontario, (1983).

[8]   **DECOMPS-GUILLOUX, A. et MOTZKIN, E.**   *Fonctions analytiques univalentes dans un corps ultramétrique complet algébriquement clos.* C.R.A.S. Paris t. 268, p. 1531-1533 (1969).

[9]   **DIARRA, B.**   *Ultraproduits ultramétriques de corps valués*, Annales Scientifiques de l'Université de Clermont II, Série Math., Fasc. 22, p.1-37, (1984).

[10]   **DWORK, B.**   *Lectures on p-adic differential equations.* Springer-Verlag, (1982).

[11]   **ESCASSUT, A.**   *Algèbres d'éléments analytiques au sens de Krasner dans un corps valué non archimédien complet algébriquement clos*, C.R.A.S. Paris, A 270 , p.758-761 (1970).

[12]   **ESCASSUT, A.**   *Compléments sur le prolongement analytique dans un corps valué non archimédien, complet algébriquement clos*, C.R.A.S.Paris, A 271 p.718-721 (1970).

[13]   **ESCASSUT, A.**   *Algèbres de Banach ultramétriques et algèbres de Krasner-Tate*, Asterisque n. 10, p.1-107, (1973).

[14]   **ESCASSUT, A.**   *Algèbres d'éléments analytiques au sens de Krasner dans un corps valué non archimédien complet algébriquement clos*, Thèse de Doctorat de spécialité, Faculté des Sciences de Bordeaux, (1970).

[15] **ESCASSUT, A.** *Algèbres d'éléments analytiques en analyse non archimédienne,* Indag. math.,t.36, p. 339-351 (1974).

[16] **ESCASSUT, A.** *Eléments analytiques et filtres percés sur un ensemble infraconnexe,* Ann. Mat. Pura Appl. t.110 p. 335-352 (1976).

[17] **ESCASSUT, A.** *T-filtres, ensembles analytiques et transformation de Fourier p-adique,* Ann. Inst. Fourier 25, n 2, p. 45-80, (1975).

[18] **ESCASSUT, A.** *Algèbres de Krasner intègres et noetheriennes,* Indag. Math. 38, p. 109-130, (1976).

[19] **ESCASSUT, A.** *Spectre maximal d'une algèbre de Krasner,* Colloquium Mathematicum, XXXVIII, fasc. 2, p. 339-357, (1978).

[20] **ESCASSUT, A.** *Eléments spectralement injectifs et générateurs universels dans une algèbre de Tate,* Collectanae Mathematica (Barcelona), Vol XXVIII, Fasc.2, p.131-148 (1977).

[21] **ESCASSUT, A.** *The ultrametric spectral theory,* Periodica Mathematica Hungarica, Vol.11, (1), p7-60, (1980).

[22] **ESCASSUT, A.** *Maximum principle for Analytic Elements and Lubin-Hensel Theorem for series with Analytic Coefficients,* Ann. Mat. Pura Appl. vol CXXXV, p. 265-278 (1983).

[23] **ESCASSUT, A.** *Derivative of Analytic Elements on Infraconnected clopen sets,* Indag. Math. 51, p.63-70, (1989).

[24] **ESCASSUT, A. and DIARRA, B.** *Analytic Elements with a null derivative on an infraconnected open set,* Journal of the London Mathematical Society, (2), 42, p.137-146 (1990).

[25] **ESCASSUT, A. and SARMANT, M.-C.** *The differential equation $y' = fy$ in algebras $H(D)$,* Collect. Math. 39, p.31-40, (1988).

[26] **ESCASSUT, A. and SARMANT, M.-C.** *The differential equation $y' = fy$ in zero residue characteristic,* The Glasgow Mathematical Journal,(33), p.149-153 (1991).

[27] **ESCASSUT, A.** *The equation $y' = fy$ in $\mathbb{C}_p$ when f is not quasi-invertible,* Rivista Matematica Pura ed Applicata, n.4,p.81-92 (1990).

[28] **ESCASSUT, A.** *The equation $y' = fy$ in $\mathbb{C}_p$ when f is quasi-invertible,* Rend. Sem. Mat. Univ. Padova, Vol. 86 p.17-27 (1991).

[29] **ESCASSUT, A. and SARMANT, M.-C.** *The p-adic differential equation $y' = \omega y$ in the closed unit disk,* Collect. Math. 41, n 1, p.165-174 (1990).

[30] **ESCASSUT, A. and SARMANT, M.-C.** *The equation $y' = \omega y$ and the meromorphic products*, p-adic functional analysis, Lecture Notes in Pure and Applied Mathematics n137, p.157-175 (1992).

[31] **ESCASSUT, A.** *Integrally closed algebras of analytic elements*, Communication in Algebra,19, p.1565-1584 (1991).

[32] **ESCASSUT, A. and DIARRA, B.** *Non integrally closed algebras $H(D)$*, p-adic functional analysis, Lecture Notes in Pure and Applied Mathematics n137, p.63-74 (1992).

[33] **ESCASSUT, A. and SARMANT, M.-C.** *Sufficient Conditions for injectivity of Analytic Elements* , Bull. Sci. Math. vol.118, n 1, p.29-46, (1994).

[34] **ESCASSUT, A. and SARMANT, M.-C.** *Mittag-Leffler series and Motzkin Products for Invertible Analytic Elements*, Rivista di Matematica Pura ed Applicata N 12, p.61-73 (1992).

[35] **ESCASSUT, A. and SARMANT, M.-C.** *Injectivity, Mittag-Leffler Series and Motzkin Products*, Annales des Sciences Mathématiques du Quebec, 16, (2), p.155-173 (1992).

[36] **ESCASSUT, A., HADDAD, L. and SARMANT, M.-C.** *Characterization of the collapsing meromorphic products*, to appear.

[37] **FENEYROL - PERRIN, Y.** *Transformations conformes dans les corps Hédériques*, Studia Scientiarum Mathematica Hungarica 24, p.219-239 (1989).

[38] **FRESNEL, J. et DE MATHAN, B.** *L'image de la transformation de Fourier p-adique*, C.R.A.S.Paris, A 278, p.653-656, (1974).

[39] **FRESNEL, J. et DE MATHAN, B.** *Algèbres L_1 p-adiques* , Bulletin de la SMF, tome 106, fasc.3, p 225-260 (1978).

[40] **GARANDEL, G.** *Les semi-normes multiplicatives sur les algèbres d'éléments analytiques au sens de Krasner*, Indag. Math., 37, n4, p.327-341, (1975).

[42] **GUENNEBAUD, B.** *Algèbres localement convexes sur les corps valués*, Bull. Sci. Math. 91, p.75-96, (1967).

[43] **GUENNEBAUD, B.** *Sur une notion de spectre pour les algèbres normées ultramétriques*, thèse Université de Poitiers, (1973).

[44] **KRASNER, M.** *Prolongement analytique uniforme et multiforme dans les corps valués complets: éléments analytiques, préliminaires du théorème d'unicité.* C.R.A.S. Paris, A 239, p.468-470, (1954).

[45] **KRASNER, M.** *Prolongement analytique dans les corps valués complets: préservation de l'analycité par des opérations rationnelles; quasiconnexité et éléments analytiques réguliers.* C.R.A.S. Paris, A 244 p.1599-1602, (1957).

[46] **KRASNER, M.** *Prolongement analytique uniforme et multiforme dans les corps valués complets: uniformité des fonctions analytiques; l'analycité des fonctions méromorphes.* C.R.A.S. Paris, A 244, p.1996-1999, (1957).

[47] **KRASNER, M.** *Prolongement analytique uniforme et multiforme dans les corps valués complets: préservation de l'analycité par la convergence uniforme, Théorème de Mittag-Leffler généralisé pour les éléments analytiques,* C.R.A.S. Paris, A 244, p.2570-2573, (1957).

[48] **KRASNER, M.** *Prolongement analytique uniforme et multiforme dans les corps valués complets. Les tendances géométriques en algèbre et théorie des nombres,* Clermont-Ferrand, p.94-141 (1964). Centre National de la Recherche Scientifique (1966), (Colloques internationaux de C.N.R.S. Paris, 143).

[49] **LAZARD, M.** *Les zéros des fonctions analytiques sur un corps valué complet,* IHES, Publications Mathématiques n14, p.47-75 (1962).

[50] **MOTZKIN, E. and ROBBA, Ph.** *Prolongement analytique en analyse p-adique.* Séminaire de theorie des nombres, année 1968-69, Faculté des Sciences de Bordeaux.

[51] **MOTZKIN, E.** *La décomposition d'un élément analytique en facteurs singuliers,* Ann. Inst. Fourier 27, n 1, p.67-82 (1977).

[52] **ROBBA, Ph.** *Fonctions analytiques sur les corps valués ultramétriques complets. Prolongement analytique et algèbres de Banach ultramétriques,* Astérisque, n.10, p. 109-220 (1973).

[53] **VAN ROOIJ, A.C.M.** *Non-Archimedean Functional Analysis,* Marcel Decker, inc. (1978).

[54] **DURIX-SARMANT, M.-C.** *Prolongement de la fonction exponentielle en dehors de son cercle de convergence,*.C.R.A.S.Paris, A 269,p.123-125, (1969).

[55] **SARMANT, M.-C. et ESCASSUT A.** *T-suites idempotentes,* Bull. Sci. Math. 106, p.289-303, (1982).

[56] **SARMANT, M.-C. et ESCASSUT A.** *Fonctions analytiques et produits croulants,* Collect. Math. XXXVI, n 2, p.199-218, (1985).

[57] **SARMANT, M.-C.** *Produits Méromorphes,* Bull. Sci. Math. t.109, p.155-178 (1985).

[58] **SARMANT, M.-C.**   *Prolongement analytique à travers un T-filtre*, Studia Scientiarum Mathematicarum Hungarica 22, p.407-444, (1987).

[59] **SARMANT, M.-C.**   *Produits Méromorphes et Prolongement Analytique*, thèse de Doctorat d'Etat, Université Pierre et Marie Curie (1987).

[60] **SARMANT, M.-C.**   *Factorisation en Produit Méromorphe d'un Elément semi-inversible*, Bull. Sci. Math. t. 115 , p.379-394 (1991).

[61] **SCHIKHOF, W.H.**   *Ultrametric calculus. An introduction to p-adic analysis*, Cambridge University Press (1984).

[62] **SCHILLING, O.**   *The Theory of valuations* , Math. Survey IV,(1950).

DEFINITIONS

NOTATIONS

Chapter 1.

$|\cdot|,\ v(\cdot),\ |D|,\ v(D)$

$U_L,\ M_L,\ \mathcal{K}$

$Max(A),\ Max_1(A)$

$SM(A),\ Mult(A),\ Mult_m(A), Mult^*(A), SM(A,T),\ Mult(A,T), Mult_m(A,T)$

$Mult^*(A,T), SM(A,\|\cdot\|), Mult(A,\|\cdot\|)),\ Mult_m(A,\|\cdot\|))\,,\ Mult^*(A,\|\cdot\|))$

$Int(x).$

Chapter 2.

$d(a,r),\ d(a,r^-)\,,\ C(a,r)$

$\Gamma(a,r_1,r_2),\ \Delta(a,r_1,r_2)$

$\mathcal{I}(\Lambda),\ \mathcal{E}(\Lambda)\,.$

$\overline{D},\ \overset{\circ}{D}\,.$

$diam(D)$

$\widetilde{D}$

$\delta(a,D)$

$\widehat{L},\ \widehat{d}(a,r)\,,\ \widehat{d}(a,r^-)),\ \widehat{D}.$

Chapter 3.

$\mathcal{C}_D(\mathcal{F}),\ \mathcal{P}_D(\mathcal{F}),\ \mathcal{C}(\mathcal{F}),\ \mathcal{P}(\mathcal{F}).$

Chapter 4.

$\overline{x},\ \|\cdot\|$

$v(h_a,\mu),\ v_a(h,\mu),\ \varphi_{\mathcal{F}}(h)).$

$N^+(P,\mu)\,,\ N^-(P,\mu)).$

Chapter 5.

$(g,h).$

Chapter 6.

$irr(a,L).$

Chapter 8.

$|\cdot|_p,\ v_p(\cdot)$

$\mathbb{Z}_p,\ \mathbb{Q}_p,\ \mathbb{C}_p$

A_s, B_s.

Chapter 9.
$R(D)$, $R_b(D)$, $R_0(D)$, $\mathcal{U}_D$, $S(D)$, $R_E(A)$
$\|f\|_D$, $\Phi(D)$, $\Phi^*(D)$.

Chapter 10.
$H(D)$, $H_b(D)$, $H_0(D)$
$\mathcal{A}$.

Chapter 11.
$\Omega(D', D)$, $\mathcal{H}om(A, B)$

Chapter 12.
$\varphi_a(f)$, $_D\varphi_{\mathcal{F}}(f)$, $_D\varphi_{a,r}$, $_D\varphi_A$, $_D\varphi_\infty$
$\mathcal{I}(\mathcal{F})$, $\mathcal{I}_0(\mathcal{F})$.

Chapter 13.
$A(d(a, r^-))$, $A_b(d(a, r^-))$, $A(K \setminus d(0, r))$, $A_b(K \setminus d(0, r))$, $A_0(K \setminus d(0, r))$, $A_{0,b}(K \setminus d(0, r))$, $A(\Gamma(0, r_1, r_2))$, $A_b(\Gamma(0, r_1, r_2))$.

Chapter 15.
$\overline{f_T}$, $\overline{f_0}$.
$\widehat{H}(A)$

Chapter 16.
$\mathcal{I}(a)$

Chapter 17.
$E^{\sim}$, $res(f, T)$

Chapter 18.
Conditions A) and B)

Chapter 20.
$v_a(f, \mu)$, $v(f, \mu)$, $N^+(f, \mu)$, $N^-(f, \mu)$

Chapter 24.
$\mathcal{Q}_n(D)$, $\mathcal{Q}(D)$